Multiphysics Modeling

Numerical Methods and Engineering Applications

Multiphysics Modeling

Numerical Methods and Engineering Applications

Qun Zhang

Song Cen

Produced in collaboration with Tsinghua University Press Limited

AMSTERDAM • BOSTON • HEIDELBERG • LONDON
NEW YORK • OXFORD • PARIS • SAN DIEGO
SAN FRANCISCO • SINGAPORE • SYDNEY • TOKYO
Academic Press is an Imprint of Elsevier

Academic Press is an imprint of Elsevier
125, London Wall, EC2Y 5AS, UK
525 B Street, Suite 1800, San Diego, CA 92101-4495, USA
225 Wyman Street, Waltham, MA 02451, USA
The Boulevard, Langford Lane, Kidlington, Oxford OX5 1GB, UK

Notices
Knowledge and best practice in this field are constantly changing. As new research and experience broaden our understanding, changes in research methods, professional practices, or medical treatment may become necessary.

Practitioners and researchers must always rely on their own experience and knowledge in evaluating and using any information, methods, compounds, or experiments described herein. In using such information or methods they should be mindful of their own safety and the safety of others, including parties for whom they have a professional responsibility.

To the fullest extent of the law, neither the Publisher nor the authors, contributors, or editors, assume any liability for any injury and/or damage to persons or property as a matter of products liability, negligence or otherwise, or from any use or operation of any methods, products, instructions, or ideas contained in the material herein.

British Library Cataloguing-in-Publication Data
A catalogue record for this book is available from the British Library

Library of Congress Cataloging-in-Publication Data
A catalog record for this book is available from the Library of Congress

ISBN: 978-0-12-407709-6

Publisher: Joe Hayton
Acquisition Editor: Simon Tian
Editorial Project Manager: Naomi Robertson
Production Project Manager: Melissa Read
Designer: Matthew Limbert

Printed and bound in the United States of America

Contents

Preface

The readers of this book are the researchers and engineers who are interested in the numerical methodology study, code implementation and engineering applications in multidisciplinary problems. This book can also be used as a textbook for graduate students and high level undergraduate students in mechanical engineering, automotive engineering, aerospace engineering, civil engineering, biomechanical engineering, and many other areas.

Research and engineering applications regarding multiphysics simulations have been given great attention over the last decade. This book describes the basic principles and methods for multiphysics modeling, covering related areas of physics, such as structure mechanics, fluid dynamics, heat transfer, electromagnetic, and acoustics fields. Although the fundamental equations for each and every physics model are presented in this book, the main focus will be on the coupling related terms and conditions, as well as the nonlinearity and stabilization issues.

In this book, the coupling problems are classified into different categories, namely: (1) essential coupling; (2) production term coupling; (3) natural boundary condition coupling; (4) constitutive equation coupling; and (5) analysis domain coupling by physical characteristics, which include strong coupling problems and weak coupling problems, by the level of the coupling. All of the possible interface coupling conditions and load transfer conditions among these five physics models are listed, and the resulted coupled equations are presented. The Direct Matrix Assembly (DMA) method, Direct Interface Coupling (DIC) method, Multipoint Constraint (MPC) equation-based, Lagrangian Multiplier (LM) based and Penalty Method (PM) based strong coupling methods are proposed for different types of strongly coupled problems. Background theories, algorithms, key technologies, and code implementation for inter-solver weak coupling methods are also covered and discussed in details.

The challenges and important topics in multiphysics simulation are also presented: the nonlinearities and numerical stabilization in spatial and time domain; multiphysics simulation of rotating machinery; moving boundary problems for nonstructural physics models; parallel computing for large scale multiphysics simulation, etc.

This book systematically discusses about the multiphysics modeling among fluid, structure, thermal, electromagnetic, and acoustics problems. The fundamental equations, numerical schemes as well as the strategies and procedures for code implementation are presented. Most of the technologies presented in this book have been implemented in general purposed multiphysics simulation software INTESIM. More than 20 valuable engineering applications in automotive, aerospace, MEMS device, rotating machinery, and biomedical engineering etc. are presented in this book.

Organization of chapters

In Chapter 1, we briefly review the fundamental equation for fluid dynamics, structure mechanic, thermal, electromagnetic analysis, and acoustics. Special emphasis is put on the coupling terms in each equation, as well as the discretization and stabilization algorithm.

In Chapter 2, we clarify the coupling types and coupling characteristics among different physics models, and also review and discuss about appropriate coupling methods for different coupling problems.

In Chapter 3, we discuss about the coupling methods, which include the strong coupling method, general weak coupling method and intersolver-based weak coupling method.

The morphing and automatic re-meshing scheme is presented in Chapter 4 for the nonstructural physics with moving boundary analysis and coupled physics simulation with deformed structure.

The stabilization method in space and time domain to cope with the coupling nonlinearity and convergence issues in each physics model, are discussed in Chapter 5.

The multiphysics coupling problems of rotating machinery are addressed in Chapter 6.

In Chapter 7, the parallel algorithm for strong and weak coupling methods is introduced.

Three general fluid–structure interaction analyses will be presented in Chapter 8.

Multiphysics simulations in automotive engineering, aerospace engineering, MEMS devices, turbine machinery and biomechanical engineering are presented in Chapter 9, Chapter 10, Chapter 11, Chapter 12, and Chapter 13, respectively.

In Chapter 14, the FSI simulation of a sensor device used in civil engineering, and an acoustic-structure coupling problem of a closed cube metal box are presented.

In Chapter 15, an overview of the commercial multiphysics software is given and the features of code implementation for multiphysics coupling is presented.

This book covers five different physics fields, namely: fluid, structure, thermal, electromagnetic, and acoustics fields. We tried our best to use common and consistent styles of statement and symbols throughout this book to make it easy to read.

This is the first edition of the book. Due to the wide coverage of the book and limitation of our knowledge, if you find any mistakes and errors in this edition, please kindly give us the feedback. We will appreciate your tolerance and efforts and correct them in the next edition.

Acknowledgments

Many of our collaborators: Prof Liu Bin and Dr Zhu Baoshan, from Tsinghua University, Prof Guan Zhenqun and Prof Jun Liu from Dalian University of Technology, Prof Li Hongguang from Shanghai Jiaotong University, and Prof Hu Qiya from the Chinese Academy of Science, for their advice and the contributions to the development of INTESIM technology.

We would like to say thanks to all our colleagues from INTESIM (Dalian) Co. Ltd. who have worked hard to prepare this book. Computational technology development and engineering examples: Dr Jiang Peng. Engineering examples: Mr Han Yepeng, Ms Huang Xiaoxiao, Mr Sun Xinghua, and Ms Xu Yu.

Many thanks to our colleagues and students who helped us in checking and verifying the English and prepare graphs, tables, etc: Ms Bai Liang, Dr Li Jianqiao, Mr Wu Xiaoming, Mr Xu Shuoyuan, Mr Zhou Xiaogang, and Mr Zhu Rubin.

We would not have the chapter on multiphysics simulation in aerospace engineering without the help of our collaborator, Prof Liu Jun from Dalian University of Technology.

Finally, many thanks to Mr Shi Lei, from Tsinghua University Press, for helping us to prepare this book.

The physics models

1

Chapter Outline

Q. Zhang & S. Cen: Multiphysics Modeling. http://dx.doi.org/10.1016/B978-0-12-407709-6.00001-8

1.1 Heat flow fundamentals

1.1.1 Basic equations

The solution for heat flow analysis is to find the temperature distribution $T(X,t) \in \mathcal{S}_T$, so that the following governing Equation (1.1) is satisfied.

$$\frac{\partial(\rho c_v T)}{\partial t} + \frac{\partial}{\partial x_i}\left(\rho c_v T v_i - k\frac{\partial T}{\partial x_i}\right) = \overleftarrow{q^B}, \text{ in } \Omega_t^T \tag{1.1}$$

The primary variable is temperature T in the infinite-dimensional space of $\mathcal{S}_T$. The first part on the left-hand side of Equation (1.1) is the time derivative term, the second term is the convection term caused by fluid flow, and the third term is the diffusion term. The source term of heat generation is on the right-hand side of Equation (1.1).

The material properties needed to be decided are; mass density, ρ; specific heat, c_v; and thermal conductivity, k. For fully incompressible flow, ρ is assumed to be constant, but for slightly compressible and low-speed compressible flow, we assume $\rho = \rho(p)$ and $\rho = \rho(p, T)$, respectively. Also, we assume specific heat c_v and conductivity k to be either constant or a function of temperature. $\boldsymbol{v}$ is the convective velocity, and q^B is the heat generation from other physics models in multiphysics simulation.

Remark 1.1: q^B is the source term of heat generation that may come from thermoelastic damping, fluid viscous heat, or electro- or electromagnetic heat in multiphysics simulation. On the other hand, temperature T as output may affect the material properties or other quantities of the corresponding physics models.

Symbol $\overleftarrow{\ }$ indicates the value that is received from other physics model, and $\overrightarrow{\ }$ represents the value transferred to other physics model in multiphysics simulation. Symbol $\overleftrightarrow{\ }$ means the value can be either sent out to or received from other physics model in multiphysics simulation.

1.1.2 Boundary conditions

Three types of boundary conditions are considered:

1. Specified temperature on Γ_g:

$$T = \overleftrightarrow{T^s} \quad \text{on} \quad \Gamma_g \tag{1.2}$$

Here, T^s is the specified temperature.

2. Specified heat flow on Γ_{q1}:

$$k \cdot \frac{(n_i \partial T)}{(\partial x_i)} = \overleftarrow{q^s} \quad \text{on} \quad \Gamma_{q1} \tag{1.3}$$

where, q^s is the specified heat flow and n_i is the component of the unit normal direction $\boldsymbol{n}$.

Remark 1.2: The boundary conditions: Equations (1.2) and (1.3) can be used in conjugate heat transfer coupling.

3. Specified convection surfaces acting over surface Γ_{q2}:

$$k \cdot \frac{(n_i\ \partial T)}{(\partial x_i)} = h_f(T_B - T_S) \quad \text{on} \quad \Gamma_{q2} \tag{1.4}$$

Where; h_f, is the convective heat transfer coefficient; T_B, is the bulk temperature of the adjacent physics model; and T_S, is the temperature at the surface of the model.

Remark 1.3: The positive specified heat flow is into the boundary.

1.1.3 Weak forms of the thermal equation

The weak form of heat flow equations is given in Equation (1.5) as follows:
find temperature $T \in \mathcal{S}_T$, such that $\forall w_T \in \mathcal{W}_T$:

$$\int_\Omega w_T \left[\frac{\partial(\rho c_v T)}{\partial t} + \frac{\partial}{\partial x_i}\left(\rho c_v T v_i - k\frac{\partial T}{\partial x_i} \right) \right] \mathrm{d}\Omega = \int_\Omega w q^B \mathrm{d}\Omega. \tag{1.5}$$

Here, w_T is the test function of temperature T in infinite-dimensional space $\mathcal{W}_T$.

The semi-discrete Galerkin finite element formulation of the heat flow problem is stated as:

find $T^h \in \mathcal{S}_T^h$, such that $\forall w_T^h \in \mathcal{W}_T^h$:

$$\int_\Omega w_T^h \left[\frac{\partial(\rho c_v T^h)}{\partial t} + \frac{\partial}{\partial x_i}\left(\rho c_v T^h v_i^h - k\frac{\partial T^h}{\partial x_i} \right) \right] \mathrm{d}\Omega = \int_\Omega w_T^h q^{B^h} \mathrm{d}\Omega \tag{1.6}$$

Where, $\mathcal{S}_T^h$ and $\mathcal{W}_T^h$ are the sets of finite-dimensional trial and test functions for temperature, respectively.

1.1.4 The shape functions for FEM

The basic shape function for the coordinate:

$$\boldsymbol{x}(\xi) = \sum_{i=1}^{n_{en}} N_i^E(\xi)\boldsymbol{x}_i = N_i^E(\xi)\boldsymbol{x}_i. \tag{1.7}$$

Here; $N_i^E(\xi)$, is the shape function at node i; ξ, is the element local coordinate; and $\boldsymbol{x}_i$, is the nodal coordinate at node i; n_{en}, is the number of nodes per element, respectively.

The interpolation for temperature variables:

$$T(\xi) = \sum_{i=1}^{n_{en}} N_i^{\mathrm{T}} T_i = N_i^{\mathrm{T}} T_i. \tag{1.8}$$

Here, N_i^T is the shape function for temperature at node i, which is consistent with N_i^E for the class of isoparametrical element. The standard isoparametrical element can be used for the temperature interpolation, for example, the low order 4-node tetrahedral element and 8-node hexahedral element, are the most commonly used for thermal analysis, although the high order 10-node tetrahedral element and 20-node hexahedral element are also useful in achieving better accuracy with the same number of nodes. The other shapes, for example, wedge and pyramid can be used for the transition region between tetrahedral and hexahedral elements. The details about these shape functions can be found in Bathe (2006) and Wang (2000).

1.1.5 Formulations in matrix form

Integrating of Equation (1.6) by part for the third term on the left-hand side, applying boundary conditions of Equations (1.2)–(1.4), and substituting the shape functions into it, one can obtain assembled global Equation (1.9) in matrix form:

$$\boldsymbol{C}\dot{\boldsymbol{T}} + \left(\boldsymbol{K}(\boldsymbol{v})^{tm} + \boldsymbol{K}^{tb} + \boldsymbol{K}^{tc}\right)\boldsymbol{T} = \boldsymbol{Q}^{\text{flux}} + \boldsymbol{Q}^{\text{conv}} + \overleftarrow{\boldsymbol{Q}^{g}\left(q^{B}\right)}. \tag{1.9}$$

Where,

$\boldsymbol{C} = \sum_{e=1}^{ne} \boldsymbol{C}_e$, $C_{e(i,j)} = \rho \int_{\Omega_e} c_v N_i^T N_j^T \, \mathrm{d}\Omega$, is the element specific heat matrix.

$\boldsymbol{K}^{tm} = \sum_{e=1}^{ne} \boldsymbol{K}_e^{tm}$, $K_{e(i,j)}^{tm} = \rho \int_{\Omega_e} c_v N_i^T v_k N_{j,k}^T \, \mathrm{d}\Omega$ is the element mass transport conductivity matrix.

$\boldsymbol{K}^{tb} = \sum_{e=1}^{ne} \boldsymbol{K}_e^{tb}$, $K_{e(i,j)}^{tb} = \int_{\Omega_e} N_{i,m}^T k_{mm} N_{j,m}^T \, \mathrm{d}\Omega$ is the element diffusion conductivity matrix.

$\boldsymbol{K}^{tc} = \sum_{e=1}^{ne} \boldsymbol{K}_e^{tc}$, $K_{e(i,j)}^{tc} = \int_{\Gamma_{q2}} h_f N_i^T N_j^T \, \mathrm{d}\Gamma$ is element convection surface conductivity matrix.

$\boldsymbol{Q}^{\text{flux}} = \sum_{f} \boldsymbol{Q}_f^{\text{flux}}$, $Q_{f(i)}^{\text{flux}} = \int_{\Gamma_{q1}} N_i^T q^s \, \mathrm{d}\Gamma$ is element surface heat flux vector.

$\boldsymbol{Q}^{\text{conv}} = \sum_{f} \boldsymbol{Q}_f^{\text{conv}}$, $Q_{f(i)}^{\text{conv}} = \int_{\Gamma_{q2}} T_B h_f N_i^T \, \mathrm{d}\Gamma$ is the element convection surface heat flow vector.

$\boldsymbol{Q}^{g} = \sum_{e} \boldsymbol{Q}_e^{g}$, $Q_{e(i)}^{g} = \int_{\Omega_e} q^B N_i^T \, \mathrm{d}\Omega$ is the element heat generation vector.

Here, $\sum_{e=1}^{ne}$ represents the assembly operator for element matrix or vector with number of *ne* elements.

$\sum_{f}$ represents the assembly operator for element surface loads.

1.1.6 The nonlinearity in thermal analysis

1.1.6.1 Material properties

The material properties depend on the temperature, that is, $k = k(T)$, $c_v = c_v(T)$.

1.1.6.2 Convection term from computational fluid dynamics (CFD) coupling

The definition of Peclet number for convective heat transfer problem is

$$\mathrm{Pe} = \frac{c_v \rho}{k} \|v\| \, h^e. \tag{1.10}$$

Where, h^e stands for element size.

For a heat flow problem with $\mathrm{Pe} \geq 2.0$, to avoid oscillation of the solution of the temperature field and to get a better quality matrix equation, the spatial stabilization method may be needed. The streamline upwind Petrov/Galerkin (SUPG) method is presented in the further section.

1.1.7 Stabilization method for convection-dominant transport equations

The SUPG method is used for the convection-dominated heat transfer problem for the case, where Peclet number is higher than 2.0. For details about implementation of the stabilization method for transport equations, please refer Section 1.2.9.3.

1.1.8 Penalty-based thermal contact

The thermal contact problem can be used for thin air gap or other thin-layered problems with different thermal conductivities.

Assuming the thickness of a thin air gap is d, and T_1 and T_2 are the temperatures on the side A and side B of the contact wall, respectively, then the heat flux from T_2 side into T_1 side is:

$$q = k\frac{T_2 - T_1}{d}. \tag{1.11}$$

Here, k is the thermal conductivity, and the heat flux from T_1 side into T_2 side is the negative of q in Equation (1.11). Here we assume the air layer is thin enough, so linear distribution assumption of temperature across the air gap is acceptable.

Then thermal conductance for this gap is:

$$h_f = \frac{k}{d} \tag{1.12}$$

where h_f, as the contact thermal conductance, can be setup as input property.

1.1.8.1 The matrix equation for thermal contact

The heat flow for side A

$$\int_{\Gamma_A} w_A \, h_f \left(T_B - T_A\right) \mathrm{d}\Gamma \tag{1.13}$$

The previous equation can be added into the left-hand side of Equation (1.5) to consider the thermal contact affect from the air gap, and the matrix form can be expressed as:

$$k_{AiAj} = \int_{\Gamma_A} N_{Ai} h_f N_{Aj} \mathrm{d}\Gamma \tag{1.14}$$

$$k_{AiBj} = -\int_{\Gamma_A} N_{Ai} \, h_f N_{Bj} \mathrm{d}\Gamma \tag{1.15}$$

where, Ai is the surface node i on side A and Bj is the surface node j on side B, respectively. $k_{AiAj} T_A + k_{AiBj} T_B$ needs to be added into the left side of Equation (1.9) for side A, and its negative value will be added in the same way for side B.

1.2 Fluid dynamics

1.2.1 Basic equations for fluid flow

We assume the viscous fluid to be isothermal and barotropic (i.e., $F(p, \rho) = 0$) and that $\partial p / \partial \rho = B / \rho$, in which B, p, and ρ are fluid bulk modulus, pressure, and fluid density, respectively. The Arbitrary Lagrangian Eulerian (ALE) formulation is usually used to handle moving boundary problems of fluid flow in coupling analysis. The fundamental equations of fluid flow (Zhang and Hisada, 2001) are expressed as:

$$\frac{1}{B} \left. \frac{\partial \vec{p}}{\partial t} \right|_{\chi} + \frac{1}{B} c_i \frac{\partial p}{\partial x_i} + \frac{\partial v_i}{\partial x_i} = 0 \quad \text{in} \quad \Omega_t^f \tag{1.16}$$

$$\rho \left. \frac{\partial \vec{v_i}}{\partial t} \right|_{\chi} + \rho c_j \frac{\partial v_i}{\partial x_j} = \frac{\partial \sigma_{ij}}{\partial x_j} + \rho g_i + \vec{f_i} \quad \text{in} \quad \Omega_t^f \tag{1.17}$$

$$\begin{aligned} \left. \frac{\partial \left(\rho c_v \vec{T} \right)}{\partial t} \right|_{\chi} &+ \frac{\partial}{\partial x_i} \left(\rho c_v T v_i - k \frac{\partial T}{\partial x_i} \right) = 2 \mu D^2 \\ &+ \frac{\partial v_i}{\partial x_i} \left(-p + \lambda \frac{\partial v_i}{\partial x_i} \right) + \overleftarrow{q^B} \quad \text{in} \quad \Omega_t^f . \end{aligned} \tag{1.18}$$

The primary variables are pressure p, velocity vector $\boldsymbol{v}$, and temperature T. Here, $\boldsymbol{\sigma}$ is the Cauchy stress tensor, $\boldsymbol{g}$ is the acceleration of gravity, and $\boldsymbol{c} = \boldsymbol{v} - \boldsymbol{v}_m$ is the relative velocity of fluid particles to the ALE coordinates with $\boldsymbol{v}_m$ the mesh velocity. Ω_t^f denotes the spatial thermal fluid domain bounded by the boundary Γ_t^f of interest at any instant t. Here the superscript f stands for the fluid component. λ is the second viscosity, and k is the thermal conductivity. And in Equation (1.18), the strain rate related energy term is expressed as:

$$D^2 = e_{ij}e_{ij} \quad \text{with} \quad e_{ij} = \frac{1}{2}(v_{i,j} + v_{j,i}).$$

Remark 1.4: The source term f_i in Equation (1.17) may come from the coupling physics, for example, the electromagnetic force. The outgoing variables are the pressure p and velocity vector $\boldsymbol{v}$ in the coupled physics simulation.

1.2.2 Boundary and initial conditions for fluid flow

The boundary is composed of Γ_t^g and Γ_t^h corresponding to Dirichlet- and Neumann-type boundary conditions, respectively.

$$v_i = \overleftarrow{g_i} \quad \text{on} \quad \Gamma_g^f \tag{1.19}$$

$$\sigma_{ji} n_j = \overrightarrow{h_i} \quad \text{on} \quad \Gamma_h^f \tag{1.20}$$

$$T = T^s \quad \text{on} \quad \Gamma_g^T \tag{1.21}$$

$$k \cdot n_i \partial T_i / \partial x_i = q^s \quad \text{on} \quad \Gamma_q^T \tag{1.22}$$

And subject to the following initial conditions:

$$v_i(0) = {}^0 v_i \quad \text{on} \quad \Omega^f \tag{1.23}$$

$$p(0) = {}^0 p \quad \text{on} \quad \Omega^f \tag{1.24}$$

$$T(0) = {}^0 T \quad \text{on} \quad \Omega^f \tag{1.25}$$

The boundary conditions for the energy equation are the same as those in previous sections.

Remark 1.5: Equations (1.19) and (1.20) represent the coupling boundaries in the fluid–structure interaction (FSI) problem with g_i received from the structure and h_i sent to the structure.

1.2.3 The constitutive equation for fluid flow

The fluid is assumed to be Newtonian, and the constitutive equation is:

$$\sigma_{ij} = -p\,\delta_{ij} + \frac{1}{2}\mu\left(v_{i,j} + v_{j,i}\right) \tag{1.26}$$

where δ_{ij} is the component of identity tensor and μ is the dynamic viscosity of the fluid.

The properties for thermal are provided in the previous section.

1.2.4 The weak forms

1.2.4.1 Galerkin formulation for N–S equations

The weak forms for the fluid equations given in Equations (1.16)–(1.18) by the Galerkin method are written as: find pressure $p \in \mathcal{S}_p$, velocity $\boldsymbol{v} \in \mathcal{S}_v$, and temperature $T \in \mathcal{S}_T$, such that $\forall w_p \in \mathcal{W}_p, \forall \boldsymbol{w}_v \in \mathcal{W}_v$, and $\forall w_T \in \mathcal{W}_T$:

$$\int_\Omega w_p \left(\frac{1}{B}\frac{\partial p}{\partial t}\bigg|_\chi + \frac{1}{B} c_i \frac{\partial p}{\partial x_i} + \nabla \cdot \boldsymbol{v} \right) \mathrm{d}\Omega = 0 \tag{1.27}$$

$$\int_\Omega \boldsymbol{w}_v \cdot \rho \left(\frac{\partial \boldsymbol{v}}{\partial t} + \boldsymbol{c} \cdot \nabla \boldsymbol{v} - \boldsymbol{f} \right) \mathrm{d}\Omega + \int_\Omega \varepsilon(\boldsymbol{w}) : \sigma(\boldsymbol{v}, p)\,\mathrm{d}\Omega = \int_{\Gamma_h} \boldsymbol{w}_v \cdot \boldsymbol{h}\,\mathrm{d}\Gamma \tag{1.28}$$

$$\begin{aligned} &\int_\Omega w_T \left[\frac{\partial(\rho c_v T)}{\partial t} + \nabla \cdot (\rho c_v T v_i - k \nabla T) \right] \mathrm{d}\Omega = \\ &\quad \int_\Omega w_T \left(2\mu D^2 + q^B + \nabla \cdot \boldsymbol{v}\left(-p + \lambda(\nabla \cdot \boldsymbol{v})\right)\right) \mathrm{d}\Omega \end{aligned} \tag{1.29}$$

Where $\mathcal{S}_p$, $\mathcal{S}_v$, and $\mathcal{S}_T$ denote the sets of infinite-dimensional trial functions for pressure, velocity, and temperature, respectively. $\mathcal{W}_p$, $\mathcal{W}_v$, and $\mathcal{W}_T$ are the sets of test functions (weighing functions) for the continuity, equilibrium, and energy equations, respectively.

The Galerkin finite element formulation of the fluid equations is stated as follows: find pressure $p^h \in \mathcal{S}_p^h$, velocity $\boldsymbol{v}^h \in \mathcal{S}_v^h$, and temperature $T^h \in \mathcal{S}_T^h$ such that $\forall w_p^h \in \mathcal{W}_p^h, \forall \boldsymbol{w}_v^h \in \mathcal{W}_v^h$, $\forall w_T^h \in \mathcal{W}_T^h$:

$$\int_\Omega w_p^h \left(\frac{1}{B}\frac{\partial p^h}{\partial t}\bigg|_\chi + \frac{1}{B} c_i^h \frac{\partial p^h}{\partial x_i} + \nabla \cdot \boldsymbol{v}^h \right) \mathrm{d}\Omega = 0 \tag{1.30}$$

$$\int_\Omega \boldsymbol{w}_v^h \cdot \rho \left(\frac{\partial \boldsymbol{v}^h}{\partial t} + \boldsymbol{c}^h \cdot \nabla \boldsymbol{v}^h - \boldsymbol{f}^h \right) \mathrm{d}\Omega + \int_\Omega \varepsilon(\boldsymbol{w}_v^h) : \sigma(\boldsymbol{v}^h, p^h)\,\mathrm{d}\Omega = \int_{\Gamma_h} \boldsymbol{w}_v^h \cdot \boldsymbol{h}^h\,\mathrm{d}\Gamma \tag{1.31}$$

$$\int_{\Omega} w_T^h \left[\frac{\partial(\rho c_v T^h)}{\partial t} + \nabla \cdot \left(\rho c_v T^h \boldsymbol{v}^h - k \nabla T^h \right) \right] \mathrm{d}\Omega = \int_{\Omega} w_T^h \left(2\mu D^2 + q^B + \nabla \cdot \boldsymbol{v}^h \left(-p^h + \lambda \left(\nabla \cdot \boldsymbol{v}^h \right) \right) \right) \mathrm{d}\Omega \tag{1.32}$$

1.2.4.1.1 The shape functions

Basic shape function for the coordinate:

$$\boldsymbol{x}(\xi) = \sum_{a=1}^{n_{en}} N_a^E \boldsymbol{x}_a = N_a^E \boldsymbol{x}_a \tag{1.33}$$

Shape function for any solution variables:

$$\boldsymbol{v}^h(\xi) = \sum_{a=1}^{n_{en}} N_a^v \boldsymbol{v}_a = N_a^v \boldsymbol{v}_a \tag{1.34}$$

$$p^h(\xi) = \sum_{a=1}^{n_{en}} N_a^p p_a = N_a^p p_a. \tag{1.35}$$

Here, N_a^v and N_a^p are the shape function for velocity and pressure at node a, respectively. The shape function for velocity and pressure can be consistent or different, and the pressure stabilized Petrov–Galerkin (PSPG) stabilization algorithm will be needed to avoid the pressure oscillation if the pressure has the same shape function as the velocity.

1.2.5 Finite element equations

By using the finite element discretization for the ALE form of Navier–Stokes Equations (1.30)–(1.32), we can obtain the finite element Equation (1.36).

$$\boldsymbol{M}^f \overset{*}{\boldsymbol{\varphi}}{}^f + \boldsymbol{C}^f \boldsymbol{\varphi}^f = \boldsymbol{F}^f \tag{1.36}$$

where

$$\boldsymbol{M}^f = \begin{bmatrix} \boldsymbol{M}^P & 0 & \boldsymbol{0} \\ \boldsymbol{0} & \boldsymbol{M} & \boldsymbol{0} \\ \boldsymbol{0} & \boldsymbol{0} & \boldsymbol{C} \end{bmatrix} \qquad \boldsymbol{C}^f = \begin{bmatrix} \boldsymbol{\Lambda}^P & \boldsymbol{G}^T & \boldsymbol{0} \\ -\boldsymbol{G} & \boldsymbol{K}_\mu + \boldsymbol{\Lambda} & \boldsymbol{0} \\ \boldsymbol{0} & \boldsymbol{0} & \boldsymbol{K}(V) \end{bmatrix}$$

$$\boldsymbol{\phi}^f = \begin{Bmatrix} \boldsymbol{p}^f \\ \boldsymbol{V}^f \\ \boldsymbol{T}^f \end{Bmatrix} \qquad \boldsymbol{F}^f = \begin{Bmatrix} \boldsymbol{0} \\ \boldsymbol{F}_v^E + \overleftarrow{\boldsymbol{F}_{EM}^v} \\ \boldsymbol{Q}^E + \overleftarrow{\boldsymbol{Q}^f(p,v)} + \overleftarrow{\boldsymbol{Q}(q^B)} \end{Bmatrix} \tag{1.37}$$

$\boldsymbol{M}^P$ and $\boldsymbol{M}$ are the generalized mass matrices for pressure and velocity, respectively. $\boldsymbol{\Lambda}^P$ and $\boldsymbol{\Lambda}$ are the generalized matrices of convective terms for pressure and velocity, respectively. $\boldsymbol{K}_\mu$ is the fluid viscosity matrix, $\boldsymbol{G}$ is the divergence operator matrix. $\boldsymbol{p}^f$ and $\boldsymbol{V}^f$ are the pressure and velocity vectors, respectively. $\boldsymbol{F}^f$ is the external force vector that consist of the directly applied external load vector $\boldsymbol{F}_v^E$ and the coupling load vector from electromagnetic field $\boldsymbol{F}_{EM}^v$. $\boldsymbol{Q}^f(p, v)$ is the heat flow vector caused by viscous flow. $\boldsymbol{Q}(q^B)$ consists of other coupling terms of heat flow vector contributed by the thermoelastic damping or electromagnetic heat. $\boldsymbol{Q}^E$ is the heat flow vector contributed by any other directly applied external heat boundaries.* indicates the time derivative in the ALE coordinates. In the present work, the secant matrices are used for nonlinear iterations, so the incremental forms of Equation (1.36) can be simply obtained.

For the continuity equation:

$$\boldsymbol{M}^P = \sum_e \boldsymbol{M}_e^P, \tag{1.38}$$

$$M_{e(I,J)}^P = \int_{\Omega_e} \frac{1}{B} N_a^P N_b^P \mathrm{d}\Omega \tag{1.39}$$

$$\boldsymbol{\Lambda}^P = \sum_e \boldsymbol{\Lambda}_e^P, \tag{1.40}$$

$$\Lambda_{e(I,J)}^P = \int_{\Omega_e} \frac{1}{B} N_a^P c_k \frac{\partial N_b^P}{\partial x_k} \mathrm{d}\Omega \tag{1.41}$$

$$\boldsymbol{G}^P = \sum_e \boldsymbol{G}_e^P, \tag{1.42}$$

$$G_{e(M,J)}^P = \int_{\Omega_e} N_b^P \frac{\partial N_c^v}{\partial x_m} \mathrm{d}\Omega. \tag{1.43}$$

Here,

$$\begin{cases} I = a \\ J = b \\ M = (c-1)\cdot \mathrm{NSD} + \mathrm{m} \end{cases} \qquad \begin{cases} 1 \le a,\, b \le \mathrm{nep} \\ 1 \le c \le \mathrm{nev} \\ 1 \le k,\, m \le \mathrm{NSD} \end{cases} . \tag{1.44}$$

Here, nep, number of pressure nodes per element; nev, number of velocity nodes per element; and NSD, spatial dimension of the analysis domain.

For equilibrium equations:

$$\boldsymbol{M} = \sum_e \boldsymbol{M}_e, \tag{1.45}$$

$$M_{e(I,J)} = \int_{\Omega_e} \delta_{ij} \rho N_a^v N_b^v \mathrm{d}\Omega \tag{1.46}$$

$$\Lambda = \sum_e \Lambda_e, \tag{1.47}$$

$$\Lambda_{e(I,J)} = \int_{\Omega_e} \delta_{ij} \rho N_a c_k \frac{\partial N_b}{\partial x_k} \mathrm{d}\Omega \tag{1.48}$$

$$\boldsymbol{F} = \sum_e \boldsymbol{F}_e, \tag{1.49}$$

$$\boldsymbol{F}_{e(I)} = \int_{\Omega_e} \rho N_a^v f_i \mathrm{d}\Omega \tag{1.50}$$

$$\boldsymbol{K}_e = \int_{\Omega_e} \boldsymbol{B}^T [\boldsymbol{D}] \boldsymbol{B} \mathrm{d}\Omega.$$

Here,

$$\begin{cases} I = (a-1) \cdot \mathrm{NSD} + i \\ J = (b-1) \cdot \mathrm{NSD} + j \end{cases} \qquad \begin{cases} 1 \le a,\, b \le \mathrm{nev} \\ 1 \le i, j, k \le \mathrm{NSD} \end{cases} \tag{1.51}$$

$$\boldsymbol{B}_a^T = \begin{bmatrix} \frac{\partial N_a}{\partial x_1} & \frac{\partial N_a}{\partial x_2} & 0 & 0 & 0 & \frac{\partial N_a}{\partial x_3} \\ 0 & \frac{\partial N_a}{\partial x_1} & \frac{\partial N_a}{\partial x_2} & 0 & \frac{\partial N_a}{\partial x_3} & 0 \\ 0 & 0 & 0 & \frac{\partial N_a}{\partial x_3} & \frac{\partial N_a}{\partial x_2} & \frac{\partial N_a}{\partial x_1} \end{bmatrix} \tag{1.52}$$

$$[\boldsymbol{D}] = \begin{bmatrix} 2\mu & 0 & 0 & 0 & 0 & 0 \\ 0 & \mu & 0 & 0 & 0 & 0 \\ 0 & 0 & 2\mu & 0 & 0 & 0 \\ 0 & 0 & 0 & 2\mu & 0 & 0 \\ 0 & 0 & 0 & 0 & \mu & 0 \\ 0 & 0 & 0 & 0 & 0 & \mu \end{bmatrix} \tag{1.53}$$

Remark 1.6: The formulations and engineering applications of the finite volume method for compressible fluid flow are given in Chapter 10.

1.2.6 The nonlinearity and numerical challenging in CFD

The nonlinear term in Navier–Stokes equations of Equation (1.17) is the convection term, and most of the numerical difficulties and stability issues for fluid flow are caused by this term.

The numerical oscillation of the solutions in space for the convection-dominated flow may happen because of the usage of an inappropriate interpolation scheme. For a high Peclet number problem (Pe > 2), the central differencing scheme usually produces numerical oscillations in space, and the upwind scheme may help to prevent from oscillation but with a compromise on the numerical accuracy.

For the fluid flow with a high Reynolds number, the flow can be turbulence with multiscale responses. Extreme fine mesh discretization must be needed for resolving the fine-scale flow phenomenal when using the direct numerical method. However, it is usually unpractical for huge computational requests. The turbulence model, for example, k-epsilon model, large eddy simulation [LES] model can be used for modeling fine-scale effects on the coarse scale mesh discretization level. Details about the stabilization method are given in the Section 1.2.7, and the turbulence models are presented in Section 1.2.8.

1.2.7 The stabilization methods

1.2.7.1 SUPG and PSPG methods

The SUPG and PSPG formulations for Navier–Stokes equations are stated as find pressure $p^h \in \mathcal{S}_p^h$, velocity $\boldsymbol{v}^h \in \mathcal{S}_v^h$ such that $\forall w_p^h \in \mathcal{W}_p^h, \forall \boldsymbol{w}_v^h \in \mathcal{W}_v^h$:

$$\begin{aligned}
&\int_{\Omega} \boldsymbol{w}_v^h \cdot \rho\left(\frac{\partial \boldsymbol{v}^h}{\partial t}+\boldsymbol{c}^h \cdot \nabla \boldsymbol{v}^h-\boldsymbol{f}^h\right) \mathrm{d}\Omega+\int_{\Omega} \varepsilon\left(\boldsymbol{w}_v^h\right): \boldsymbol{\sigma}\left(\boldsymbol{v}^h, p^h\right) \mathrm{d}\Omega-\int_{\Gamma_h} \boldsymbol{w}_v^h \cdot \boldsymbol{h} \mathrm{d}\Gamma \\
&+\int_{\Omega} w_p^h\left(\left.\frac{1}{B} \frac{\partial p^h}{\partial t}\right|_{\chi}+\frac{1}{B} c_i^h \frac{\partial p^h}{\partial x_i}+\nabla \cdot \boldsymbol{v}^h\right) \mathrm{d}\Omega \\
&+\sum_{e=1}^{n_{el}} \int_{\Omega^e} \tau_{\mathrm{SUPG}}\left(\boldsymbol{v}^h \cdot \nabla w_v^h\right) \cdot \boldsymbol{r}^e\left(\boldsymbol{v}^h, p^h\right) \mathrm{d}\Omega \\
&+\sum_{e=1}^{n_{el}} \int_{\Omega^e} \tau_{\mathrm{PSPG}}\left(\frac{1}{\rho} \nabla w_q^h\right) \cdot \boldsymbol{r}^e\left(\boldsymbol{v}^h, p^h\right) \mathrm{d}\Omega=0.
\end{aligned} \tag{1.54}$$

Here the residual vector of the momentum equation is:

$$\boldsymbol{r}^e\left(\boldsymbol{v}^h, p^h\right)=\rho\left(\frac{\partial \boldsymbol{v}^h}{\partial t}+\boldsymbol{c}^h \cdot \nabla \boldsymbol{v}^h-\boldsymbol{f}^h\right)-\nabla \cdot \boldsymbol{\sigma}^h. \tag{1.55}$$

The last two terms in Equation(1.54) are SUPG and PSPG stabilization terms, respectively.

The commonly used definitions of τ_{SUPG} and $\tau_{\mathrm{PSPG}} = \tau_{\mathrm{SUPG}}$ are given by Bazilevs et al. (2013):

$$\tau_{\mathrm{SUPG}} = \left(\frac{1}{\tau_{\mathrm{SUPG}_1}^2} + \frac{1}{\tau_{\mathrm{SUPG}_2}^2} + \frac{1}{\tau_{\mathrm{SUPG}_3}^2} \right)^{-\frac{1}{2}}, \tag{1.56}$$

where

$$\tau_{\mathrm{SUPG}_1} = \left(\sum_{i=1}^{n_{en}} \left| \boldsymbol{v}^h \cdot \nabla N_i \right| \right)^{-1}, \tag{1.57}$$

$$\tau_{\mathrm{SUPG}_2} = \frac{\Delta t}{2}, \tag{1.58}$$

$$\tau_{\mathrm{SUPG}_3} = \frac{h_r^2}{4\nu}. \tag{1.59}$$

Here, h_r is the generalized element size,

$$h_r = 2.0 \left(\sum_{a=1}^{n_{en}} \left| \boldsymbol{j} \cdot \nabla \boldsymbol{N}_a \right| \right)^{-1}, \tag{1.60}$$

and $\boldsymbol{j}$ is the unit velocity of the gradient of the norm of the convective velocity,

$$\boldsymbol{j} = \frac{\nabla \left\| \boldsymbol{v}^h \right\|}{\left\| \boldsymbol{v}^h \right\|}. \tag{1.61}$$

1.2.7.2 Discontinuity capturing operator (Tezduyard, 2012)

The additional term to overcome the undershoot and overshoot problem in convection dominant flow is the discontinuity capturing operator:

$$\boldsymbol{S}_{\mathrm{DCDD}} = \sum_{e=1}^{n_{el}} \int_{\Omega^e} \rho \nabla w_\theta^h : \left(\nu_{\mathrm{DCDD}} \boldsymbol{\kappa} \cdot \nabla \boldsymbol{v}^h \right) \mathrm{d}\Omega, \tag{1.62}$$

where

$$\kappa = \boldsymbol{I} \tag{1.63}$$

If the Y, Z, β method is used (Tezduyar and Senga, 2007),

$$\nu_{\mathrm{DCDD}} = \left| Y^{-1} Z \right| \left(\sum_{i=1}^{nsd} \left| Y^{-1} \frac{\partial \left\| v^h \right\|}{\partial x_i} \right|^2 \right)^{\frac{\beta}{2}-1} \left(\frac{h_r}{2} \right)^{\beta}. \tag{1.64}$$

Some details about the stabilization method are presented in Chapter 5.

1.2.7.3 Underrelaxation method and solution capping

The underrelaxation scheme and solution capping schemes are convenient and useful for stabilizing the solution during the nonlinear iteration process when facing inconsistent convergence or even divergence phenomena. The underrelaxation methods can be used for updating the major primary variable, for example, velocity, pressure, temperature. Details about the underrelaxation scheme and solution capping scheme are presented in Sections 5.4 and 5.5, respectively.

1.2.8 Turbulence model in CFD

Turbulence causes eddies of flow with a wide range of spatial length and time scales that interact in a dynamically complex way. The research about turbulence model is one of the major subjects for development of CFD. The ways to solve turbulence flow can be classified into three categories: turbulence modes for Reynolds-averaged Navier–Stokes (RANS) equation, LES, and direct numerical simulation (DNS).

1.2.8.1 k-Epsilon turbulence model

For the RANS model, the effects of the turbulence on the time-averaged mean flow are modeled with classical turbulence models. The most popular ones, the k-epsilon and k-omega models, are presented in this section. Compared with LES and DNS, the computing resources needed for reasonable predictions of flow field are modest. Two extra transport equations need to be solved for the k-epsilon and k-omega models.

1.2.8.1.1 Basic equations for the k-epsilon model

The governing equation for turbulence kinetic energy k:

$$\frac{\partial(\rho k)}{\partial t}+\frac{\partial(\rho k)}{\partial x_k}v_k=\frac{\partial}{\partial x_k}\left[\rho\left(v+\frac{v_t}{\sigma_k}\right)\frac{\partial k}{\partial x_k}\right]+P_k-\rho\varepsilon \tag{1.65}$$

The similar transport equation for the rate of viscous dissipation ε:

$$\frac{\partial(\rho\varepsilon)}{\partial t}+\frac{\partial(\rho\varepsilon)}{\partial x_k}v_k=\frac{\partial}{\partial x_k}\left[\rho\left(v+\frac{v_t}{\sigma_\varepsilon}\right)\frac{\partial\varepsilon}{\partial x_k}\right]+C_{\varepsilon 1}\frac{\varepsilon}{k}P_k-\rho C_{\varepsilon 2}\frac{\varepsilon^2}{k} \tag{1.66}$$

The turbulence production term attributable to viscous forces is,

$$P_k=2\mu_t S_{ij}\cdot S_{ij}-\frac{2}{3}\frac{\partial v_k}{\partial x_k}\left(3\mu_t\frac{\partial v_k}{\partial x_k}+\rho k\right). \tag{1.67}$$

Here,

$$S_{ij}=\frac{1}{2}\left(v_{i,j}+v_{j,i}\right). \tag{1.68}$$

If the flow is incompressible, then the second term is vanishing and $P_k = 2\mu_t S_{ij} \cdot S_{ij}$. The effective viscosity

$$\nu_{\text{eff}} = \nu + \nu_t, \tag{1.69}$$

where the eddy viscosity,

$$\nu_t = C_\mu \frac{k^2}{\varepsilon}. \tag{1.70}$$

The parameters C_μ, $C_{\varepsilon 1}$, $C_{\varepsilon 2}$, σ_k, and σ_ε are constant values, the standard k-epsilon model takes the value: $C_\mu = 0.09$, $C_{\varepsilon 1} = 1.44$, $C_{\varepsilon 2} = 1.92$, $\sigma_k = 1.0$, and $\sigma_\varepsilon = 1.3$.

Rewriting Equations (1.65) and (1.66), we have:

$$\frac{\partial(\rho k)}{\partial t} + \frac{\partial(\rho k)}{\partial x_k} v_k - \frac{\partial}{\partial x_k}\left[\rho\left(\nu + \frac{\nu_t}{\sigma_k}\right)\frac{\partial k}{\partial x_k}\right] + \rho\varepsilon = P_k \tag{1.71}$$

$$\frac{\partial(\rho \varepsilon)}{\partial t} + \frac{\partial(\rho \varepsilon)}{\partial x_k} v_k - \frac{\partial}{\partial x_k}\left[\rho\left(\nu + \frac{\nu_t}{\sigma_\varepsilon}\right)\frac{\partial \varepsilon}{\partial x_k}\right] + C_{\varepsilon 2}\rho\frac{\varepsilon^2}{k} = C_{\varepsilon 1}\frac{\varepsilon}{k}P_k \tag{1.72}$$

1.2.8.1.2 Equations in weak form

$$\int_\Omega \left\{\frac{\partial(\rho k)}{\partial t} + \frac{\partial(\rho k)}{\partial x_k} v_k - \frac{\partial}{\partial x_k}\left[\rho\left(\nu + \frac{\nu_t}{\sigma_k}\right)\frac{\partial k}{\partial x_k}\right] + \rho\varepsilon\right\}\delta k \mathrm{d}\Omega = \int_\Omega P_k \delta k \mathrm{d}\Omega \tag{1.73}$$

$$\begin{aligned} &\int_\Omega \left\{\frac{\partial(\rho \varepsilon)}{\partial t} + \frac{\partial(\rho \varepsilon)}{\partial x_k} v_k - \frac{\partial}{\partial x_k}\left[\rho\left(\nu + \frac{\nu_t}{\sigma_\varepsilon}\right)\frac{\partial \varepsilon}{\partial x_k}\right] + C_{\varepsilon 2}\rho\frac{\varepsilon^2}{k}\right\}\delta\varepsilon \mathrm{d}\Omega \\ &= \int_\Omega \left\{C_{\varepsilon 1}\frac{\varepsilon}{k}P_k\right\}\delta\varepsilon \mathrm{d}\Omega \end{aligned} \tag{1.74}$$

Integration by part for third term on the left of Equation (1.73):

$$\begin{aligned} &\int_\Omega \left\{\frac{\partial}{\partial x_k}\left[\rho\left(\nu + \frac{\nu_t}{\sigma_k}\right)\frac{\partial k}{\partial x_k}\right]\right\}\delta k \mathrm{d}\Omega = \int_\Omega \frac{\partial}{\partial x_k}\left\{\left[\rho\left(\nu + \frac{\nu_t}{\sigma_k}\right)\frac{\partial k}{\partial x_k}\right]\delta k\right\}\mathrm{d}\Omega \\ &- \int_\Omega \left\{\rho\left(\nu + \frac{\nu_t}{\sigma_k}\right)\frac{\partial k}{\partial x_k}\frac{\partial \delta k}{\partial x_k}\right\}\mathrm{d}\Omega, \end{aligned} \tag{1.75}$$

By divergence theorem, we have

$$\int_\Omega \frac{\partial}{\partial x_k}\left\{\left[\rho\left(\nu + \frac{\nu_t}{\sigma_k}\right)\frac{\partial k}{\partial x_k}\right]\delta k\right\}\mathrm{d}\Omega = \int_\Gamma \left[\rho\left(\nu + \frac{\nu_t}{\sigma_k}\right)\frac{\partial k}{\partial n}\right]\mathrm{d}\Gamma \tag{1.76}$$

where $\boldsymbol{n}$ is the normal of the boundary.

1.2.8.1.3 Boundary conditions

The boundary types for turbulence term consist of the inlet, outlet, free stream, and solid walls.

Inlet: distribution of k and ε must be given.

Outlet, symmetry: $\frac{\partial k}{\partial n} = 0$, $\frac{\partial \varepsilon}{\partial n} = 0$;

Free stream: distribution of k and ε must be given as $\frac{\partial k}{\partial n} = 0$ and $\frac{\partial \varepsilon}{\partial n} = 0$;

Solid wall: approach depends on Reynolds number.

The boundary layer consists of near wall viscous sublayer, log law layer, turbulence region, and inertial dominated region. To have a good resolution with reasonable mesh, the wall function for the logarithmic layer is normally used.

The logarithmic relation for the near wall velocity is given by:

$$v^+ = \frac{V_t}{v_\tau} = \frac{1}{\kappa} \ln\left(Ey^+\right). \tag{1.77}$$

Here, dimensionless wall distance:

$$y^+ = \frac{\Delta y v_\tau}{v}. \tag{1.78}$$

The friction velocity

$$v_\tau = \left(\frac{\tau_w}{\rho}\right)^{1/2} \tag{1.79}$$

Now, v^+ is the near wall velocity; V_t is the known velocity tangent to the wall at distance of Δy from the wall; τ_w is the wall shear stress; κ is the von Karman constant; and E is law of the wall constant. The value of κ and E are 0.4 and 0.9, respectively, and the latter corresponds to a smooth wall condition.

Based on the known value of V_t and the near wall distance Δy, by using the log law equation, the wall shear stress τ_w can be obtained, and so is the calculation of dynamic viscosity:

$$\mu_w = \Delta y \frac{\tau_w}{V_t} \tag{1.80}$$

The wall element viscosity value is the larger one of the laminar viscosity and the calculated value from the Equation (1.80).

The adjacent value of the turbulence kinetic energy is obtained from the k-epsilon model. The near wall value of dissipation rate is dominated by the length scale and is given by equation:

$$\varepsilon_w = \frac{\left(C_\mu\right)^{3/4}\left(k_w\right)^{3/2}}{\kappa \Delta y}. \tag{1.81}$$

1.2.8.1.4 Equations in matrix form

The shape functions of k and ε:

$$k = N_i^k k_i \qquad i = 1, \text{ne}, \tag{1.82}$$

$$\varepsilon = N_i^\varepsilon \varepsilon_i \qquad i = 1, \text{ne}. \tag{1.83}$$

Substituting Equations (1.82) and (1.83) into the weak form of the k-epsilon Equations (1.73) and (1.74), one has:

$$\boldsymbol{M}^k \dot{\boldsymbol{k}} + \boldsymbol{C}^k \boldsymbol{k} + \boldsymbol{K}^K \boldsymbol{k} = \boldsymbol{F}^k - \boldsymbol{G}^K \boldsymbol{\varepsilon} \tag{1.84}$$

$$\boldsymbol{M}^\varepsilon \dot{\boldsymbol{\varepsilon}} + \boldsymbol{C}^\varepsilon \boldsymbol{\varepsilon} + \boldsymbol{K}^\varepsilon \boldsymbol{\varepsilon} + \boldsymbol{G}^\varepsilon \boldsymbol{\varepsilon} = \boldsymbol{F}^\varepsilon. \tag{1.85}$$

Where the matrices and vectors in Equations (1.84) and (1.85) can be obtained by a standard assembly process of the element matrices and vectors. The components in each matrix or vector at element level are given further.

For k equation at element level:

$$M_{e(i,j)}^k = \int_{\Omega^e} \rho N_i^k N_j^k \mathrm{d}\Omega \tag{1.86}$$

$$C_{e(i,j)}^k = \int_{\Omega^e} \rho N_i^k N_{j,k}^k v_k \mathrm{d}\Omega \tag{1.87}$$

$$K_{e(i,j)}^k = \int_{\Omega^e} \rho \left(v + \frac{v_t}{\sigma_k} \right) N_{i,k}^k N_{j,k}^k \mathrm{d}\Omega \tag{1.88}$$

$$G_{e(i,j)}^k = \int_{\Omega^e} \rho N_i^k N_j^\varepsilon \mathrm{d}\Omega \tag{1.89}$$

$$F_{e(i)}^k = \int_{\Omega^e} P_k N_i^k \mathrm{d}\Omega \tag{1.90}$$

For epsilon equation at element level:

$$M_{e(i,j)}^\varepsilon = \int_{\Omega^e} \rho N_i^\varepsilon N_j^\varepsilon \mathrm{d}\Omega \tag{1.91}$$

$$C_{e(i,j)}^\varepsilon = \int_{\Omega^e} \rho N_i^\varepsilon N_{j,k}^\varepsilon v_k \mathrm{d}\Omega \tag{1.92}$$

$$K^{\varepsilon}_{e(i,j)} = \int_{\Omega^e} \rho\left(v + \frac{v_t}{\sigma_\varepsilon}\right) N^{\varepsilon}_{i,k} N^{\varepsilon}_{j,k} \mathrm{d}\Omega \tag{1.93}$$

$$G^{\varepsilon}_{e(i,j)} = \int_{\Omega^e} \rho C_{\varepsilon 2} \frac{\varepsilon}{k} N^{\varepsilon}_{i} N^{\varepsilon}_{j} \mathrm{d}\Omega \tag{1.94}$$

$$F^{\varepsilon}_{e(i)} = \int_{\Omega^e} \left\{ C_{\varepsilon 1} \frac{\varepsilon}{k} P_k \right\} N^{\varepsilon}_{i} \mathrm{d}\Omega \tag{1.95}$$

1.2.8.2 Wilcox k-omega turbulence model

1.2.8.2.1 Basic equations for k-omega model

The transport equation for k and ω for turbulence flow:

$$\frac{\partial(\rho k)}{\partial t} + \frac{\partial(\rho k)}{\partial x_k} v_k = \frac{\partial}{\partial x_k}\left[\rho\left(v + \frac{v_t}{\sigma_k}\right)\frac{\partial k}{\partial x_k}\right] + P_k - \beta^* \rho k \omega \tag{1.96}$$

where

$$P_k = 2\mu_t S_{ij} \cdot S_{ij} - \frac{2}{3}\frac{\partial v_k}{\partial x_k}\left(3\mu_t \frac{\partial v_k}{\partial x_k} + \rho k\right) \tag{1.97}$$

$$\frac{\partial(\rho \omega)}{\partial t} + \frac{\partial(\rho \omega)}{\partial x_k} v_k = \frac{\partial}{\partial x_k}\left[\rho\left(v + \frac{v_t}{\sigma_\omega}\right)\frac{\partial \omega}{\partial x_k}\right] + \gamma_1 \frac{\omega}{k} P_k - \beta_1 \rho \omega^2 \tag{1.98}$$

Here, β_1, β^*, γ_1, σ_k, and σ_ω are the constants, $\beta_1 = 0.075$, $\beta^* = 0.09$, $\gamma_1 = 0.553$, $\sigma_k = 2.0$, and $\sigma_\omega = 2.0$.

If we use this variable, the length scale is $\ell = \sqrt{k}/\omega$, and the eddy dynamic viscosity:

$$\mu_t = \rho \frac{k}{\omega}. \tag{1.99}$$

The Reynolds stress

$$\tau_{ij} = -\overline{v_i' \, v_j'} = 2\mu_t S_{ij} - \frac{2}{3}\rho k \delta_{ij} \tag{1.100}$$

1.2.8.2.2 Boundary conditions

Inlet: distribution of k and ω must be given.

Outlet: symmetry: $\frac{\partial k}{\partial n} = 0, \frac{\partial \omega}{\partial n} = 0;$

Free stream: small none zero ω must be specified.

Solid wall: in boundary conditions on solid wall for $\boldsymbol{k}$, the k-omega model of Wilcox has the advantage that an analytical expression is known for ω in the viscous sublayer. The main idea behind the present formulation is to blend the wall value of ω between the logarithmic and near wall formulation.

The algebraic expression for ω in the logarithmic region:

$$\omega_l = \frac{v^*}{a_l \kappa \Delta y} = \frac{1}{a_l \kappa v} \frac{v^*}{y^+} \tag{1.101}$$

And the corresponding expression in the viscous sublayer:

$$\omega_s = \frac{6v}{\beta (\Delta y)^2} \tag{1.102}$$

where Δy is the distance between the first and the second mesh point. To achieve smooth blending and to avoid cyclic convergence behavior, the following formation is adopted.

$$\omega_\omega = \omega_s \sqrt{1 + \left(\frac{\omega_l}{\omega_s}\right)}. \tag{1.103}$$

1.2.8.2.3 Weak forms of k-omega model

The weak forms for $k - \omega$ can be obtained as the procedure for $k - \varepsilon$ equation, as well as for the finite element method (FEM) equations in matrix form.

1.2.8.2.4 Equations in matrix form

The shape functions for k and ω:

$$k = N_i^k k_i \qquad i = 1, \text{ne}, \tag{1.104}$$

$$\omega = N_i^\omega \omega_i \qquad i = 1, \text{ne}, \tag{1.105}$$

The assembled FEM equations:

$$\boldsymbol{M}^k \dot{\boldsymbol{k}} + \boldsymbol{C}^k \boldsymbol{k} + \boldsymbol{K}^{\mathrm{K}} \boldsymbol{k} + \boldsymbol{G}^K \boldsymbol{k} = \boldsymbol{F}^k, \tag{1.106}$$

$$\boldsymbol{M}^\omega \dot{\boldsymbol{\omega}} + \boldsymbol{C}^\omega \boldsymbol{\omega} + \boldsymbol{K}^\omega \boldsymbol{\omega} + \boldsymbol{G}^\omega \boldsymbol{\omega} = \boldsymbol{F}^\omega. \tag{1.107}$$

Where for k equation at element level:

$$M_{e(i,j)}^k = \int_{\Omega^e} \rho N_i^k N_j^k \mathrm{d}\Omega \tag{1.108}$$

$$C^k_{e(i,j)} = \int_{\Omega^e} \rho N^k_i N^k_{j,k} v_k \mathrm{d}\Omega \tag{1.109}$$

$$K^k_{e(i,j)} = \int_{\Omega^e} \rho \left(v + \frac{v_t}{\sigma_k} \right) N^k_{i,k} N^k_{j,k} \mathrm{d}\Omega \tag{1.110}$$

$$G^k_{e(i,j)} = \int_{\Omega^e} \beta^* \rho \omega N^k_i N^k_j \mathrm{d}\Omega \tag{1.111}$$

$$F^k_{e(i)} = \int_{\Omega^e} P_k N^k_i \mathrm{d}\Omega \tag{1.112}$$

For the omega equation at element level:

$$M^\omega_{ij} = \int_{\Omega^e} \rho N^\omega_i N^\omega_j \mathrm{d}\Omega \tag{1.113}$$

$$C^\omega_{ij} = \int_{\Omega^e} \rho N^\omega_i N^\omega_{j,k} v_k \mathrm{d}\Omega \tag{1.114}$$

$$K^\omega_{ij} = \int_{\Omega^e} \rho \left(v + \frac{v_t}{\sigma_\omega} \right) N^\omega_{i,k} N^\omega_{j,k} \mathrm{d}\Omega \tag{1.115}$$

$$G^\omega_{ij} = \int_{\Omega^e} \rho \beta_1 \omega N^\omega_i N^\omega_j \mathrm{d}\Omega \tag{1.116}$$

$$F^\omega_i = \int_{\Omega^e} \left\{ \gamma_1 \frac{\omega}{k} P_k \right\} N^\omega_i \mathrm{d}\Omega \tag{1.117}$$

1.2.8.3 Procedure for solving the k-epsilon/k-omega turbulence model

Time loop
  PMA predictor
  Iteration loop
    Solve N–S equations
    Solve k-epsilon equation
      Solve *k* equation
      Solve epsilon/omega equation
    Loop till *k* and epsilon/omega equation converge
    PMA corrector
    Solve mesh motion
  End of iteration loop
End of time loop

1.2.8.4 Large eddy simulation

Instead of time averaging, LES uses a spatial filtering operation to separate the larger and smaller eddies. The effects of unresolved small eddies on the mesh level are included by means of subgrid scale (SGS) models. Transient flow equations must be solved, so much larger demands on computer storage and computing speed are requested compared with the RANS model.

Spatial filtering of unsteady Navier–Stoke equations: in LES, we define a spatial filtering operation by means of a filter function $G(\boldsymbol{x}, \boldsymbol{x}', \Delta)$

$$\bar{\phi}(\boldsymbol{x},t)=\int_{-\infty}^{\infty}\int_{-\infty}^{\infty}\int_{-\infty}^{\infty}G(\boldsymbol{x},\boldsymbol{x}',\Delta)\,\phi(\boldsymbol{x}',t)\,\mathrm{d}x_1'\,\mathrm{d}x_2'\,\mathrm{d}x_3' \tag{1.118}$$

where

$\bar{\phi}(\boldsymbol{x},t)$ = filtered function
$\phi(\boldsymbol{x},t)$ = original (unfiltered) function
Δ = filter cutoff width

In this section, the over bar indicates that the spatial filtering is not time averaging. The choices for the filtering (Versteeg and Malalasekera, 2007):

- Box filter:

$$G(\mathbf{x},\mathbf{x}',\Delta)=\begin{cases}\dfrac{1}{\Delta^3} & |\boldsymbol{x}-\boldsymbol{x}'|\le\Delta/2\\ 0 & |\boldsymbol{x}-\boldsymbol{x}'|>\Delta/2\end{cases} \tag{1.119}$$

- Gaussian filter:

$$G(\mathbf{x},\mathbf{x}',\Delta)=\left(\frac{\gamma}{\pi\Delta^2}\right)^{3/2}\exp\left(-\gamma\frac{|\mathbf{x}-\mathbf{x}'|^2}{\Delta^2}\right) \tag{1.120}$$

γ is the parameter, and the typical value of it is 6.0.

- Spectral cutoff

$$G(\boldsymbol{x},\boldsymbol{x}',\Delta)=\prod_{i=1}^{3}\frac{\sin\left[(x_i-x_i')/\Delta\right]}{x_i-x_i'} \tag{1.121}$$

The Box filter is normally used in finite volume implementations of the LES method. The Gaussian filter is used in finite element and finite differences implementation. The spectral cutoff filter gives a sharp cutoff in the energy spectrum at the wavelength of Δ/π, but it cannot be used in general-purpose CFD.

The filter cutoff width can be defined as

$$\Delta=\sqrt{\Delta_x^2+\Delta_y^2+\Delta_z^2}\quad\text{or}\quad\Delta=\sqrt[3]{\Delta_x\Delta_y\Delta_z}, \tag{1.122}$$

for three-dimensional meshes. Here, Δx, Δy, Δz are the element sizes in x, y, z directions, respectively.

Filtered unsteady Navier–Stokes equations

The filtered unsteady Navier–Stokes equations can be expressed as:

$$\rho\left(\frac{\partial \bar{v}}{\partial t}+\bar{c}\cdot\nabla\bar{v}\right)=-\nabla\bar{p}+\rho\nabla\cdot(\nu\nabla\bar{v}-\tau)+\rho f. \tag{1.123}$$

Here, the SGS stress tensor τ can be written as:

$$\tau_{ij}=\overline{v_i v_j}-\bar{v}_i\bar{v}_j=L_{ij}+C_{ij}+R_{ij} \tag{1.124}$$

As expressed in the previous equation, the SGS stresses consist of three parts:

Leonard stresses L_{ij}: $L_{ij}=\rho\overline{\bar{v}_i\bar{v}_j}-\rho\bar{v}_i\bar{v}_j$

Cross-stresses C_{ij}: $C_{ij}=\rho\overline{\bar{v}_i v'_j}+\rho\overline{v'_i\bar{v}_j}$

LES Reynolds stresses: $R_{ij}=\rho\overline{v'_i\, v'_j}$

Smagorinsky–Lilly SGS model

Here we assume $L_{ij}+C_{ij}\approx 0$, $\tau_{ij}\approx R_{ij}$, and the SGS stresses become LES Reynolds stresses R_{ij}.

In the Smagorinsky's SGS model, the local LES stresses R_{ij} are taken to be proportional to the local rate of strain of the resolved flow:

$$\bar{S}_{ij}=\frac{1}{2}\left(\bar{v}_{i,j}+\bar{v}_{j,i}\right), \tag{1.125}$$

$$R_{ij}=-2\rho\nu_{\mathrm{SGS}}\bar{S}_{ij}+\frac{1}{3}R_{ij}\delta_{ij}, \tag{1.126}$$

$$\tau^*_{ij}=\tau_{ij}-\frac{1}{3}\delta_{ij}\tau_{kk}=-2\rho\nu_{\mathrm{SGS}}\bar{S}_{ij}, \tag{1.127}$$

where

$$\nu_{\mathrm{SGS}}=\left(C_s\Delta\right)^2\left|\bar{S}\right|,\quad \left|\bar{S}\right|=\sqrt{2\bar{S}_{ij}\bar{S}_{ij}}, \tag{1.128}$$

Then the N–S equations with Smagorinsky modes become:

$$\rho\left(\frac{\partial \bar{v}}{\partial t}+\bar{c}\cdot\nabla\bar{v}\right)=-\nabla\left(p+\frac{1}{3}\tau_{kk}\right)+\rho\nabla\cdot\left(\left(\nu+\nu_{\mathrm{SGS}}\right)\nabla\bar{v}\right)+\rho\mathbf{g}, \tag{1.129}$$

The following Van Driest function Equation (1.130) is generally used for the wall correction of C_s:

$$f=1.0-\exp\left(-\frac{y^+}{A^+}\right) \tag{1.130}$$

where

$$y^+ = \frac{v_\tau y}{v}, \quad v_\tau = \sqrt{\frac{\tau_w}{\rho}}, \quad A^+ = 25 \tag{1.131}$$

$$\tau_w = \mu \frac{\partial V}{\partial y}\bigg|_{y=0}. \tag{1.132}$$

1.2.9 The general transport equations

1.2.9.1 The governing equation of the transport equation

The scalar, time-dependent form of the advection diffusion equation is

$$\rho \frac{\partial \phi}{\partial t} + \rho \frac{\partial \phi}{\partial x_k} v_k - \frac{\partial}{\partial x_k}\left(\mu \frac{\partial \phi}{\partial x_k} \right) = \rho f. \tag{1.133}$$

Dividing the equation by ρ, we could obtain the classic form of the advection diffusion equation.

$$\frac{\partial \phi}{\partial t} + \frac{\partial \phi}{\partial x_k} v_k - \frac{\partial}{\partial x_k}\left(v \frac{\partial \phi}{\partial x_k} \right) = f \tag{1.134}$$

$$v = \frac{\mu}{\rho}. \tag{1.135}$$

1.2.9.2 The weak form of advection diffusion equation

The weak form of the advection–diffusion equation is to find $\phi \in \boldsymbol{\mathcal{S}}_\phi$, such that $\forall w \in \boldsymbol{W}_\phi$:

$$\int_\Omega w \cdot \left(\frac{\partial \phi}{\partial t} + \boldsymbol{c} \cdot \nabla \phi \right) \mathrm{d}\Omega + \int_\Omega \nabla w \cdot v \nabla \phi = \int_{\Gamma_h} w h \mathrm{d}\Gamma + \int_\Omega w f \mathrm{d}\Omega. \tag{1.136}$$

The Galerkin finite element formation of the advection–diffusion equation is stated as follows: find $\phi^h \in \boldsymbol{\mathcal{S}}_\phi^h$, such that $\forall w^h \in \boldsymbol{W}_\phi^h$

$$\int_\Omega w^h \cdot \left(\frac{\partial \phi}{\partial t} + \boldsymbol{c}^h \cdot \nabla \phi^h \right) \mathrm{d}\Omega + \int_\Omega \nabla w^h \cdot v \nabla \phi^h = \int_{\Gamma_h} w^h h \mathrm{d}\Gamma + \int_\Omega w^h f \mathrm{d}\Omega \tag{1.137}$$

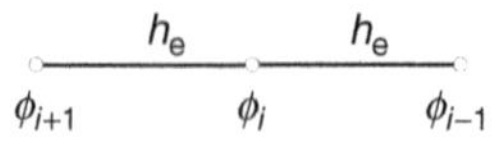

Figure 1.1 One-dimensional demonstration of the central differencing approach.

Where $\boldsymbol{S}_\phi^h$ and $\boldsymbol{W}_\phi^h$ are sets of finite-dimensional trial and test functions for scalar ϕ, respectively.

1.2.9.3 The SUPG stabilization for the advection-dominated advection–diffusion equation

Characteristics of the unstable equation system: for the advection-dominated advection–diffusion equation with Peclet number greater than 2.0, the traditional central differencing scheme may produce the spatial oscillation in the result. Therefore, a numerical stabilization method may be needed to stabilize the solution field.

1.2.9.3.1 Central differencing approach

The central differencing approach has second order accuracy, but it can produce spatial oscillation, when the Peclet number is greater than 2.0. It can be expressed as:

$$\left.\frac{\partial \phi}{\partial x}\right|_i = \frac{\phi_{i+1} - \phi_{i-1}}{2h_e}. \tag{1.138}$$

Where, ϕ_i is the primary scalar variable at node i and h_e is the element size (Figure 1.1).

$$\left.\frac{\partial \phi}{\partial x}\right|_i = \frac{\phi_{i+1} - \phi_{i-1}}{2h^e} \tag{1.139}$$

1.2.9.3.2 Upwind method for convection-dominant transport equations (first-order accuracy)

The upwind differencing scheme is the most stable and unconditionally bounded scheme, but it is based on the backward differencing formula, so the accuracy is first order on the basis of the Taylor series truncation error. It introduces a high level of false diffusion. The upwind differencing takes the flow direction into account when the value of ϕ is determined at a cell face; the convective value of ϕ at a cell face is taken to be equal to the value at the upstream node.

$$\left.\frac{\partial \phi}{\partial x}\right|_i = \begin{cases} \dfrac{\phi_{i+1} - \phi_i}{h^e} \text{flow direction} : i+1 \rightarrow i \\ \dfrac{\phi_i - \phi_{i-1}}{h^e} \text{flow direction} : i-1 \rightarrow i \end{cases}. \tag{1.140}$$

The SUPG method is used for convection-dominated heat transfer problem, when the Peclet number is higher than 2.0. The weak form of SUPG formulation for the transport equation is given in Equation (1.141). For details about the implementation of stabilization method for transport, please refer Chapter 5.

$$\begin{aligned}&\int_{\Omega} w^h \cdot \rho\left(\frac{\partial \phi^h}{\partial t}+\boldsymbol{c}^h \cdot \nabla \phi^h - f^h\right)\mathrm{d}\Omega + \int_{\Omega} \nabla w^h \cdot \nabla \phi^h \mathrm{d}\Omega - \int_{\Gamma_h} w^h h^h \mathrm{d}\Gamma \\ &+\sum_{e=1}^{n_{el}} \int_{\Omega^e} \tau_{\mathrm{SUPG}}(\boldsymbol{c}^h \cdot \nabla \phi^h) \cdot r^e(\phi^h)\mathrm{d}\Omega.\end{aligned} \tag{1.141}$$

Here, the residual of the transport equation is:

$$r^e\left(\phi^h\right) = \rho\left(\frac{\partial \phi^h}{\partial t}+\boldsymbol{c}^h \cdot \nabla \phi^h - f^h\right) - \nabla \cdot \left(\nu \nabla \phi^h\right) \tag{1.142}$$

The commonly used definitions of τ_{SUPG} are (Tezduyar and Sathe, 2003)

$$\tau_{\mathrm{SUPG}} = \left(\frac{1}{\tau_{\mathrm{SUPG}_1}^2}+\frac{1}{\tau_{\mathrm{SUPG}_2}^2}+\frac{1}{\tau_{\mathrm{SUPG}_3}^2}\right)^{-\frac{1}{2}} \tag{1.143}$$

where

$$\tau_{\mathrm{SUPG}_1} = \left(\sum_{i=1}^{nen} \left|\boldsymbol{c}^h \cdot \nabla N_i\right|\right)^{-1} \tag{1.144}$$

$$\tau_{\mathrm{SUPG}_2} = \frac{\Delta t}{2} \tag{1.145}$$

$$\tau_{\mathrm{SUPG}_3} = \frac{h_r^2}{4\nu} \tag{1.146}$$

$$h_r = 2.0\left(\sum_{a=1}^{n_{en}} \left|\boldsymbol{j} \cdot \nabla \boldsymbol{N}_a\right|\right)^{-1} \tag{1.147}$$

$$\boldsymbol{j} = \frac{\nabla \phi^h}{\left\|\nabla \phi^h\right\|} \tag{1.148}$$

1.2.9.4 Discontinuity capturing operator for the advection–diffusion equation

The additional term to overcome the undershoot and overshoot problem in the advection–diffusion equation is the discontinuity capturing operator:

$$S_{\text{DCDD}} = \sum_{e=1}^{n_{el}} \int_{\Omega^e} \rho \nabla \boldsymbol{w}^h : \left(\nu_{\text{DCDD}} \boldsymbol{\kappa} \cdot \nabla \boldsymbol{v}^h \right) d\Omega, \tag{1.149}$$

where

$$\boldsymbol{\kappa} = \boldsymbol{I} \tag{1.150}$$

For the Y, Z, β method,

$$\nu_{\text{DCDD}} = \left| Y^{-1} Z \right| \left(\sum_{i=1}^{n_{sd}} \left| Y^{-1} \frac{\partial \left\| \boldsymbol{v}^h \right\|}{\partial x_i} \right|^2 \right)^{\frac{\beta}{2}-1} \left(\frac{h_r}{2} \right)^{\beta} \tag{1.151}$$

For the governing equations of compressible flow, please refer Chapter 10.

1.3 Structural mechanics

1.3.1 Governing equations for structure analysis

The governing equations for structural mechanics is:

$$\rho \left. \frac{d^2 \overrightarrow{u_L}}{d^2 t} \right|_X = \frac{\partial \sigma_{ij}}{\partial x_j} + \rho g_i + \bar{f}_i \quad \text{in} \quad \Omega_t^S. \tag{1.152}$$

Boundary conditions

$$u_i = \overrightarrow{g_i} \qquad \text{on} \quad \Gamma_t^g \tag{1.153}$$

$$\sigma_{ji} n_j = \bar{h}_i \qquad \text{on} \quad \Gamma_t^h \tag{1.154}$$

Constitutive equation

$$\sigma_{ij} = c_{ijkl} \left(\varepsilon_{kl} - \frac{\alpha_{ij} \overleftarrow{\Delta T}}{T} \right) \quad \text{in} \quad \Omega_t^s. \tag{1.155}$$

Where, u_i is the component of the displacement vector $\boldsymbol{u}$; σ_{ij} is the component of the stress tensor $\boldsymbol{\sigma}$; c_{ijkl} is the component of the constitutive coefficient tensor $\boldsymbol{c}$; ε_{kl} is the component of the strain tensor $\boldsymbol{\varepsilon}$; α_{ij} is the component of the expansion coefficient tensor $\boldsymbol{\alpha}$; and ΔT is the temperature change. The strain components are:

$$\varepsilon_{ij} = \frac{1}{2}\left(u_{i,j} + u_{j,i}\right) \quad \text{in } \Omega_t^S. \tag{1.156}$$

Remark 1.7: The source term f_i represents the external body force that may come from the coupling physics model, for example, the Lorentz force. h_i in Equation (1.154) the surface force that may include fluid force, electrostatic force, and magnetic force. The displacement vector $\boldsymbol{u}$ will influence the simulation domain of any other nonstructural physics models involved in multiphysics simulation.

1.3.2 The equation in matrix form

1.3.2.1 The weak form by virtual work (Bathe, 2006)

Total Lagrange method-based virtual work:

$$\int_{\Omega_0} \delta \boldsymbol{u} \cdot \rho_0 \left({}^t\ddot{\boldsymbol{u}} + \ddot{\boldsymbol{u}}\right) \mathrm{d}V + \int_{\Omega_0} {}^{t+\Delta t}_{0}\boldsymbol{S} : \delta\, {}^{t+\Delta t}_{0}\boldsymbol{E} \mathrm{d}\Omega = \delta\, {}^{t+\Delta t}\mathfrak{R}. \tag{1.157}$$

Where ${}^{t+\Delta t}_{0}\boldsymbol{S}$ is the second Piola–Kirchhoff of stress tensor at time $= t + \Delta t$ on the original configuration ${}^{t+\Delta t}_{0}\boldsymbol{E}$, is the Green–Lagrange strain tensor; $\delta\, {}^{t'}\mathfrak{R}$, is the virtual work by external loads; ${}^t\ddot{\boldsymbol{u}}$, is the acceleration vector at time t; $\ddot{\boldsymbol{u}}$, is the increment of acceleration vector from time point $t + \Delta t$ to time point t; and δ, is the variation operator.

Updated Lagrange method based virtual work can be expressed as:

$$\int_{\Omega_t} \delta \boldsymbol{u} \cdot \rho_t \left({}^t\ddot{\boldsymbol{u}} + \ddot{\boldsymbol{u}}\right) \mathrm{d}V + \int_{\Omega_t} {}^{t+\Delta t}_{t}\boldsymbol{S} : \delta\, {}^{t+\Delta t}_{t}\boldsymbol{E} \mathrm{d}\Omega = \delta\, {}^{t+\Delta t}\mathfrak{R}. \tag{1.158}$$

Different from the total Lagrange formulation, the variables in the updated Lagrange method in Equation (1.158) are about the state at time t.

1.3.2.1.1 The solution space and shape function

For the displacement-based structure element, we need to do a spatial discretization for displacement vector space:

$$\boldsymbol{u}^h(\xi) = \sum_{i=1}^{n_{en}} N_i^u(\xi)\boldsymbol{u}_i = N_i^u(\xi)\boldsymbol{u}_i. \tag{1.159}$$

Here, $\boldsymbol{u}^h(\xi)$, is the displacement vector at any point ξ of the element; $N_i^u(\xi)$, is the corresponding shape function; $\boldsymbol{u}_i$, is the displacement vector at node i; and n_{en}, is the number of nodes of the structure element.

Remark 1.8: The mixed formulation may be needed to avoid numerical locking problems such as mixed interpolation shell element (Section 1.3.4.2) or displacement-pressure based element for incompressible hyperelasticity (Section 1.3.3.3). Although the mixed formulation is capable of avoiding the numerical locking problem, but as a cost, the inf-sup condition needs to be satisfied to have a stable and unique solution and the extra degrees of freedom (DOFs) will be added to the displacement system with more computational costs.

1.3.2.2 The equations in matrix form

Substituting the interpolation Equation (1.159) into the weak form of virtual work, one can obtain the equilibrium equations in matrix form:

$$\boldsymbol{M}\ddot{\boldsymbol{d}}+\boldsymbol{Q}(\boldsymbol{u})+\boldsymbol{K}_{ut}^{S}\Delta\boldsymbol{T}=\boldsymbol{F}^{E}+\overleftarrow{\boldsymbol{F}_{f}^{s}}+\overleftarrow{\boldsymbol{F}_{EM}^{S}}. \tag{1.160}$$

The primary variable for structure/solid analysis is usually displacement (displacement + pressure for mixed element). The material properties for linear elasticity are; Young's modulus, E; Poisson's ratio, ν; and mass density, ρ.

Where, $\boldsymbol{M}$, is the assembled mass matrix; $\boldsymbol{Q}(\boldsymbol{u})$, is the internal force vector, in the linear case; $\boldsymbol{Q}(\boldsymbol{u})=\boldsymbol{K}\boldsymbol{u}$; and $\boldsymbol{K}_{ut}^{S}$, is thermoelastic stiffness matrix,

$$\boldsymbol{K}_{ut}^{S}=\sum_{e}\boldsymbol{K}_{ut}^{S}=\sum_{e}\left(-\int_{\Omega^e}\boldsymbol{B}^{\mathrm{T}}\beta\boldsymbol{N}^{\mathrm{T}}\mathrm{d}\Omega\right). \tag{1.161}$$

$\boldsymbol{F}^{E}$, is the directly applied element external force vector; $\boldsymbol{F}_{f}^{s}$, is the element load vector caused by fluid force in FSI analysis; and $\boldsymbol{F}_{EM}^{S}$, is the force vector from the electromagnetic physics model in multiphysics simulation (e.g., electrostatic force, magnetic force, and Lorentz force).

1.3.3 The general nonlinearity in structural dynamics

1.3.3.1 Nonlinear material properties

The nonlinear behavior of material properties in solid/structure analysis can be classified into the following categories: (1) the nonlinear stress–strain relations, for example, the nonlinear elasticity, hyperelasticity, nonlinear elastoplastic material, nonlinear viscoelastic material, and nonlinear viscoplastic material, etc.; (2) material properties that depend on other properties or solutions, for example, temperature-dependent material properties and time-dependent material properties; (3) the nonlinearities caused by material failure, for example, material crack or damage. Besides the incremental formulation and the use of the appropriate tangential stiffness matrix, for the material nonlinear problem, especially for the problem with discontinuity behavior in the stiffness or strength, special numerical treatment is usually needed to obtain a converged solution during nonlinear iteration process (e.g., returning mapping method for plasticity material, homogenization method for crack development problems). The material nonlinearity increases the nonlinearity (Bathe, 2006) and numerical difficulties for a coupled physics problem. Since our major focus in this book is multiphysics simulation, only the nonlinear hyperelasticity material that is used in the engineering examples is presented here. For more details about the formulation and implementation of material nonlinear analysis in the FEM, we recommend books by Simo and Hughes (2006) and Belytschko (2013).

1.3.3.2 Geometrical nonlinearity

One source that causes geometrical nonlinearity and discontinuity is the large deformation problems, which can be large deflection, large rotation, post buckling, or large strain, and it is likely for all of them to happen at once for a single object under certain circumstances. On the other hand, the contact and impact problem is another resource that causes geometrical nonlinearity and discontinuity, not just by the large deformation of one object but also by the geometric interaction behavior of multiple bodies. The geometrical nonlinearity in structural physics is a major resource for nonlinearity and numerical difficulty of a coupling problem, such as FSI problems with large domain changes (Zhang and Hisada, 2001) in which spatial treatments for each physics solver and coupling characteristics will be generally needed to obtain a converged solution with good accuracy and reasonable computational amounts (refer Chapter 3 for some details). The contact element and large rotation and deformation in thin shell structure analysis are discussed and presented in the further section.

1.3.4 Incompressible hyperelastic material for the rubber (Watanabe, 1995)

1.3.4.1 High Mooney–Rivlin model

The hyperelastic material satisfies the following Equation (1.162):

$$S_{ij} = \frac{\partial W}{\partial E_{ij}} \tag{1.162}$$

where $\boldsymbol{E}$, is the Green–Lagrange strain tensor; $\boldsymbol{S}$, is the second Piola–Kirchhoff stress tensor; and W, is the elastic potential function. The high Mooney–Rivlin model in reduced invariants form can be expressed as:

$$\begin{aligned} W_R^H = & c_1\left(\tilde{I}c-3\right)+c_2\left(\widetilde{II}c-3\right)+c_3\left(\tilde{I}c-3\right)^2+c_4\left(\tilde{I}c-3\right)\left(\widetilde{II}c-3\right)^2 \\ & +c_5\left(\widetilde{II}c-3\right)^2+c_6\left(\tilde{I}c-3\right)^3+c_7\left(\tilde{I}c-3\right)^2\left(\widetilde{II}c-3\right) \\ & +c_8\left(\tilde{I}c-3\right)\left(\widetilde{II}c-3\right)^2+c_9\left(\widetilde{II}c-3\right)^3 \end{aligned} \tag{1.163}$$

$$\tilde{I}_C = \frac{I_C}{III_C^{1/3}}, \quad \tilde{II}_C = \frac{I_C}{III_C^{2/3}} \tag{1.164}$$

where I_c, II_c , and III_c are the main constants of the right Cauchy–Green strain tensor $\boldsymbol{C}(\boldsymbol{E} = (\boldsymbol{C} - \boldsymbol{I}/2))$. The parameters of $C_1 \sim C_9$ can be determined by curve fitting of experimental data. We use the mixed FEM for the spatial discretization of the hyperelastic solids.

1.3.4.2 Mixed interpolation for incompressible hyperelasticity

To model the incompressible hyperelasticity material, mixed interpolation for the displacement and pressure is usually used, which must satisfy the inf-sup condition to avoid the pressure oscillation (Bathe, 2006).

1.3.4.2.1 Finite element equation

The equilibrium equations of solid in matrix form can be expressed as follows:

$$\boldsymbol{M}^s \ddot{\boldsymbol{d}}^s + \boldsymbol{Q}^s\left(\boldsymbol{U}^s, \boldsymbol{P}^s\right) = \boldsymbol{F}^s \tag{1.165}$$

where,

$$\boldsymbol{M}^s = \begin{bmatrix} \boldsymbol{M}_u^s & \\ & \boldsymbol{0} \end{bmatrix}, \quad \boldsymbol{d}^s = \begin{Bmatrix} \boldsymbol{U}^s \\ \boldsymbol{p}^s \end{Bmatrix}, \quad \boldsymbol{Q}^s = \begin{Bmatrix} \boldsymbol{Q}_u^s \\ \boldsymbol{Q}_p^s \end{Bmatrix}, \quad \boldsymbol{F}^s = \begin{Bmatrix} \boldsymbol{F}_u^s \\ \boldsymbol{0} \end{Bmatrix}. \tag{1.166}$$

$\boldsymbol{M}^s$, is the mass matrix; $\boldsymbol{U}^s$ and $\boldsymbol{p}^s$, are the displacement and pressure vectors, respectively; and $\boldsymbol{Q}^s$ and $\boldsymbol{F}^s$, are the internal and external force vectors, respectively. Subscripts u and p, represent the value corresponding to velocity and pressure, respectively. $\ddot{}$ indicates the second order time derivative in the Lagrangian coordinates. The tangent matrices are used for nonlinear iterations of the incremental form of Equation (1.165).

$$\boldsymbol{M}^s \Delta \ddot{\boldsymbol{d}}^s + {}^t\boldsymbol{K}^s \Delta \boldsymbol{d}^s = {}^{t+\Delta t}\boldsymbol{F}^s - {}^t\boldsymbol{Q}^s \tag{1.167}$$

Here, ${}^t\boldsymbol{K}^s$ is the tangent stiffness matrix of solid at time t and Δ stands for the increments of variable vector of solid. Hyperelasticity material is used for the rubber modeling in the hydraulic engine mount analysis in Section 7.1.

The linear elasticity material with geometrical nonlinearity is used for the large deflection analysis of the wing structure in the Section 7.2.

1.3.5 Formulations for thin structure with large deformation

1.3.5.1 Basic equations for thin shell structure

The virtual work by the total Lagrange method is given in Equation (1.157). To have the incremental form of the virtual work equation, we first divide the stress and strain tensor into the total term at time point t and the incremental term from time point t to time point $t + \Delta t$.

$$ {}^{t+\Delta t}_{0}\boldsymbol{S} = {}^{t}_{0}\boldsymbol{S} + {}_{0}\boldsymbol{S} \tag{1.168}$$

$$ {}^{t+\Delta t}_{0}S^{ij} = {}^{t}_{0}S^{ij} + {}_{0}S^{ij} \tag{1.169}$$

$$ \delta\, {}^{t+\Delta t}_{0}E_{ij} = \delta({}^{t}_{0}E_{ij} + {}_{0}E_{ij}) = \delta\, {}_{0}E_{ij} = \delta\, {}_{0}e_{ij} + \delta\, {}_{0}\eta_{ij} \tag{1.170}$$

Here, ${}_{0}e_{ij}$ and ${}_{0}\eta_{ij}$, are the linear and nonlinear terms of ${}_{0}E_{ij}$, respectively.

Substituting Equations (1.169) and (1.170) into Equation (1.157), will give the incremental form of the virtual work equation.

$$ \begin{aligned} &\int_{\Omega_0} \delta u_i \rho_0 \ddot{u}_i \mathrm{d}\Omega + \int_{\Omega_0} {}_{0}S^{ij} : \delta\, {}_{0}E_{ij} \mathrm{d}\Omega + \int_{\Omega_0} {}^{t}_{0}S^{ij} : \delta\, {}_{0}\eta_{ij} \mathrm{d}\Omega = \\ &\quad \delta\, {}^{t+\Delta t}\mathfrak{R} - \int_{\Omega_0} {}^{t}_{0}S^{ij} : \delta\, {}_{0}e_{ij} \mathrm{d}\Omega - \int_{\Omega_0} \delta u_i \rho_0\, {}^{t}\ddot{u}_i \mathrm{d}\Omega. \end{aligned} \tag{1.171}$$

The Green–Lagrange strain tensor at time $t + \Delta t$ in the original configuration,

$$ {}^{t+\Delta t}_{0}\boldsymbol{E} = \frac{1}{2}\left({}^{t+\Delta t}g_i \cdot {}^{t+\Delta t}g_j - G_i \cdot G_j\right) G^i \otimes G^j \tag{1.172}$$

For the incremental term of ${}_{0}\boldsymbol{E}$,

$$ {}_{0}\boldsymbol{E} = \frac{1}{2}\left({}^{t+\Delta t}\boldsymbol{g}_i \cdot {}^{t+\Delta t}\boldsymbol{g}_j - {}^{t}\boldsymbol{g}_i \cdot {}^{t}\boldsymbol{g}_j\right) \boldsymbol{G}^i \otimes \boldsymbol{G}^j \tag{1.173}$$

where $\boldsymbol{G}_i$, ${}^{t}\boldsymbol{g}_i$, and ${}^{t+\Delta t}\boldsymbol{g}_i$ are the cobase vectors at time = 0, t and $t + \Delta t$, respectively.

Based on the definition,

$$ {}^{t+\Delta t}\boldsymbol{g}_i = \frac{\partial\, {}^{t+\Delta t}\boldsymbol{x}}{\partial r^i} = \frac{\partial({}^{t}\boldsymbol{x} + \boldsymbol{u})}{\partial r^i} = {}^{t}\boldsymbol{g}_i + \frac{\partial \boldsymbol{u}}{\partial r^i} \tag{1.174}$$

substituting Equation (1.174) into Equation (1.173), we have

$$ {}_{0}\dot{e}_{ij} = \frac{1}{2}\left(\frac{\partial \dot{\boldsymbol{u}}}{\partial r^i} \cdot {}^{t}\boldsymbol{g}_j + {}^{t}\boldsymbol{g}_i \cdot \frac{\partial \dot{\boldsymbol{u}}}{\partial r^j}\right) \tag{1.175}$$

$$\delta_0 e_{ij} = \frac{1}{2}\left(\frac{\partial \boldsymbol{u}}{\partial r^i}\cdot {}^t\boldsymbol{g}_j + {}^t\boldsymbol{g}_i \cdot \frac{\partial \boldsymbol{u}}{\partial r^j}\right) \tag{1.176}$$

$$\delta_0 \dot{\eta}_{ij} = \frac{1}{2}\left(\frac{\partial \delta \boldsymbol{u}}{\partial r^i}\cdot \frac{\partial \dot{\boldsymbol{u}}}{\partial r^j} + \frac{\partial \dot{\boldsymbol{u}}}{\partial r^i}\cdot \frac{\partial \delta \boldsymbol{u}}{\partial r^j}\right) \tag{1.177}$$

1.3.5.1.1 The transformation of the constitutive law

$${}_0\boldsymbol{C} = {}_0C^{ijkl}\boldsymbol{G}^i \otimes \boldsymbol{G}^j \otimes \boldsymbol{G}^k \otimes \boldsymbol{G}^l = {}_0\tilde{C}_{mnop}\tilde{\boldsymbol{e}}_m \otimes \tilde{\boldsymbol{e}}_n \otimes \tilde{\boldsymbol{e}}_o \otimes \tilde{\boldsymbol{e}}_p \tag{1.178}$$

$${}_0C^{ijkl} = {}_0\tilde{C}_{mnop}(\boldsymbol{G}^i \cdot \tilde{\boldsymbol{e}}_m)(\boldsymbol{G}^k \cdot \tilde{\boldsymbol{e}}_o)(\boldsymbol{G}^l \cdot \tilde{\boldsymbol{e}}_p)(\boldsymbol{G}^j \cdot \tilde{\boldsymbol{e}}_n) \tag{1.179}$$

Here, $\tilde{\boldsymbol{e}}_m$ is the unit orthogonal base vector in the local coordinate of the shell element.

1.3.6 Mixed interpolation tensorial component shell element (Bathe, 2006)

Assumptions for mixed interpolation tensorial components (MITC) shell element:

1. Isoparametric degenerated shell assumption.
2. Redefine the out of plane shear strain by direct interpolation from those values on the sample points.
3. Use the covariant and contravariant base vectors to define the strain and stress components in the natural coordinate system.

1.3.6.1 Interpolation and shape function for MITC shell element

The coordinate and displacement of any point in the shell element can be expressed as the interpolation of related nodal values.

Coordinates of any point at time = 0,

$$\boldsymbol{X} = \sum_{a=1}^{a=ne} N_a(r^1, r^2)\boldsymbol{X}^a + \frac{r^3}{2}\sum_{a=1}^{a=ne} N_a(r^1, r^2)\,{}^0\boldsymbol{V}_3^a \tag{1.180}$$

Coordinates of any point at time = t,

$${}^t\boldsymbol{x} = \sum_{a=1}^{a=ne} N_a(r^1, r^2)\,{}^t\boldsymbol{x}^a + \frac{r^3}{2}t\sum_{a=1}^{a=ne} N_a(r^1, r^2)\,{}^t\boldsymbol{V}_3^a \tag{1.181}$$

where, N_a is the shape function at node a, so the displacement vector at time = t and the displacement increment from time point t to time point $t + \Delta t$,

$${}^t\boldsymbol{u} = {}^t\boldsymbol{x} - \boldsymbol{X}, \boldsymbol{u} = {}^{t+\Delta t}\boldsymbol{x} - {}^t\boldsymbol{x} \tag{1.182}$$

$$ {}^{t}\boldsymbol{u} = \sum_{a=1}^{a=ne} N_a(r^1, r^2)\, {}^{t}\boldsymbol{u}^a + \frac{r^3}{2} t \sum_{a=1}^{a=ne} N_a(r^1, r^2)({}^{t}\boldsymbol{V}_3^a - {}^{0}\boldsymbol{V}_3^a) \tag{1.183} $$

$$ \boldsymbol{u} = \sum_{a=1}^{a=ne} N_a(r^1, r^2)\boldsymbol{u}^a + \frac{r^3}{2} t \sum_{a=1}^{a=ne} N_a(r^1, r^2)({}^{t+\Delta t}\boldsymbol{V}_3^a - {}^{t}\boldsymbol{V}_3^a) \tag{1.184} $$

Finite rotation matrix ${}^{t+\Delta t}_{t}\boldsymbol{R}$ indicates the rotation from time point t to time point $t + \Delta t$,

$$ {}^{t+\Delta t}\boldsymbol{V}_3^a = {}^{t+\Delta t}_{t}\boldsymbol{R}\, {}^{t}\boldsymbol{V}_3^a \tag{1.185} $$

$$ \boldsymbol{u} = \sum_{a=1}^{a=ne} N_a(r^1, r^2)\boldsymbol{u}^a + \frac{r^3}{2} t \sum_{a=1}^{a=ne} N_a(r^1, r^2)({}^{t+\Delta t}_{t}\boldsymbol{R} - \boldsymbol{I})\ {}^{t}\boldsymbol{V}_3^a \tag{1.186} $$

By introducing the finite rotation, Equation (1.186) changes to:

$$ \begin{aligned} \boldsymbol{u} = & \sum_{a=1}^{a=ne} N_a(r^1, r^2)\boldsymbol{u}^a + \frac{r^3}{2} t \sum_{a=1}^{a=ne} N_a(r^1, r^2)\left(-{}^{t}\boldsymbol{V}_2^a \alpha^a + {}^{t}\boldsymbol{V}_1^a \beta^a\right) \\ & + \frac{r^3}{2} t \sum_{a=1}^{a=ne} N_a(r^1, r^2) \frac{-1}{2!}\left\{(\alpha^a)^2 + (\beta^a)^2\right\} {}^{t}\boldsymbol{V}_3^a \end{aligned} \tag{1.187} $$

Here, the first two terms in Equation (1.187) express for small rotational degenerated shell element, although the last term indicates the finite rotation effect.

The interpolation for out of plane shear strain,

$$ E_{13} = \frac{1}{2}(1 + r^2)E_{13}^A + \frac{1}{2}(1 - r^2)E_{13}^C \tag{1.188} $$

$$ E_{23} = \frac{1}{2}(1 + r^1)E_{23}^D + \frac{1}{2}(1 - r^1)E_{23}^B \tag{1.189} $$

where, E_{13} and E_{23} are the shear strains of the shell element at arbitrary points and $E^{A \sim D}$ is the shear strain of the shell element at the sample points.

1.3.6.2 Finite element equations

Based on Equation (1.171) and combined with Equations (1.187)–(1.189), the finite element-based equilibrium equation can be achieved in incremental form:

$$ {}^{t}\boldsymbol{M}\Delta\ddot{\boldsymbol{U}} + {}^{t}(\boldsymbol{K}_L + \boldsymbol{K}_{NL})\Delta\boldsymbol{U} = {}^{t+\Delta t}\boldsymbol{F} - {}^{t}\boldsymbol{Q} \tag{1.190} $$

where $\boldsymbol{U}$ is the displacement vector of the shell structure, ${}^{t}\boldsymbol{M}$ is the mass matrix of the shell structure at time t, ${}^{t}(\boldsymbol{K}_L + \boldsymbol{K}_{NL})$ is the summation of linear and nonlinear stiffness matrices of the shell structure, ${}^{t+\Delta t}\boldsymbol{F}$ is the external load vector at time $t + \Delta t$ of the shell structure, ${}^{t}\boldsymbol{Q}$ is the internal nodal force vector of the shell structure at time t, $\Delta\ddot{\boldsymbol{U}}$ is the increment of the acceleration vector of the shell structure from time t to time $t + \Delta t$, and $\Delta\boldsymbol{U}$ is the increment of the displacement vector of the shell structure from time t to time $t + \Delta t$.

1.3.6.3 The added stiffness matrix by considering finite rotation

Green–Lagrange strain by considering the second order of nonlinear terms in Equations (1.175)–(1.177),

$$ {}_0\dot{e}_{ij} = \frac{1}{2}\left(\frac{\partial\dot{\boldsymbol{u}}_s}{\partial r^i}\cdot{}^t\boldsymbol{g}_j + {}^t\boldsymbol{g}_i\cdot\frac{\partial\dot{\boldsymbol{u}}_s}{\partial r^j}\right) \tag{1.191} $$

$$ \delta_0 e_{ij} = \frac{1}{2}\left(\frac{\partial\boldsymbol{u}_s}{\partial r^i}\cdot{}^t\boldsymbol{g}_j + {}^t\boldsymbol{g}_i\cdot\frac{\partial\boldsymbol{u}_s}{\partial r^j}\right) \tag{1.192} $$

$$ \begin{aligned} \delta_0\dot{\eta}_{ij} &= \frac{1}{2}\left(\frac{\partial\delta\boldsymbol{u}_s}{\partial r^i}\cdot\frac{\partial\dot{\boldsymbol{u}}_s}{\partial r^j} + \frac{\partial\dot{\boldsymbol{u}}_s}{\partial r^i}\cdot\frac{\partial\delta\boldsymbol{u}_s}{\partial r^j}\right) \\ &\quad + \frac{1}{2}\left(\frac{\partial(\delta\boldsymbol{u}_{ex})^{\cdot}}{\partial r^i}\cdot{}^t\boldsymbol{g}_j + {}^t\boldsymbol{g}_i\cdot\frac{\partial\dot{\boldsymbol{u}}_s}{\partial r^j}\cdot\frac{\partial(\delta\boldsymbol{u}_{ex})^{\cdot}}{\partial r^j}\right) \end{aligned} \tag{1.193} $$

Here in Equation (1.193), the time derivative of the finite rotation related displacement can be expressed as:

$$ \delta\dot{u}_{ex} = -\frac{r^3}{2}t\sum_{a=1}^{a=ne}N_a(r^1,r^2)(\delta\alpha^a\dot{\alpha}^a + \delta\beta^a\dot{\beta}^a)\ {}^t\boldsymbol{V}_3^a. \tag{1.194} $$

And the final stiffness matrix K can be obtained by adding the finite rotation term $\boldsymbol{K}_{NL,ex}$ to the standard term $\boldsymbol{K}_s$ (Equation (1.95).

$$ \boldsymbol{K} = \boldsymbol{K}_s + \boldsymbol{K}_{NL,ex}, \tag{1.195} $$

1.3.6.4 Stabilization method for the MITC shell element

The modified mass matrix for rotational DOFs is introduced for thin shell structure to stabilize the dynamics solution with reasonable time step size (refer Section 5.3.3 for details). The engineering applications of MITC shell elements in structure analysis and coupling physics simulations are given in Sections 5.3.4 and 13.1, respectively.

1.3.7 Generalized conforming flat shell element

The generalized conforming element method (Long et al., 2009; Long and Xin, 1989; Long et al., 1995) was originally proposed to solve the difficulty of C1 continuity required by thin plate elements. It opens a new way between conforming and nonconforming elements. On the one hand, the shortcomings, that sometimes conforming elements are overstiff and even difficult to be constructed, are overcome. On the other hand, fatal weakness that nonconforming elements may not be convergent is also eliminated. It obtains first success in constructions of thin plate elements, and various high-performance thin plate element models of different types are then successfully constructed.

At the same time, the idea of generalized conforming method has also been successfully generalized into many other areas, such as thick plate elements (Soh et al., 2001), laminated composite plate elements (Cen et al., 2002; Cen et al., 2007), piezoelectric laminated composite plate elements (Cen et al., 2002), isoparametric membrane elements (Long and Huang, 1988; Soh et al., 1999), membrane elements with rotational freedom(Long and Xu, 1994a, 1994b), generalized conforming flat (GCF) shell element (Chen et al., 2003), generalized conforming curved shallow shell element (Sun et al., 2001), and so on.

This section mainly introduces research works on the applications of generalized conforming element methods for flat shell element. Thus, the universal significance of generalized conforming theory can be exhibited. Flat shell element, which is composed of plate bending element and plane membrane element (Zienkiewicz and Taylor, 1991), is the simplest shell element model, and it is widely used in the linear and nonlinear problems. Reviews on general formulations and characters of the flat shell element can be found in Zienkiewicz and Taylor (1991) and Long and Cen (2001). The appearance of the new generalized conforming membrane element with drilling freedoms and new generalized conforming thin plate element makes it possible to construct high-performance flat shell elements.

Long et al. (2009) have introduced the concept of drilling freedom (the additional in-plane rigid vertex rotational freedom) and proposed the formulations of new generalized conforming rectangular membrane element GR12 and triangular membrane elements GT9 and GT9M8 with drilling freedoms. Furthermore, in Long et al. (1995), generalized conforming rectangular thin plate element GPL-R12 and triangular thin plate element GPL-T9, were constructed. Then in Long and Xu (1993, 1994), Xu et al. (1994), Xu and Long (1996), and Xu et al. (1999), by starting with the generalized conforming theory and degenerated potential energy principle, above-plane membrane and plate-bending elements are used to formulate several GCF shell elements for analysis of cylindrical and arbitrary shells:

1. Generalized conforming rectangular flat shell element GCR24 (Long and Xu, 1994), developed by the combination of generalized conforming rectangular membrane element GR12 and rectangular thin plate element GPL-R12
2. Generalized conforming triangular flat shell element GST18 (Long and Xu, 1993), developed by the combination of generalized conforming triangular membrane element GT9 and triangular thin plate element GPL-T9, and one-point reduced integration scheme is used for GT9
3. Generalized conforming triangular flat shell element GST18M (Xu et al., 1999), developed by the combination of generalized conforming triangular membrane element GT9M8 and triangular thin plate element GPL-T9

Furthermore, Chen et al. (2003) developed a generalized conforming triangular thick/thin flat shell element GMST18. First, the formulation of generalized conforming triangular membrane element GT9 is used as membrane component of the shell element. Both a one-point reduced integration scheme and a corresponding stabilization matrix proposed by Fish and Belytschko (1992) are adopted for avoiding membrane locking and hourglass phenomenon. Second, bending component of the element comes from a new generalized conforming thick/thin plate element TSL-T9, which is derived from rational shear interpolation (Soh et al., 2001; Cen et al., 2006) and SemiLoof conforming scheme (Long, 1992, 1993). In this section, as an example element GMST18 (Chen et al., 2003) is used to describe the construction procedure of GCF shell element.

Using the updated Lagrangian (UL) formulation with arc-length method, application of the GT9 shell element in the postbuckling and geometric nonlinear analysis of spatial shell structure and stiffened shell structure was first studied by Zhang et al. (1998; Zhang, Ku, and Kuang, 1999). The geometrical nonlinear formulation of GMST18 was derived and used for larger deformation simulation of shell structure (Chen et al., 2003).

1.3.7.1 Formulations of generalized conforming flat shell element

As shown in Figure 1.2, the triangular flat shell element in the local coordinate system xyz is assembled by plane membrane and plate-bending element.

The element nodal displacement vector $\boldsymbol{q}^e$ in the local coordinate system xyz is composed of the vertex freedoms:

$$\boldsymbol{q}^e = \begin{Bmatrix} \boldsymbol{q}_1^e \\ \boldsymbol{q}_2^e \\ \boldsymbol{q}_3^e \end{Bmatrix}, \boldsymbol{q}_i^e = [\, u_i \;\; v_i \;\; w_i \;\; \theta_{xi} \;\; \theta_{yi} \;\; \theta_{zi} \,]^{\mathrm{T}}, (i = 1, 2, 3). \tag{1.196}$$

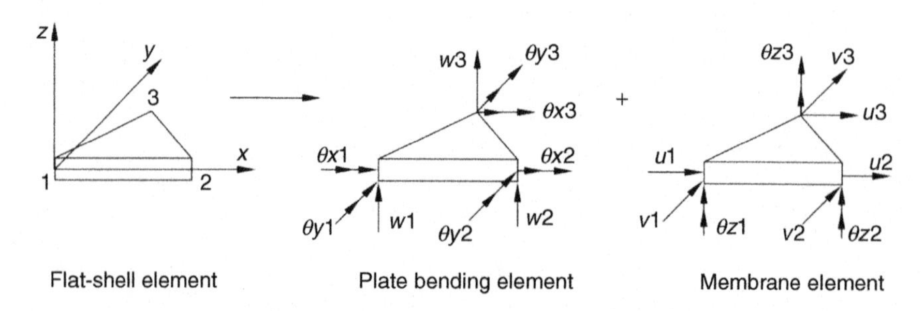

Figure 1.2 Flat shell element in the local coordinate system xyz.

Let $\boldsymbol{q}_m^e$, be the nodal displacement vector related to the membrane element and $\boldsymbol{q}_p^e$, be the nodal displacement vector related to the plate element. Then we have,

$$\boldsymbol{q}_m^e = \begin{Bmatrix} \boldsymbol{q}_{m1}^e \\ \boldsymbol{q}_{m2}^e \\ \boldsymbol{q}_{m3}^e \end{Bmatrix}, \boldsymbol{q}_{mi}^e = \begin{Bmatrix} u_i \\ v_i \\ \theta_{zi} \end{Bmatrix}, \boldsymbol{q}_p^e = \begin{Bmatrix} \boldsymbol{q}_{p1}^e \\ \boldsymbol{q}_{p2}^e \\ \boldsymbol{q}_{p3}^e \end{Bmatrix}, \boldsymbol{q}_{pi}^e = \begin{Bmatrix} w_i \\ \theta_{xi} \\ \theta_{yi} \end{Bmatrix}, (i = 1, 2, 3). \quad (1.197)$$

1.3.7.2 Membrane part – triangular membrane element GT9

The membrane element used here is the 3-node triangular generalized conforming membrane element GT9 (Long and Xu, 1994) with rigid rotational freedoms, and its displacement field is given by,

$$\begin{Bmatrix} u \\ v \end{Bmatrix} = \sum_{i=1}^{3} \begin{bmatrix} L_i & 0 & N_{u\theta i} \\ 0 & L_i & N_{v\theta i} \end{bmatrix} \boldsymbol{q}_{mi}^e \quad (1.198)$$

where, L_i (i = 1,2,3) are the triangular area coordinates; $N_{u\theta i}$ and $N_{v\theta i}$, are the shape functions for rigid rotation freedoms θ_{zi} (Figure 1.2) in which

$$N_{u\theta i} = \frac{1}{2} L_i (b_k L_j - b_j L_k), N_{v\theta i} = \frac{1}{2} L_i (c_k L_j - c_j L_k), (i, j, k = \overrightarrow{1,2,3}) \quad (1.199)$$

with

$$b_i = y_j - y_k, \quad c_i = x_j - x_k \quad (1.200)$$

Differentiation of Equation (1.198) yields the strain matrix of element,

$$\boldsymbol{B}_m = \begin{bmatrix} \boldsymbol{B}_{m1} & \boldsymbol{B}_{m2} & \boldsymbol{B}_{m3} \end{bmatrix} \quad (1.201)$$

where,

$$\boldsymbol{B}_{mi} = \frac{1}{4A} \begin{bmatrix} 2b_i & 0 & b_i (b_k L_j - b_j L_k) \\ 0 & 2c_i & c_i (c_k L_j - c_j L_k) \\ 2c_i & 2b_i & (b_i c_k + b_k c_i) L_j - (b_i c_j + b_j c_i) L_k \end{bmatrix}, (i, j, k = \overrightarrow{1,2,3}) \quad (1.202)$$

where A denotes the area of the element.

Then the element stiffness matrix of GT9 in the local coordinate system can be obtained as follows:

$$\boldsymbol{K}_m^e = \int_{A^e} \boldsymbol{B}_m^{\mathrm{T}} \boldsymbol{D}_m \boldsymbol{B}_m \mathrm{d}A \tag{1.203}$$

where $\boldsymbol{D}_m$ is the elasticity matrix:

$$\boldsymbol{D}_m = \frac{Eh}{1-\nu^2}\begin{bmatrix} 1 & \nu & 0 \\ \nu & 1 & 0 \\ 0 & 0 & \dfrac{1-\nu}{2} \end{bmatrix} \tag{1.204}$$

where, E is Young's modulus; ν is Poisson's ratio; and h is the thickness of the element.

If $\theta_{z1} = \theta_{z2} = \theta_{z3}$, there exists an extra zero energy mode in addition to the conventional rigid body movement. This extra zero energy mode can be easily suppressed by setting one of the nodal rotational degrees to be zero in one element of the mesh.

1.3.7.2.1 Matrix stabilization method for GT9

To avoid membrane locking in the calculation of shells, one-point reduced integration is often used for the membrane component with rotational freedoms. But unfortunately, extra zero energy modes of the element will appear, and the hourglass phenomenon may occur. Fish and Belytschko (1992) suggested a method of adding a stabilization matrix to overcome this shortcoming. According to their approach, the stabilization matrix of the element GT9 is given as follows

$$\boldsymbol{K}_{m\,stab}^e = \varpi \begin{bmatrix} 0 & 0 & 0 & 0 & 0 & 0 & 0 & 0 & 0 \\ 0 & 0 & 0 & 0 & 0 & 0 & 0 & 0 & 0 \\ 0 & 0 & 2 & 0 & 0 & -1 & 0 & 0 & -1 \\ 0 & 0 & 0 & 0 & 0 & 0 & 0 & 0 & 0 \\ 0 & 0 & 0 & 0 & 0 & 0 & 0 & 0 & 0 \\ 0 & 0 & -1 & 0 & 0 & 2 & 0 & 0 & -1 \\ 0 & 0 & 0 & 0 & 0 & 0 & 0 & 0 & 0 \\ 0 & 0 & 0 & 0 & 0 & 0 & 0 & 0 & 0 \\ 0 & 0 & -1 & 0 & 0 & -1 & 0 & 0 & 2 \end{bmatrix} \tag{1.205}$$

with

$$\varpi = \frac{1}{3}\chi\left[\boldsymbol{K}_m^1(3,3)^e + \boldsymbol{K}_m^1(6,6)^e + \boldsymbol{K}_m^1(9,9)^e\right] \tag{1.206}$$

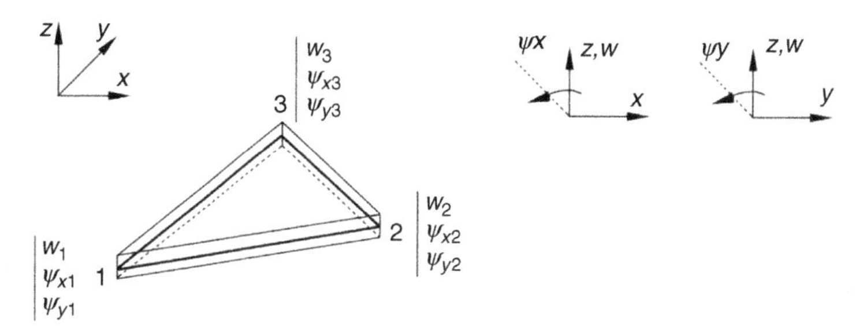

Figure 1.3 Triangular plate-bending element TSL-T9.

where $\boldsymbol{K}_m^{1e}$, denotes the element stiffness matrix of GT9 using one-point integration; and χ, is a perturbation factor. From numerical experiments, it is found that when χ is not less than 10^{-6}, the rank and eigenvalues of the new shell element are correct. So $\chi = 10^{-6}$ is adopted throughout this section.

Thus, the element stiffness matrix of GT9 in the local coordinate system can be modified as

$$\boldsymbol{K}_m^e = \boldsymbol{K}_m^{1e} + \boldsymbol{K}_{m\,stab}^e \tag{1.207}$$

1.3.7.3 Plate bending part – triangular thick/thin plate element TSL-T9

The triangular plate-bending element TSL-T9 in the local coordinate system xyz is shown in Figure 1.3. The element nodal displacement vector is composed of deflection w and normal slopes ψ_x and ψ_y of the midsurface

$$\bar{\boldsymbol{q}}^e = \begin{bmatrix} w_1 & \psi_{x1} & \psi_{y1} & w_2 & \psi_{x2} & \psi_{y2} & w_3 & \psi_{x3} & \psi_{y3} \end{bmatrix}^{\mathrm{T}}. \tag{1.208}$$

Note that, since the definitions of the rotations (ψ_x, ψ_y and θ_x, θ_y) are different, there exists the following relations between $\bar{\boldsymbol{q}}^e$ and $\boldsymbol{q}_p^e$ in Equation (1.197):

$$\bar{\boldsymbol{q}}^e = \boldsymbol{L}\boldsymbol{q}_p^e,\ \boldsymbol{L} = \begin{bmatrix} \boldsymbol{I} & \boldsymbol{0} & \boldsymbol{0} \\ \boldsymbol{0} & \boldsymbol{I} & \boldsymbol{0} \\ \boldsymbol{0} & \boldsymbol{0} & \boldsymbol{I} \end{bmatrix},\ \boldsymbol{I} = \begin{bmatrix} 1 & 0 & 0 \\ 0 & 0 & -1 \\ 0 & 1 & 0 \end{bmatrix},\ \boldsymbol{0} = \begin{bmatrix} 0 & 0 & 0 \\ 0 & 0 & 0 \\ 0 & 0 & 0 \end{bmatrix} \tag{1.209}$$

If the construction procedure of the element shear strain fields is the same as that described in (Soh et al., 1999), then the final element shear strain field is

$$\gamma = \boldsymbol{B}_s\boldsymbol{q}_p^e = \boldsymbol{H}\boldsymbol{\Delta}'\tilde{\boldsymbol{G}}\boldsymbol{L}\boldsymbol{q}_p^e \tag{1.210}$$

where $\boldsymbol{B}_s$ is the element shear strain matrix. $\boldsymbol{H}$, $\boldsymbol{\Delta}'$, and $\tilde{\boldsymbol{G}}$ are given by

$$\boldsymbol{H} = \frac{1}{2A}\begin{bmatrix} b_3L_2 - b_2L_3 & b_1L_3 - b_3L_1 & b_2L_1 - b_1L_2 \\ c_3L_2 - c_2L_3 & c_1L_3 - c_3L_1 & c_2L_1 - c_1L_2 \end{bmatrix} \tag{1.211}$$

$$\boldsymbol{\Delta}' = \begin{bmatrix} \delta_1 & 0 & 0 \\ 0 & \delta_2 & 0 \\ 0 & 0 & \delta_3 \end{bmatrix} \text{with } \delta_i = \frac{(h/d_i)^2}{\frac{5}{6}(1-\nu)+2(h/d_i)^2} \tag{1.212}$$

where, h is the thickness and d_i is the edge length.

$$\tilde{\boldsymbol{G}} = \begin{bmatrix} 0 & 0 & 0 & -2 & -c_1 & b_1 & 2 & -c_1 & b_1 \\ 2 & -c_2 & b_2 & 0 & 0 & 0 & -2 & -c_2 & b_2 \\ -2 & -c_3 & b_3 & 2 & -c_3 & b_3 & 0 & 0 & 0 \end{bmatrix} \tag{1.213}$$

The element deflection field is assumed to be the same as

$$w = \boldsymbol{F}_\lambda \boldsymbol{\lambda} \tag{1.214}$$

where

$$\boldsymbol{\lambda} = [\,\lambda_1\ \lambda_2\ \lambda_3\ \lambda_4\ \lambda_5\ \lambda_6\ \lambda_7\ \lambda_8\ \lambda_9\ \lambda_{10}\ \lambda_{11}\ \lambda_{12}\,]^{\mathrm{T}} \tag{1.215}$$

$$\begin{aligned} \boldsymbol{F}_\lambda = [\, & L_1\ L_2\ L_3\ L_2L_3\ L_3L_1\ L_1L_2\ L_2L_3(L_2-L_3) \\ & \times L_3L_1(L_3-L_1)\ \ L_1L_2(L_1-L_2)\ L_1^2L_2L_3\ L_2^2L_3L_1\ L_3^2L_1L_2\,] \end{aligned} \tag{1.216}$$

According to Mindlin–Reissner plate theory, the element rotation fields are

$$\boldsymbol{\psi} = \begin{Bmatrix} \psi_x \\ \psi_y \end{Bmatrix} = \begin{Bmatrix} \frac{\partial w}{\partial x} - \gamma_{xz} \\ \frac{\partial w}{\partial y} - \gamma_{yz} \end{Bmatrix} = \begin{bmatrix} \boldsymbol{F}_{\lambda,x} \\ \boldsymbol{F}_{\lambda,y} \end{bmatrix} \boldsymbol{\lambda} - \boldsymbol{B}_s \boldsymbol{q}_p^e \tag{1.217}$$

where $\boldsymbol{F}_{\lambda,x}$ and $\boldsymbol{F}_{\lambda,y}$ denote the derivative matrices of $\boldsymbol{F}_\lambda$ with respect to x and y, respectively.

Along the element sides, deflection $\tilde{w}$ is interpolated according to the thick beam theory, and the normal slope $\tilde{\psi}_n$ is assumed to be linearly distributed.

SemiLoof point conforming conditions are introduced in the following equations:

$$(w-\tilde{w})_i = 0 \qquad (i = 1,2,3) \tag{1.218}$$

$$(w-\tilde{w})_j = 0 \quad (j=4,5,6) \tag{1.219}$$

$$(\psi_n - \tilde{\psi}_n)_k = 0 \qquad (k = A_1, B_1, A_2, B_2, A_3, B_3) \tag{1.220}$$

Equations (1.218) and (1.219) are the point conforming conditions about deflections at corner nodes (nodes 1, 2, and 3) and midside points (points 4, 5, and 6), respectively. Equation (1.220) denotes the point conforming conditions about normal slopes at Gauss points on element side (points A_1, B_1, A_2, B_2, A_3, and B_3).

Then, $\lambda_1, \ldots, \lambda_{12}$ can be obtained, in which the last three coefficients are equal to each other, that is, $\lambda_{10} = \lambda_{11} = \lambda_{12}$. Therefore, Equation (1.214) can be rewritten as

$$w = \boldsymbol{F}'_\lambda \boldsymbol{\lambda}' \tag{1.221}$$

where

$$\boldsymbol{\lambda}' = [\ \lambda_1\ \ \lambda_2\ \ \lambda_3\ \ \lambda_4\ \ \lambda_5\ \ \lambda_6\ \ \lambda_7\ \ \lambda_8\ \ \lambda_9\ \ \lambda_{10}\ ]^{\mathrm{T}} \tag{1.222}$$

$$\begin{aligned}\boldsymbol{F}'_\lambda = [\,& L_1\ \ L_2\ \ L_3\ \ L_2L_3\ \ L_3L_1\ \ L_1L_2\ \ L_2L_3(L_2-L_3) \\ & \times L_3L_1(L_3-L_1)\ \ \ L_1L_2(L_1-L_2)\ \ \ L_1L_2L_3\,]\end{aligned} \tag{1.223}$$

$\boldsymbol{\lambda}'$ can be expressed in terms of the element nodal displacement vector

$$\boldsymbol{\lambda}' = \boldsymbol{C}\overline{\boldsymbol{q}}^e \tag{1.224}$$

where

$$\boldsymbol{C} = [\ \boldsymbol{C}_1\ \ \boldsymbol{C}_2\ \ \boldsymbol{C}_3\ ] \tag{1.225}$$

$$\boldsymbol{C}_1 = \begin{bmatrix} 1 & 0 & 0 \\ 0 & 0 & 0 \\ 0 & 0 & 0 \\ 0 & 0 & 0 \\ 0 & -\frac{1}{2}c_2 & \frac{1}{2}b_2 \\ 0 & \frac{1}{2}c_3 & -\frac{1}{2}b_3 \\ \begin{array}{l}\frac{1}{2}(r_2+r_3)-(\frac{1}{3}+r_2)\delta_2 \\ \quad +(\frac{1}{3}-r_3)\delta_3\end{array} & \begin{array}{l}\frac{1}{12}(c_1-3r_2c_2+3r_3c_3)+\frac{1}{6}(1+3r_2)c_2\delta_2 \\ \quad +\frac{1}{6}(1-3r_3)c_3\delta_3\end{array} & \begin{array}{l}-\frac{1}{12}(b_1-3r_2b_2+3r_3b_3)-\frac{1}{6}(1+3r_2)b_2\delta_2 \\ \quad -\frac{1}{6}(1-3r_3)b_3\delta_3\end{array} \\ \begin{array}{l}-\frac{1}{2}(3+r_3)+\frac{8}{3}\delta_2 \\ \quad +(\frac{1}{3}+r_3)\delta_3\end{array} & \begin{array}{l}-\frac{1}{12}(3r_3c_3+c_3-8c_2)-\frac{4}{3}c_2\delta_2 \\ \quad +\frac{1}{6}(1+3r_3)c_3\delta_3\end{array} & \begin{array}{l}\frac{1}{12}(3r_3b_3+b_3-8b_2)+\frac{4}{3}b_2\delta_2 \\ \quad -\frac{1}{6}(1+3r_3)b_3\delta_3\end{array} \\ \begin{array}{l}\frac{1}{2}(3-r_2)-(\frac{1}{3}-r_2)\delta_2 \\ \quad -\frac{8}{3}\delta_3\end{array} & \begin{array}{l}\frac{1}{12}(3r_2c_2-c_2+8c_3)+\frac{1}{6}(1-3r_2)c_2\delta_2 \\ \quad -\frac{4}{3}c_3\delta_3\end{array} & \begin{array}{l}-\frac{1}{12}(3r_2b_2-b_2+8b_3)-\frac{1}{6}(1-3r_2)b_2\delta_2 \\ \quad +\frac{4}{3}b_3\delta_3\end{array} \\ (r_3-r_2)+2(r_2\delta_2-r_3\delta_3) & \frac{1}{2}(r_2c_2+r_3c_3)-r_2c_2\delta_2-r_3c_3\delta_3 & -\frac{1}{2}(r_2b_2+r_3b_3)+r_2b_2\delta_2+r_3b_3\delta_3 \end{bmatrix} \tag{1.226}$$

$$\boldsymbol{C}_2 = \begin{bmatrix} 0 & 0 & 0 \\ 1 & 0 & 0 \\ 0 & 0 & 0 \\ 0 & \frac{1}{2}c_1 & -\frac{1}{2}b_1 \\ 0 & 0 & 0 \\ 0 & -\frac{1}{2}c_3 & \frac{1}{2}b_3 \\ \frac{1}{2}(3-r_3)-(\frac{1}{3}-r_3)\delta_3 - \frac{8}{3}\delta_1 & \frac{1}{12}(3r_3c_3-c_3+8c_1)+\frac{1}{6}(1-3r_3)c_3\delta_3 - \frac{4}{3}c_1\delta_1 & -\frac{1}{12}(3r_3b_3-b_3+8b_1)-\frac{1}{6}(1-3r_3)b_3\delta_3 + \frac{4}{3}b_1\delta_1 \\ \frac{1}{2}(r_3+r_1)-(\frac{1}{3}+r_3)\delta_3 + (\frac{1}{3}-r_1)\delta_1 & \frac{1}{12}(c_2-3r_3c_3+3r_1c_1)+\frac{1}{6}(1+3r_3)c_3\delta_3 + \frac{1}{6}(1-3r_1)c_1\delta_1 & -\frac{1}{12}(b_2-3r_3b_3+3r_1b_1)-\frac{1}{6}(1+3r_3)b_3\delta_3 - \frac{1}{6}(1-3r_1)b_1\delta_1 \\ -\frac{1}{2}(3+r_1)+\frac{8}{3}\delta_3 + (\frac{1}{3}+r_1)\delta_1 & -\frac{1}{12}(3r_1c_1+c_1-8c_3)-\frac{4}{3}c_3\delta_3 + \frac{1}{6}(1+3r_1)c_1\delta_1 & \frac{1}{12}(3r_1b_1+b_1-8b_3)+\frac{4}{3}b_3\delta_3 - \frac{1}{6}(1+3r_1)b_1\delta_1 \\ (r_1-r_3)+2(r_3\delta_3-r_1\delta_1) & \frac{1}{2}(r_3c_3+r_1c_1)-r_3c_3\delta_3-r_1c_1\delta_1 & -\frac{1}{2}(r_3b_3+r_1b_1)+r_3b_3\delta_3+r_1b_1\delta_1 \end{bmatrix} \tag{1.227}$$

$$\boldsymbol{C}_3 = \begin{bmatrix} 0 & 0 & 0 \\ 0 & 0 & 0 \\ 1 & 0 & 0 \\ 0 & -\frac{1}{2}c_1 & \frac{1}{2}b_1 \\ 0 & \frac{1}{2}c_2 & -\frac{1}{2}b_2 \\ 0 & 0 & 0 \\ -\frac{1}{2}(3+r_2)+\frac{8}{3}\delta_1 + (\frac{1}{3}+r_2)\delta_2 & -\frac{1}{12}(3r_2c_2+c_2-8c_1)-\frac{4}{3}c_1\delta_1 + \frac{1}{6}(1+3r_2)c_2\delta_2 & \frac{1}{12}(3r_2b_2+b_2-8b_1)+\frac{4}{3}b_1\delta_1 - \frac{1}{6}(1+3r_2)b_2\delta_2 \\ \frac{1}{2}(3-r_1)-(\frac{1}{3}-r_1)\delta_1 - \frac{8}{3}\delta_2 & \frac{1}{12}(3r_1c_1-c_1+8c_2)+\frac{1}{6}(1-3r_1)c_1\delta_1 - \frac{4}{3}c_2\delta_2 & -\frac{1}{12}(3r_1b_1-b_1+8b_2)-\frac{1}{6}(1-3r_1)b_1\delta_1 + \frac{4}{3}b_2\delta_2 \\ \frac{1}{2}(r_1+r_2)-(\frac{1}{3}+r_1)\delta_1 + (\frac{1}{3}-r_2)\delta_2 & \frac{1}{12}(c_3-3r_1c_1+3r_2c_2)+\frac{1}{6}(1+3r_1)c_1\delta_1 + \frac{1}{6}(1-3r_2)c_2\delta_2 & -\frac{1}{12}(b_3-3r_1b_1+3r_2b_2)-\frac{1}{6}(1+3r_1)b_1\delta_1 - \frac{1}{6}(1-3r_2)b_2\delta_2 \\ (r_2-r_1)+2(r_1\delta_1-r_2\delta_2) & \frac{1}{2}(r_1c_1+r_2c_2)-r_1c_1\delta_1-r_2c_2\delta_2 & -\frac{1}{2}(r_1b_1+r_2b_2)+r_1b_1\delta_1+r_2b_2\delta_2 \end{bmatrix} \tag{1.228}$$

in which

$$r_i = \frac{d_j^2 - d_k^2}{d_i^2} \quad (i = \overrightarrow{1,2,3}; \quad j = \overrightarrow{2,3,1} \quad k = \overrightarrow{3,1,2}) \tag{1.229}$$

Substituting Equation (1.224) into Equation (1.221) yields

$$w = \boldsymbol{F}_\lambda' \boldsymbol{C}\overline{\boldsymbol{q}}^e = \boldsymbol{F}_\lambda' \boldsymbol{C}\boldsymbol{L}\boldsymbol{q}_p^e = \boldsymbol{N}^e \boldsymbol{q}_p^e, \quad \boldsymbol{N}^e = \boldsymbol{F}_\lambda' \boldsymbol{C}\boldsymbol{L} \tag{1.230}$$

where $\boldsymbol{N}^e$ is the shape function matrix of the deflection w. It can be verified that this deflection field and the rotation fields given by Equation (1.218) determined by this deflection field satisfy the following generalized conforming condition:

$$\oint_{\partial A^e} [Q_n(w-\tilde{w}) - M_n(\psi_n - \tilde{\psi}_n) - M_{ns}(\psi_s - \tilde{\psi}_s)] = 0 \tag{1.231}$$

where Q_n, M_n, and M_{ns} denote the shear force, normal bending moment, and tangent bending moment along element boundary ∂A^e, respectively. Hence, the element derived here is a generalized conforming element, and its convergence can be ensured.

The rotation field $\boldsymbol{\psi}$ in Equation (1.217) can be rewritten as

$$\boldsymbol{\psi} = \begin{Bmatrix} \psi_x \\ \psi_y \end{Bmatrix} = \begin{Bmatrix} \frac{\partial w}{\partial x} - \gamma_{xz} \\ \frac{\partial w}{\partial y} - \gamma_{yz} \end{Bmatrix} = \left(\begin{bmatrix} \boldsymbol{F}'_{\lambda,x} \\ \boldsymbol{F}'_{\lambda,y} \end{bmatrix} \boldsymbol{C} - \boldsymbol{B}_s \right) \boldsymbol{L}\boldsymbol{q}_p^e \tag{1.232}$$

Then the curvature field $\boldsymbol{\kappa}$ of the plate element is

$$\boldsymbol{\kappa} = \begin{Bmatrix} \kappa_x \\ \kappa_y \\ 2\kappa_{xy} \end{Bmatrix} = \begin{Bmatrix} -\frac{\partial \psi_x}{\partial x} \\ -\frac{\partial \psi_y}{\partial y} \\ -\frac{\partial \psi_x}{\partial y} - \frac{\partial \psi_y}{\partial x} \end{Bmatrix} = \left(- \begin{bmatrix} \boldsymbol{F}'_{\lambda,xx} \\ \boldsymbol{F}'_{\lambda,yy} \\ 2\boldsymbol{F}'_{\lambda,xy} \end{bmatrix} \boldsymbol{C}\boldsymbol{L} \right) \boldsymbol{q}_p^e = \boldsymbol{B}_b \boldsymbol{q}_p^e \tag{1.233}$$

in which $\boldsymbol{B}_b$ is the bending strain matrix.

Thus, the element stiffness matrix of the thick/thin-bending element TSL-T9 can be obtained

$$\boldsymbol{K}_p^e = \int_{A^e} \boldsymbol{B}_b^{\mathrm{T}} \boldsymbol{D}_b \boldsymbol{B}_b \mathrm{d}A + \int_{A^e} \boldsymbol{B}_s^{\mathrm{T}} \boldsymbol{D}_s \boldsymbol{B}_s \mathrm{d}A \tag{1.234}$$

Numerical results show that element TSL-T9 possesses excellent performance for both thin and thick plate bending problems. And its stress solutions are also improved by the hybrid-enhanced postprocessing procedure in reference (Chen et al., 2003). We will not explain this in detail. Readers who are interested can refer Chen et al. (2003).

1.3.7.4 Stiffness matrix of the flat shell element GMST18

Assembling Equations (1.207) and (1.234) according to the DOF's sequence given by Equation (1.196), we obtain element stiffness matrix $\boldsymbol{K}^e$ of the flat shell element GMST18 in the local coordinates. And after transforming $\boldsymbol{K}^e$ to the global coordinates by a standard procedure, the element can be used to calculate shell structures.

1.3.7.5 Geometrically nonlinear analysis

On the basis of generalized conforming thick/thin triangular flat shell element, the UL formulations of GT9 element and GMST18 element are derived by Zhang et al. (1998)

and Chen et al. (2003), respectively. The UL formulations were used for analysis of geometrically nonlinear problems and demonstrated good performance for numerical examples.

In the incremental method, all the physical components of a structure from time 0 to time t are assumed to have been obtained. What we are interested in is the increment that occurs from time t to time $t + \Delta t$. The reference configuration is the configuration at time t. The principle of virtual displacement expressed by the UL method can be written as

$$\int_V ({}^{t+\Delta t}\sigma_{ij})\, \delta({}^{t+\Delta t}\varepsilon_{ij})\mathrm{d}V = {}^{t+\Delta t}W \tag{1.235}$$

where ${}^{t+\Delta t}\sigma_{ij}$ and ${}^{t+\Delta t}\varepsilon_{ij}$ are the modified Kirchhoff stress tensor and the modified Green strain tensor, respectively; ${}^{t+\Delta t}W$ is the virtual work done by external loadings at the time $t + \Delta t$.

$${}^{t+\Delta t}\sigma_{ij} = \sigma_{ij}^{E} + \Delta\sigma_{ij} \tag{1.236}$$

where σ_{ij}^{E} is the Cauchy stress tensor at the time t and $\Delta\sigma_{ij}$ is the Kirchhoff stress tensor increment from time t to time $t + \Delta t$.

$$\left.\begin{array}{l} {}^{t+\Delta t}\varepsilon_{ij} = \Delta\varepsilon_{ij} = \Delta e_{ij} + \Delta\eta_{ij} \\ \Delta e_{ij} = \dfrac{1}{2}(\Delta u_{i,j} + \Delta u_{j,i}), \quad \Delta\eta_{ij} = \dfrac{1}{2}\Delta u_{k,i}\Delta u_{k,j} \end{array}\right\} \tag{1.237}$$

where Δe_{ij} and $\Delta\eta_{ij}$ are the linear and nonlinear Green strain tensor increment from time t to time $t + \Delta t$, respectively. Δu_i is the displacement increment from time t to time $t + \Delta t$.

If Δt is small enough, the following relationship can be established

$$\Delta\sigma_{ij} = D_{ijkl}\Delta\varepsilon_{kl} \tag{1.238}$$

where D_{ijkl} is the elastic tensor.

Substitution of Equations (1.236)–(1.238) into Equation (1.235) yields (the higher order terms have been neglected)

$$I_1 + I_2 = ({}^{t+\Delta t}W) - I_3 \tag{1.239}$$

with

$$I_1 = \int_V D_{ijkl}\Delta e_{kl}\,\delta\Delta e_{ij}\,\mathrm{d}V, \quad I_2 = \int_V \sigma_{ij}^{E}\,\delta\Delta\eta_{ij}\,\mathrm{d}V, \quad I_3 = \int_V \sigma_{ij}^{E}\,\delta\Delta e_{ij}\,\mathrm{d}V \tag{1.240}$$

where I_1 is the linear increment of virtual work; I_2 is the incremental virtual work relevant to the initial stresses; and I_3 is the incremental virtual work done by the internal forces.

For the flat shell element in the local coordinates, I_1, I_2, and I_3 in Equation (1.240) can be rewritten in the following discrete form:

$$\left.\begin{aligned} I_1 &= \int_{A^e} (\delta\Delta\varepsilon_m^{\mathrm{T}} \boldsymbol{D}_m \Delta\varepsilon_m + \delta\Delta\kappa^{\mathrm{T}} \boldsymbol{D}_b \Delta\kappa + \delta\Delta\gamma^{\mathrm{T}} \boldsymbol{D}_s \Delta\gamma)\mathrm{d}A \\ I_2 &= \int_{A^e} \delta\Delta w'^{\mathrm{T}} \bar{\boldsymbol{N}}^E \Delta w' \mathrm{d}A \\ I_3 &= \int_{A^e} (\delta\Delta\varepsilon_m^{\mathrm{T}} N_m^E + \delta\Delta\kappa^{\mathrm{T}} \boldsymbol{M}^E + \delta\Delta\gamma^{\mathrm{T}} \boldsymbol{Q}^E)\mathrm{d}A \end{aligned}\right\} \tag{1.241}$$

where Δ means the increment of relevant variables; $\Delta\boldsymbol{\varepsilon}_m$ is the linear increment of the membrane strain and is given by

$$\Delta\varepsilon_m = \left[\frac{\partial\Delta u}{\partial x} \quad \frac{\partial\Delta v}{\partial y} \quad \frac{\partial\Delta u}{\partial y} + \frac{\partial\Delta v}{\partial x} \right]^{\mathrm{T}} = \boldsymbol{B}_m \Delta\boldsymbol{q}_m^e \tag{1.242}$$

$\Delta\boldsymbol{\kappa}$ is the linear increment of the curvature vector given in Equation (1.233); $\Delta\boldsymbol{\gamma}$ is the increment of the transverse shear strain vector given in Equation (1.210).

$$\Delta\boldsymbol{w}' = \begin{Bmatrix} \dfrac{\partial\Delta w}{\partial x} \\ \dfrac{\partial\Delta w}{\partial y} \end{Bmatrix} = \begin{bmatrix} \boldsymbol{F}'_{\lambda,x} \\ \boldsymbol{F}'_{\lambda,y} \end{bmatrix} \boldsymbol{CL}\Delta\boldsymbol{q}_p^e = \boldsymbol{B}_G \Delta\boldsymbol{q}_p^e, \quad \boldsymbol{B}_G = \begin{bmatrix} \boldsymbol{F}'_{\lambda,x} \\ \boldsymbol{F}'_{\lambda,y} \end{bmatrix} \boldsymbol{CL} \tag{1.243}$$

$$\bar{\boldsymbol{N}}^E = \begin{bmatrix} \int_{-h/2}^{h/2} \sigma_x^E \, dz & \int_{-h/2}^{h/2} \tau_{xy}^E \, dz \\ \int_{-h/2}^{h/2} \tau_{xy}^E \, dz & \int_{-h/2}^{h/2} \sigma_y^E \, dz \end{bmatrix} = \begin{bmatrix} N_x^E & N_{xy}^E \\ N_{xy}^E & N_y^E \end{bmatrix} \tag{1.244}$$

$\boldsymbol{N}_m^E$, $\boldsymbol{M}^E$, and $\boldsymbol{Q}^E$ are the membrane force, bending moment, and shear force vectors at the time t, respectively,

$$\boldsymbol{N}_m^E = [\, N_x^E \;\; N_y^E \;\; N_{xy}^E \,]^{\mathrm{T}}, \quad \boldsymbol{M}^E = [\, M_x^E \;\; M_y^E \;\; M_{xy}^E \,]^{\mathrm{T}}, \quad \boldsymbol{Q}^E = [\, Q_x^E \;\; Q_y^E \,]^{\mathrm{T}} \tag{1.245}$$

Substitution of the geometric relation Equations (1.210), (1.233), and (1.242) into Equation (1.241) yields

$$\left.\begin{aligned} I_1 &= \delta\Delta\boldsymbol{q}_m^{e\mathrm{T}} \left(\int_{A^e} \boldsymbol{B}_m^{\mathrm{T}} \boldsymbol{D}_m \boldsymbol{B}_m \,\mathrm{d}A \right) \Delta\boldsymbol{q}_m^e \\ &\quad + \delta\Delta\boldsymbol{q}_p^{e\mathrm{T}} \left(\int_{A^e} (\boldsymbol{B}_b^{\mathrm{T}} \boldsymbol{D}_b \boldsymbol{B}_b + \boldsymbol{B}_s^{\mathrm{T}} \boldsymbol{D}_s \boldsymbol{B}_s) \mathrm{d}A \right) \Delta\boldsymbol{q}_p^e \\ I_2 &= \delta\Delta\boldsymbol{q}_p^{e\mathrm{T}} \left(\int_{A^e} \boldsymbol{B}_G^{\mathrm{T}} \bar{\boldsymbol{N}}^E \boldsymbol{B}_G \,\mathrm{d}A \right) \Delta\boldsymbol{q}_p^e \\ I_3 &= \delta\Delta\boldsymbol{q}_m^{e\mathrm{T}} \left(\int_{A^e} \boldsymbol{B}_m^{\mathrm{T}} \boldsymbol{N}_m^E \,\mathrm{d}A \right) \Delta\boldsymbol{q}_m^e \\ &\quad + \delta\Delta\boldsymbol{q}_p^{e\mathrm{T}} \left(\int_{A^e} (\boldsymbol{B}_b^{\mathrm{T}} \boldsymbol{M}^E + \boldsymbol{B}_s^{\mathrm{T}} \boldsymbol{Q}^E) \mathrm{d}A \right) \Delta\boldsymbol{q}_p^e \end{aligned}\right\} \tag{1.246}$$

And ${}^{t+\Delta t}W$ can be rewritten as

$$ {}^{t+\Delta t}W = \delta\Delta \boldsymbol{q}_m^{e\mathrm{T}}\ {}^{t+\Delta t}\boldsymbol{R}_m^e + \delta\Delta \boldsymbol{q}_p^{e\mathrm{T}}\ {}^{t+\Delta t}\boldsymbol{R}_p^e \tag{1.247} $$

where ${}^{t+\Delta t}\boldsymbol{R}_m^e$ and ${}^{t+\Delta t}\boldsymbol{R}_p^e$ are the equivalent nodal force vectors at the time $t + \Delta t$ of the membrane element and plate bending element, respectively.

Since $\delta\Delta \boldsymbol{q}_m^e$ and $\delta\Delta \boldsymbol{q}_p^e$ are arbitrary, according to the variational principle and Equations (1.239), (1.246), and (1.247), we can obtain the element incremental equations in the local coordinates:

$$ \begin{bmatrix} \boldsymbol{K}_m^e & \boldsymbol{0} \\ \boldsymbol{0} & \boldsymbol{K}_p^e + \boldsymbol{K}_\sigma^e \end{bmatrix} \begin{Bmatrix} \Delta \boldsymbol{q}_m^e \\ \Delta \boldsymbol{q}_p^e \end{Bmatrix} = \begin{Bmatrix} {}^{t+\Delta t}\boldsymbol{R}_m^e \\ {}^{t+\Delta t}\boldsymbol{R}_p^e \end{Bmatrix} - \begin{Bmatrix} \boldsymbol{\Psi}_m^e \\ \boldsymbol{\Psi}_p^e \end{Bmatrix} \tag{1.248} $$

where $\boldsymbol{K}_m^e$ is the linear stiffness matrix of the membrane element and given by Equation (1.207); $\boldsymbol{K}_p^e$ is the linear stiffness matrix of the plate bending element and given by Equation (1.234); and $\boldsymbol{K}_\sigma^e$ is the geometric stiffness matrix,

$$ \boldsymbol{K}_\sigma^e = \int_{A^e} \boldsymbol{B}_G^{\mathrm{T}} \bar{\boldsymbol{N}}^E \boldsymbol{B}_G \,\mathrm{d}A \tag{1.249} $$

$\boldsymbol{\Psi}_m^e$ is the equivalent nodal internal force vector of the membrane element; $\boldsymbol{\Psi}_p^e$ is the equivalent nodal internal force vector of the plate bending element,

$$ \boldsymbol{\Psi}_m^e = \int_{A^e} \boldsymbol{B}_m^{\mathrm{T}} \boldsymbol{N}_m^E \,\mathrm{d}A, \boldsymbol{\Psi}_p^e = \int_{A^e} (\boldsymbol{B}_b^{\mathrm{T}} \boldsymbol{M}^E + \boldsymbol{B}_s^{\mathrm{T}} \boldsymbol{Q}^E)\mathrm{d}A \tag{1.250} $$

Rewriting Equation (1.248) according to the DOF's sequence yields

$$ (\boldsymbol{K}^e + \boldsymbol{K}_\sigma^e)\Delta \boldsymbol{q}^e = {}^{t+\Delta t}\boldsymbol{R}^e - \boldsymbol{\Psi}^e \tag{1.251} $$

After transforming Equation (1.251) to the global coordinates by standard procedure, elements GT9 and GMST18 can be used to analyze the geometrically nonlinear problem of shells.

1.3.8 Contact problems in structural mechanics

1.3.8.1 Contact problem

For a contact problem, both kinetic conditions and dynamic conditions need to be satisfied on the contact interface, and impenetrability condition is the most important one for mechanical contact problems. For a general contact problem in the structure analysis, the contact region may change during analysis, for example, slide, separation, and rebound. Besides an appropriate contact searching algorithm, the discontinuity of the interface velocity as well as contact forces make the contact problem one of the most difficult nonlinear problems in structural dynamics. Some special treatments may need to be used to improve the robustness, efficiency, and accuracy. The contact interface can be frictionless or frictional, and friction laws can be applied on the frictional contact interface. The constraint conditions on the contact interface can be satisfied by the following methods; the comparisons between these methods are given in Chapter 3.

- Penalty method
- Lagrange multiplier method
- Augmented Lagrangian method
- Multipoint constraint method

The type of contact includes node-to-surface, surface-to-surface, and self-contact. In further section, penalty method based integration point-to-surface contact will be presented.

1.3.8.2 Contact equations

The contact interface is the surface intersection of the two bodies that in contact (Figure 1.4),

$$\Gamma_c = \Gamma^A \cap \Gamma^B \tag{1.252}$$

where Γ^A is the boundary face of the contactor A; Γ^B is the boundary face of the target B; and Γ_C is the in-contact part of Γ^A and Γ^B.

It is convenient to describe contact conditions in the local coordinates of the contact surface ${}^t\Gamma^B$. Here, the superscript t indicates the current time. The three orthogonal unit vectors are ${}^t\boldsymbol{e}_1$, ${}^t\boldsymbol{e}_2$, and ${}^t\boldsymbol{e}_3$, where ${}^t\boldsymbol{e}_1$ and ${}^t\boldsymbol{e}_2$ are the unit vectors in the tangential plane of Γ_B, and ${}^t\boldsymbol{e}_3$ is the outward normal vector (Figure 1.5).

$${}^t\boldsymbol{n}^B = {}^t\boldsymbol{e}_3 = {}^t\boldsymbol{e}_1 \times {}^t\boldsymbol{e}_2 \tag{1.253}$$

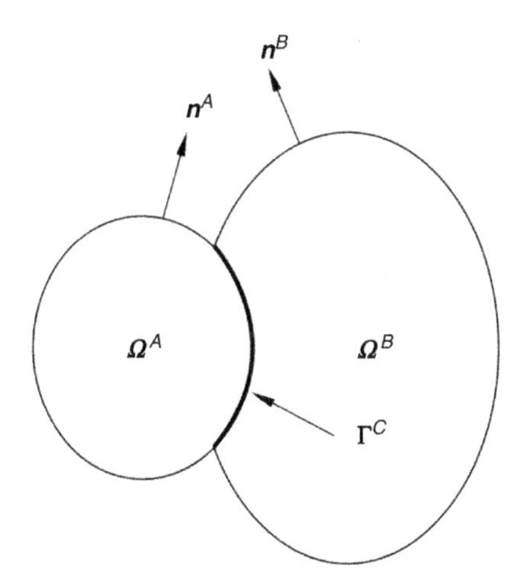

Figure 1.4 Contact between the contactor body A and body target B with contact interface Γ_c.

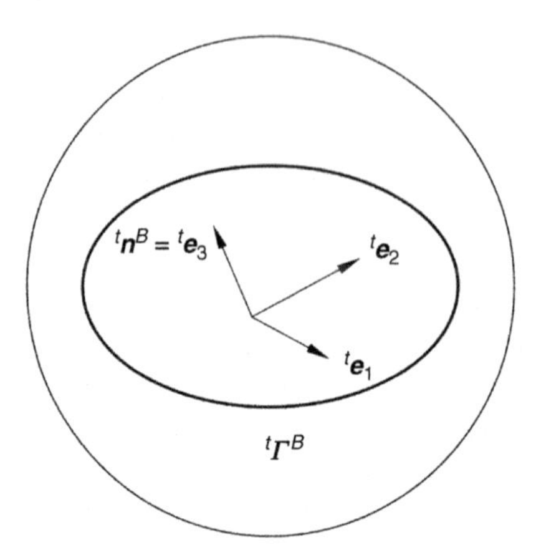

Figure 1.5 The local coordinate of the contact interface.

If a point P at Γ_A is in contact with point Q at Γ_B, then we call P and Q a contact pair. The force vector at P and Q are ${}^t\boldsymbol{F}_P^A$ and ${}^t\boldsymbol{F}_Q^B = -{}^t\boldsymbol{F}_P^A$. Here we can omit the subscripts P and Q.

$$ {}^t\boldsymbol{F}^r = {}^tF_N^r\,{}^t\boldsymbol{n}^B + {}^tF_1^r\,{}^t\boldsymbol{e}_1 + {}^tF_1^r\,{}^t\boldsymbol{e}_2 = {}^t\boldsymbol{F}_N^r + {}^t\boldsymbol{F}_T^r, \quad r = (A, B) \tag{1.254} $$

$$ {}^t\boldsymbol{F}_N^r = {}^tF_N^r\,{}^t\boldsymbol{n}^B \tag{1.255} $$

$$ {}^t\boldsymbol{F}_T^r = {}^tF_1^r\,{}^t\boldsymbol{e}_1 + {}^tF_1^r\,{}^t\boldsymbol{e}_2 \quad r = (A, B) \tag{1.256} $$

As the law of reaction force is equal to the negative of action force,

$$ {}^t\boldsymbol{F}^B = -{}^t\boldsymbol{F}^A \tag{1.257} $$

From the previous equation:

$$ {}^t\boldsymbol{F}_N^B = -{}^t\boldsymbol{F}_N^A \tag{1.258} $$

$$ {}^t\boldsymbol{F}_T^B = -{}^t\boldsymbol{F}_T^A \tag{1.259} $$

$$ {}^tF_N^B = -{}^tF_N^A \tag{1.260} $$

$$ {}^tF_1^B = -{}^tF_1^A \tag{1.261} $$

$$ {}^tF_2^B = -{}^tF_2^A \tag{1.262} $$

Similarly, for the velocity vector in the local coordinates:

$$ {}^t\boldsymbol{v}^r = {}^tv_N^r\,{}^t\boldsymbol{n}^B + {}^tv_1^r\,{}^t\boldsymbol{e}_1 + {}^tv_1^r\,{}^t\boldsymbol{e}_2 = {}^t\boldsymbol{v}_N^r + {}^t\boldsymbol{v}_T^r, \quad r = (A, B) \tag{1.263} $$

where ${}^t\boldsymbol{v}_N^r$ and ${}^t\boldsymbol{v}_T^r$ is the normal and tangential component, respectively (Figure 1.6).

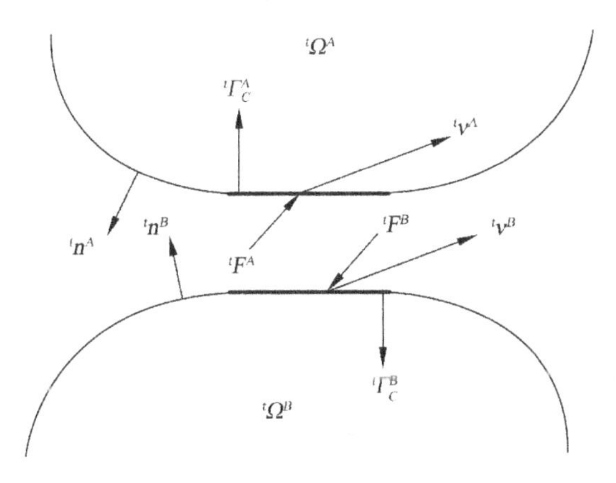

Figure 1.6 Contact forces and velocity at contact point.

$$ {}^{t}\boldsymbol{v}_{N}^{r} = {}^{t}v_{N}^{r}\,{}^{t}\boldsymbol{n}^{B} \tag{1.264} $$

$$ {}^{t}\boldsymbol{v}_{T}^{r} = {}^{t}v_{1}^{r}\,{}^{t}\boldsymbol{e}_{1} + {}^{t}v_{1}^{r}\,{}^{t}\boldsymbol{e}_{2} \quad r = (A, B) \tag{1.265} $$

1.3.8.2.1 The contact conditions in normal direction

1. The impenetrability contact condition in normal direction

$$ {}^{t}g_{N} = g\left({}^{t}\boldsymbol{x}_{P}^{A}, t\right) = \min_{{}^{t}\boldsymbol{x}^{B} \in \Gamma_{B}} \left\| {}^{t}\boldsymbol{x}_{P}^{A} - {}^{t}\boldsymbol{x}^{B} \right\| \tag{1.266} $$

where ${}^{t}g_{N}$ is the minimum distance of ${}^{t}\boldsymbol{x}_{P}^{A}$ on $\Gamma^{\boldsymbol{B}}$.
If Γ^{B} is smooth surface, then ${}^{t}g_{N}$ must be in the direction of ${}^{t}\boldsymbol{n}^{B}$.

$$ {}^{t}g_{N} = g({}^{t}\boldsymbol{x}_{P}^{A}, t) = ({}^{t}\boldsymbol{x}_{P}^{A} - {}^{t}\boldsymbol{x}_{Q}^{B})^{t} \cdot \boldsymbol{n}_{Q}^{B} \geq 0 \tag{1.267} $$

${}^{t}g_{N} > 0$ indicates that point P separates from $\Gamma^{\boldsymbol{B}}$, ${}^{t}g_{N} = 0$ means that point P is in contact with $\Gamma^{\boldsymbol{B}}$, and ${}^{t}g_{N} < 0$ stands for that P is in penetration with surface $\Gamma^{\boldsymbol{B}}$.
Since this condition is applicable for all points on contact surface, it can be expressed in a general form (Figure 1.7):

$$ {}^{t}g_{N} = g({}^{t}\boldsymbol{x}^{A}, t) = ({}^{t}\boldsymbol{x}^{A} - {}^{t}\boldsymbol{x}^{B})^{t} \cdot \boldsymbol{n}^{B} \geq 0 \tag{1.268} $$

2. The contact normal force must be pressured
Without considering the cohesive condition, the contact force in the normal direction must be pressured.

$$ {}^{t}F_{N}^{B} = -{}^{t}F_{N}^{A} \leq 0 \tag{1.269} $$

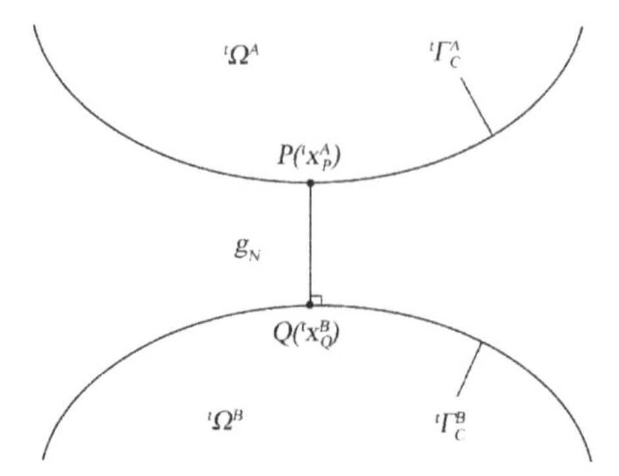

Figure 1.7 Contact pair and the normal distance.

1.3.8.2.2 The contact conditions in tangential direction

1. Frictionless contact
   If the frictions on the contact surface are negligible, then the frictionless contact condition can be adopted, that is,

$$ {}^t\boldsymbol{F}_T^A = {}^t\boldsymbol{F}_T^B \equiv \boldsymbol{0} \tag{1.270}$$

   or in the component form:

$$ {}^tF_J^A = {}^tF_J^B \equiv 0 \quad (J=1,2) \tag{1.271}$$

2. Frictional contact
   If the friction must be considered on the contact interface, then the frictional contact model needs to be used. The most general friction model is the Coulomb model:

$$ \left\| {}^t\boldsymbol{F}_T^A \right\| = \left(({}^tF_1^A)^2 + ({}^tF_2^A)^2 \right)^{1/2} \le \mu \left\| {}^t\boldsymbol{F}_N^A \right\|. \tag{1.272}$$

   When

$$ \left\| {}^t\boldsymbol{F}_T^A \right\| < \mu \left\| {}^t\boldsymbol{F}_N^A \right\|, \tag{1.273}$$

   there is no relative sliding on the contact interfaces: ${}^t\overline{\boldsymbol{v}}_T = {}^t\boldsymbol{v}_T^A - {}^t\boldsymbol{v}_T^B = \boldsymbol{0}$.

   On the other hand, for the relative sliding condition,

$$ {}^t\overline{\boldsymbol{v}}_T = {}^t\boldsymbol{v}_T^A - {}^t\boldsymbol{v}_T^B \neq \boldsymbol{0} \text{ when } \left\| {}^t\boldsymbol{F}_T^A \right\| = \mu \left\| {}^t\boldsymbol{F}_N^A \right\| \tag{1.274}$$

And the relative sliding velocity ${}^t\overline{\boldsymbol{v}}_T$ is in the opposite direction of frictional force acting on the contactor ${}^t\boldsymbol{F}_T^A$:

$$ {}^t\overline{\boldsymbol{v}}_T \,{}^t\!\cdot \boldsymbol{F}_T^A < 0 \tag{1.275}$$

1.3.8.3 The incremental form of contact conditions

For a nonlinear contact problem, the incremental form is generally adopted. It assumes that the solution and contact status have been known at time t and the system equation with contact is to obtain the solution and new contact status at time $t + \Delta t$.

1. The impenetrability at $t + \Delta t$

$$ {}^{t+\Delta t}g_N = ({}^{t+\Delta t}\boldsymbol{x}^A - {}^{t+\Delta t}\boldsymbol{x}^B)\cdot {}^{t+\Delta t}\boldsymbol{n}^B \tag{1.276} $$

where

$$ {}^{t+\Delta t}\boldsymbol{x}^A = {}^{t}\boldsymbol{x}^A + \boldsymbol{u}^A \tag{1.277} $$

$$ {}^{t+\Delta t}\boldsymbol{x}^B = {}^{t}\boldsymbol{x}^B + \boldsymbol{u}^B \tag{1.278} $$

$\boldsymbol{u}^A$ and $\boldsymbol{u}^B$ are the increments of displacement from time t to time $t + \Delta t$.

$$ \boldsymbol{u}^A = {}^{t+\Delta t}\boldsymbol{u}^A - {}^{t}\boldsymbol{u}^A \tag{1.279} $$

$$ \boldsymbol{u}^B = {}^{t+\Delta t}\boldsymbol{u}^B - {}^{t}\boldsymbol{u}^B \tag{1.280} $$

Substituting Equations (1.277) and (1.278) into Equation (1.276), we get:

$$ {}^{t+\Delta t}g_N = (\boldsymbol{u}^A - \boldsymbol{u}^B)\cdot {}^{t+\Delta t}\boldsymbol{n}^B + ({}^{t}\boldsymbol{x}^A - {}^{t}\boldsymbol{x}^B)\cdot {}^{t+\Delta t}\boldsymbol{n}^B = u_N^A - u_N^B + {}^{t}\overline{g}_N \geq 0 \tag{1.281} $$

where

$$ u_N^A = \boldsymbol{u}^A \cdot {}^{t+\Delta t}\boldsymbol{n}^B,\ u_N^B = \boldsymbol{u}^B \cdot {}^{t+\Delta t}\boldsymbol{n}^B, \tag{1.282} $$

$$ {}^{t}\overline{g}_N = ({}^{t}\boldsymbol{x}^A - {}^{t}\boldsymbol{x}^B)\cdot {}^{t+\Delta t}\boldsymbol{n}^B \tag{1.283} $$

Remark 1.9: Here ${}^{t}\overline{g}_N \approx {}^{t}g_N$ for small displacement analysis, and in nonlinear iterations, we normally treat ${}^{t+\Delta t}\boldsymbol{n}^B$ as constant value without consideration of its vibrations, but we use the updated ${}^{t+\Delta t}\boldsymbol{n}^B$ for the next iteration.

2. For no relative sliding conditions for stick contact

$$ \overline{\boldsymbol{u}}_T = \boldsymbol{u}_T^A - \boldsymbol{u}_T^B = \boldsymbol{0}, \quad \text{when} \ \left\| {}^{t+\Delta t}\boldsymbol{F}_T^A \right\| < \mu \left\| {}^{t+\Delta t}\boldsymbol{F}_N^A \right\|, \tag{1.284} $$

In the component form:

$$ \overline{u}_J = u_J^A - u_J^B = 0, \quad \text{when} \left\| {}^{t+\Delta t}\boldsymbol{F}_T^A \right\| < \mu \left\| {}^{t+\Delta t}\boldsymbol{F}_N^A \right\|, \tag{1.285} $$

here, $\boldsymbol{u}_T^A$ and $\boldsymbol{u}_T^B$ are the tangential components of the incremental displacement from $\boldsymbol{t}$ to time $\boldsymbol{t} + \boldsymbol{\Delta t}$.

3. For the relative sliding contact interface

$$\bar{\boldsymbol{u}}_T = \boldsymbol{u}_T^A - \boldsymbol{u}_T^B \neq \boldsymbol{0}, \quad \text{when} \left\| {}^{t+\Delta t}\boldsymbol{F}_T^A \right\| = \mu \left\| {}^{t+\Delta t}\boldsymbol{F}_N^A \right\| \tag{1.286}$$

and

$$\bar{\boldsymbol{u}}_T \cdot {}^{t+\Delta t}\boldsymbol{F}_T^A < 0, \tag{1.287}$$

in the component form:

$${}^{t+\Delta t}F_J^A + \mu\, {}^{t+\Delta t}F_N^A \bar{u}_J / \bar{u}_T \quad (J = 1,2), \tag{1.288}$$

where $\bar{u}_T = \|\bar{\boldsymbol{u}}_T\|$.

1.3.8.4 *Verification conditions and the definite solution conditions for the contact problem*

Although the equations are indefinite, we solve these indefinite equations by definite condition with the verification and adjustment of the contact status.

1.3.8.4.1 The stick conditions

Definite conditions

$$(1)\ {}^{t+\Delta t}g_N = 0 \tag{1.289}$$

$$(2)\ \boldsymbol{v}_T^A - \boldsymbol{v}_T^B = \boldsymbol{0} \tag{1.290}$$

Or in the incremental form:

$$(1)\ u_N^A - u_N^B + {}^t\bar{g}_N = 0 \tag{1.291}$$

$$(2)\ \boldsymbol{u}_T^A - \boldsymbol{u}_T^B = \boldsymbol{0} \tag{1.292}$$

Verification of the contact status

(1) ${}^{t+\Delta t}F_N^A > 0$ if not satisfied then change to separation
(2) $\left\|{}^t\boldsymbol{F}_T^A\right\| < \mu \left\|{}^t\boldsymbol{F}_N^A\right\|$ if not satisfied then change to sliding

1.3.8.4.2 Sliding condition

Definite conditions

$$(1)\ {}^{t+\Delta t}g_N = 0 \tag{1.293}$$

$$(2) \left\| {}^{t+\Delta t}\boldsymbol{F}_T^A \right\| = \mu \left\| {}^{t+\Delta t}\boldsymbol{F}_N^A \right\| \text{or}\ {}^{t+\Delta t}F_J^A + \mu\, {}^{t+\Delta t}F_N^A \bar{v}_J / \bar{v}_T \quad (J = 1,2) \tag{1.294}$$

Or in the incremental form:

$$(1)\ u_N^A - u_N^B + {}^t\overline{g}_N = 0 \tag{1.295}$$

$$(2)\left\|{}^{t+\Delta t}\boldsymbol{F}_T^A\right\| = \mu\left\|{}^{t+\Delta t}\boldsymbol{F}_N^A\right\| \text{or } {}^{t+\Delta t}F_J^A + \mu\, {}^{t+\Delta t}F_N^A \overline{v}_J / \overline{v}_T = 0 \quad (J = 1, 2) \tag{1.296}$$

Verification of the contact status on the incremental form:

$$\begin{aligned} &(1)\ {}^{t+\Delta t}F_N^A > 0, \text{ if not satisfied then change to separation} \\ &(2)\ {}^{t+\Delta t}\boldsymbol{F}_T^{\text{trial}} = {}^t\boldsymbol{F}_T + \alpha_T({}^{t+\Delta t}\boldsymbol{u}_T - {}^t\boldsymbol{u}_T) \end{aligned} \tag{1.297}$$

The return mapping is completed by setting:

$${}^{t+\Delta t}\boldsymbol{F}_T = \begin{cases} {}^{t+\Delta t}\boldsymbol{F}_T^{\text{trial}} & \text{(stick) if} \left\|{}^{t+\Delta t}\boldsymbol{F}_T^{\text{trial}}\right\| \le \mu\, {}^{t+\Delta t}F_n \\ \dfrac{\mu\, {}^{t+\Delta t}F_n}{\left\|{}^{t+\Delta t}\boldsymbol{F}_T^{\text{trial}}\right\|}\, {}^{t+\Delta t}\boldsymbol{F}_T^{\text{trial}} & \text{(slide) if} \left\|{}^{t+\Delta t}\boldsymbol{F}_T^{\text{trial}}\right\| > \mu\, {}^{t+\Delta t}F_n \end{cases} \tag{1.298}$$

Here,

$${}^t\boldsymbol{u}_T = \boldsymbol{u}_T^A - \boldsymbol{u}_T^B \tag{1.299}$$

If condition (2) is not satisfied, then change to sick. If it is satisfied, then search for new contact pairs.

1.3.8.4.3 Separation condition

Definite conditions

$${}^{t+\Delta t}\boldsymbol{F}^A = {}^{t+\Delta t}\boldsymbol{F}^B \equiv \boldsymbol{0}, \tag{1.300}$$

Verification of the contact status

${}^{t+\Delta t}g_N > \varepsilon_d$, if not satisfied, then change to stick.

Remark 1.10: ε_d is also a specified small value of normal tolerance for checking the normal distance of the contact surface.

1.3.8.5 Penalty-based contact method

1.3.8.5.1 The principle of virtual work for contact

The principle of virtual work for a contact system with body A and B in current configuration is:

$${}^{t+\Delta t}W = {}^{t+\Delta t}W_{\text{internal}} - {}^{t+\Delta t}W_L - {}^{t+\Delta t}W_{\text{inertial}} - {}^{t+\Delta t}W_C \tag{1.301}$$

Here, ${}^{t+\Delta t}W_{\text{internal}}$ is the virtual work by internal force, ${}^{t+\Delta t}W_L$ is the virtual work by external loads, ${}^{t+\Delta t}W_{\text{inertial}}$ is the virtual work by inertial force, and ${}^{t+\Delta t}W_C$ is the virtual work by contact forces. Since all other terms have been discussed in the previous sections, we have only given the details about the virtual work by contact forces here.

$$ {}^{t+\Delta t}W_c = \sum_{r=}^{A,B} \int_{{}^{t+\Delta t}\Gamma^r} {}^{t+\Delta t}F_i^r \, \delta u_i^r \, \mathrm{d}\Gamma \tag{1.302} $$

$$ = \int_{{}^{t+\Delta t}\Gamma^A} {}^{t+\Delta t}F_i^A \delta u_i^A \mathrm{d}S + \int_{{}^{t+\Delta t}\Gamma^B} {}^{t+\Delta t}F_i^B \delta u_i^B \mathrm{d}\Gamma \tag{1.303} $$

$$ = \int_{{}^{t+\Delta t}\Gamma^A} {}^{t+\Delta t}F_J^A \delta u_J^A \mathrm{d}S + \int_{{}^{t+\Delta t}\Gamma^B} {}^{t+\Delta t}F_J^B \delta u_J^B \mathrm{d}\Gamma \tag{1.304} $$

$$ = \int_{{}^{t+\Delta t}\Gamma^A} {}^{t+\Delta t}F_J^A \, (\delta u_J^A - \delta u_J^B) \mathrm{d}\Gamma. \tag{1.305} $$

The subscript i indicates the value in the global Cartesian coordinate, $i = 1, 2, 3$, and the subscript J indicates the value in the contact local coordinate.

1.3.8.5.2 The virtual work by the penalty-based contact method

The potential function of a contact system by the penalty method is

$$ \Pi = \Pi_u + \Pi_{\text{CP}} \tag{1.306} $$

where Π_{CP} is the added potential by the contact condition of penalty method.

For the stick contact case: based on the definite conditions in Section 1.3.8.4:

$$ \Pi_{\text{CP}} = \int_{{}^{t+\Delta t}\Gamma_C} \alpha_N (u_N^A - u_N^B + {}^t\overline{g}_N)^2 + \alpha_1 (u_1^A - u_1^B)^2 + \alpha_2 (u_2^A - u_2^B)^2 \, \mathrm{d}\Gamma \tag{1.307} $$

$$ \begin{aligned} \delta\Pi_{\text{CP}} = \int_{{}^{t+\Delta t}\Gamma_C} & \alpha_N (u_N^A - u_N^B + {}^t\overline{g}_N)(\delta u_N^A - \delta u_N^B) \\ & + \alpha_1 (u_1^A - u_1^B)(\delta u_1^A - \delta u_1^B) + \alpha_2 (u_2^A - u_2^B)(\delta u_2^A - \delta u_2^B) \, \mathrm{d}\Gamma \end{aligned} \tag{1.308} $$

From $\delta\Pi = \delta\Pi_u + \delta\Pi_{\text{CP}}$ and Equation (1.308):

$$ \begin{aligned} {}^{t+\Delta t}W_C = -\delta\Pi_{\text{CP}} = \int_{{}^{t+\Delta t}\Gamma_C} & -\alpha_N (u_N^A - u_N^B + {}^t\overline{g}_N)(\delta u_N^A - \delta u_N^B) \\ & -\alpha_1 (u_1^A - u_1^B)(\delta u_1^A - \delta u_1^B) - \alpha_2 (u_2^A - u_2^B)(\delta u_2^A - \delta u_2^B) \, \mathrm{d}\Gamma \end{aligned} \tag{1.309} $$

Compared with Equation (1.286), the contact force can be expressed as:

$$ {}^{t+\Delta t}F_N^A = -{}^{t+\Delta t}F_N^B = -\alpha_N(u_N^A - u_N^B + {}^t\overline{g}_N) = -\alpha_N\,{}^{t+\Delta t}g_N \tag{1.310} $$

$$ {}^{t+\Delta t}F_J^A = -{}^{t+\Delta t}F_J^B = -\alpha_J(u_J^A - u_J^B) \quad (J=1,2) \tag{1.311} $$

For a frictionless contact interface:

$$ {}^{t+\Delta t}F_J^A = -{}^{t+\Delta t}F_J^B = \alpha_J = 0 \quad (J=1,2) \tag{1.312} $$

and the virtual work becomes:

$$ {}^{t+\Delta t}W_C = -\delta\Pi_{CP} = \int_{{}^{t+\Delta t}\Gamma_C} -\alpha_N(u_N^A - u_N^B + {}^t\overline{g}_N)(\delta u_N^A - \delta u_N^B)\,\mathrm{d}\Gamma \tag{1.313} $$

For a frictional sliding contact interface:

$$ {}^{t+\Delta t}F_J^A = -{}^{t+\Delta t}F_J^B = -\alpha_J(u_J^A - u_J^B) = \mu\alpha_N(u_N^A - u_N^B + {}^t\overline{g}_N)\overline{u}_J / \overline{u}_T \tag{1.314} $$

and the virtual work is

$$ \begin{aligned} {}^{t+\Delta t}W_C &= -\delta\Pi_{CP} \\ &= \int_{{}^{t+\Delta t}\Gamma_C} -\alpha_N(u_N^A - u_N^B + {}^t\overline{g}_N)\left[(\delta u_N^A - \delta u_N^B) - \mu\overline{u}_J / \overline{u}_T(\delta u_N^A - \delta u_N^B)\right]\mathrm{d}\Gamma \end{aligned} \tag{1.315} $$

The transformations of the force and displacement vector between local and global coordinates are presented as

$$ u_i = e_{Ji}u_J, \quad u_J = e_{Ji}u_i, \tag{1.316} $$

$$ F_i = e_{Ji}F_J, \quad F_J = e_{Ji}F_i, \tag{1.317} $$

$$ F_i u_i = F_J u_J. \tag{1.318} $$

1.3.8.5.3 The FEM matrix equations

1. Stick contact state

   We use the matrix form for the contact forces in Equations (1.310) and (1.311), that is,

$$ {}^{t+\Delta t}\boldsymbol{F}_c^A = -\boldsymbol{\alpha}_{st}\cdot(\boldsymbol{u}^A - \boldsymbol{u}^B) - \alpha_N\,{}^t\overline{\boldsymbol{g}} \tag{1.319} $$

where

$$\alpha_{st} = \begin{bmatrix} \alpha_1 & & \\ & \alpha_2 & \\ & & \alpha_N \end{bmatrix} {}^t\bar{g} = \begin{bmatrix} 0 \\ 0 \\ {}^t\bar{g}_N \end{bmatrix} \tag{1.320}$$

The contact force in the global coordinate system:

$${}^{t+\Delta t}Q_{ci}^A = e_{Ji}\,{}^tF_J^A = e_{Ji}(-\alpha_{stJ}(e_{Jk}u_k^A - e_{Jk}u_k^B)) - e_{3i}\alpha_N\,{}^t\bar{g}_N \quad (1.321$$

$${}^{t+\Delta t}Q_{ci}^B = -{}^{t+\Delta t}Q_{ci}^A \tag{1.322}$$

or

$${}^{t+\Delta t}Q_{ci}^A = K_{ik}^A u_k^A + K_{ij}^{AB} u_j^B - e_{3i}\alpha_N\,{}^t\bar{g}_N \tag{1.323}$$

where

$$K_{ik}^{AA} = -\alpha_{stJ}e_{Ji}e_{Jk},\, K_{ik}^{AB} = +\alpha_{stJ}e_{Ji}e_{Jk} \tag{1.324}$$

2. The frictional sliding interface

$${}^{t+\Delta t}\boldsymbol{F}_c^A = -\alpha_{st}(u_N^A - u_N^B + {}^t\bar{g}_N) \tag{1.325}$$

$$\alpha_{st} = \alpha\begin{bmatrix} -\mu\dfrac{\bar{u}_1}{\bar{u}_T} & -\mu\dfrac{\bar{u}_2}{\bar{u}_T} & 1 \end{bmatrix}^{\mathrm{T}} \tag{1.326}$$

$${}^{t+\Delta t}Q_{ci}^A = e_{Ji}\,{}^tF_J^A = e_{Ji}(-\alpha_{stJ}(u_N^A - u_N^B + {}^t\bar{g}_N)) \tag{1.327}$$

$$= e_{Ji}(-\alpha_{stJ}(e_{3k}u_k^A - e_{3k}u_k^B + {}^t\bar{g}_N)) \tag{1.328}$$

$$= -\alpha_{stJ}e_{Ji}e_{3k}u_k^A + \alpha_{stJ}e_{Ji}e_{3k}u_k^B - \alpha_{stJ}e_{Ji}\,{}^t\bar{g}_N \tag{1.329}$$

Or in the matrix form:

$${}^{t+\Delta t}Q_{ci}^A = K_{ik}^{AA}u_k^A + K_{ik}^{AB}u_k^B - \alpha_{stJ}e_{Ji}\,{}^t\bar{g}_N = -\alpha_{stJ}e_{Ji}\,{}^{t+\Delta t}\bar{g}_N \tag{1.330}$$

where

$$K_{ik}^{AA} = -\alpha_{stJ}e_{Ji}e_{3k}, \quad K_{ik}^{AB} = \alpha_{stJ}e_{Ji}e_{3k} \tag{1.331}$$

3. The frictionless sliding interface
For the frictionless sliding interface, substituting $\mu = 0$ into Equations (1.330) and (1.331), we have:

$$K_{ik}^{AA} = -\alpha e_{3i}e_{3k}, \quad K_{ik}^{AB} = \alpha e_{3i}e_{3k}, \tag{1.332}$$

and

$$^{t+\Delta t}Q_{ci}^{A} = -\alpha e_{3i} \, ^{t+\Delta t}\overline{g}_N \tag{1.333}$$

The element matrix and nodal force can be obtained by surface integration for these cases:

$$K_{e(I,J)}^{A_a A_b} = \int_{^{t+\Delta t}\Gamma_C^{A_e}} K_{ij}^{AA} N_S^{A_a} N_S^{A_b} \mathrm{d}\Gamma \tag{1.334}$$

$$K_{e(I,K)}^{A_a B_b} = \int_{^{t+\Delta t}\Gamma_C^{A_e}} K_{ik}^{AB} N_S^{A_a} N_S^{B_b} \mathrm{d}\Gamma \tag{1.335}$$

$$Q_{e(I)}^{A_a} = \int_{^{t+\Delta t}\Gamma_C^{A_e}} {}^{t+\Delta t}Q_{ci}^{A} N_S^{A_a} \mathrm{d}\Gamma \tag{1.336}$$

where $N_S^{A_a}, N_S^{A_b}$ are the shape functions of a contact surface element on contactor A at nodes a and b, respectively. $N_S^{B_b}$ is the shape function of node b of target surface of the body B. Subscript e stands for the element value.

$$\begin{cases} I = (a-1)\cdot \mathrm{NSD} + i \\ J = (b-1)\cdot \mathrm{NSD} + j \\ K = (b-1)\cdot \mathrm{NSD} + \mathrm{b} \end{cases} \qquad \begin{cases} 1 \le A_a, A_b \le \mathrm{ns}_A \\ 1 \le B_b \le \mathrm{ns}_B \\ 1 \le i, j, k \le \mathrm{NSD} \end{cases}$$

where ns_A and ns_B are the numbers of nodes per surface element on contactor A and on target surface B, respectively.

Remark 1.11: The previous equations consider body A as the contactor and body B as the target, and surface integration is performed at the contactor's surface. All the contributions of contact forces also only merge to the contactors surface. For numerical convenience, we can do the reversion for the contactor and target side and repeat the same calculation for the new contactor (body B). Then the contact force for both the contactor and target is included.

1.3.8.6 The simulation procedure for the penalty-based contact by implicit method

For the time transient contact problems, starting time is 0, and end time is T, and i is the iteration number. The return mapping method is used with incremental form to carry out the simulation for the contact process.

1. Set $t = 0, i = 0$.
2. Calculate and assemble the effective matrices without contact terms.
3. Calculate the residual vectors without contact terms.

4. Search for contact pair at integration points of each contact surface on side A.
5. Compute and assemble the contact matrix $K_{c\alpha}^{AA}, K_{c\alpha}^{AB}$ for each and every integration point.
6. Compute and assemble the contact force vector Q_c^A, Q_c^B for each integration points.
7. Reverse the sides A and B, and repeat the same calculation from step 4 to step 6 for side B.
8. Solve the contact system.
9. Update the solution vectors.
10. Update the stress and the contact force ${}^{t+\Delta t}\boldsymbol{F}_c^{A(l)}$, ${}^{t+\Delta t}\boldsymbol{F}_c^{B(l)}$.
    Note: the incremental form of return mapping method is used for updating the tangential contact force.

$$ {}^{t+\Delta t}F_n = \alpha_n \langle {}^{t+\Delta t}g_N \rangle \tag{1.337} $$

$$ {}^{t+\Delta t}\boldsymbol{F}_T^{\text{trail}} = {}^{t}\boldsymbol{F}_T + \alpha_T({}^{t+\Delta t}\boldsymbol{u}_T - {}^{t}\boldsymbol{u}_T) \tag{1.338} $$

The return mapping is completed by setting:

$$ {}^{t+\Delta t}\boldsymbol{F}_T = \begin{cases} {}^{t+\Delta t}\boldsymbol{F}_T^{\text{trial}} & (\text{stick}) \text{ if } \left\| {}^{t+\Delta t}\boldsymbol{F}_T^{\text{trial}} \right\| \le \mu\, {}^{t+\Delta t}F_n \\ \dfrac{\mu\, {}^{t+\Delta t}F_n}{\left\| {}^{t+\Delta t}\boldsymbol{F}_T^{\text{trial}} \right\|}\, {}^{t+\Delta t}\boldsymbol{F}_T^{\text{trial}} & (\text{slide}) \text{ if } \left\| {}^{t+\Delta t}\boldsymbol{F}_T^{\text{trial}} \right\| > \mu\, {}^{t+\Delta t}F_n \end{cases} \tag{1.339} $$

11. Verify the contact conditions: if dissatisfied. go to step 4; if yes, go to step 12.
12. Convergence check: if yes, go to next time step; if not, set $i = i + 1$ and go back to step 2.
13. $t = t + \Delta t$
14. if $t < T$, then go to next step. If not, finish the simulation.

1.3.8.7 *Searching algorithm for contact problems*

As mentioned in the previous section, contact status checks of the integration points on contactor and target surfaces need to be carried out for all the nonlinear contact iterations. An appropriate searching algorithm is able to report the normal distance, mapping results (the mapped facet of the surface element and local coordinate on that facet), and the tangential and normal unit vectors of mapped point on the target face. The bucket search method for surface presented in Chapter 3 is an efficient and robust mapping method for the contact problem with a good accuracy.

1.3.8.8 *Other methods for contact problems*

Other methods include the Lagrange multiplier method and the augmented Lagrange method (Belytschko et al., 2013) are commonly used for contact problems.

Examples of FSI problems with contact are presented in Section 4.6.4.

Remark 1.12: (1)The stabilization method for the constitutive law of viscous condition is presented in Section 5.2.5. (2) Surface heat generation on the sliding viscous contact surface is another heat source in structural and thermal simulation (e.g., the brake disc). (3) The pinched flow problem with structure contact is one of the challenges in the FSI analysis, refer to Section 4.6.4 for details.

1.4 Electromagnetic field

1.4.1 Fundamental equations

Maxwell's equations are the basic descriptions of the laws of electromagnetic fields in mathematic form, which define relations for electromagnetic field quantities and the source terms. Maxwell's equations are the starting point for simulating all electromagnetic problems.

1.4.1.1 Governing equations

The governing equations for electromagnetic problems are:

$$\nabla \times \boldsymbol{H} = \boldsymbol{J} + \frac{\partial \boldsymbol{D}}{\partial t} = \boldsymbol{J}_s + \boldsymbol{J}_e + \boldsymbol{J}_v + \frac{\partial \boldsymbol{D}}{\partial t} \quad \text{on} \quad \Omega^{em} \tag{1.340}$$

$$\nabla \times \boldsymbol{E} = -\frac{\partial \boldsymbol{B}}{\partial t} \quad \text{on} \quad \Omega^{em} \tag{1.341}$$

$$\nabla \cdot \boldsymbol{B} = 0 \quad \text{on} \quad \Omega^{em} \tag{1.342}$$

$$\nabla \cdot \boldsymbol{D} = \rho \quad \text{on} \quad \Omega^{em} \tag{1.343}$$

Here, $\nabla\times$ is the curl operator; $\nabla\cdot$ is the divergence operator; $\boldsymbol{H}$ is the magnetic field intensity vector; $\boldsymbol{J}$ is the total current density vector; $\boldsymbol{J}_s$ is the applied source current density vector; $\boldsymbol{J}_e$ is the induced current density vector; $\boldsymbol{J}_v$ is the velocity current density vector; $\boldsymbol{D}$ is the electric flux density vector; $\boldsymbol{E}$ is the electric field intensity vector; $\boldsymbol{B}$ is the magnetic flux density vector; ρ is the electric charge density; and Ω^{em} denotes the spatial electromagnetic domain.

1.4.1.2 Constitutive equations

For problems considering saturated material for soft materials (without permanent magnets), the constitutive relation for magnetic field is

$$\boldsymbol{B} = [\mu]\boldsymbol{H} \tag{1.344}$$

Here, $[\boldsymbol{\mu}]$ is the magnetic permeability matrix (inverse of the magnetic reluctivity matrix $[\boldsymbol{\nu}]$), in general, depending on the magnetic field vector, temperature, and so on.

When hard materials (with permanent magnets) are considered, the constitutive relation becomes:

$$\boldsymbol{B} = [\mu]\boldsymbol{H} + [\mu_0]\boldsymbol{M}_0 \tag{1.345}$$

where $[\boldsymbol{\mu}_0]$ is the permeability of free space; $\boldsymbol{M}_0$ is the permanent intrinsic magnetization vector, and $[\boldsymbol{\mu}_0]\boldsymbol{M}_0$ represents residual flux density $\boldsymbol{B}_r$.

The constitutive relations for electric field are

$$J_e = [\sigma]E \tag{1.346}$$

$$J_v = [\sigma]\bar{v} \times B \tag{1.347}$$

$$D = [\varepsilon]E \tag{1.348}$$

where $[\sigma]$ is the electrical conductivity matrix (inverse of the electrical resistivity matrix), $[\varepsilon]$ is the permittivity matrix, and $\bar{v}$ is the incoming velocity vector.

1.4.2 Classification of Maxwell's equations and potential formulations

Most general equations describing electromagnetic phenomenon are presented in Section 1.4.1. In the simplest case, time variation of the field quantities can be neglected, which is denoted as the static field, corresponding to the magnetic field, electrostatic field, and electric field. In this case, these static fields can be regarded independently without considering the interactions between them. By contrast, for time-varying case, the magnetic field and electric field are coupled, and the interaction between them cannot be neglected, which results in the eddy current in conducting regions or wave propagation in free space and so on.

In this section, we will make a classification of Maxwell's equations according to whether we neglect specific terms in Maxwell's equations or not (Kuczmann, 2009). Based on different types of electromagnetic equations in terms of electromagnetic field quantities, potential equations are established correspondingly.

Before we get start, we first give a brief graph to introduce the applicable regions for electromagnetic problems, as shown later (Figure 1.8). It will be referred to with each electromagnetic type discussed further.

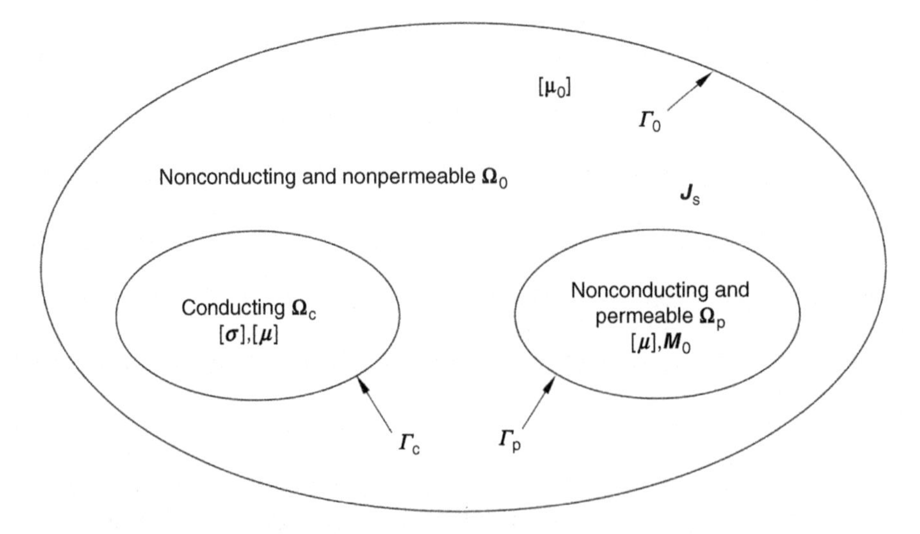

Figure 1.8 Electromagnetic field regions.

Here in Figure 1.8, Ω_0 is the free space region; Ω_c is the conducting region; Ω_p is the nonconducting permeable region; and:

$$\Omega^{em} = \Omega_0 \cup \Omega_c \cup \Omega_p.$$

1.4.2.1 *Classification of Maxwell's equations*

For static fields, as we have discussed earlier, time varying effects are ignored, and Maxwell's equations with constitutive relations can be separated into magnetic, electrostatic, and electric cases.

1.4.2.1.1 Magnetic field

The exciting source is time-independent coil current; generating time-independent magnetic field intensity, $\boldsymbol{H}$; and time-independent magnetic flux density, $\boldsymbol{B}$. Maxwell's equations are thus reduced to the following form:

$$\nabla \times \boldsymbol{H} = \boldsymbol{J}_s \quad \text{in} \quad \Omega_0 \cup \Omega_p \tag{1.349}$$

$$\nabla \cdot \boldsymbol{B} = \boldsymbol{0} \quad \text{in} \quad \Omega_0 \cup \Omega_p \tag{1.350}$$

associated with boundary conditions Γ_B and Γ_H.

Where Γ_B is defined by

$$\boldsymbol{n} \cdot \boldsymbol{B} = 0 \tag{1.351}$$

and Γ_H is defined by

$$\boldsymbol{n} \times \boldsymbol{H} = \boldsymbol{0} \tag{1.352}$$

$$\Gamma_0 \cup \Gamma_p = \Gamma_B \cup \Gamma_H \tag{1.353}$$

1.4.2.1.2 Electrostatic field

The exciting source is time-independent charge, generating time-independent electric field intensity $\boldsymbol{E}$, and electric flux density $\boldsymbol{D}$. Maxwell's equations are reduced to the following form:

$$\nabla \times \boldsymbol{E} = 0 \quad \text{in} \quad \Omega_0 \cup \Omega_p \tag{1.354}$$

$$\nabla \cdot \boldsymbol{D} = \rho \quad \text{in} \quad \Omega_0 \cup \Omega_p \tag{1.355}$$

associated with boundary conditions Γ_D and Γ_E.

Where Γ_D is defined by

$$\boldsymbol{n} \cdot \boldsymbol{D} = 0 \tag{1.356}$$

and Γ_E is defined by

$$\boldsymbol{n}\times\boldsymbol{E}=\boldsymbol{0} \tag{1.357}$$

$$\Gamma_0\cup\Gamma_p=\Gamma_D\cup\Gamma_E \tag{1.358}$$

1.4.2.1.3 Electric field

Time-independent current caused by time-independent electric field intensity and Maxwell's equations are reduced to the following form:

$$\nabla\times\boldsymbol{E}=\boldsymbol{0} \quad \text{in} \quad \Omega_c \tag{1.359}$$

$$\nabla\cdot\boldsymbol{J}_e=0 \quad \text{in} \quad \Omega_c \tag{1.360}$$

associated with boundary conditions Γ_J and Γ_E.
where Γ_J is defined by

$$\boldsymbol{n}\cdot\boldsymbol{J}_e=0 \tag{1.361}$$

and Γ_E is defined by

$$\boldsymbol{n}\times\boldsymbol{E}=\boldsymbol{0} \tag{1.362}$$

$$\Gamma_c=\Gamma_J\cup\Gamma_E \tag{1.363}$$

It should be noted that Equation (1.360) is derived from Equation (1.340) by taking the divergence and omitting the terms $\boldsymbol{J}_s$, $\boldsymbol{J}_v$, and $\frac{\partial \boldsymbol{D}}{\partial t}$.

For the time-varying case, the time-varying effects could not be ignored, and the magnetic and electric fields are coupled together. Maxwell's equations with constitutive relations can be separated into eddy current field and wave propagation of electrodynamics.

1.4.2.1.4 Eddy current field

The magnetic and electric fields are coupled, but the displacement current can be neglected in the usual frequency range because $|\boldsymbol{J}| \gg |\partial \boldsymbol{D}/\partial t|$ in this situation. Currents in coils generate magnetic field in the conducting region, and the time-varying magnetic field induces electric field, which causes eddy currents in the conducting region. In the nonconducting region, governing equations could be described by magnetic Equations (1.349) and (1.350); in the conducting region, governing equations could be described by "quasistatic" equations as follows:

$$\nabla\times\boldsymbol{H}=\boldsymbol{J}_e+\boldsymbol{J}_v \quad \text{in} \quad \Omega_c \tag{1.364}$$

$$\nabla\cdot(\boldsymbol{J}_e+\boldsymbol{J}_v)=0 \quad \text{in} \quad \Omega_c \tag{1.365}$$

$$\nabla\times\boldsymbol{E}=-\frac{\partial \boldsymbol{B}}{\partial t} \quad \text{in} \quad \Omega_c \tag{1.366}$$

$$\nabla\times\boldsymbol{B}=\boldsymbol{0}\quad\text{in}\quad\Omega_c \tag{1.367}$$

It should be noted that Equation (1.365) is derived from Equation (1.340) by taking the divergence and omitting the imposed current term J_s.

This approximation is applicable in electromagnetic devices such as motors, relays, and transformers with a low frequency below a few tens of kilohertz.

1.4.2.1.5 Wave propagation

In this case, electric flux density could not be neglected because of high-frequency effects, and high-frequency equations should thus be the full form of Maxwell's equations. Specially, wave equations in a vacuum are expressed as follows:

$$\nabla\times\boldsymbol{H}=-\frac{\partial\boldsymbol{D}}{\partial t}\quad\text{in}\quad\Omega_0 \tag{1.368}$$

$$\nabla\times\boldsymbol{E}=-\frac{\partial\boldsymbol{B}}{\partial t}\quad\text{in}\quad\Omega_0 \tag{1.369}$$

$$\nabla\times\boldsymbol{B}=\boldsymbol{0}\quad\text{in}\quad\Omega_0 \tag{1.370}$$

$$\nabla\cdot\boldsymbol{D}=\rho\quad\text{in}\quad\Omega_0 \tag{1.371}$$

In this book, we narrow our attention mainly into low-frequency electromagnetic problems (i.e., static field and eddy current field problems).

1.4.2.2 *Potential formulations for different types of equations*

The solution of magnetic field problems is commonly obtained by using potential functions. In this book, $\boldsymbol{A}$, $\boldsymbol{A}-V$ form is adopted to describe Maxwell's equations, which is widely used in low-frequency electromagnetic simulation.

By introducing magnetic vector potential $\boldsymbol{A}$ and electric scalar potential V, the magnetic field density $\boldsymbol{B}$ and the electric field $\boldsymbol{E}$ could be expressed as:

$$\boldsymbol{B}=\nabla\times\boldsymbol{A} \tag{1.372}$$

$$\boldsymbol{E}=-\frac{\partial\boldsymbol{A}}{\partial t}-\nabla V \tag{1.373}$$

Equations (1.372) and (1.373) will guarantee that Equations (1.342) and (1.341) could be satisfied automatically. By substituting Equations (1.372) and (1.373) into governing equations of magnetic fields, electrostatic fields, and so on, we could obtain their potential formulations, respectively.

1.4.2.2.1 Magnetic field

Substitute Equation (1.372) into Equation (1.349), noting that Equation (1.350) is satisfied automatically, we get the magnetic field equation in curl–curl form:

$$\nabla\times[\nu]\nabla\times\boldsymbol{A}=\boldsymbol{J}_s+\boldsymbol{H}_c\quad\text{in}\quad\Omega_0\cup\Omega_p \tag{1.374}$$

associated with boundary conditions in potential form:

$$\boldsymbol{n}\times\boldsymbol{A}=\boldsymbol{0} \quad \text{on}\ \Gamma_B \tag{1.375}$$

$$\boldsymbol{n}\times\nabla\times\boldsymbol{A}=\boldsymbol{0} \quad \text{on}\ \Gamma_H \tag{1.376}$$

where

[ν] is the reluctivity matrix (inverse of the magnetic permeability matrix).
$\boldsymbol{J}_s$ is the source current density vector.
$\boldsymbol{H}_c$ is the coercive force vector, which is related to the permanent intrinsic magnetization vector, represented as $\boldsymbol{H}_c=\frac{1}{\nu_0}[\nu]\boldsymbol{M}_0$.

ν_0 is the reluctivity of free space.

1.4.2.2.2 Electrostatic field

Substitute Equation (1.373) into Equation (1.355) by omitting $\boldsymbol{A}$ and note that Equation (1.354) could be satisfied automatically; we get the Poisson form equation in terms of electric scalar potential:

$$-\nabla\cdot([\varepsilon]\nabla V)=\rho \quad \text{in} \quad \Omega_0\cup\Omega_p \tag{1.377}$$

associated with boundary conditions in potential form:

$$\boldsymbol{n}\cdot\nabla V=\boldsymbol{0} \quad \text{on}\ \Gamma_D \tag{1.378}$$

$$V=V_0=\text{constant} \quad \text{on}\ \Gamma_E \tag{1.379}$$

1.4.2.2.3 Electric field

Substitute Equation (1.373) into Equation (1.360) by omitting $\boldsymbol{A}$ and note that Equation (1.359) could be satisfied automatically; we get

$$-\nabla\cdot([\sigma]\nabla V)=0 \quad \text{in} \quad \Omega_0\cup\Omega_p \tag{1.380}$$

associated with boundary conditions in potential form:

$$\boldsymbol{n}\cdot\nabla V=\boldsymbol{0} \quad \text{on}\ \Gamma_J \tag{1.381}$$

$$V=V_0=\text{constant} \quad \text{on}\ \Gamma_E \tag{1.382}$$

1.4.2.2.4 Eddy current field

In this case, as we have discussed in Section 1.4.2.1 that the governing equations should be separated into two parts according to their conducting properties.

In the nonconducting region, the potential formulation is the same as magnetic field equations, but in the conducting region, by substituting Equations (1.372) and (1.373) into eddy current field equations, the potential formulation of "quasistatic" equations are thus derived and expressed as follows:

$$\nabla \cdot \left\{ [\sigma] \left\{ \left(-\frac{\partial \boldsymbol{A}}{\partial t} - \nabla V \right) + \vec{v} \times \nabla \times \boldsymbol{A} \right\} \right\} = 0 \quad \text{in } \Omega_C \tag{1.383}$$

$$\nabla \times [\nu] \nabla \times \boldsymbol{A} + [\sigma] \left(\frac{\partial \boldsymbol{A}}{\partial t} + \nabla V \right) - [\sigma] (\vec{v} \times \nabla \times \boldsymbol{A}) = 0 \quad \text{in } \Omega_C \tag{1.384}$$

associated with boundary conditions containing Equations (1.375), (1.376), (1.381), and (1.382).

Nonuniqueness of magnetic vector potential

It should be noted that the introduction of magnetic vector potential $\boldsymbol{A}$ leads to nonuniqueness of the solution. We will learn that discretization of the differential curl–curl operator will give rise to a singular FEM matrix. It is easily seen by adding the gradient of a scalar function to $\boldsymbol{A}$, which leads to a modified vector $\boldsymbol{A}^*$:

$$\boldsymbol{A}^* = \boldsymbol{A} + \nabla \phi \tag{1.385}$$

We conclude that

$$\boldsymbol{B} = \nabla \times \boldsymbol{A} = \nabla \times \boldsymbol{A}^* \tag{1.386}$$

based on the zero property of curl of the gradient operator. The nonuniqueness of the magnetic vector potential will lead to a singular FEM matrix, as we discretize the differential curl–curl operator by the edge element method.

To ensure the uniqueness of magnetic vector potential $\boldsymbol{A}$, more constraints should be enforced on $\boldsymbol{A}$, and a common approach is to impose a Coulomb-type gauge:

$$\nabla \cdot \boldsymbol{A} = 0 \tag{1.387}$$

1.4.3 FEM discretization of potential formulations

The numerical analysis of electromagnetic field problems using FEM has been one of the main directions of research in computational electromagnetic (Bossavit, 1998; Meunier, 2010; Kuczmann, 2009; Bíró, 1999; Jin, 2014). It is used to approximate the solution of the $\boldsymbol{A}$, $\boldsymbol{A}$–V formulation derived from Maxwell's equations. In this formulation, the edge element method and nodal element method are used for approximating $\boldsymbol{A}$ and V, respectively. We will give a brief view of the edge-based discretization method and introduce different types of FEM equations in the matrix form.

1.4.3.1 Weak forms for electromagnetic equations

To obtain the weak form of electromagnetic equations, Sobolev spaces are introduced (Meunier, 2010):

$$\boldsymbol{H}(\text{curl};\Omega)=\left\{\boldsymbol{u}\in L^2(\Omega)^3;\text{curl}\boldsymbol{u}\in L^2(\Omega)^3\right\} \tag{1.388}$$

$$\boldsymbol{H}_0(\text{curl};\Omega)=\left\{\boldsymbol{u}\in \boldsymbol{H}(\text{curl};\Omega);\boldsymbol{u}\times\boldsymbol{n}=0 \text{ on } \partial\Omega\right\} \tag{1.389}$$

$\boldsymbol{H}^s(\Omega)$ $(s\geq 0)$ is the standard Sobolev space, and $H_0^s(\Omega)$ $(s\geq 0)$ is the standard Sobolev space with boundary conditions. In addition, $\boldsymbol{H}^s(\Omega)$ and $\boldsymbol{H}_0^s(\Omega)$ are standard Sobolev spaces defined on $L^2(\Omega)^3$.

To establish weak formulations with respect to electromagnetic equations associated with appropriate boundary conditions, Green formulae should be used:

$$(\boldsymbol{u},\ \text{grad}v)+(\text{div}\boldsymbol{u},\ v)=\langle v,\ \boldsymbol{n}\cdot\boldsymbol{u}\rangle|_{\partial\Omega}\quad \boldsymbol{u}\in \boldsymbol{H}^1(\Omega),\ v\in H^1(\Omega) \tag{1.390}$$

$$(\boldsymbol{u},\ \text{curl}\boldsymbol{v})-(\text{curl}\boldsymbol{u},\ \boldsymbol{v})=\langle \boldsymbol{n}\times\boldsymbol{u},\ \boldsymbol{v}\rangle|_{\partial\Omega}\quad \boldsymbol{u},\ \boldsymbol{v}\in \boldsymbol{H}^1(\Omega) \tag{1.391}$$

where $(,)$ denotes inner product on volume Ω and $\langle,\rangle$ denotes inner product on surface $\partial\Omega$.

By using Green formulae and associated with boundary conditions, the weak forms of electromagnetic field equations with different types are obtained.

1.4.3.1.1 Magnetic field

Applying Equations (1.391), (1.375), and (1.376) into Equation (1.374), we get the weak form of magnetic field equations:

Find magnetic vector potential $\boldsymbol{A}\in \boldsymbol{H}_0(\text{curl};\Omega)$, such that

$$\left([\nu]\ \text{curl}\ \boldsymbol{A},\ \text{curl}\ \boldsymbol{w}\right)=\left(\boldsymbol{f},\boldsymbol{w}\right)\quad \forall \boldsymbol{w}\in \boldsymbol{H}_0(\text{curl};\Omega) \tag{1.392}$$

where, $\boldsymbol{f}$ is $\boldsymbol{J}_s+\boldsymbol{H}_c$; Ω denotes $\Omega_0\cup\Omega_p$.

1.4.3.1.2 Electrostatic field

Applying Equation (1.390),(1.378), and (1.379) into Equation (1.377), we get the weak form of electrostatic field equations:

Find electric scalar potential $V\in H_0^1(\Omega)$, such that

$$\left([\varepsilon]\ \text{grad}V,\ \text{grad}q\right)=\left(\rho,q\right)\quad \forall q\in H_0^1(\Omega) \tag{1.393}$$

Where Ω denotes $\Omega_0\cup\Omega_p$.

1.4.3.1.3 Electric field

Similar to the electrostatic field case, we could also get the weak form of electric field equations:

Find the electric scalar potential $V \in H_0^1(\Omega)$, such that

$$([\sigma] \text{ grad} V, \text{ grad } q) = (I_{in}, q) \quad \forall q \in H_0^1(\Omega) \tag{1.394}$$

where Ω denotes Ω_c.

1.4.3.1.4 Eddy current field

By using Equations (1.390) and (1.391), the weak form of eddy current field potential equations Equations (1.383) and (1.384) with boundary conditions, Equations (1.375), (1.376), (1.381), and (1.382) in the conducting region can be obtained:

Find $(\boldsymbol{A}, V) \in \boldsymbol{H}_0(\text{curl};\Omega) \times H_0^2(\Omega)$, such that

$$\begin{cases} ([\nu] \text{ curl } \boldsymbol{A}, \text{ curl} w) + ([\sigma]\text{curl } \dot{\boldsymbol{A}}, \boldsymbol{w}) + ([\sigma]V, \boldsymbol{w}) + ([\sigma]\bar{\nu} \times \text{curl}\boldsymbol{A}, \boldsymbol{w}) \\ \qquad = (\boldsymbol{f}, \boldsymbol{w}) \quad \forall \boldsymbol{w} \in \boldsymbol{H}_0(\text{curl};\Omega) \\ ([\sigma]\boldsymbol{A}, \text{grad}q) + ([\sigma]V, \text{grad}q) + ([\sigma]\bar{\nu} \times \text{curl}\boldsymbol{A}, \text{ grad}q) \\ \qquad = (I_{\text{in}}, q) \quad \forall q \in H_0^2(\Omega) \end{cases} \tag{1.395}$$

1.4.3.2 Shape functions for magnetic vector potential

The magnetic vector potential in elements could be interpolated by values along the element edges:

$$\boldsymbol{A} = \boldsymbol{W}^{\mathrm{T}} \boldsymbol{A}^{e} = \sum_{i=1}^{M} \boldsymbol{W}_i^{\mathrm{T}} A_{zi} \tag{1.396}$$

Where $\boldsymbol{W}$ is the matrix of element vector edge-based shape functions:

$$\boldsymbol{W} = \begin{bmatrix} W_{1x} & W_{1y} & W_{1z} \\ W_{2x} & W_{2y} & W_{2z} \\ W_{3x} & W_{3y} & W_{3z} \\ \vdots & \vdots & \vdots \\ W_{Mx} & W_{My} & W_{Mz} \end{bmatrix}, \boldsymbol{W}_i = \begin{bmatrix} W_{ix} & W_{iy} & W_{iz} \end{bmatrix} \tag{1.397}$$

where M is the number of element edges.

In addition, the curl of shape functions could be expressed by:

$$\nabla \times \mathbf{W}_i^{\mathrm{T}} = \begin{bmatrix} \boldsymbol{i} & \boldsymbol{j} & \boldsymbol{k} \\ \dfrac{\partial}{\partial x} & \dfrac{\partial}{\partial y} & \dfrac{\partial}{\partial z} \\ W_{ix} & W_{iy} & W_{iz} \end{bmatrix} = \begin{bmatrix} \dfrac{\partial W_{iz}}{\partial y} - \dfrac{\partial W_{iy}}{\partial z} \\ \dfrac{\partial W_{ix}}{\partial z} - \dfrac{\partial W_{iz}}{\partial x} \\ \dfrac{\partial W_{iy}}{\partial x} - \dfrac{\partial W_{ix}}{\partial y} \end{bmatrix} \tag{1.398}$$

Consider the term $\dfrac{\partial W_{iz}}{\partial y}$, for example:

$$\begin{aligned} \frac{\partial W_{iz}}{\partial y} &= \frac{\partial (1-s)(1-t)\,(\partial r/\partial z)}{\partial y} \\ &= -\frac{\partial s}{\partial y}(1-t)\frac{\partial r}{\partial z} + (1-s)\left(-\frac{\partial t}{\partial y}\right)\frac{\partial r}{\partial z} + (1-s)\,(1-t)\frac{\partial^2 r}{\partial z \partial y} \end{aligned} \tag{1.399}$$

$$\begin{aligned} \frac{\partial}{\partial z}\left(\frac{\partial r}{\partial x}\right) &= \frac{\partial}{\partial r}\left(\frac{\partial r}{\partial x}\right)\frac{\partial r}{\partial z} + \frac{\partial}{\partial s}\left(\frac{\partial r}{\partial x}\right)\frac{\partial s}{\partial z} + \frac{\partial}{\partial t}\left(\frac{\partial r}{\partial x}\right)\frac{\partial t}{\partial z} \\ &= \frac{\partial^2 r}{\partial r \partial x}\frac{\partial r}{\partial z} + \frac{\partial^2 r}{\partial s \partial x}\frac{\partial s}{\partial z} + \frac{\partial^2 r}{\partial t \partial x}\frac{\partial t}{\partial z} = 0 \end{aligned} \tag{1.400}$$

$$\frac{\partial}{\partial z}\left(\frac{\partial r}{\partial y}\right) = 0 \tag{1.401}$$

$$\frac{\partial}{\partial z}\left(\frac{\partial t}{\partial z}\right) = 0 \tag{1.402}$$

1.4.3.2.1 Shape functions for hex element

Shape functions for the hexahedral Nedelec element (Figure 1.9) are expressed as follows:

$$\overrightarrow{\boldsymbol{W}}_r^e = \frac{h_r}{8}(1 \pm s)(1 \pm t)\ \nabla r \quad \text{parallel to } r\text{-axis: } Q,\ S,\ W,\ U \tag{1.403}$$

$$\overrightarrow{\boldsymbol{W}}_s^e = \frac{h_s}{8}(1 \pm t)(1 \pm r)\nabla s \quad \text{parallel to } s\text{-axis: } R,\ V,\ X,\ T \tag{1.404}$$

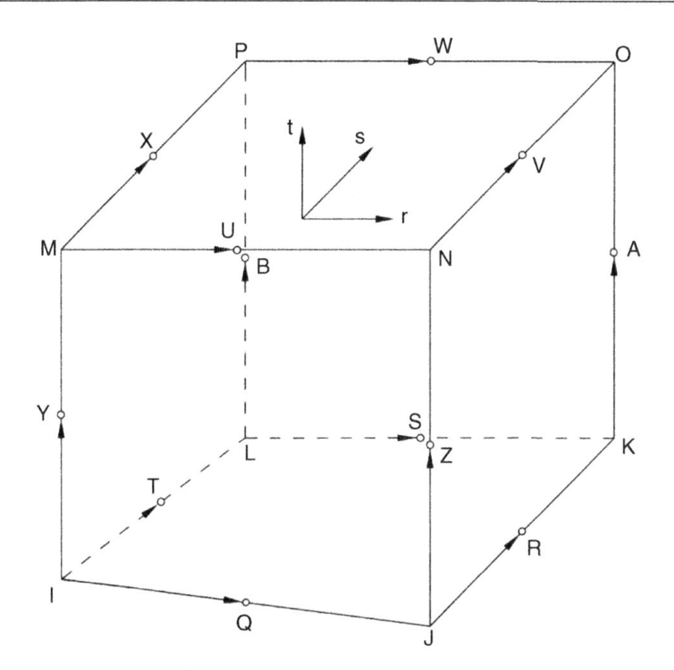

Figure 1.9 Configurations of first-order hexahedral Nedelec element.

$$\vec{W}_t^e = \frac{h_t}{8}(1 \pm r)(1 \pm s)\ \nabla t \quad \text{parallel to } t\text{-axis: } Y,\ Z,\ A,\ B \tag{1.405}$$

Where h_r, h_s, and h_t represent the length of the element edge and $\nabla r, \nabla s$, and ∇t represent the gradient of local r, s and t, respectively. Note that because we use the length with direction to define the shape function for A_z DOF (Equation (1.396)), to guarantee the continuity conditions during the element assembly, the shared edges of the neighboring elements should have the same direction defined for the magnetic flux density vector.

1.4.3.2.2 Shape functions for tet element

Shape functions for tetrahedral Nedelec element (Figure 1.10) are expressed as follows:

$$\vec{W}_{IJ} = h_{IJ}\left(\lambda_I \nabla \lambda_J - \lambda_J \nabla \lambda_I\right) \tag{1.406}$$

$$\vec{W}_{JK} = h_{JK}\left(\lambda_J \nabla \lambda_K - \lambda_K \nabla \lambda_J\right) \tag{1.407}$$

$$\vec{W}_{KI} = h_{KI}\left(\lambda_K \nabla \lambda_I - \lambda_I \nabla \lambda_K\right) \tag{1.408}$$

$$\vec{W}_{IL} = h_{IL}\left(\lambda_I \nabla \lambda_L - \lambda_L \nabla \lambda_I\right) \tag{1.409}$$

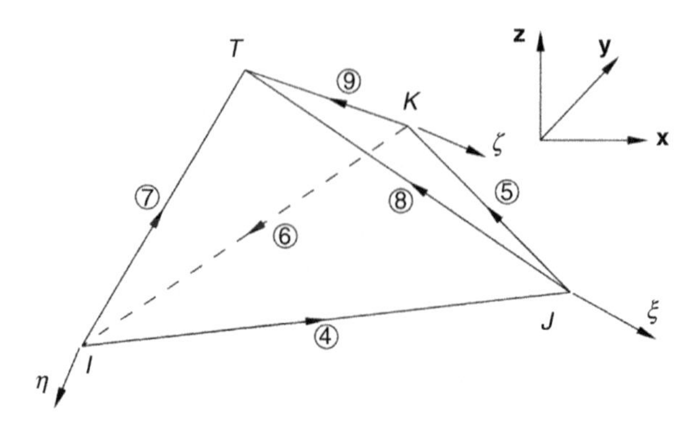

Figure 1.10 Configurations of first-order tetrahedral Nedelec element.

$$\vec{W}_{JL} = h_{JL}\left(\lambda_J \nabla\lambda_L - \lambda_L \nabla\lambda_J\right) \tag{1.410}$$

$$\vec{W}_{KL} = h_{KL}\left(\lambda_K \nabla\lambda_L - \lambda_L \nabla\lambda_K\right) \tag{1.411}$$

Where: h_{ij}, represents edge length between node I and J; $\lambda_I, \lambda_J, \lambda_K$, and λ_L, represent volume coordinates $\left(\lambda_K = 1 - \lambda_I - \lambda_J - \lambda_L\right)$; and $\nabla\lambda_I$, $\nabla\lambda_J$, $\nabla\lambda_K$, and $\nabla\lambda_L$, represent the gradient of volume coordinates. Note that the tet element is same as the hex element in which the length with direction is used to define the shape function.

1.4.3.3 *Shape functions for electric scalar potential*

The interpolation for electric scalar potential variables has the same form as the interpolation for temperature variables, Equation (1.8), and both the low- and high-order elements are appropriate for the approximation of electric scalar potential.

The electric scalar potential in elements could be interpolated by values at nodes:

$$V = \boldsymbol{N}^{\mathrm{T}} \boldsymbol{V}_e = \sum_{i=1}^{N} N_i V_i \tag{1.412}$$

where $\boldsymbol{N}$ is the matrix of element nodal shape functions:

$$\boldsymbol{N} = \begin{bmatrix} N_1 \\ N_2 \\ N_3 \\ \vdots \\ N_N \end{bmatrix} \tag{1.413}$$

where N is the number of element nodes.

Additionally, the gradient of nodal shape functions that will be used later has the following form:

$$\nabla N^{\mathrm{T}} = \begin{bmatrix} \dfrac{\partial}{\partial x} \\ \dfrac{\partial}{\partial y} \\ \dfrac{\partial}{\partial z} \end{bmatrix} [N_1\ N_2\ N_3 \ldots N_N] = \begin{bmatrix} \dfrac{\partial N_1}{\partial x} & \dfrac{\partial N_2}{\partial x} & \dfrac{\partial N_3}{\partial x} & \ldots & \dfrac{\partial N_N}{\partial x} \\ \dfrac{\partial N_1}{\partial y} & \dfrac{\partial N_2}{\partial y} & \dfrac{\partial N_3}{\partial y} & \ldots & \dfrac{\partial N_N}{\partial y} \\ \dfrac{\partial N_1}{\partial z} & \dfrac{\partial N_2}{\partial z} & \dfrac{\partial N_3}{\partial z} & \ldots & \dfrac{\partial N_N}{\partial z} \end{bmatrix} \tag{1.414}$$

1.4.3.4 Equations in matrix form after FEM discretization

1.4.3.4.1 Magnetic field

Substituting the edge-based shape functions into weak form (Equation (1.392) in finite element space, one can obtain assembled global equation (Equation. (1.415) in matrix form:

$$\boldsymbol{K}^{AA}\boldsymbol{A} = \boldsymbol{J}^{S} + \boldsymbol{J}^{PM} \tag{1.415}$$

where $\boldsymbol{A}$ represents the finite element approximation for the magnetic vector potential in solution domain.

Magnetic reluctivity matrix:

$$\boldsymbol{K}^{AA} = \sum\nolimits_e \boldsymbol{K}_e^{AA}, \boldsymbol{K}_e^{AA} = \int_{\Omega^e} \left(\nabla \times \boldsymbol{W}^{\mathrm{T}}\right)^{\mathrm{T}} [\nu] \left(\nabla \times \boldsymbol{W}^{\mathrm{T}}\right) \mathrm{d}\Omega \tag{1.416}$$

Here, $[\boldsymbol{\nu}]$ is a constant matrix. Nonlinear analysis is given in Section 1.4.3.5.

Source current density vector is:

$$\boldsymbol{J}^{S} = \sum\nolimits_e \boldsymbol{J}_e^{s}, \boldsymbol{J}_e^{S} = \int_{\Omega^e} \boldsymbol{W}^{\mathrm{T}} \boldsymbol{J}_s \mathrm{d}\Omega \tag{1.417}$$

Remanent magnetization load vector is:

$$\boldsymbol{J}^{PM} = \sum\nolimits_e \boldsymbol{J}_e^{PM}, \boldsymbol{J}_e^{PM} = \int_{\Omega^e} \left(\nabla \times \boldsymbol{W}^{\mathrm{T}}\right)^{\mathrm{T}} \boldsymbol{H}_c \,\mathrm{d}\Omega \tag{1.418}$$

1.4.3.4.2 Electrostatic field

Using standard nodal elements for discretization of weak form Equation (1.393), we obtain the following FE equations in assembled matrix form:

$$\boldsymbol{C}^{VV}\boldsymbol{V} = \boldsymbol{Q} \tag{1.419}$$

where V represents the finite element approximations for the electric scalar potential solution domain.

$$\boldsymbol{C}^{VV} = \sum_e \boldsymbol{C}_e^{VV}, \boldsymbol{C}_e^{VV} = \int_{\Omega^e} \left(\nabla \boldsymbol{N}^{\mathrm{T}}\right)^{\mathrm{T}} [\varepsilon] \left(\nabla \boldsymbol{N}^{\mathrm{T}}\right) \mathrm{d}\Omega \tag{1.420}$$

$$\boldsymbol{Q} = \sum_e \boldsymbol{Q}_e \tag{1.421}$$

$$\boldsymbol{Q}_e = \boldsymbol{Q}_e^n + \boldsymbol{Q}_e^c + \boldsymbol{Q}_e^s \tag{1.422}$$

$$\boldsymbol{Q}_e^c = \int_{\Omega^e} \boldsymbol{N} \rho \, \mathrm{d}\Omega \tag{1.423}$$

$$\boldsymbol{Q}_e^s = \int_{\Gamma^e} \boldsymbol{N} \rho_s \, \mathrm{d}\Gamma \tag{1.424}$$

Where $\boldsymbol{Q}_e^n$, is the concentrated charge on the element nodes; $\boldsymbol{Q}_e^c$, is the volume charge load vector; $\boldsymbol{Q}_e^s$, is the surface charge load vector; ρ, is the volume charge density; and ρ_s, is the surface charge density.

1.4.3.4.3 Electric field

Similarly, the discretization method is the same as electrostatic case for the approximation of electric scalar potential, and electric FE equations derived from weak form (1.394) in assembled matrix form are shown as follows:

$$\boldsymbol{K}^{VV}\boldsymbol{V}^e = \boldsymbol{I}_{\mathrm{in}} \tag{1.425}$$

$$\boldsymbol{K}^{VV} = \sum_e \boldsymbol{K}_e^{VV}, \boldsymbol{K}_e^{VV} = \int_{\Omega^e} \left(\nabla \boldsymbol{N}^{\mathrm{T}}\right)^{\mathrm{T}} [\sigma] \left(\nabla \boldsymbol{N}^{\mathrm{T}}\right) \mathrm{d}\Omega \tag{1.426}$$

where $\boldsymbol{I}_{\mathrm{in}}$ is the inject current load vector.

1.4.3.4.4 Eddy current field

In the nonconducting region, we adopt the same form as Equation (1.415); in the conducting region, we adopt the edge-based discretization method for $\boldsymbol{A}$ and the standard nodal discretization method for V. The final FEM equations derived from Equation (1.395) in assembled matrix form are expressed as follow:

$$\begin{bmatrix} \boldsymbol{K}^{AA} + \boldsymbol{K}_v^{AA} & \boldsymbol{0} \\ \boldsymbol{K}_v^{VA} & \boldsymbol{0} \end{bmatrix} \begin{Bmatrix} \boldsymbol{A} \\ - \end{Bmatrix} + \begin{bmatrix} \boldsymbol{C}^{AA} & \boldsymbol{K}^{AV} \\ \left(\boldsymbol{K}^{AV}\right)^{\mathrm{T}} & \boldsymbol{K}^{VV} \end{bmatrix} \begin{Bmatrix} \dot{\boldsymbol{A}} \\ \boldsymbol{V} \end{Bmatrix} = \begin{Bmatrix} \boldsymbol{J}^S + \boldsymbol{J}^{PM} \\ \boldsymbol{I}_{in} \end{Bmatrix} \tag{1.427}$$

where $\boldsymbol{A}$ and $\boldsymbol{V}$ are the finite element approximation for the magnetic vector potential vector and electric scalar potential vector, respectively.

Magnetic reluctivity matrix due to velocity:

$$\boldsymbol{K}_v^{AA} = \sum_e \boldsymbol{K}_{ve}^{AA},\ \boldsymbol{K}_{ve}^{AA} = -\int_{\Omega^e} \boldsymbol{W}[\sigma]\boldsymbol{v}\times\left(\nabla\times\boldsymbol{W}^{\mathrm{T}}\right)\mathrm{d}\Omega \tag{1.428}$$

Electric conductivity matrix is:

$$\boldsymbol{K}^{VV} = \sum_e \boldsymbol{K}_e^{VV}, \boldsymbol{K}_e^{VV} = \int_{\Omega^e} \left(\nabla\boldsymbol{N}^{\mathrm{T}}\right)^{\mathrm{T}} [\sigma]\left(\nabla\boldsymbol{N}^{\mathrm{T}}\right)\mathrm{d}\Omega \tag{1.429}$$

Magneto-electric coupling matrix is:

$$\boldsymbol{K}^{AV} = \sum_e \boldsymbol{K}_e^{AV}, \boldsymbol{K}_e^{AV} = \int_{\Omega^e} \boldsymbol{W}[\sigma]\left(\nabla\boldsymbol{N}^{\mathrm{T}}\right)\mathrm{d}\Omega \tag{1.430}$$

Electric-magneto coupling matrix due to velocity is:

$$\boldsymbol{K}_v^{VA} = \sum_e \boldsymbol{K}_{ve}^{VA},\ \boldsymbol{K}_{ve}^{VA} = -\int_{\Omega^e} \left(\nabla\boldsymbol{N}^{\mathrm{T}}\right)^{\mathrm{T}} [\sigma]\boldsymbol{v}\times\left(\nabla\times\boldsymbol{W}^{\mathrm{T}}\right)\mathrm{d}\Omega \tag{1.431}$$

Eddy current damping matrix is:

$$\boldsymbol{C}^{AA} = \sum_e \boldsymbol{C}_e^{AA}, \boldsymbol{C}_e^{AA} = \int_{\Omega^e} \boldsymbol{W}[\sigma]\boldsymbol{W}^T\,\mathrm{d}\Omega \tag{1.432}$$

1.4.3.5 Nonlinearity in electromagnetic analysis

All of the preceding problems involve only linear materials. However, it is often necessary to deal with nonlinear problems in electromagnetic simulation because of the existence of material nonlinearity. In this section, we will first give a brief introduction to the nonlinear behavior laws of magnetic materials and then narrow our attention to the implementation of resolution algorithms for nonlinear materials modeling.

1.4.3.5.1 Assumptions for material nonlinearity

Magnetic behavior of materials is generally nonlinear, that is, the permeability of material $[\mu]$ is a function of the magnetic field (Figures 1.11 and 1.12). It is noted that the

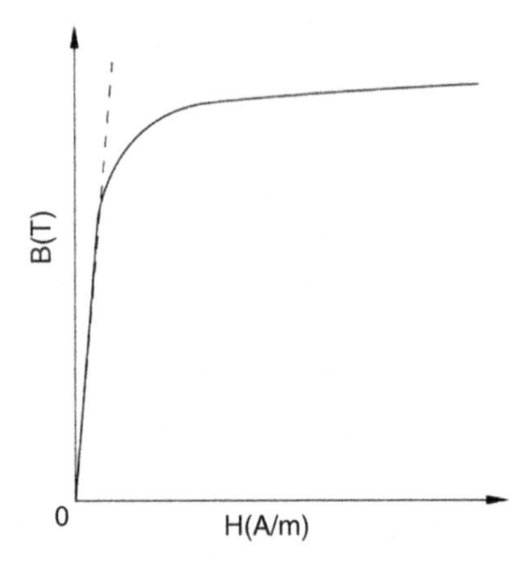

Figure 1.11 A typical *B–H* curve for soft materials, represented by the solid line; the dotted line represents a linear model.

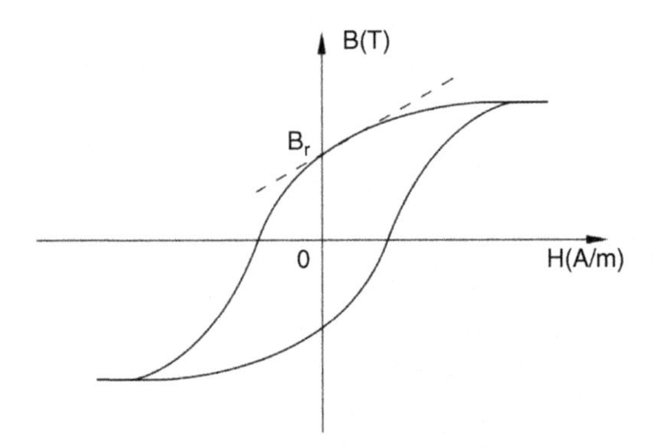

Figure 1.12 A typical *B–H* curve for hard materials, represented by the solid line; the dotted line represents a linear model.

exact *B–H* curve for soft materials involves a hysteresis loop, as shown in Figure 1.12 for hard materials. However, it is appropriate to use a single curve, called the normal magnetization curve, as shown in Figure 1.11, to represent the loop because soft materials usually have a narrow loop.

In many cases, a linear model is adequate for modeling the material behavior if the magnetic saturation does not occur in the problem and the required accuracy could be achieved (Figures 1.11 and 1.12). The constitutive relation thus has the following forms:

For soft materials:

$$\boldsymbol{B} = [\mu]\boldsymbol{H} \tag{1.433}$$

For hard materials (permanent magnets):

$$B=[\mu]H+[\mu_0]M_0 \tag{1.434}$$

The permeability matrix $[\mu]$ is assumed to be constant in Equations (1.433) and (1.434).

However, it is limited if a more accurate and realistic model is pursued. In order to take into account the magnetic saturation of soft materials, which widely exists in electric equipment, a nonlinear isotropic model is at least necessary. Specifically, without considering anisotropy and hysteresis, the magnetic field intensity $\boldsymbol{H}$ and the magnetic flux density $\boldsymbol{B}$ are assumed to be collinear and the amplitudes of $\boldsymbol{B}$ and $\boldsymbol{H}$ has the following relation for soft materials:

$$\|\boldsymbol{B}\|=\mu(\|\boldsymbol{H}\|)\|\boldsymbol{H}\| \tag{1.435}$$

Where $\mu(\|\boldsymbol{H}\|)$ (also simply denoted as μ_h), represents a constant, and the inverse of $\nu(\|\boldsymbol{H}\|)$ (also denoted as ν_h); $\|\;\|$ represents the amplitude of the field. It is noted that the permeability of material is reduced to a scalar function of the amplitude of the magnetic field intensity $\boldsymbol{H}$. Also, the nonlinear model of soft materials is built on their curve of the first magnetization because the coercive force is relatively very low. By contrast, the hard materials (permanent magnets) should be modeled by their major cycle, providing that the materials used are operated in the zone of reversibility of the cycle (safe zone).

Due to the introduction of nonlinear constitutive relation, the responses of magnetic field are nonlinear to the applied source. To obtain the solution, an iterative procedure should be established. In this section, the Newton–Raphson and fixed-point methods, which are common methods in solving nonlinear electromagnetic problems, are both discussed.

1.4.3.5.2 Newton–Raphson method

When applying the Newton–Raphson method into nonlinear electromagnetic problems, there are two types of nonlinear formulations used for iterative calculation, both appropriate for the Newton–Raphson method (Meunier, 2010).

The first type of iterative formulation

Based on the assumptions of material nonlinearity, we could derive finite element equations in nonlinear form, just simply replacing the linear terms in FE equations by nonlinear terms. To illustrate this, taking magnetic field, for example, without considering permanent magnets, we get the FE equation:

$$\boldsymbol{K}^{AA}\boldsymbol{A}=\boldsymbol{J}^S \tag{1.436}$$

Obviously, the magnetic reluctivity matrix in the FE equations is nonlinear:

$$\boldsymbol{K}^{AA}=\int_{\Omega^e}\left(\nabla\times\boldsymbol{W}^{\mathrm{T}}\right)^{\mathrm{T}}[\nu]\left(\nabla\times\boldsymbol{W}^{\mathrm{T}}\right)\mathrm{d}\Omega \tag{1.437}$$

where $[\nu]$ is depending on the current magnetic field and $\boldsymbol{K}^{AA}$ thus also could be denoted as $\boldsymbol{K}^{AA}(\boldsymbol{B})$ or $\boldsymbol{K}^{AA}(\boldsymbol{A})$.

Our objective is to find the $\boldsymbol{A}'$ (corresponding to field $\boldsymbol{H}'$ and $\boldsymbol{B}'$) that satisfying Equation (1.436). Suppose we have a current or temporary solution, $\boldsymbol{B}$ and $\boldsymbol{H}$, and the difference between $\boldsymbol{A}$ and $\boldsymbol{A}'$ will be:

$$\delta \boldsymbol{A} = \boldsymbol{A}' - \boldsymbol{A} \tag{1.438}$$

the difference between $\boldsymbol{B}$ and $\boldsymbol{B}'$:

$$\delta \boldsymbol{B} = \boldsymbol{B}' - \boldsymbol{B} \tag{1.439}$$

the difference between $\boldsymbol{H}$ and $\boldsymbol{H}'$:

$$\delta \boldsymbol{H} = \boldsymbol{H} - \boldsymbol{H}' \tag{1.440}$$

Substituting Equations (1.438) and (1.439) into Equation (1.436), we obtain:

$$\boldsymbol{K}^{AA}(\boldsymbol{A} + \delta \boldsymbol{A})(\boldsymbol{A} + \delta \boldsymbol{A}) = \boldsymbol{J}^{S} \tag{1.441}$$

Expanding Equation (1.441) into a Taylor series at $\boldsymbol{A}$, intercepting the first-order term, we obtain:

$$\boldsymbol{K}^{AA}(\boldsymbol{A})\delta \boldsymbol{A} + \boldsymbol{A}\frac{\mathrm{d}\boldsymbol{K}^{AA}(\boldsymbol{A})}{\mathrm{d}\boldsymbol{A}}\Big|_{\boldsymbol{A}} \delta \boldsymbol{A} = \boldsymbol{J}^{S} - \boldsymbol{K}^{AA}(\boldsymbol{A})\boldsymbol{A} \tag{1.442}$$

Consider the second term in Equation (1.442),

$$\boldsymbol{A}\frac{\mathrm{d}\boldsymbol{K}^{AA}(\boldsymbol{A})}{\mathrm{d}\boldsymbol{A}}\Big|_{\boldsymbol{A}} = \int_{\Omega} \left(\nabla \times \boldsymbol{W}^{\mathrm{T}}\right)^{\mathrm{T}} \frac{\mathrm{d}[\nu]}{\mathrm{d}\boldsymbol{A}}\Big|_{\boldsymbol{A}} \boldsymbol{A}\left(\nabla \times \boldsymbol{W}^{\mathrm{T}}\right) \mathrm{d}\Omega \tag{1.443}$$

where $\boldsymbol{W}$ denotes a test function in the whole solution domain and

$$\frac{\mathrm{d}[\nu]}{\mathrm{d}\boldsymbol{A}}\Big|_{\boldsymbol{A}}\boldsymbol{A} = \frac{\mathrm{d}[\nu]}{\mathrm{d}\boldsymbol{B}}\Big|_{\boldsymbol{B}} \frac{\mathrm{d}\boldsymbol{B}}{\mathrm{d}\boldsymbol{A}}\Big|_{\boldsymbol{A}}\boldsymbol{A} = \frac{\mathrm{d}[\nu]}{\mathrm{d}\boldsymbol{B}}\Big|_{\boldsymbol{B}} \nabla \times \boldsymbol{A} = \frac{\mathrm{d}[\nu]}{\mathrm{d}\boldsymbol{B}}\Big|_{\boldsymbol{B}}\boldsymbol{B} = 2\boldsymbol{B}\frac{\mathrm{d}[\nu]}{\mathrm{d}(\boldsymbol{B}^2)}\Big|_{\boldsymbol{B}}\boldsymbol{B} \tag{1.444}$$

According to assumption in material nonlinearity, we obtain:

$$\frac{\mathrm{d}[\nu]}{\mathrm{d}\boldsymbol{A}}\Big|_{\boldsymbol{A}} \boldsymbol{A} = 2\boldsymbol{B}\frac{\mathrm{d}[\nu]}{\mathrm{d}(\boldsymbol{B}^2)}\Big|_{\boldsymbol{B}} \boldsymbol{B} = 2\boldsymbol{B}\frac{\mathrm{d}\nu_h}{\mathrm{d}(\|\boldsymbol{B}\|^2)}\boldsymbol{B} \tag{1.445}$$

Substituting Equation (1.445) into Equation (1.443), we get the nonlinear part of tangent matrix

$$\boldsymbol{K}_N^{AA} = \boldsymbol{A}\frac{\mathrm{d}\boldsymbol{K}^{AA}(\boldsymbol{A})}{\mathrm{d}\boldsymbol{A}}|_A = 2\int_{\Omega}\left(\boldsymbol{B}^{\mathrm{T}}\left(\nabla\times\boldsymbol{W}^{\mathrm{T}}\right)\right)^{\mathrm{T}}\frac{\mathrm{d}\nu_h}{\mathrm{d}\left(\|\boldsymbol{B}\|^2\right)}\left(\boldsymbol{B}^{\mathrm{T}}\left(\nabla\times\boldsymbol{W}^{\mathrm{T}}\right)\right)\mathrm{d}\Omega \quad (1.446)$$

and the iterative formulation derived from (1.442) could thus be rewritten as

$$\left(\boldsymbol{K}^{AA}(\boldsymbol{A})+\boldsymbol{K}_N^{AA}\right)\delta\boldsymbol{A} = \boldsymbol{J}^S - \boldsymbol{K}^{AA}(\boldsymbol{A})\boldsymbol{A} \quad (1.447)$$

where $\boldsymbol{K}_N^{AA}$ is in assembled matrix form:

$$\begin{aligned}\boldsymbol{K}_N^{AA} &= \sum_e \boldsymbol{K}_{Ne}^{AA}, \boldsymbol{K}_{Ne}^{AA} = \\ &2\int_{\Omega^e}\left(\boldsymbol{B}^{\mathrm{T}}\left(\nabla\times\boldsymbol{W}^{\mathrm{T}}\right)\right)^{\mathrm{T}}\frac{\mathrm{d}\nu_h}{\mathrm{d}\left(\|\boldsymbol{B}\|^2\right)}|_B\left(\boldsymbol{B}^{\mathrm{T}}\left(\nabla\times\boldsymbol{W}^{\mathrm{T}}\right)\right)\mathrm{d}\Omega\end{aligned} \quad (1.448)$$

By updating $\delta\boldsymbol{A}$ successively, we could get the final solution, which satisfies the convergence criteria.

The second type of iterative formulation: by contrast to the first one, we start from the governing equations in nonlinear form:

$$\nabla\times\boldsymbol{H} = \boldsymbol{J}_s \quad (1.449)$$

Where $\boldsymbol{H}$ is depending on $\boldsymbol{A}$ nonlinearly, denoted as $\boldsymbol{H}(\boldsymbol{A})$.

We adopt the same assumptions as in the first type iterative formulation. By substitute Equation (1.438) into Equation (1.449), we get:

$$\nabla\times\boldsymbol{H}\left(\boldsymbol{A}+\delta\boldsymbol{A}\right) = \boldsymbol{J}_s \quad (1.450)$$

Expand Equation (1.450) into a Taylor series at $\boldsymbol{A}'$ and intercept the first-order term, and we obtain:

$$\nabla\times\left(\frac{\mathrm{d}\boldsymbol{H}}{\mathrm{d}\boldsymbol{A}}|_A\ \delta\boldsymbol{A}\right) = \boldsymbol{J}_s - \nabla\times\boldsymbol{H}(\boldsymbol{A}) \quad (1.451)$$

By using the equation

$$\frac{\mathrm{d}\boldsymbol{H}}{\mathrm{d}\boldsymbol{A}}|_A = \frac{\mathrm{d}\boldsymbol{H}}{\mathrm{d}\boldsymbol{B}}|_B\ \frac{\mathrm{d}\boldsymbol{B}}{\mathrm{d}\boldsymbol{A}}|_A \quad (1.452)$$

and

$$\mathrm{d}\boldsymbol{B} = \mathrm{d}\left(\nabla\times\boldsymbol{A}\right) = \nabla\times\mathrm{d}\boldsymbol{A} \quad (1.453)$$

we obtain the following derived from Equation (1.451):

$$\nabla \times \frac{\mathrm{d}\boldsymbol{H}}{\mathrm{d}\boldsymbol{B}}|_{B} \nabla \times \delta \boldsymbol{A} = \boldsymbol{J}_{s} - \nabla \times \boldsymbol{H}(\boldsymbol{A}) \tag{1.454}$$

Based on the nonlinear assumption in the previous section, the magnetic field intensity $\boldsymbol{H}$ and the magnetic flux density $\boldsymbol{B}$ are assumed to be collinear, Equation (1.454) could be rewritten as:

$$\nabla \times \frac{\mathrm{d}\|\boldsymbol{H}\|}{\mathrm{d}\|\boldsymbol{B}\|}|_{B} \nabla \times \delta \boldsymbol{A} = \boldsymbol{J}_{s} - \nabla \times \boldsymbol{H}(\boldsymbol{A}) \tag{1.455}$$

Where, the term $\frac{\mathrm{d}\|\boldsymbol{H}\|}{\mathrm{d}\|\boldsymbol{B}\|}$ is, depending on $\boldsymbol{B}$ nonlinearly, obtained from the *B–H* curves.

By using the standard edge element discretization procedure mentioned in the previous section, we get an iterative formulation for updating $\boldsymbol{A}$:

$$\boldsymbol{K}_{\tau}^{AA} \delta \boldsymbol{A} = \boldsymbol{J}^{S} - \boldsymbol{K}^{AA}(\boldsymbol{B})\boldsymbol{A} \tag{1.456}$$

where the tangent matrix $\boldsymbol{K}_{\tau}^{AA}$ is expressed as:

$$\boldsymbol{K}_{\tau}^{AA} = \sum_{e} \boldsymbol{K}_{\tau e}^{AA}, \boldsymbol{K}_{\tau e}^{AA} = \int_{\Omega^{e}} \left(\nabla \times \boldsymbol{W}^{\mathrm{T}}\right)^{\mathrm{T}} \frac{\mathrm{d}\|\boldsymbol{H}\|}{\mathrm{d}\|\boldsymbol{B}\|}|_{B} \left(\nabla \times \boldsymbol{W}^{T}\right) \mathrm{d}\Omega \tag{1.457}$$

By updating $\delta \boldsymbol{A}$ successively, we could get the final solution, which satisfies the convergence criteria.

The steps of Newton–Raphson method

The steps to apply Newton–Raphson method into nonlinear calculation could be summarized as follows:

1. Make an initial approximation for $\boldsymbol{A}$, denoted as $\boldsymbol{A}^{\mathbf{0}}$.
2. Derive the current fields $\boldsymbol{B}$ and $\boldsymbol{H}$ from the approximation of $\boldsymbol{A}^{i} (i \geq \mathbf{0})$.
3. Obtain $\frac{\mathrm{d}v_{h}}{\mathrm{d}\left(\|\boldsymbol{B}\|^{2}\right)}$ or $\frac{\mathrm{d}\|\boldsymbol{H}\|}{\mathrm{d}\|\boldsymbol{B}\|}$ based on the current fields and the *B–H* relation curves.
4. Calculate the corresponding tangent matrix mentioned in Equation (1.448) or Equation (1.457).
5. Calculate the residual based on the current approximation of $\boldsymbol{A}^{i} (i \geq \mathbf{0})$.
6. Solve the iterative equation (1.442) or (1.456).
7. Update $\boldsymbol{A}^{i+1} = \boldsymbol{A}^{i} + \delta \boldsymbol{A}$.
8. Loop step 2 to step 7 until the convergence criteria are satisfied.

Remark 1.13:

1. The convergence of the Newton–Raphson method is quadratic if the iterative process starts from an initial guess close to the exact solution. However, this condition is not always

satisfied, and the Newton–Raphson method may fail to converge. Therefore, the relaxation technique is often used to improve the convergence. The magnetic vector potential is updated in an iterative process (Yao, 2010):

$$A^{i+1} = A^{i} + \alpha \delta A \tag{1.458}$$

Where α is a relaxation that should be specified by users.

2. These two kinds of iterative formulations for the Newton–Raphson method are both available for solving nonlinear electromagnetic problems, and they have been both implemented in our procedure. However, considering computational efficiency, only the second formulation is used in INTESIM product in practice.

1.4.3.5.3 Fixed-point method

The fixed-point method was another method used in electrical engineering to solve nonlinear electromagnetic problems. The method, also known as the Gauss–Seidel method, is based on Picard–Banach fixed-point theory. For more details, please refer to Meunier (2010) and Kuczmann (2009).

The steps to apply the fixed-point method into nonlinear calculation can be summarized as follows:

1. Make an initial approximation for $\boldsymbol{A}$, denoted as $\boldsymbol{A}^0$.
2. Derive the current field intensity $\boldsymbol{H}$ from the approximation of $\boldsymbol{A}^i$ $(i \geq \boldsymbol{0})$.
3. Obtain $[\nu]$ from the *B–H* relation curves based on the current fields and calculate Equation (1.437).
4. Calculate the source terms (current or permanent magnet).
5. Solve the Equation (1.436) with the imposed boundary conditions.
6. Update solution with $\boldsymbol{A}^{i+1}$.
7. Loop step 2 to step 6 until the convergence criteria are satisfied.

Remark 1.14:

As an alternative method to the Newton–Raphson method, the fixed-point method has the following advantages:

1. It is more robust than the Newton–Raphson method, which means that if only *B–H* curves is monotonic, a convergent solution always could be obtained. For some highly nonlinear *B–H* curves (Figure 1.13), the Newton–Raphson method maybe fail to approach a convergent solution, but the fixed-point method always holds.
2. Preliminary operations and smoothness of *B–H* curves are not necessary for the method.
3. The fixed-point method is simple and easy to implement in numerical procedures.

However, the main disadvantage of the fixed-point method is obvious with slow convergence. To improve the convergence, an optimization for $[\nu]$ is usually adopted.

To summarize, the Newton–Raphson method based on the second type of iterative formulation is generally used for solving nonlinear electromagnetic problems when the level of nonlinearity of materials is not too high. Otherwise, the fixed-point method is advised from the consideration of computational stability.

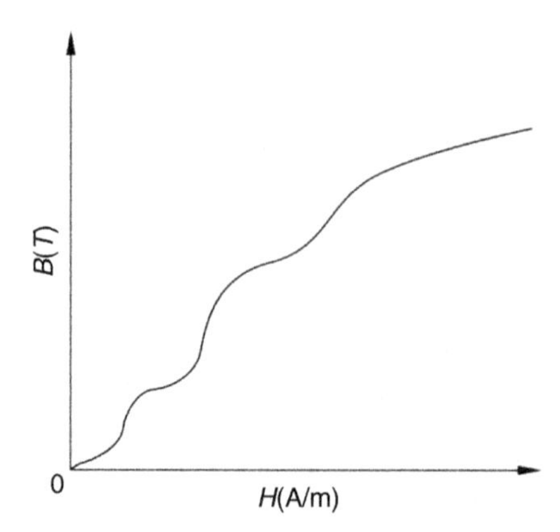

Figure 1.13 The Newton–Raphson method may fail with highly nonlinear *B–H* curves in some complex electromagnetic problems, but the fixed-point method always holds.

1.4.4 Gauge methods for electromagnetic elements

The singular system matrix arising from the discretization of the differential curl–curl operator often leads to numerical difficulty of FEM in simulating magnetic as well as eddy current field (Jin, 2014). To overcome this, instead of solving the singular system matrix directly, we can regularize the system matrix by imposing a gauge condition. One common approach is by enforcing the tree gauge (Albanese and Rubinacci, 1990). Another commonly used approach is solving the singular matrix by an iterative solver without enforcing gauge condition. As an efficient alternative to achieve uniqueness of magnetic vector potentials, incorporating Coulomb gauge into edge element system matrix by using the Lagrange multiplier method (Chen et al., 2000; Hu and Zou, 2004) is also provided in this section.

1.4.4.1 Tree gauge

1.4.4.1.1 The definition of spanning tree

The tree gauge method could be described as selecting a tree in the graph defined by the edges of the finite element mesh and setting the DOFs corresponding to those edges to zero. It is known that the nullity of the singular system matrix resulted from discretization of the differential curl–curl operator equals to the number of the edges in the spanning tree of a finite element mesh. By eliminating the unknowns corresponding to the tree edges, the redundant linear equations in the system matrix can be eliminated, and uniqueness is achieved.

To apply tree gauge to the edge-based FEM, the key point is how to construct a minimum spanning tree on the given finite element mesh, in which the edges corresponding to minimum spanning tree should connect every element node but should not form a loop. This kind of special requirement in electromagnetic simulation could be easily achieved by means of available technologies in network theory (Meunier, 2010; Yao, 2010).

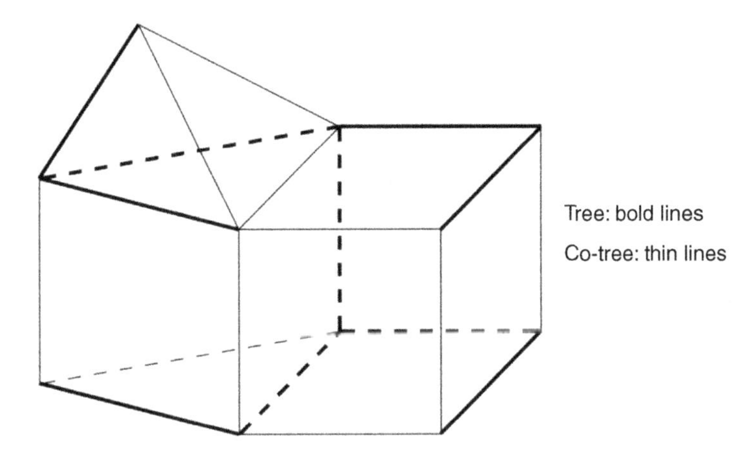

Figure 1.14 Example of a tree and co-tree.

An algorithm for finding the tree edges is presented as follows (Lee, 2009), and sample results of applying the tree scheme are shown in Figure 1.14 (Meunier, 2010). The tree edges are highlighted by the thick lines.

1.4.4.1.2 The advantages and shortcomings of the tree gauge method

Advantages:

1. There are available technologies about how to construct a spanning tree in network theory.
2. It avoids imposing coulomb gauge condition, although it is apparently shown to be a weak-sense enforcement of the Coulomb gauge (Manges and Cendes, 1995).
3. It largely decreases the number of DOFs to be solved because of eliminating the unknowns corresponding to the tree edges.

Disadvantages:

1. It is still a difficult issue about how to construct a proper spanning tree regarding the computational accuracy.
2. It is time consuming or even fails to construct a spanning tree for complex or large-scale mesh.
3. It will make the CG/ICCG solver slow.

1.4.4.2 Lagrange matrix gauge (Hu, 2004)

Another way to achieve uniqueness of the magnetic vector potential is to incorporate the Coulomb gauge into edge-based finite element formulation by using LM method or penalty method. In this section, we will give a brief introduction to the Lagrange matrix gauge method in an electromagnetic field (Chen et al., 2000; Hu and Zou, 2004).

We start by considering the following system associated with the magnetic field problem:

$$\begin{cases} \nabla\times[\boldsymbol{\nu}]\ \nabla\times\boldsymbol{A}=\boldsymbol{J}_s+\boldsymbol{H}_c & \text{in }\Omega \\ \nabla\cdot\boldsymbol{A}=0 & \text{in }\Omega \end{cases} \tag{1.459}$$

with the following boundary condition:

$$\boldsymbol{A}\times\boldsymbol{n}=\boldsymbol{0} \qquad \text{on }\partial\Omega \tag{1.460}$$

Here, $\boldsymbol{A}$ refers to magnetic vector potential and $[\boldsymbol{\nu}]$ refers to the magnetic reluctivity matrix. The second equation of Equation (1.459) is a Coulomb-type condition.

By introducing a Lagrange multiplier p and integration by parts, we achieve the following saddle-point problem corresponding to Equation (1.459).

Find $(\boldsymbol{A}, p)$ such that:

$$\begin{cases} ([\boldsymbol{\nu}]\ \text{curl}\ \boldsymbol{A}, \text{curl}\ \boldsymbol{v})+(\text{grad}\ p,\boldsymbol{v})=(\boldsymbol{f},\boldsymbol{v})\ \text{in}\ \Omega,\ \forall\boldsymbol{v}\in H_0(\text{curl};\Omega) \\ (\boldsymbol{A}, \text{grad}\ q)=\boldsymbol{0},\ \text{in}\ \Omega,\ \forall\boldsymbol{q}\in H_0^1(\Omega) \end{cases} \tag{1.461}$$

Where $\boldsymbol{f}$ is $\boldsymbol{J}_S+\boldsymbol{H}_C$.

By substituting the shape functions into weak form Equation (1.461) in finite element space, one can obtain assembled global equation Equation (1.462) in matrix form:

$$\begin{bmatrix} \boldsymbol{K}^{AA} & \boldsymbol{G}^{A\lambda} \\ \boldsymbol{G}^{A\lambda^{\mathrm{T}}} & \boldsymbol{0} \end{bmatrix} \begin{Bmatrix} \boldsymbol{A} \\ \boldsymbol{\lambda} \end{Bmatrix} = \begin{Bmatrix} \boldsymbol{J}^{S}+\boldsymbol{J}^{PM} \\ \boldsymbol{0} \end{Bmatrix} \tag{1.462}$$

where $\boldsymbol{A}$ is the edge element approximation for magnetic vector potential vector and $\boldsymbol{\lambda}$ is node-based element approximation for Lagrange multiplier.

Magneto–LM coupling matrix:

$$\boldsymbol{G}^{A\lambda}=\sum\nolimits_e \boldsymbol{G}_e^{A\lambda}, \boldsymbol{G}_e^{A\lambda}=\int_{\Omega^e}\boldsymbol{W}\left(\nabla\boldsymbol{N}^{\mathrm{T}}\right)\mathrm{d}\Omega \tag{1.463}$$

Similar to the magnetic field case, by introducing the Lagrange multiplier, we could also get weak form of eddy current field equations, and by using the finite element discretization method, we get the eddy current field FE equations:

$$\begin{bmatrix} \boldsymbol{K}^{AA}+\boldsymbol{K}_v^{AA} & \boldsymbol{0} & \boldsymbol{G}^{A\lambda} \\ \boldsymbol{K}_v^{VA} & \boldsymbol{0} & \boldsymbol{0} \\ \left(\boldsymbol{G}^{A\lambda}\right)^{\mathrm{T}} & \boldsymbol{0} & \boldsymbol{0} \end{bmatrix} \begin{Bmatrix} \boldsymbol{A} \\ - \\ \boldsymbol{\lambda} \end{Bmatrix} + \begin{bmatrix} \boldsymbol{C}^{AA} & \boldsymbol{K}^{AV} & \boldsymbol{0} \\ \left(\boldsymbol{K}^{AV}\right)^{\mathrm{T}} & \boldsymbol{K}^{VV} & \boldsymbol{0} \\ \boldsymbol{0} & \boldsymbol{0} & \boldsymbol{0} \end{bmatrix} \begin{Bmatrix} \dot{\boldsymbol{A}} \\ \boldsymbol{V} \\ \dot{\boldsymbol{\lambda}} \end{Bmatrix} = \begin{Bmatrix} \boldsymbol{J}^{S}+\boldsymbol{J}^{PM} \\ \boldsymbol{I}_{in} \\ \boldsymbol{0} \end{Bmatrix} \tag{1.464}$$

Remark 1.15: To obtain the unique solution of the magnetic vector potential, the tree gauge method or CG/ICCG method is widely used in low-frequency electromagnetic simulation (e.g., some commercial software, such as ANSYS and JMAG adopt one of them). Generally speaking, tree gauge is appropriate for a direct solver. When a large-scale problem needs to be solved, an iterative solver (CG/ICCG) without imposing gauge is more appropriate. They are both implemented in INTESIM product, provided to users for electromagnetic simulation. Additionally, LM gauge is implemented in our procedure, mainly used for tackling special issues such as sliding interface, periodic boundary conditions, and so on.

1.4.5 Output results

1.4.5.1 Electromagnetic fields

$$\boldsymbol{B} = \nabla \times \boldsymbol{A} \tag{1.465}$$

$$\boldsymbol{H} = [\nu]\boldsymbol{B} - \frac{1}{[\nu_0]}[\nu]\boldsymbol{M}_0 \tag{1.466}$$

$$\boldsymbol{E} = -\frac{\partial \boldsymbol{A}}{\partial t} - \nabla V \tag{1.467}$$

$$\boldsymbol{D} = [\varepsilon]\boldsymbol{E} \tag{1.468}$$

1.4.5.2 Lorentz forces

Magnetic forces density in current-carrying conductors:

$$\boldsymbol{F}_l = \boldsymbol{J} \times \boldsymbol{B} \tag{1.469}$$

1.4.5.3 Maxwell forces

The Maxwell stress tensor is used to determine the magnetic or electrostatic force density,

$$f_i = T_{ij} n_j \tag{1.470}$$

where

$$T_{ij} = \frac{1}{\mu_0}\left(B_i B_j - \frac{1}{2}\delta_{ij}|\boldsymbol{B}|^2 \right) \tag{1.471}$$

is the stress tensor of the magnetic field.

$$T_{ij} = \varepsilon_0 \left(E_i E_j - \frac{1}{2}\delta_{ij} |\boldsymbol{E}|^2 \right) \tag{1.472}$$

is the stress tensor of the electrostatic field.

The force vector can be obtained by surface integration of force density f_i of Equation (1.470) or by volumetric integration:

$$\boldsymbol{F}_{MX} = \int_{\Omega_e} [\boldsymbol{B}]^T [\boldsymbol{T}] \mathrm{d}\Omega \tag{1.473}$$

where $\boldsymbol{F}_{MX}$ is the element Maxwell force vector; $[B]$ is the strain displacement matrix; and $[\boldsymbol{T}]$ is the Maxwell stress tensor.

1.4.5.4 Joule heat

The Joule heat generated in a nonzero resistivity and a nonzero current density region can be obtained.

In static or transient analysis

$$q^j = \rho\, \boldsymbol{J} \cdot \boldsymbol{J} \tag{1.474}$$

Here, ρ is resistivity matrix and $\boldsymbol{J}$ is total current density.

In harmonic analysis,

$$q^j = \mathrm{Re}\left(\frac{1}{2} \rho\, \boldsymbol{J} \cdot \boldsymbol{J}^* \right) \tag{1.475}$$

Here, $\boldsymbol{J}$ is complex total current density in the element and $\boldsymbol{J}^*$ is complex conjugate of $\boldsymbol{J}$.

1.4.5.5 Iron losses

Iron losses consist of three components namely eddy currents loss, hysteresis loss, and anomalous loss. To calculate total iron losses, we need to add up all element losses.

$$W_{\mathrm{loss}} = \sum_{i=1}^{ne} \left(W_{h_i} + W_{e_i} + W_{a_i} \right) \tag{1.476}$$

Where W_{h_i} is hysteresis loss of element i; W_{e_i} is eddy current loss of element i; W_{a_i} is anomalous current loss of element i; and ne is number of elements.

The hysteresis losses are considered to be related to the magnetic domains movement and rotation, as well as material grain's composition and size (Bastos and Sadowski, 2003). As shown in Figure 1.15, the curve of ***B–H*** forms a hysteresis cycle after the first magnetization. As the magnetic field $\boldsymbol{B}$ and $\boldsymbol{H}$ vary along this cycle periodically, the energy in device is consumed.

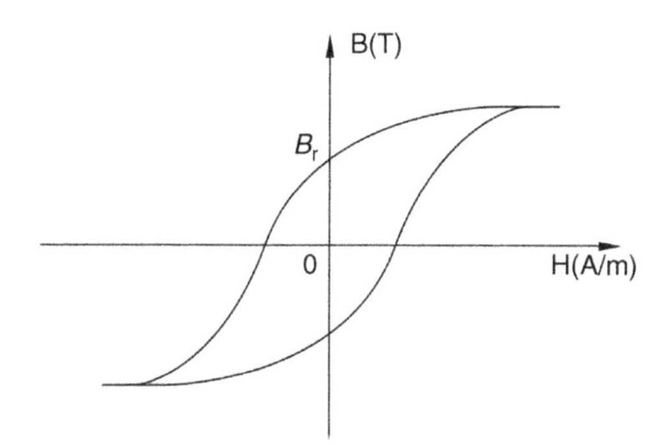

Figure 1.15 Hysteresis curve.

According to the formulation of the volumetric magnetic energy density:

$$W_h = \int_0^B H \,\mathrm{d}B \tag{1.477}$$

We conclude that the consumed energy in one loop is proportional to the cycle area shown in the Figure 1.15.

Iron losses are generally calculated by using a posteriori approach, that is, they are estimated by postprocessing.

There are several empirical methods to calculate hysteresis losses, but here we only present the commonly used FFT and Steinmetz methods:

FFT method: by applying FFT frequency analysis to the time–series data of magnetic flux density, the amplitudes of magnetic flux density of each component could be calculated. Then hysteresis loss is estimated by summing up all the contributions from those components. The FFT loss calculation method has been implemented in electromagnetic analysis software JMAG developed in Japan.

Steinmetz method: Steinmetz's equation, a kind of empirical relationship established by Steinmetz in 1892, is used for calculate the hysteresis loss, which was obtained by experimental results for values of B_m in the range of $0.2T$ and $1.5T$, is expressed as:

$$W_h = \eta B_m^{\alpha} \tag{1.478}$$

where η and α are both obtained based on experimental results.

However, Equation (1.478) is only appropriate to sinusoidal induction variations. Lavers et al. (1978) made an extension on Equation (1.478) to calculate hysteresis loss for nonsinusoidal induction waveforms, that is, presenting reversals ΔB as shown in Figure 1.16:

$$W_h = \eta B_m^{\alpha} \left[1 + \frac{0.65}{B_m} \sum_{i=1}^{n} \Delta B_i \right] \tag{1.479}$$

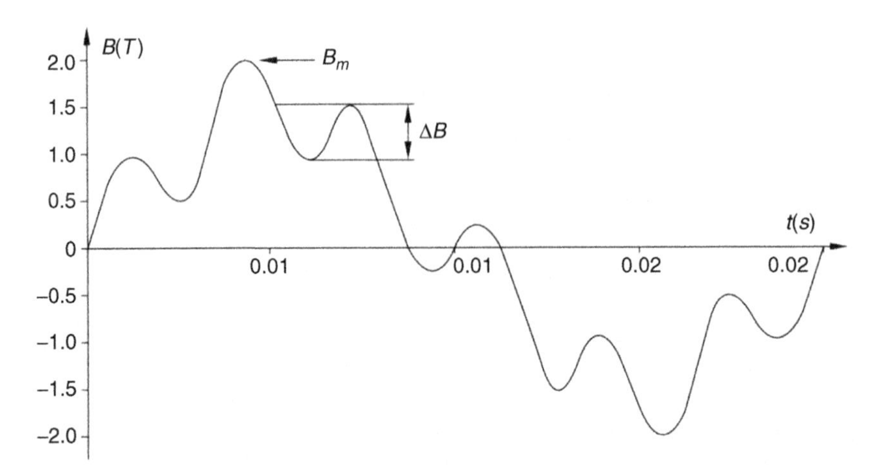

Figure 1.16 Nonsinusoidal induction as a function of time and associated hysteresis loop.

The FFT and Steinmetz methods have been implemented in INTESIM software and used in the engineering simulation in the application chapters.

A more precise method: the best way to calculate hysteresis losses is to evaluate precisely the hysteresis loop surface. For this method, the mathematical model for describing *B–H* curve is needed. It is more accurate than an empirical relationship-based method, but the shortcoming of this method is its lack of computation efficiency (Bastos and Sadowski, 2003).

Apart from the calculation method referred to in Equations (1.474) and (1.475), eddy current loss can also be estimated by the FFT method or maximum value method, in which the time–series data of magnetic flux density is used.

Remark 1.16: The resulting Lorentz and Maxwell forces work as coupling load vectors for mechanical models in multiphysics simulations, and the Joule and iron loss heats work as source terms for thermal analysis. The incoming data for electromagnetic analysis are the displacement or velocity from the mechanical field.

1.4.6 Sliding interface coupling in electromagnetic problems

1.4.6.1 Primary variables for electromagnetic problems

The DOF for edge-based element is the magnetic vector potential **A** in the tangential direction along the edge, that is, AZ.

In this section, we will limit our discussion to the most commonly used edge-based magnetic vector potential element.

1.4.6.2 Coupling methods for edge elements

Constraint equation method: since the directional characteristics of coupling variable along the edge requires Ladder shaped regular meshes across the interface for using the multipoint constraint method.

Lagrange multiplier (or mortar element method): for coupling by the Lagrange multiplier method, the following aspects need to be considered: (1) the need to generate control surface meshes for surface integration; (2) shape functions and additional DOFs are introduced for the Lagrange multiplier; and (3) the need to evaluate the tangential component of the magnetic vector potential on the coupling interface.

Penalty method: It needs control surface meshes for surface integration, although no additional interface DOFs are introduced. The continuity conditions on the coupling interface are satisfied approximately.

Discontinuous Galerkin FEM with FEM: It needs to couple discontinuous Galerkin FEM with FEM.

1.4.6.3 Interface conditions

Continuity condition of the tangential component of or magnetic vector potential A:

$$\boldsymbol{u}^A \Lambda\, \boldsymbol{n}_\Gamma |_\Gamma = \boldsymbol{u}^B \Lambda\, \boldsymbol{n}_\Gamma |_\Gamma . \quad (1.480)$$

Here, $\boldsymbol{u}$ represents $\boldsymbol{A}$ or any other vector of electromagnetic field.

1.4.6.4 Interface condition in weak form for Lagrange multiplier method

$$\int_{\Gamma c} (\boldsymbol{u}^A \ \Lambda \boldsymbol{n}_\Gamma - \boldsymbol{u}^B \Lambda\, \boldsymbol{n}_\Gamma) \cdot \boldsymbol{\lambda}\, \mathrm{d}\Gamma = \boldsymbol{0} \quad (1.481)$$

1.4.6.5 Weak form for the Lagrange multiplier method

Here $\boldsymbol{\lambda}$ is the Lagrange multiplier that needs to be defined on the coupling interface.

$$\int_{\Gamma_c} \left(\boldsymbol{u}^A \Lambda\, \boldsymbol{n}_\Gamma - \boldsymbol{R}^{AB} \boldsymbol{u}^B \Lambda\, \boldsymbol{n}_\Gamma\right) \cdot \boldsymbol{\lambda}\, \mathrm{d}\Gamma, \quad (1.482)$$

the variational form of Equation(1.482)

$$\begin{aligned} &\int_{\Gamma_c} \left(\boldsymbol{u}^A \Lambda\, \boldsymbol{n}_\Gamma - \boldsymbol{R}^{AB} \boldsymbol{u}^B \Lambda\, \boldsymbol{n}_\Gamma\right) \cdot \delta\boldsymbol{\lambda} \mathrm{d}\Gamma + \\ &\quad \int_{\Gamma_c} \left(\delta \boldsymbol{u}^A \Lambda\, \boldsymbol{n}_\Gamma - \delta\left(\boldsymbol{R}^{AB} \boldsymbol{u}^B\right) \Lambda\, \boldsymbol{n}_\Gamma\right) \cdot \boldsymbol{\lambda} \mathrm{d}\Gamma = 0. \end{aligned} \quad (1.483)$$

R^{AB} is the transformation matrix.

1.4.6.6 The interface continuity conditions in matrix form

Shape function for Lagrange multiplier space:

$$\boldsymbol{\lambda} = \boldsymbol{W}_\lambda \boldsymbol{\lambda}^e . \quad (1.484)$$

The way to define $\boldsymbol{W}_\lambda$ that satisfies the inf-sup condition is the key factor for the Lagrange multiplier-based coupling method (Section 1.4.6.7).

From Equation (1.483):

$$\boldsymbol{K}_{\Gamma}^{\lambda A} = \int_{\Gamma_c} \boldsymbol{W}_{\lambda}^{\mathrm{T}} \left(\boldsymbol{W}^{A^{\mathrm{T}}} \Lambda \boldsymbol{n}_{\Gamma} - \boldsymbol{R}^{AB} \boldsymbol{W}^{B^{\mathrm{T}}} \Lambda \boldsymbol{n}_{\Gamma} \right) \mathrm{d}\Gamma. \tag{1.485}$$

$$\boldsymbol{K}_{\Gamma}^{A\lambda} = \int_{\Gamma_c} \left(\boldsymbol{W}^{A} \Lambda \boldsymbol{n}_{\Gamma} - \boldsymbol{W}^{B} \boldsymbol{R}^{AB^{\mathrm{T}}} \Lambda \boldsymbol{n}_{\Gamma} \right) \boldsymbol{W}_{\lambda} \, \mathrm{d}\Gamma. \tag{1.486}$$

Note: the shape function of $\boldsymbol{W}_{\lambda}$ is not equal to $\boldsymbol{W}^{A}$, and there are three options. Γ is the interface meshes on one side (say domain A) of the domain or the interface meshes that is independent from coupling surface mesh.

The component form of the interface continuity conditions:

$$\begin{aligned}
&\int_{\Gamma_c} \left((\boldsymbol{u}^{A} \cdot \boldsymbol{\tau}_1) \boldsymbol{\tau}_1 - \left((\boldsymbol{R}^{AB} \boldsymbol{u}^{B}) \cdot \boldsymbol{\tau}_1 \right) \boldsymbol{\tau}_1 \right) \cdot (\boldsymbol{\lambda} \cdot \boldsymbol{\tau}_1) \boldsymbol{\tau}_1 \, \mathrm{d}\Gamma + \\
&\int_{\Gamma_c} \left((\boldsymbol{u}^{A} \cdot \boldsymbol{\tau}_2) \boldsymbol{\tau}_2 - \left((\boldsymbol{R}^{AB} \boldsymbol{u}^{B}) \cdot \boldsymbol{\tau}_2 \right) \boldsymbol{\tau}_2 \right) \cdot (\boldsymbol{\lambda} \cdot \boldsymbol{\tau}_2) \boldsymbol{\tau}_2 \, \mathrm{d}\Gamma = \\
&\int_{\Gamma_c} \left((\boldsymbol{u}^{A} \cdot \boldsymbol{\tau}_1) - \left((\boldsymbol{R}^{AB} \boldsymbol{u}^{B}) \cdot \boldsymbol{\tau}_1 \right) \right) (\boldsymbol{\lambda} \cdot \boldsymbol{\tau}_1) \mathrm{d}\Gamma + \\
&\int_{\Gamma_c} \left((\boldsymbol{u}^{A} \cdot \boldsymbol{\tau}_2) - \left((\boldsymbol{R}^{AB} \boldsymbol{u}^{B}) \cdot \boldsymbol{\tau}_2 \right) \right) (\boldsymbol{\lambda} \cdot \boldsymbol{\tau}_2) \mathrm{d}\Gamma.
\end{aligned} \tag{1.487}$$

In matrix form ($\delta\boldsymbol{\lambda}$),

$$\begin{aligned}
\boldsymbol{K}_{\Gamma}^{\lambda A} = &\int_{\Gamma_c} \left(\boldsymbol{W}_{\lambda}^{\mathrm{T}} \cdot \boldsymbol{\tau}_1 \right) \left(\boldsymbol{\tau}_1 \cdot \boldsymbol{W}^{A^{\mathrm{T}}} - \boldsymbol{\tau}_1 \cdot \boldsymbol{R}^{AB} \boldsymbol{W}^{B^{\mathrm{T}}} \right) \mathrm{d}\Gamma \\
&+ \int_{\Gamma_c} \left(\boldsymbol{W}_{\lambda}^{\mathrm{T}} \cdot \boldsymbol{\tau}_2 \right) \left(\boldsymbol{\tau}_2 \cdot \boldsymbol{W}^{A^{\mathrm{T}}} - \boldsymbol{\tau}_2 \cdot \boldsymbol{R}^{AB} \boldsymbol{W}^{B^{\mathrm{T}}} \right) \mathrm{d}\Gamma.
\end{aligned} \tag{1.488}$$

In matrix form ($\delta\boldsymbol{u}$),

$$\begin{aligned}
\boldsymbol{K}_{\Gamma}^{A\lambda} = &\int_{\Gamma_c} \left(\boldsymbol{W}^{A} \cdot \boldsymbol{\tau}_1 - \boldsymbol{W}^{B} \boldsymbol{R}^{AB^{\mathrm{T}}} \cdot \boldsymbol{\tau}_1 \right) \left(\boldsymbol{\tau}_1 \cdot \boldsymbol{W}_{\lambda} \right) \mathrm{d}\Gamma \\
&+ \int_{\Gamma_c} \left(\boldsymbol{W}^{A} \cdot \boldsymbol{\tau}_2 - \boldsymbol{W}^{B} \boldsymbol{R}^{AB^{\mathrm{T}}} \cdot \boldsymbol{\tau}_2 \right) \left(\boldsymbol{\tau}_2 \cdot \boldsymbol{W}_{\lambda} \right) \mathrm{d}\Gamma.
\end{aligned} \tag{1.489}$$

$\boldsymbol{R}^{AB} = \boldsymbol{R}^{A^{\mathrm{T}}} \boldsymbol{R}^{B}$: Rotation matrix from B system to A system,

$$\boldsymbol{R}^{AB} = \boldsymbol{R}_{xp\_X}{}^{\mathrm{T}} \boldsymbol{R}\left(\theta_{\mathrm{thefta}}\right) \boldsymbol{R}_{xp\_X} \tag{1.490}$$

θ_{thefta} is the rotation angle from frame B (global stationary) to frame A (body-attached system). For details about multiple frames of reference in multiphysics simulation, please refer to Chapter 6.

The unit vectors in tangential directions satisfy:

$$\boldsymbol{\tau}_1 \times \boldsymbol{\tau}_2 = \boldsymbol{n} \tag{1.491}$$

Three ways to choose the control surface: Γ_c

1. The master (e.g., stator) interface mesh
2. Build control surface from mesh A and mesh B
3. Build an independent interface mesh (a general method)

The final format for tree gauge with LM-based interface coupling:

$$\begin{bmatrix} \boldsymbol{K}^{AA}+\boldsymbol{K}_{v}^{AA} & \mathbf{0} & \boldsymbol{K}_{\Gamma}^{A\lambda} \\ \boldsymbol{K}_{v}^{VA} & \mathbf{0} & \mathbf{0} \\ \left(\boldsymbol{K}_{\Gamma}^{A\lambda}\right)^{\mathrm{T}} & \mathbf{0} & \mathbf{0} \end{bmatrix} \begin{Bmatrix} \boldsymbol{A}^{e} \\ - \\ \lambda_{\Gamma} \end{Bmatrix} + \begin{bmatrix} \boldsymbol{C}^{AA} & \boldsymbol{K}^{AV} & \mathbf{0} \\ \left(\boldsymbol{K}^{AV}\right)^{\mathrm{T}} & \boldsymbol{K}^{VV} & \mathbf{0} \\ \mathbf{0} & \mathbf{0} & \mathbf{0} \end{bmatrix} \begin{Bmatrix} \dot{\boldsymbol{A}}^{e} \\ \boldsymbol{V}^{e} \\ - \end{Bmatrix}$$
$$+ \begin{bmatrix} \boldsymbol{M}^{AA} & \boldsymbol{C}^{AV} & \mathbf{0} \\ \left(\boldsymbol{C}^{AV}\right)^{\mathrm{T}} & \boldsymbol{C}^{VV} & \mathbf{0} \\ \mathbf{0} & \mathbf{0} & \mathbf{0} \end{bmatrix} \begin{Bmatrix} \ddot{\boldsymbol{A}}^{e} \\ \dot{\boldsymbol{V}}^{e} \\ - \end{Bmatrix} = \begin{Bmatrix} \boldsymbol{J}_{e}^{S}+\boldsymbol{J}_{e}^{\mathrm{pm}} \\ \boldsymbol{I}^{e} \\ \mathbf{0} \end{Bmatrix} \tag{1.492}$$

The final format for tree gauge with LM-based interface coupling for the static case:

$$\begin{bmatrix} \boldsymbol{K}^{AA} & \mathbf{0} & \boldsymbol{K}_{\Gamma}^{A\lambda} \\ \left(\boldsymbol{G}^{AV}\right)^{\mathrm{T}} & \mathbf{0} & \mathbf{0} \\ \left(\boldsymbol{K}_{\Gamma}^{A\lambda}\right)^{\mathrm{T}} & \mathbf{0} & \mathbf{0} \end{bmatrix} \begin{Bmatrix} \boldsymbol{A}^{e} \\ - \\ \lambda_{\Gamma} \end{Bmatrix} + \begin{bmatrix} \mathbf{0} & \boldsymbol{G}^{AV} & \mathbf{0} \\ \mathbf{0} & \boldsymbol{K}^{VV} & \mathbf{0} \\ \mathbf{0} & \mathbf{0} & \mathbf{0} \end{bmatrix} \begin{Bmatrix} \dot{\boldsymbol{A}}^{e} \\ \boldsymbol{V}^{e} \\ - \end{Bmatrix} = \begin{Bmatrix} \boldsymbol{J}_{e}^{S}+\boldsymbol{J}_{e}^{\mathrm{pm}} \\ \boldsymbol{I}^{e} \\ \mathbf{0} \end{Bmatrix} \tag{1.493}$$

where $\boldsymbol{\lambda}_{\Gamma}$ is the Lagrange multiplier vector for interface coupling.

1.4.6.7 *Interface conditions in weak form for the penalty method*

In component form:

$$\int_{\Gamma_c} \frac{1}{2}\beta\left(\left(\boldsymbol{u}^{A}\cdot\boldsymbol{\tau}_{1}\right)-\left(\boldsymbol{u}^{B}\cdot\boldsymbol{\tau}_{1}\right)\right)^{2}\mathrm{d}\Gamma+\int_{\Gamma_c} \frac{1}{2}\beta\left(\left(\boldsymbol{u}^{A}\cdot\boldsymbol{\tau}_{2}\right)-\left(\boldsymbol{u}^{B}\cdot\boldsymbol{\tau}_{2}\right)\right)^{2}\mathrm{d}\Gamma \tag{1.494}$$

Take variation/disturbance $\delta\boldsymbol{u}^{A}$, $\delta\boldsymbol{u}^{B}$

$$\begin{aligned} &\int_{\Gamma_c} \beta\left(\left(\delta\boldsymbol{u}^{A}\cdot\boldsymbol{\tau}_{1}\right)-\left(\delta\boldsymbol{u}^{B}\cdot\boldsymbol{\tau}_{1}\right)\right)\left(\left(\boldsymbol{\tau}_{1}\cdot\boldsymbol{u}^{A}\right)-\left(\boldsymbol{\tau}_{1}\cdot\boldsymbol{u}^{B}\right)\right)\mathrm{d}S+ \\ &\int_{\Gamma_c} \beta\left(\left(\delta\boldsymbol{u}^{A}\cdot\boldsymbol{\tau}_{2}\right)-\left(\delta\boldsymbol{u}^{B}\cdot\boldsymbol{\tau}_{2}\right)\right)\left(\left(\boldsymbol{\tau}_{2}\cdot\boldsymbol{u}^{A}\right)-\left(\boldsymbol{\tau}_{2}\cdot\boldsymbol{u}^{B}\right)\right)\mathrm{d}\Gamma \end{aligned} \tag{1.495}$$

In matrix form ($\delta\boldsymbol{u}^A$,$\delta\boldsymbol{u}^B$):

$$\begin{aligned}\boldsymbol{K}_{\Gamma\ \tau_1}^{AA} &= \int_{\Gamma_c} \beta\left(\boldsymbol{W}^A\cdot\boldsymbol{\tau}_1\right)\left(\boldsymbol{\tau}_1\cdot\boldsymbol{W}^{A^{\mathrm{T}}}\right)\mathrm{d}\Gamma - \int_{\Gamma_c} \beta\left(\boldsymbol{W}^B\cdot\boldsymbol{\tau}_1\right)\left(\boldsymbol{\tau}_1\cdot\boldsymbol{W}^{A^{\mathrm{T}}}\right)\mathrm{d}\Gamma \\ &- \int_{\Gamma_c} \beta\left(\boldsymbol{W}^A\cdot\boldsymbol{\tau}_1\right)\left(\boldsymbol{\tau}_1\cdot\boldsymbol{W}^{B^{\mathrm{T}}}\right)\mathrm{d}\Gamma + \int_{\Gamma_c} \beta\left(\boldsymbol{W}^B\cdot\boldsymbol{\tau}_1\right)\left(\boldsymbol{\tau}_1\cdot\boldsymbol{W}^{B^{\mathrm{T}}}\right)\mathrm{d}\Gamma\end{aligned} \tag{1.496}$$

$$\begin{aligned}\boldsymbol{K}_{\Gamma\ \tau_2}^{AA} &= \int_{\Gamma_c} \beta\left(\boldsymbol{W}^A\cdot\boldsymbol{\tau}_2\right)\left(\boldsymbol{\tau}_2\cdot\boldsymbol{W}^{A^{\mathrm{T}}}\right)\mathrm{d}\Gamma - \int_{\Gamma_c} \beta\left(\boldsymbol{W}^B\cdot\boldsymbol{\tau}_2\right)\left(\boldsymbol{\tau}_2\cdot\boldsymbol{W}^{A^{\mathrm{T}}}\right)\mathrm{d}\Gamma \\ &- \int_{\Gamma_c} \beta\left(\boldsymbol{W}^A\cdot\boldsymbol{\tau}_2\right)\left(\boldsymbol{\tau}_2\cdot\boldsymbol{W}^{B^{\mathrm{T}}}\right)\mathrm{d}\Gamma + \int_{\Gamma_c} \beta\left(\boldsymbol{W}^B\cdot\boldsymbol{\tau}_2\right)\left(\boldsymbol{\tau}_2\cdot\boldsymbol{W}^{B^{\mathrm{T}}}\right)\mathrm{d}\Gamma\end{aligned} \tag{1.497}$$

$$\boldsymbol{K}_{\Gamma}^{AA} = \boldsymbol{K}_{\Gamma\ \tau_1}^{AA} + \boldsymbol{K}_{\Gamma\ \tau_2}^{AA} \tag{1.498}$$

where, β is the penalty factor

1.4.6.8 Procedures to perform the interface coupling calculation

Steps to calculate the coupling matrices and load vectors:

1. Loop over the interface elements on the slave mesh (e.g., the rotator in rotating machine, the slave mesh in a coupling problem).
2. Loop over the integration points on the slave element (ξ, η).
3. Calculate the coordinates of the current integration point (x, y, z).
4. Get the matching element index and face index using the mapping tool.
5. Get the local coordinates in the master mesh (ζ, η).
6. Get the tangential and normal direction ($\boldsymbol{\tau}_1$, $\boldsymbol{\tau}_2$, $\mathbf{n}$) on the slave mesh (using the mapping tool).
7. Calculate the coupling matrices and the vectors.
8. Assemble the coupling matrices and the vectors to the global equations.
9. End of the loop for integration points
10. End of the loop for interface elements

1.4.6.9 Integration surface

We choose the interface mesh on the slave side as the control surface (integration surface), and the Lagrange multiplier is also interpolated on this mesh (Figures 1.17 and 1.18).

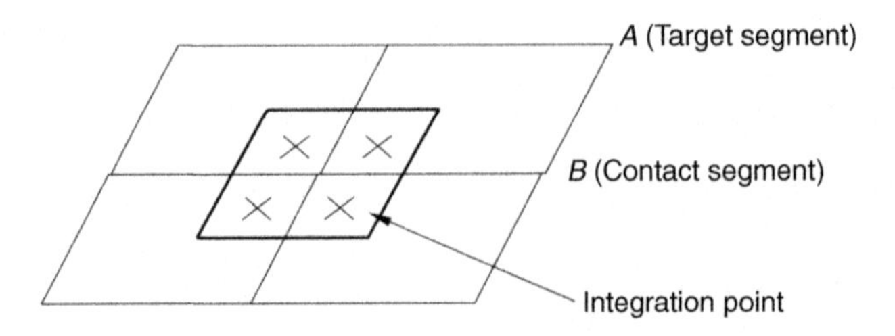

Figure 1.17 Master and slave mesh on coupling interfaces.

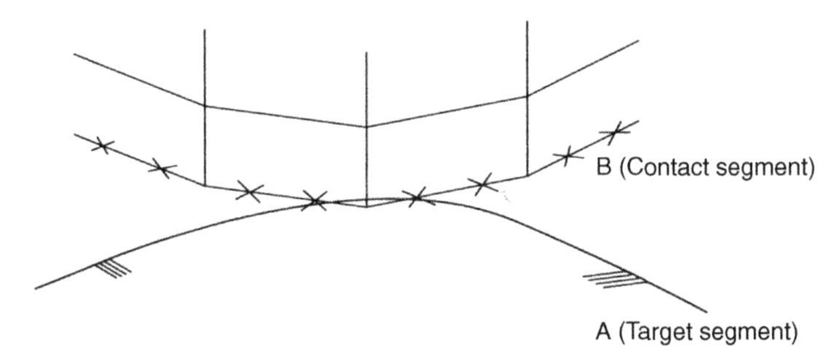

Figure 1.18 Contact detection points located at the Gauss integration point.

Note: here $\Gamma_c = \Gamma_B$ are in contact.

Remark 1.17: The DOFs correlations between the overlapping element need to be updated during simulation. Need to build DOF spaces for Lagrange multiplier.

1.4.6.10 The Lagrange multiplier space

The motor edge element methods have been proposed for multidomain simulation of the edge-based electromagnetic element, in which Nedelec's curl-conforming edge element method is used quite often. This section provides only some conceptual description.

1.4.6.10.1 Nedelec's curl-conforming edge elements

For a tetrahedron $T \in \mathcal{T}_i$ $(d = 3)$ resp, a triangle $T \in \mathcal{T}_{ij}$ $(d = 2)$ and the lowest order edge element $Nd_1(T)$ is defined by means of

$$Nd_1(T) := \{\boldsymbol{q} := \boldsymbol{a} + \boldsymbol{b}\Lambda\boldsymbol{x} \mid \boldsymbol{a}, \boldsymbol{b} \in \boldsymbol{R}^3,\ \boldsymbol{x} \in T\},\ d = 3 \tag{1.499}$$

$$Nd_1(T) := \left\{\boldsymbol{q} := \boldsymbol{a} + b\begin{pmatrix} -x_2 \\ x_1 \end{pmatrix} \mid \boldsymbol{a}, \boldsymbol{b} \in \boldsymbol{R}^2,\ \boldsymbol{x} \in T\right\},\ d = 2 \tag{1.500}$$

where $\mathcal{T}_i$ and $\mathcal{T}_{ij}$ are the triangulation of the subdomain Ω_i by tetrahedral element and Γ_{ij} by triangular element, respectively.

With the DOF given by the zero-order moments of tangential components with respect to edges $e \in \mathcal{E}_h(T)$

$$\ell_e(\boldsymbol{q}) := \int_e \boldsymbol{t}_e \cdot \boldsymbol{q}\, \mathrm{d}\Gamma, \qquad e \in \mathcal{E}_h(T) \tag{1.501}$$

where $\boldsymbol{t}_e$ stands for the tangential unit vector along e.

Alternatively denoted by $x_e^{(M)}$, the midpoint of $e \in \mathcal{E}_h(T)$, the DOF may be chosen as

$$\ell_e(\boldsymbol{q}) := (\boldsymbol{t}_e \cdot \boldsymbol{q})\left(x_e^{(M)}\right), \qquad e \in \mathcal{E}_h(T) \tag{1.502}$$

The edge element spaces $Nd_1(\Omega_i; \mathcal{T}_i)$ and $Nd_1(\Gamma_{ij}; \mathcal{T}_{ij})$ are then given as follows.

$$Nd_1(\Omega_i; \mathcal{T}_i) := \left\{ \boldsymbol{q}_h \in L^2(\Omega_i)^3 \,|\, \boldsymbol{q}_h|_T \in Nd_1(T),\ T \in \mathcal{T}_i \right\} \tag{1.503}$$

$$Nd_1(\Gamma_{ij}; \mathcal{T}_{ij}) := \left\{ \boldsymbol{q}_h \in L^2(\Gamma_{ij})^2 \,|\, \boldsymbol{q}_h|_T \in Nd_1(T),\ T \in \mathcal{T}_{ij} \right\} \tag{1.504}$$

We refer to $Nd_{1,\Gamma}(\Omega_i; \mathcal{T}_i)$ and $Nd_{1,0}(\Gamma_{ij}; \mathcal{T}_{ij})$ as its subspaces of vanishing tangential components on $\Gamma \cap \partial\Omega_i$ and $\partial\Gamma_{ij}$, respectively.

Remark 1.18: It is necessary to fix $\ell_e(\boldsymbol{q})$ as 0 on $\partial\Gamma_{ij}$ for Lagrange multiplier DOFs.

Based on these definitions, the product space is considered as:

$$\tilde{V}_h := \left\{ \boldsymbol{q}_h \in L^2(\Omega)^3 \,|\, \boldsymbol{q}_h|_{\Omega_i} \in Nd_{1,0}(\Omega_i; \mathcal{T}_i), 0 \le i \le n \right\} \tag{1.505}$$

The Lagrange multiplier space $M_h(\Gamma_{ij})$ will be constructed using the trace on Γ_{ij} of the edge element space with respect to the nonmortar side, that is, in terms of basis fields of $Nd_1(\Gamma_{ij}; \mathcal{T}_{ij})$ such that:

$$\dim M_h(\Gamma_{ij}) = \dim Nd_{1,0}(\Gamma_{ij}; \mathcal{T}_{ij}) \tag{1.506}$$

We now define $M_h(\Gamma_{ij})$ by an extension of the basis functions $\boldsymbol{q}^e \in Nd_{1,0}(\Gamma_{ij}; \mathcal{T}_{ij})$ associated with the edges in Γ_{ij} that have at least one neighboring edge on $\partial\Gamma_{ij}$. To be more precise,

1. For an interior edge $e \in \mathcal{E}_h(\Gamma_{ij})$, we denote it by

$$\mathcal{E}_h^{\partial\Gamma_{ij}}(e) := \left\{ f \in \mathcal{E}_h(\partial\Gamma_{ij}) \,|\, f \subset \text{supp } \boldsymbol{q}^e \right\} \tag{1.507}$$

the set of the neighboring edges of edge e on $\partial\Gamma_{ij}$.

2. For a boundary edge $f \in \varepsilon_h(\partial\Gamma_{ij})$, we refer to

$$\mathcal{E}_h^{\Gamma_{ij}}(f) := \left\{ e \in \mathcal{E}_h(\Gamma_{ij}) \,|\, e \subset \text{supp } \boldsymbol{q}_f \right\} \tag{1.508}$$

as the set of neighboring edges of edge f in the interior of $\delta_{m(j)}$.

Finally, we define

$$\mathcal{E}_h^{\Gamma_{ij}}\left(\partial\Gamma_{ij}\right) := \bigcup_{f\in\mathcal{E}_h\left(\partial\Gamma_{ij}\right)} \mathcal{E}_h^{\Gamma_{ij}}(f) \tag{1.509}$$

as the set of interior edges with a neighboring edge on $\partial\Gamma_{ij}$.

Then for $e \in \mathcal{E}_h^{\Gamma_{ij}}\left(\partial\Gamma_{ij}\right)$, we choose appropriate weighting factors $\lambda_{e,f} \in R$, $f \in \mathcal{E}_h^{\partial\Gamma_{ij}}(e)$ and define the basis field $\tilde{\boldsymbol{q}}_e$, $e \in \mathcal{E}_h\left(\Gamma_{ij}\right)$, according to:

$$\tilde{\boldsymbol{q}}^e = \begin{cases} \boldsymbol{q}^e, & e \in \mathcal{E}_h\left(\Gamma_{ij}\right) \setminus \mathcal{E}_h^{\Gamma_{ij}}\left(\partial\Gamma_{ij}\right) \\ \boldsymbol{q}^e + \sum\limits_{e\in\mathcal{E}_h^{\Gamma_{ij}}(f)} \lambda_{e,f}\boldsymbol{q}_f, & e \in \mathcal{E}_h^{\Gamma_{ij}}\left(\partial\Gamma_{ij}\right) \end{cases} \tag{1.510}$$

Where the weighting factors are assumed to satisfy,

$$\lambda_{e,f} \ge 0, \quad \sum_{e\in\mathcal{E}_h^{\Gamma_{ij}}(f)} \lambda_{e,f} = 1, \quad f \in \mathcal{E}_h\left(\partial\Gamma_{ij}\right) \tag{1.511}$$

Thus, specified basis fields define

$$M_h\left(\delta_{m(j)}\right) := \operatorname{span}\left\{\tilde{\boldsymbol{q}}^e \mid e \in \mathcal{E}_h\left(\delta_{m(j)}\right)\right\}. \tag{1.512}$$

This Lagrange multiplier space in Equation (1.512) has been implemented and used in the rotating electromachine simulation in this book.

1.4.6.11 The choices of element Gauge for interface coupling

As discussed in Section 1.4.6, the spanning tree gauge, Lagrange multiplier-based gauge, and penalty-based gauge are available for the magnetic vector potential-based edge element, but only the Lagrange multiplier-based gauge and penalty-based gauge are applicable to interface coupling problems.

1.5 Acoustic analysis

1.5.1 Governing equations

The basic assumptions for the acoustic fluid are as follows:

1. The fluid is compressible.
2. The fluid is inviscid or nonviscous.
3. There is no mean flow of the fluid.
4. The mean density and pressure are uniform throughout the fluid.

With these assumptions, the basic governing equations for acoustic analysis become Continuity equation is:

$$\frac{\partial \rho}{\partial t} + \rho_0 \frac{\partial v_i}{\partial x_i} = 0. \tag{1.513}$$

1.5.2 Momentum equation

$$\rho_0 \frac{\partial v_i}{\partial t} = -\frac{\partial p}{\partial x_i}. \tag{1.514}$$

The state equation is:

$$p = c_0^2 \rho. \tag{1.515}$$

The previous equations can be simplified to get the acoustic wave equation as follows:

$$\frac{1}{c^2}\frac{\partial^2 p}{\partial t^2} - \frac{\partial^2 p}{\partial x_i^2} = 0, \tag{1.516}$$

where, c is the speed of sound $\left(\sqrt{\frac{k}{\rho_0}}\right)$ in fluid medium.

1.5.3 The boundary conditions

The commonly used boundary conditions for acoustic analysis include the pressure-specified boundary condition, absorbing boundary condition (the third term in Equation (1.520), normal velocity or acceleration conditions on the domain boundary (the fourth term in Equation (1.520)), and so on.

1.5.4 The equations in weak form

Now we multiply equation by a virtual change in pressure and integrate it over the volume of domain and then get the following equation:

$$\int_{\Omega_f} \frac{1}{c^2} \delta p \frac{\partial^2 p}{\partial t^2} \mathrm{d}\Omega + \int_{\Omega_f} \frac{\partial(\delta p)}{\partial x_i} \frac{\partial p}{\partial x_i} \mathrm{d}\Omega = \int_{\Gamma_0} n_i \, \delta p \frac{\partial p}{\partial x_i} \mathrm{d}\Gamma, \tag{1.517}$$

where, Ω_f, volume of domain; Γ_0, elastic boundary surface where the pressure is applied and the dissipation may occur; δp, a virtual change in pressure; and n, unit normal to the interface Γ_0.

Accounting for the dissipation of energy due to damping, a dissipation term is added to previous equation. Now, we get equation:

$$\begin{aligned}&\int_{\Omega_f}\frac{1}{c^2}\delta p\frac{\partial^2 p}{\partial t^2}\mathrm{d}\Omega+\int_{\Omega_f}\frac{\partial(\delta p)}{\partial x_i}\frac{\partial p}{\partial x_i}\mathrm{d}\Omega+\int_{\Gamma_0}\delta p\left(\frac{r}{\rho_0 c}\right)\frac{1}{c}\frac{\partial p}{\partial t}\mathrm{d}\Gamma\\&=\int_{\Gamma_0}n_i\,\delta p\frac{\partial p}{\partial x_i}\mathrm{d}\Gamma,\end{aligned}\tag{1.518}$$

where r, is the absorption at the boundary.

On the fluid–structure interface, the following relationships exist between the normal pressure gradient of the fluid and the normal acceleration of the structure:

$$n_i\frac{\partial p}{\partial x_i}=-\rho_0 n_i\frac{\partial^2 u_i}{\partial t^2},\tag{1.519}$$

where, $\boldsymbol{u}$, displacement component of the structure at the interface.

Substituting Equation (1.519) into Equation (1.518), we get the following equation:

$$\begin{aligned}&\int_{\Omega_f}\frac{1}{c^2}\delta p\frac{\partial^2 p}{\partial t^2}\mathrm{d}\Omega+\int_{\Omega_f}\frac{\partial(\delta p)}{\partial x_i}\frac{\partial p}{\partial x_i}\mathrm{d}\Omega+\int_{\Gamma_0}\delta p\left(\frac{r}{\rho_0 c}\right)\frac{1}{c}\frac{\partial p}{\partial t}\mathrm{d}\Gamma+\\&\int_{\Gamma_0}\rho_0\,\delta p\;n_i\frac{\partial^2 u_i}{\partial t^2}\mathrm{d}\Gamma=0.\end{aligned}\tag{1.520}$$

1.5.5 Acoustics Equations in Matrix Form

The shape functions for the acoustic pressure and structure displacement vector:

$$p=\boldsymbol{N}_p^{\mathrm{T}}\boldsymbol{p}^e,\tag{1.521}$$

$$\boldsymbol{u}=\boldsymbol{N}_u^{\mathrm{T}}\boldsymbol{u}^e,\tag{1.522}$$

where, $\boldsymbol{N}_p$, element shape function for pressure; $\boldsymbol{N}_u$, element shape function for displacements; $\boldsymbol{p}^e$, element nodal pressure vector; and $\boldsymbol{u}^e$, element nodal displacement vector.

$$\frac{\partial^2 p}{\partial t^2}=\boldsymbol{N}_p{}^{\mathrm{T}}\ddot{\boldsymbol{p}}^{e\,\mathrm{T}}=\ddot{\boldsymbol{p}}^{e\,\mathrm{T}}\boldsymbol{N}_p,\tag{1.523}$$

$$\frac{\partial p}{\partial t}=\boldsymbol{N}_p{}^{\mathrm{T}}\dot{\boldsymbol{p}}^{e}=\dot{\boldsymbol{p}}^{e\,\mathrm{T}}\boldsymbol{N}_p,\tag{1.524}$$

$$\frac{\partial^2}{\partial t^2}\boldsymbol{u} = \boldsymbol{N}_u^{\mathrm{T}}\ddot{\boldsymbol{u}}^e, \tag{1.525}$$

$$\frac{\partial p}{\partial x_i} = \frac{\partial \boldsymbol{N}_p^{\mathrm{T}}\boldsymbol{p}^e}{\partial x_i} = \frac{\partial \boldsymbol{N}_p^{\mathrm{T}}}{\partial x_i}\boldsymbol{p}^e = \boldsymbol{B}_p\boldsymbol{p}^e, \tag{1.526}$$

where

$\boldsymbol{B}_{ji} = \dfrac{\partial \boldsymbol{N}_j}{\partial x_i}$, j is the node index of element.

Substituting Equations (1.526)–(1.529) into the weak form Equation (1.523) and then taking terms that do not vary over the element out of the integral sign, we eliminate virtual change term $\{\delta p\}^{\mathrm{T}}$ from equation, which will yield and the following equation will be obtained, which is written in matrix notation:

$$\boldsymbol{M}_e^f\ddot{\boldsymbol{p}}^e + \boldsymbol{C}_e^f\dot{\boldsymbol{p}}^e + \boldsymbol{K}_e^f\boldsymbol{p}^e + \rho_0\boldsymbol{R}^{e\,\mathrm{T}}\ddot{\boldsymbol{u}}^e = 0 \tag{1.527}$$

where,

$\boldsymbol{M}_e^f = \dfrac{1}{c^2}\int_{\Omega_f}\boldsymbol{N}_p\boldsymbol{N}_p^{\mathrm{T}}\,\mathrm{d}\Omega = \text{acoustic mass matrix}.$

$\boldsymbol{C}_e^f = \left(\dfrac{r}{\rho_0 c}\right)\dfrac{1}{c}\int_{\Gamma_0}\boldsymbol{N}_p\boldsymbol{N}_p^{\mathrm{T}}\,\mathrm{d}\Gamma = \text{acoustic damping matrix}.$

$\boldsymbol{K}_e^f = \int_{\Omega_f}\boldsymbol{B}_p^{\mathrm{T}}\boldsymbol{B}_p\,\mathrm{d}\Omega = \text{acoustic stiffness matrix}.$

$\boldsymbol{R}_e$ is the transformation matrix from the element nodal pressure vector to the element force vector:

$$\boldsymbol{R}_e = \int_{\Gamma_0}\boldsymbol{N}_u\boldsymbol{n}\boldsymbol{N}_p^{\mathrm{T}}\,\mathrm{d}\Gamma. \tag{1.528}$$

with the component of

$$[R_e]_{ai,b} = \int_{\Gamma_0} N_u^a n_i N_p^b\,\mathrm{d}\Gamma.$$

Remark 1.19: In the acoustic structure coupling simulation, the acoustic pressure loads are applied to the structure model. The resulting displacement and acceleration are sent to the acoustic physics model.

Physics coupling phenomena and formulations

2

Chapter Outline

Q. Zhang & S. Cen: Multiphysics Modeling. http://dx.doi.org/10.1016/B978-0-12-407709-6.00002-X

2.1 Introduction to coupling problems

The key issue for coupled physics simulations is to satisfy the conservative and continuity conditions across the coupling interfaces.

Multiphysics problems, which include, but are not limited to coupling among thermal, structural, fluidic, electromagnetic, and acoustic problems, have been given great attention over the years. We will discuss couplings among thermal-, structural-, fluidic-, electromagnetic-, and acoustic-physics models. The coupling may happen between two physics models, for example, fluid–structure interaction (FSI) analysis (fluid flow and structural field), thermal-stress analysis (thermal-structure), magneto-structure interaction (magnetic-structure), and electrostatic–structural interaction; among three physics models, namely, magneto-thermal-structural coupling (electric machine), electric-thermal-structural coupling (Joule heating), high-frequency electromagnetic-thermal-structural coupling (RF heating), harmonic-electromagnetic-thermal-fluidic coupling (induction stirring) etc.; or coupling among four physics models (electromagnetic-thermal-structural-fluid), for example, RF thermal probe and electric motor. Physically, we classify coupling into (1) essential coupling; (2) production-term coupling; (3) natural boundary-condition coupling; (4) constitutive-equation coupling; and (5) analysis-domain coupling. Mathematically, the coupling equations can be classified into matrix coupling, load–vector coupling, and integration–domain coupling. Based on the level of nonlinearity caused by the coupling, these coupling problems can be classified into strong coupling (SC) and weak coupling problems.

This chapter discusses coupling phenomena among the physics models listed and presents the coupling equation. Details about coupling methods are given in Chapter 3.

2.2 General coupling equations

2.2.1 Coupling equations in matrix form

The coupled algebraic system of equations between two physical domains (denoted by Ω_1 and Ω_2) are shown as:

$$\boldsymbol{Q}_1(\boldsymbol{X}_1, \boldsymbol{X}_2) = \boldsymbol{F}_1(\boldsymbol{X}_2) \tag{2.1}$$

$$\boldsymbol{Q}_2(\boldsymbol{X}_1, \boldsymbol{X}_2) = \boldsymbol{F}_2(\boldsymbol{X}_1). \tag{2.2}$$

Where, $\boldsymbol{X}_1$ and $\boldsymbol{X}_2$ are the primary variable vectors, $\boldsymbol{Q}_1$ and $\boldsymbol{Q}_2$ are the internal force vectors, and $\boldsymbol{F}_1$ and $\boldsymbol{F}_2$ are the external force vectors. Equations (2.1) and (2.2) reveal a balanced relation in the coupled system involving two physical models, and of course, they can be extended to three or more physical models.

Figure 2.1 shows the relationships between two physical domains in the coupled system; they may share coupling volume Ω_c, or coupling interface Γ_c.

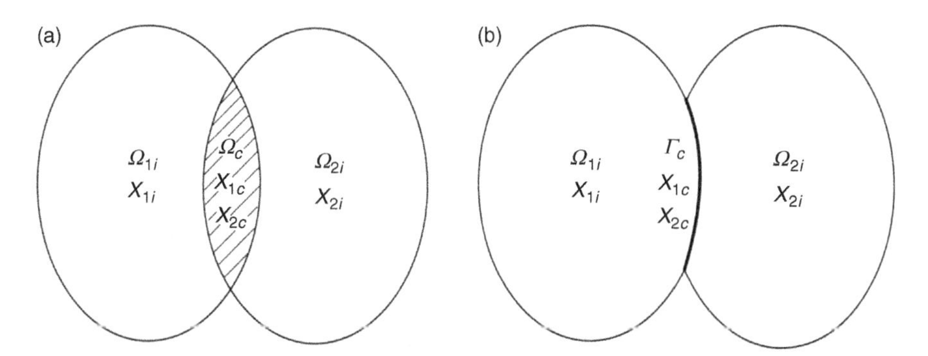

Figure 2.1 Volumetric or surface coupling between two physics models. (a) Volumetric coupling and (b) surface coupling.

Where subscripts i and c indicate the internal components of each physics model and the components on the coupling domain or interface, respectively.

$$\Omega_c = \Omega_1 \cap \Omega_2; \tag{2.3}$$

$$\Omega_{1i} = \Omega_1 - \Omega_c; \tag{2.4}$$

$$\Omega_{2i} = \Omega_2 - \Omega_c; \tag{2.5}$$

- $\boldsymbol{X}_{1i}$ is the primary variable vector of physical model 1 in domain Ω_{1i}.
- $\boldsymbol{X}_{1c}$ is the primary variable vector of physical model 1 in the coupled domain Ω_c.
- $\boldsymbol{X}_{2i}$ is the primary variable vector of physical model 2 in domain Ω_{2i}.
- $\boldsymbol{X}_{2c}$ is the primary variable vector of physical model 2 in domain Ω_c.

2.2.2 Constraint equations on interface

If the two physics domains have common primary variable on the coupling interface Γ_c, then the continuity condition across the coupling interface can be written as:

$$\boldsymbol{X}_1 = \boldsymbol{X}_2, \text{on} \quad \Gamma_c. \tag{2.6}$$

This condition is classified as an essential continuity condition of primary variables.

If the coupling between two physics models with the flux/force conservation or force balance involved, then the coupling equations are in the form of derivative of the primary variables Equation (2.7).

$$\text{flux}(\boldsymbol{X}_1) + \text{flux}(\boldsymbol{X}_2) = 0, \text{ on } \Gamma_c. \tag{2.7}$$

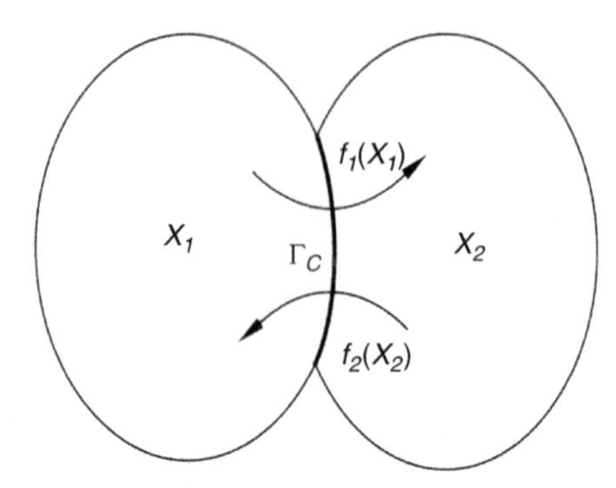

Figure 2.2 General coupling on the interface.

Where flux ($\boldsymbol{X}_1$) and flux ($\boldsymbol{X}_2$) are the flux values across the interface Γ_c from Ω_1 to Ω_2 and from Ω_2 to Ω_1, respectively. The two physical domains may not always have a common primary variable on the coupling interface Γ_c, for example, in the case of acoustic-structure interaction.

Equations (2.3) and (2.4) correspond to the equilibrium equation on fluid–structure coupling interface and flux balance across the fluid–solid interface in the conjugate heat transfer problem, respectively.

A more general coupling formulation of primary variables (Figure 2.2) can be written as:

$$f_1(\boldsymbol{X}_1)+f_2(\boldsymbol{X}_2)=0, \text{ on } \Gamma_c. \tag{2.8}$$

Here f_1 and f_2 are generic functions.

An example for this type of constraint Equation (2.8) in the coupling interface, can be found in an electromagnetic simulation with magnetic vector potential as the primary variable, but the corresponding continuity is forced by Equation (1.480).

2.2.3 Coupling example 1: DIC with common interface DOFs

The continuity condition on the coupling interface is:

$$\boldsymbol{X}_1=\boldsymbol{X}_2 \text{ on } \Gamma_c. \tag{2.9}$$

In this case, as shown in Figure 2.3a,

$$\begin{cases} \Omega_1 \cap \Omega_2 = \Gamma_c \\ \boldsymbol{X}_1 \cap \boldsymbol{X}_2 = \boldsymbol{X}_c = \boldsymbol{X}_{1c} = \boldsymbol{X}_{2c} \end{cases}.$$

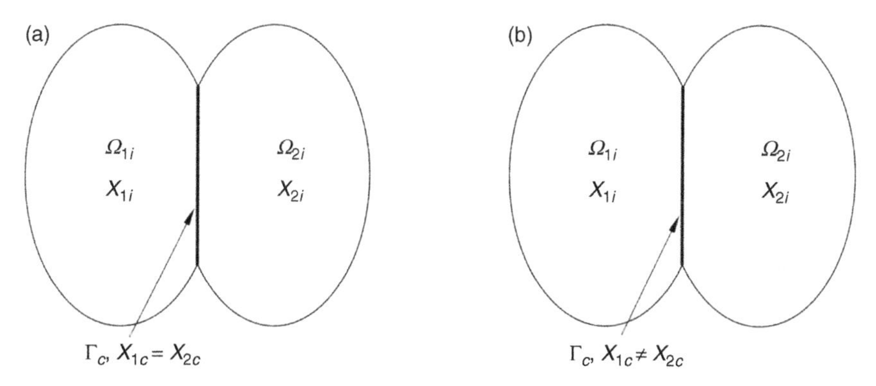

Figure 2.3 Interface coupling with or without common interface DOFs. (a) With common DOFs and (b) without common DOFs.

For numerical analysis, the internal force vectors $\boldsymbol{Q}_1$, $\boldsymbol{Q}_2$ in Equations (2.1) and (2.2) can be usually written in a matrix form as:

$$\boldsymbol{Q}_1(\boldsymbol{X}_1,\boldsymbol{X}_2)=\boldsymbol{K}_{11}^{1}(\boldsymbol{X}_1,\boldsymbol{X}_2)\boldsymbol{X}_1+\boldsymbol{K}_{12}^{12}(\boldsymbol{X}_1)\boldsymbol{X}_2, \tag{2.10}$$

$$\boldsymbol{Q}_2(\boldsymbol{X}_1,\boldsymbol{X}_2)=\boldsymbol{K}_{21}^{21}(\boldsymbol{X}_2)\boldsymbol{X}_1+\boldsymbol{K}_{22}^{2}(\boldsymbol{X}_1,\boldsymbol{X}_2)\boldsymbol{X}_2. \tag{2.11}$$

In the case of $\Gamma_c \neq 0$ (e.g., FSI problem, conjugate heat transfer problem), using the continuity condition, Equation (2.9), Equations (2.10) and (2.11) can be combined as:

$$\begin{bmatrix} \boldsymbol{K}_{1i,1i}^{1}(\boldsymbol{X}_1,\boldsymbol{X}_2) & \boldsymbol{K}_{1i,c}^{1}(\boldsymbol{X}_1,\boldsymbol{X}_2) & \boldsymbol{K}_{1i,2i}^{12}(\boldsymbol{X}_{1i}) \\ \boldsymbol{K}_{c,1i}^{1}(\boldsymbol{X}_1,\boldsymbol{X}_2) & \boldsymbol{K}_{cc}^{1}(\boldsymbol{X}_1,\boldsymbol{X}_2)+\boldsymbol{K}_{cc}^{2}(\boldsymbol{X}_1,\boldsymbol{X}_2) & \boldsymbol{K}_{c,2i}^{2}(\boldsymbol{X}_1,\boldsymbol{X}_2) \\ \boldsymbol{K}_{2i,1i}^{21}(\boldsymbol{X}_{2i}) & \boldsymbol{K}_{2i,c}^{2}(\boldsymbol{X}_1,\boldsymbol{X}_2) & \boldsymbol{K}_{2i,2i}^{2}(\boldsymbol{X}_1,\boldsymbol{X}_2) \end{bmatrix} \begin{Bmatrix} \boldsymbol{X}_{1i} \\ \boldsymbol{X}_c \\ (\boldsymbol{X}_{2i}) \end{Bmatrix} = \begin{Bmatrix} \boldsymbol{F}_{1i}(\boldsymbol{X}_2) \\ (\boldsymbol{F}_c) \\ \boldsymbol{F}_{2i}(\boldsymbol{X}_1) \end{Bmatrix}. \tag{2.12}$$

This type of coupling can be established by the direct interface coupling (DIC) method, multiple constraint equations-based coupling method, Lagrange multiplier (LM)-based coupling, or penalty method-based strong coupling (SC). The weak coupling method is also applicable if nonlinearity of the coupling system is not high (please refer Chapter 3 for further details). The FSI and conjugate heat transfer problems belong to this type of a coupling problem.

2.2.4 Coupling example 2: DIC without common interface DOFs

For some interface coupling problems, such as acoustic–structure coupling analysis, there is no common interface DOFs on the coupling interface (Figure 2.3b),

$$\begin{cases} \Omega_1 \cap \Omega_2 = \Gamma_c \\ X_1 \cap X_2 = 0, \text{ or } X_{1c} \neq X_{2c} \end{cases}.$$

As there are no common primary variables on the coupling interface, Equation (2.9) reduces to:

$$\begin{bmatrix} K_{11}^{1}(X_1, X_2) & K_{12}^{12}(X_1, X_2) \\ K_{21}^{21}(X_1, X_2) & K_{22}^{2}(X_1, X_2) \end{bmatrix} \begin{Bmatrix} X_1 \\ X_2 \end{Bmatrix} = \begin{Bmatrix} F_1(X_2) \\ F_2(X_1) \end{Bmatrix}. \tag{2.13}$$

For this type of coupling problem, the direct matrix assembly SC method and weak coupling can be applied. Acoustic-structure, thermal-stress, and electro-thermal coupling belong to these type of coupling problem.

2.3 Types of coupling interfaces

2.3.1 Types of coupling interface regions

The interface regions for coupling problems can be classified into:

- Surface–to–surface interaction (Figure 2.1b).
- Volumetric–to–volumetric interaction (Figure 2.1a).
- Porous media.
- General interdimensional coupling, for example, 1D–2D, 2D–3D, and 1D–3D.

2.3.2 Updates and matching of the interface region

- Bounded interface: for the bounded interface coupling, coupling regions of two physics models are tied together and do not change during the simulation. The interface mapping needs to be done only once for unmatching interface meshes in the initialization process.
- Sliding interface: for the sliding interface, two physics models across the interface slide against each other during simulation, but there are no interface separations, so interface mapping needs to be done at every time step during the simulation.
- General contact: for a general contact problem in structure analysis, the contact region may change during the analysis, for example, slide, separation, and rebound. The mapping needs to be done for each and every nonlinear iteration.

Remark 2.1: The mapping needs to be done for every single coupling iteration in sliding interface problems and general contact cases, to update the matching pair in the simulation process.

2.3.3 General connections

For a standard coupling interface, we consider that the two physics models are solved in the same coordinate system, and there is no frame change for a rotating coupling case, also these two physics models share a common interface without pitch changes or a periodic boundary. The treatment for dealing with periodic boundaries, frame changes, and pitch changes for multiphysics simulations of the rotating machinery, have been given in Chapter 6.

2.4 Classification of coupling phenomena

2.4.1 Coupling types by physics models

- Essential coupling: the coupled fields have a common primary variable on the shared interface, and they need to be continuous across the coupling interface.
- Production-term coupling: the coupled fields do not have common primary variable, but there are production terms of the primary variables from different physics models in the coupled equation.
- Natural boundary condition coupling: the coupled fields do not have the same primary variable, but one field is considered as an output that provides natural boundary conditions or source terms for the others.
- Constitutive-equation coupling: the primary variable or output of one field influences the material properties or the constitutive equations of the other physics models.
- Analysis-domain coupling: a field, usually the structural field, provides displacement boundary conditions for the nonstructural field to influence their analysis domain.

2.4.2 Coupling types by nonlinearity: strong and weak coupling

Based on the level of nonlinearity caused by the coupling, coupling problems can be classified into SC problems with high nonlinearity and weak coupling problems with low nonlinearity.

For a coupled system with high correlation between physics models, such as FSI problems with incompressible fluid enclosed by soft structure, FSI problems with structural buckling, or with discontinuous shock wave in CFD field, the nonlinearity of the system is high. This kind of coupling is referred to as an SC problem.

If the dependency between physics models is weak and the nonlinearity of a coupled system is not critical, it is considered as a weak coupling problem, for example, standard thermal-stress coupling problem without highly nonlinear temperature dependent material behavior, fluid–structure problem with high structure stiffness and low Reynolds number flow in an open domain, etc.

The nonlinearity for a coupled system may come from the nature of coupling and the nonlinear behavior of each individual physics model.

2.4.3 Unidirectional and bidirectional coupling

All coupling problems are essentially bidirectional. If the influence of a physics model A on physics model B is negligible, but not vice versa, then we consider this type of coupling problem as unidirectional.

2.5 The coupling matrices among physics models

In this chapter, the following coupling phenomena will be discussed.

- Thermal–stress coupling
- FSI
- Conjugate heat transfer problem
- Acoustic–structure coupling
- Piezoelectric analysis
- Electrostatic–structure coupling
- Magneto–structure coupling
- Magneto–fluid coupling
- Electro–thermal coupling
- Magnetic–thermal coupling.

The possible load transfers among electric, electrostatic, magnetostatic, electromagnetic, thermal, structure, fluid, and acoustics models are demonstrated in Figure 2.4. The coupling region can be on the surface region, on the volumetric domain, or on both, and the load transfers can be bidirectional or unidirectional.

2.6 Thermal–stress coupling

2.6.1 Constitutive equations of thermos elasticity

The coupled thermos-elastic constitutive equations (Ney,1957) are:

$$\varepsilon = [D]^{-1}\sigma + \alpha\Delta T, \tag{2.14}$$

$$S = \alpha^{T}\sigma + \frac{\rho C_{P}}{T_{0}}\Delta T. \tag{2.15}$$

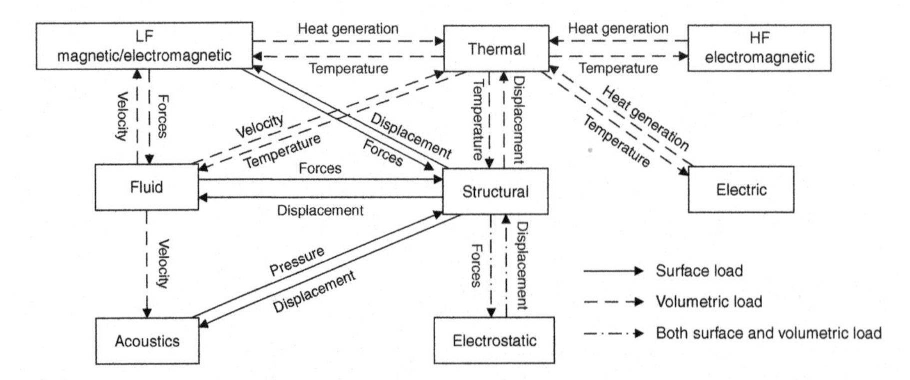

Figure 2.4 The coupling and load transfers among physics models.

Where, $\boldsymbol{\varepsilon}$ is the total strain vector, $\varepsilon=\{\varepsilon_x,\varepsilon_y,\varepsilon_z,\varepsilon_{xy},\varepsilon_{yz},\varepsilon_{xz}\}^T$; S is entropy density; $\boldsymbol{\sigma}$ is the stress vector, $\sigma=\{\sigma_x,\sigma_y,\sigma_z,\sigma_{xy},\sigma_{yz},\sigma_{xz}\}^T$; T_0 is the absolute reference temperature, $T_0=T_{\text{ref}}+T_{\text{off}}$; $\Delta T=T-T_{\text{ref}}$; T is the current temperature; T_{ref} is the reference temperature; T_{off} is the offset temperature from absolute zero to zero; $[\boldsymbol{D}]$ is the elastic stiffness matrix; $\boldsymbol{\alpha}$ is the vector of coefficients of thermal expansion, $\alpha=\{\alpha_x,\alpha_y,\alpha_z,0,\ 0,\ 0\}^T$; ρ is density; and C_P is specific heat at constant stress or pressure.

From Equation (2.14), we have:

$$\sigma=[\boldsymbol{D}]\ \varepsilon-\beta\Delta T \tag{2.16}$$

where, $\boldsymbol{\beta}$ is vector of thermo-elastic coefficient, $\boldsymbol{\beta}=[\boldsymbol{D}]\boldsymbol{\alpha}$.

Using $\boldsymbol{\varepsilon}$ and ΔT as independent variables, replacing the entropy density S in Equation (2.15) by heat density Q, and using the second law of thermodynamics for a reversible change, we have:

$$Q=T_0S, \tag{2.17}$$

$$Q=T_0\beta^T\varepsilon+\rho C_V\Delta T. \tag{2.18}$$

C_V, specific heat at constant strain or volume $= C_P-\frac{T_0}{\rho}\alpha^T\beta$.

Substituting Q from Equation (2.18) into the heat flow Equation (1.1) produces,

$$\frac{\partial Q}{\partial t}=T_0\beta^T\frac{\partial\varepsilon}{\partial t}+\rho C_V\frac{\partial\Delta T}{\partial t}-k\nabla^2T. \tag{2.19}$$

2.6.2 Coupled equations in matrix form

Applying the variational principle to stress equation of motion and the heat flow conservation equation (they are coupled by the thermo-elastic constitutive equations) produces the following finite element matrix equation:

$$\begin{bmatrix}\boldsymbol{M}^s & \boldsymbol{0}\\ \boldsymbol{0} & \boldsymbol{0}\end{bmatrix}\begin{Bmatrix}\ddot{\boldsymbol{u}}\\ -\end{Bmatrix}+\begin{bmatrix}\boldsymbol{0} & \boldsymbol{0}\\ \boldsymbol{C}_{tu}^T & \boldsymbol{C}_{tt}^T(\boldsymbol{u})\end{bmatrix}\begin{Bmatrix}\dot{\boldsymbol{u}}\\ \dot{\boldsymbol{T}}\end{Bmatrix}+\begin{bmatrix}\boldsymbol{K}^s(\boldsymbol{u},\boldsymbol{T}) & \boldsymbol{K}_{ut}^s\\ \boldsymbol{0} & \boldsymbol{K}_{tt}^T(\boldsymbol{u},\boldsymbol{T})\end{bmatrix}\begin{Bmatrix}\boldsymbol{u}\\ \boldsymbol{T}\end{Bmatrix}=\begin{Bmatrix}\boldsymbol{F}^s\\ \boldsymbol{Q}^{\text{T}}\end{Bmatrix}. \tag{2.20}$$

Where, $\boldsymbol{M}^S$, mass matrix; $\boldsymbol{K}_{ut}^S$, thermo–elastic stiffness matrix, $\boldsymbol{K}_{ut}^s=-\int_\Omega \boldsymbol{B}_u^{\text{T}}\beta\left(\boldsymbol{N}_t^{\text{T}}\right)\text{d}\Omega$; $\boldsymbol{C}_{tu}$, thermoelastic-damping matrix, where, $\boldsymbol{C}_{tu}=-T_0\boldsymbol{K}_{ut}^{s\ \text{T}}$.

It is obvious that there are no common DOFs between thermal and stress analysis. The major coupling term in this equation is thermal strain force vector,

$\boldsymbol{F}_T(T) = -\boldsymbol{K}_{ut}^s \Delta \boldsymbol{T}$, and we call it the natural boundary coupling (load–vector coupling). The temperature term $\boldsymbol{T}$ in matrices $\boldsymbol{K}^s(\boldsymbol{u},T)$ and $\boldsymbol{K}_{tt}^T(\boldsymbol{u},T)$ demonstrates the temperature-dependent material properties, and $\boldsymbol{K}^s(\boldsymbol{u},T)$ represents constitutive law coupling (matrix coupling). The displacement term $\boldsymbol{u}$ in $\boldsymbol{C}_{tt}^T(\boldsymbol{u})$ and $\boldsymbol{K}_{tt}^T(\boldsymbol{u},T)$ represents the effects of the integration domain change on the thermal field. It is called the integration-domain coupling (integration domain and view factor changes in radiation simulation). Thermal–stress coupling is a bidirectional coupling problem.

2.6.3 SC method for thermal–stress coupling

We call it the SC method if we solve Equation (2.20) directly. Although the source term is in the form of $F_1 = \boldsymbol{K}_{12}\boldsymbol{X}_2$, the SC method in thermal stress analysis is straightforward to implement.

2.6.4 Weak coupling method for thermal–stress coupling

If we move the coupling source terms in Equation (2.20) to the right-hand side of the equation and solve the thermal and structural equations separately, we obtain Equation (2.21), and the weak coupling method is selected. Figure 2.5 shows the volumetric data transfer between thermal and structural analyses, that is, the temperature from thermal to structure model and the displacement and heat from structure to thermal model.

$$\begin{bmatrix} \boldsymbol{M}^s & \boldsymbol{0} \\ \boldsymbol{0} & \boldsymbol{0} \end{bmatrix} \begin{Bmatrix} \ddot{\boldsymbol{u}} \\ - \end{Bmatrix} + \begin{bmatrix} \boldsymbol{0} & \boldsymbol{0} \\ \boldsymbol{0} & \boldsymbol{C}_{tt}^T(\boldsymbol{u}) \end{bmatrix} \begin{Bmatrix} - \\ \dot{\boldsymbol{T}} \end{Bmatrix} + \begin{bmatrix} \boldsymbol{K}^s(\boldsymbol{u},T) & \boldsymbol{0} \\ \boldsymbol{0} & \boldsymbol{K}_{tt}^T(\boldsymbol{u},T) \end{bmatrix} \begin{Bmatrix} \boldsymbol{u} \\ \boldsymbol{T} \end{Bmatrix} = \begin{Bmatrix} \boldsymbol{F}^S + \boldsymbol{F}^S(T) \\ \boldsymbol{Q}^T + \boldsymbol{Q}_{\text{ted}}^T \end{Bmatrix}. \tag{2.21}$$

Where, the coupling source term $\boldsymbol{Q}_{\text{ted}}^T$ is the heat-generating rate vector for thermoselastic damping $\boldsymbol{Q}_{\text{ted}}^T = -\boldsymbol{C}_{tu}^T \dot{\boldsymbol{u}}$ and $\boldsymbol{F}^S(\boldsymbol{T}) = -\boldsymbol{K}_{ut}^s \Delta \boldsymbol{T}$ is thermal strain force vector due to temperature changes.

If we assume the structure material is independent of temperature and the displacement effect on thermal and heat generate rate vector for thermo-elastic damping is negligible, then we have the general thermal stress problem:

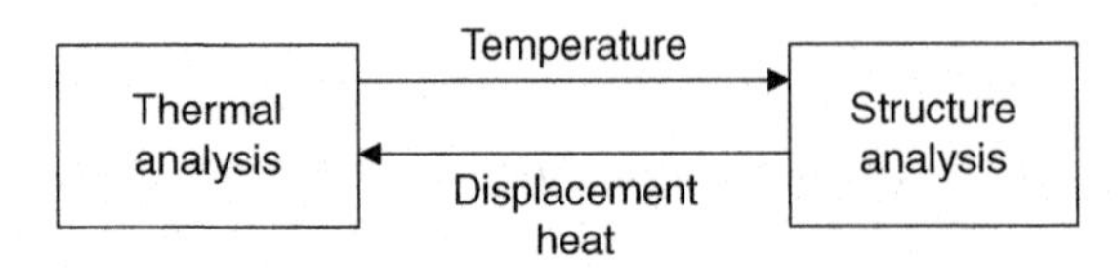

Figure 2.5 The data transfers between thermal- and structural analysis.

$$\begin{bmatrix} \boldsymbol{M}^s & \boldsymbol{0} \\ \boldsymbol{0} & \boldsymbol{0} \end{bmatrix} \begin{Bmatrix} \ddot{\boldsymbol{u}} \\ - \end{Bmatrix} + \begin{bmatrix} \boldsymbol{0} & \boldsymbol{0} \\ \boldsymbol{0} & \boldsymbol{C}^T \end{bmatrix} \begin{Bmatrix} - \\ \dot{\boldsymbol{T}} \end{Bmatrix} + \begin{bmatrix} \boldsymbol{K}^s(\boldsymbol{u}) & \boldsymbol{0} \\ \boldsymbol{0} & \boldsymbol{K}_{tt}^T \end{bmatrix} \begin{Bmatrix} \boldsymbol{u} \\ \boldsymbol{T} \end{Bmatrix} = \begin{Bmatrix} \boldsymbol{F}^S + \boldsymbol{F}^S(T) \\ \boldsymbol{Q}^T \end{Bmatrix}. \tag{2.22}$$

This is a unidirectional coupling problem and the thermal problem always has to be solved first and the temperature vector has to be transferred to the structure solution.

2.7 Fluid–structure interaction

2.7.1 Interface conditions

For a fully coupled fluid–structure system (Figure 2.6), the geometrical compatibility conditions and equilibrium conditions on the interface must be satisfied, that is, $\Gamma_c \neq 0$. In this case, the nonslip conditions are assumed on fluid and solid interfaces, that is, the compatibility conditions,

$$v_i^f = v_i^s \quad (i = 1,2,3) \quad \text{on} \quad \Gamma_c. \tag{2.23}$$

And the equilibrium conditions:

$$\sigma_{ji}^f n_j^f + \sigma_{ji}^s n_j^s = 0 \quad (i = 1,2,3) \quad \text{on} \quad \Gamma_c. \tag{2.24}$$

Where, n_j is the outward normal of the coupling interface. Superscripts f and s indicate the value corresponding to the fluid and the structure, respectively, and the Γ_c is the interface of the fluid and the structure.

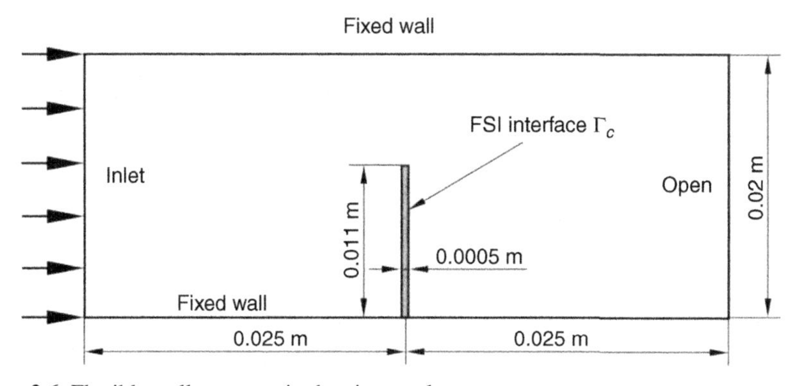

Figure 2.6 Flexible wall structure in the air tunnel.

2.7.2 Coupled equations

Since the continuum elements can be used for both fluid and structure, the geometrical compatibility on Γ_c is satisfied by simply defining the shared interface nodes and the same shape functions for the fluid and structure on the interface (Figure 2.7). The equilibrium conditions are satisfied automatically in the sense of the weak form of Equation (2.24). This approach is called Direct Interface Coupling (DIC) method; details are presented in Chapter 3. The Multipoint Constraint (MPC) and the Lagrange Multiplier (LM) method can also be used for the interface coupling of fluid–structure problems, and both methods allow nonconforming mesh discretization across the coupling interface. For the sake of simplicity, we only present the DIC method in this section; other coupling approaches are discussed in Chapter 3.

The Arbitrary Lagrangian Eulerian formulation for the fluid flows becomes a pure Lagrangian description due to the stick fluid–structure interface assumption for viscous flow. This result in a consistent Lagrangian description is used for both of the fluid and the solid on the interface, and then the variable vector of the coupled system can be divided into three parts,

$$\boldsymbol{\Phi}^{fs} = \begin{Bmatrix} \boldsymbol{\Phi}_i^f \\ \boldsymbol{\Phi}_c^{fs} \\ \boldsymbol{\Phi}_i^s \end{Bmatrix}, \ \boldsymbol{U}^s = \begin{Bmatrix} - \\ \boldsymbol{u}_c^{fs} \\ \boldsymbol{u}_i^s \end{Bmatrix}. \tag{2.25}$$

Here, $\boldsymbol{\Phi}_i^f$, is the pressure and internal velocity vector of fluid, $\left(\boldsymbol{\Phi}_i^f = \left\{\boldsymbol{P}^f,\ \boldsymbol{V}_i^f\right\}^T\right)$; $\boldsymbol{\Phi}_c^{fs}$, is the coupled velocity vector, $\left(\boldsymbol{\Phi}_c^{fs} = \boldsymbol{V}_c^{fs}\right)$; $\boldsymbol{\Phi}_i^s$, is the internal velocity vector of the solid domain (the time derivatives of the pressure vector are also included if mixed u–p is used for solid domain), $\left(\boldsymbol{\Phi}_i^s = \boldsymbol{V}_i^s\right)$; $\boldsymbol{u}_c^{fs}$, is the coupled displacement/mesh

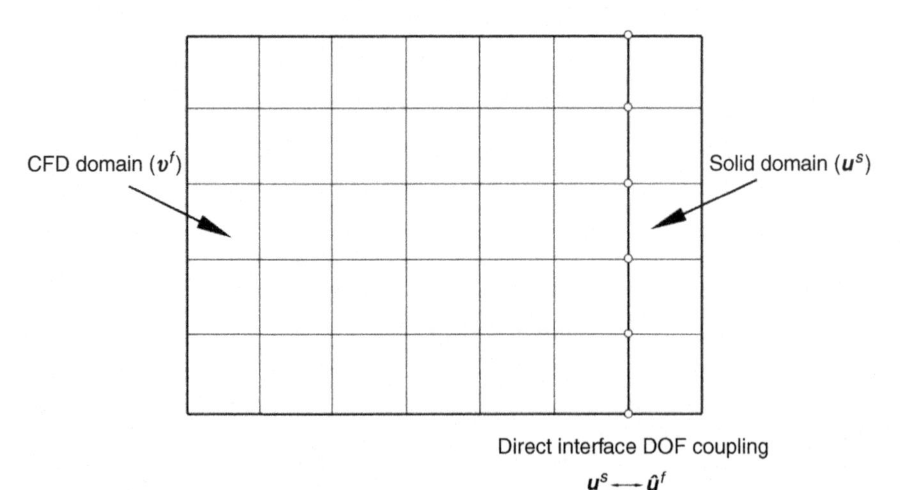

Figure 2.7 DIC DOF.

displacement vector of structure/fluid; and $\boldsymbol{u}_i^s$, is the internal displacement vector of solid (pressure vector is also included if mixed u–p is used for solid domain). Based on the definition of the variables of the coupled system, the coupled Equations (1.36) and (1.160) can be expressed as,

$$\boldsymbol{Q}^{fs} = \boldsymbol{F}^{fs}. \tag{2.26}$$

Here, the internal force vector, $\boldsymbol{Q}^{fs}$, includes the inertial force of the coupled system,

$$\boldsymbol{Q}^{fs} = \boldsymbol{M}^{fs} \cdot \overset{*}{\boldsymbol{\Phi}}{}^{fs} + \boldsymbol{C}^{f} \cdot \boldsymbol{\Phi}^{fs} + \boldsymbol{Q}^{s}\left(\boldsymbol{u}^{s}\right). \tag{2.27}$$

$\boldsymbol{F}^{fs}$, is the external force vector. In Equation (2.27), * denotes the time derivative in Arbitrary Lagrangian Eulerian coordinate for fluid and time derivative in Lagrangian coordinate for interface and solid, respectively; $\boldsymbol{M}^{fs}$, is the mass matrix composed of those for the fluid and the structure; and $\boldsymbol{C}^{f}$, consists of the divergence, viscous, and convective terms of fluid. The details of these coupled matrices and vectors can be expressed as follows.

$$\boldsymbol{M}^{fs} = \begin{bmatrix} \boldsymbol{M}^{p} & \mathbf{0} & \mathbf{0} & \mathbf{0} \\ \mathbf{0} & \boldsymbol{M}_{ii}^{f} & \boldsymbol{M}_{ic}^{f} & \mathbf{0} \\ \mathbf{0} & \boldsymbol{M}_{ci}^{f} & \boldsymbol{M}_{cc}^{f} + \boldsymbol{M}_{cc}^{s} & \boldsymbol{M}_{ci}^{f} \\ \mathbf{0} & \mathbf{0} & \boldsymbol{M}_{ic}^{s} & \boldsymbol{M}_{ii}^{s} \end{bmatrix};$$
$$\boldsymbol{C}^{f} = \begin{bmatrix} \boldsymbol{\Lambda}^{p} & \boldsymbol{G}_{i}^{\mathrm{T}} & \boldsymbol{G}_{c}^{\mathrm{T}} & \mathbf{0} \\ -\boldsymbol{G}_{i} & \boldsymbol{\Lambda}_{ii} + \boldsymbol{K}_{\mu ii} & \boldsymbol{\Lambda}_{ic} + \boldsymbol{K}_{\mu ic} & \mathbf{0} \\ \boldsymbol{G}_{c} & \boldsymbol{\Lambda}_{ci} + \boldsymbol{K}_{\mu ci} & \boldsymbol{\Lambda}_{cc} + \boldsymbol{K}_{\mu cc} & \mathbf{0} \\ \mathbf{0} & \mathbf{0} & \mathbf{0} & \mathbf{0} \end{bmatrix} \tag{2.28}$$

$$\boldsymbol{\Phi}^{s} = \begin{Bmatrix} - \\ - \\ \boldsymbol{Q}_{c}^{s} \\ \boldsymbol{Q}_{i}^{s} \end{Bmatrix}, \quad \boldsymbol{F}^{s} = \begin{Bmatrix} 0 \\ \boldsymbol{F}_{i}^{f} \\ \boldsymbol{F}_{c}^{fs} \\ \boldsymbol{F}_{i}^{s} \end{Bmatrix}. \tag{2.29}$$

In these equations, subscripts c and i denote the coupling parts and internal parts of variables, respectively.

It is obvious that the fluid structure is a bidirectional fully coupled system (essential boundary coupling, matrix coupling, load–vector coupling, integration-domain coupling).

The incremental form of the coupled Equation (2.26) can be obtained by taking linearization.

$$ {}^{t}\boldsymbol{M}^{fs} \cdot \Delta\overset{*}{\boldsymbol{\Phi}}{}^{fs} + {}^{t}\boldsymbol{C}^{f} \cdot \Delta\boldsymbol{\Phi}^{fs} + {}^{t}\boldsymbol{K}^{s}\Delta\boldsymbol{u}^{s} = {}^{t+\Delta t}\boldsymbol{F}^{fs} - {}^{t}\boldsymbol{Q}^{fs}. \tag{2.30} $$

Here, ${}^{t}\boldsymbol{K}^{s}$ is the tangent stiffness matrix and the coefficient matrices, ${}^{t}\boldsymbol{M}^{fs}$ and ${}^{t}\boldsymbol{C}^{f}$, are secant matrices. Although the linearization of ${}^{t}\boldsymbol{M}^{fs}$ and ${}^{t}\boldsymbol{C}^{f}$ has been discussed by Zhang (1999), in practice, if an effective time step control algorithm is used to control the time step Δt, the secant matrices of ${}^{t}\boldsymbol{M}^{fs}$ and ${}^{t}\boldsymbol{C}^{f}$ frequently work well. Using the implicit time integration method (Newmark-β) for the linearized coupled Equation (2.30), we obtain the following linear equation,

$$ \left[{}^{t}\boldsymbol{M}^{fs} + \gamma\Delta t\, {}^{t}\boldsymbol{C}^{f} + \beta\Delta t^{2\,t}\boldsymbol{K}^{s} \right] \Delta\overset{*}{\boldsymbol{\Phi}}{}^{fs} = {}^{t+\Delta t}\boldsymbol{F}^{fs} - {}^{t}\boldsymbol{Q}^{fs}. \tag{2.31} $$

Where, γ and β are parameters that can be chosen so as to obtain integration accuracy and stability and $\overset{*}{\boldsymbol{\Phi}}{}^{fs}$ is the increment of the variable vector of the coupled system. Furthermore, Equation (2.31) can be expressed in the simplest form as,

$$ \boldsymbol{\Lambda}^{fs}\Delta\overset{*}{\boldsymbol{\Phi}}{}^{fs} = \boldsymbol{R}^{fs}. \tag{2.32} $$

Where, $\boldsymbol{\Lambda}^{fs}$ is the effective mass matrix of the coupled system and $\boldsymbol{R}^{fs}$ as the residual vector of the coupled system.

2.7.3 SC method for FSI (Zhang and Hisada, 2001)

For the SC method, the increment of the variable vector of the coupled system is obtained simultaneously by solving the linearized coupled Equation (2.32), and the variable vectors of the coupled system are updated simultaneously (Zhang and Hisada, 2004). For notational simplicity, we omit the subscripts, t and $t + \Delta t$, and the iteration number i. The most recently updated variable vectors are used to calculate matrices and vectors, which appear in the following equations.

$$ \begin{bmatrix} \boldsymbol{\Lambda}_{ii}^{fs} & \boldsymbol{\Lambda}_{ci}^{f} & 0 \\ \boldsymbol{\Lambda}_{ci}^{f} & \boldsymbol{\Lambda}_{cc}^{f} + \boldsymbol{\Lambda}_{cc}^{s} & \boldsymbol{\Lambda}_{ci}^{s} \\ 0 & \boldsymbol{\Lambda}_{ic}^{s} & \boldsymbol{\Lambda}_{ii}^{s} \end{bmatrix} \begin{Bmatrix} \Delta\overset{*}{\boldsymbol{\Phi}}{}_{i}^{fs} \\ \Delta\overset{*}{\boldsymbol{\Phi}}{}_{c}^{fs} \\ \Delta\overset{*}{\boldsymbol{\Phi}}{}_{i}^{s} \end{Bmatrix} = \begin{Bmatrix} \boldsymbol{R}_{i}^{f} \\ \boldsymbol{R}_{c}^{fs} \\ \boldsymbol{R}_{i}^{s} \end{Bmatrix}. \tag{2.33} $$

The iteration procedure for solving coupled equation with automatic mesh moving is carried out using the predictor–multicorrector algorithm. The calculation procedure is shown as follows.

Begin time step loop

1. Increment time step $t+\Delta t$
2. Form predictors
3. Solve mesh motion
4. Begin nonlinear loop
   a. Solve the coupled Equation (2.33)
   b. Solve the mesh motion
   c. Update the solution variables
5. End nonlinear loop

End time step

2.7.4 Weak coupling method for FSI problems

2.7.4.1 The load transfer between fluid and structure

In contrast to the SC method, in the weak coupling method, the structure dynamic solver and fluid dynamic solver are carried out in a staggered manner.

Figure 2.8 demonstrates the load transfer between fluid analysis and structure analysis, in which the fluid forces are transferred from the fluid to the structure, as return the displacement and velocity are transferred from the structure to the fluid model (Zhang and Hisada, 2004).

2.7.4.2 Calculation procedure for weak coupling method

The computational procedure of the weak coupling method is shown as follows.

1. Solve the fluid components:

$$\Lambda_{ii}^{f}\Delta\overset{*}{\Phi}_{i}^{f}=\boldsymbol{R}_{i}^{f}. \tag{2.34}$$

   Here, $\Delta\overset{*}{\Phi}_{c}^{fs}=0$ is assumed.
2. Update the fluid components by $\Delta\overset{*}{\Phi}_{i}^{f}$, transfer the updated fluid force vector and added matrix to the structure.
3. Solve the structure components with interfaces,

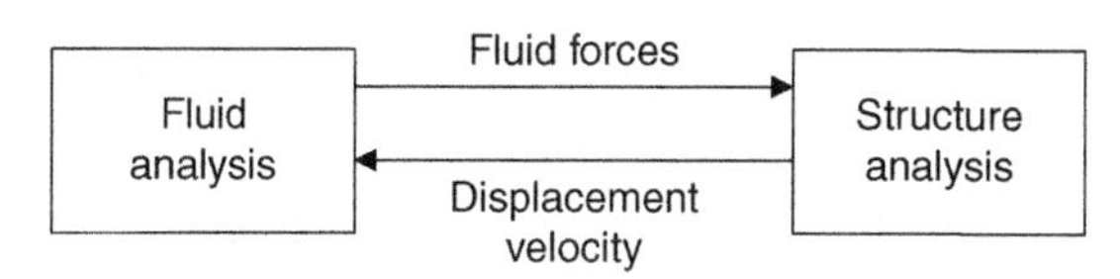

Figure 2.8 The data transfers for FSI.

$$\left[\Lambda^f_{\text{add}} + \Lambda^s_{cc}\right] \Delta\overset{*}{\Phi}{}^{fs}_c + \Lambda^s_{ci}\, \Delta\overset{*}{\Phi}{}^s_i = \boldsymbol{R}^{fs}_c, \tag{2.35}$$

$$\Lambda^s_{ic} \Delta\overset{*}{\Phi}{}^{fs}_c + \Lambda^s_{ii} \Delta\overset{*}{\Phi}{}^s_i = \boldsymbol{R}^s_i. \tag{2.36}$$

Here, Λ^f_{add} is the added matrix from the fluid; and the residual vector $\boldsymbol{R}^{fs}_c$ includes the components of the updated fluid force.

4. Transfer the updated velocity and displacement to the fluid and go to step one, if the solution is not converged.

Remark 2.2 In this method, the fluid force vector and added matrix, Λ^f_{add}, are transferred from the FD solver to the SD solver. The choice of the added matrix, Λ^f_{add}, and the fluid force vector, which should be transferred to the SD solver, is a key issue for FSI problem. It will affect the convergence rate, accuracy, and also the efficiency of the calculation of a coupled system; some advanced weak coupling approaches will be discussed in Chapter 3.

2.8 Conjugate heat transfer problem

When the thermal fluid flow is coupled with a thermal stress problem, the continuity conditions of temperature and conservative conditions of heat flow need to be satisfied across the coupling interface. This kind of coupling is called the conjugate heat transfer problem. We need to solve this conjugate heat transfer problem with temperature being transferred from one field to another and also the heat flux. Similar to the fluid–structure-coupling problem, this is also a fully coupled (essential coupling, i.e., $\Gamma_c \neq 0$) problem.

The fully coupled matrix is shown as Equation (2.37).

$$\begin{bmatrix} \boldsymbol{C}^f_{ii} & \boldsymbol{C}^f_{ic} & \boldsymbol{0} \\ \boldsymbol{C}^f_{ci} & \boldsymbol{C}^f_{cc}+\boldsymbol{C}^s_{cc} & \boldsymbol{C}^s_{ci} \\ \boldsymbol{0} & \boldsymbol{C}^s_{2c} & \boldsymbol{C}^s_{ii} \end{bmatrix} \begin{Bmatrix} \dot{\boldsymbol{T}}^f_i \\ \dot{\boldsymbol{T}}_c \\ \dot{\boldsymbol{T}}^s_i \end{Bmatrix} + \begin{bmatrix} \boldsymbol{K}^f_{ii} & \boldsymbol{K}^f_{ic} & \boldsymbol{0} \\ \boldsymbol{K}^f_{ci} & \boldsymbol{C}^f_{cc}+\boldsymbol{C}^s_{cc} & \boldsymbol{K}^s_{ci} \\ \boldsymbol{0} & \boldsymbol{K}^s_{ic} & \boldsymbol{K}^s_{ii} \end{bmatrix} \begin{Bmatrix} \boldsymbol{T}^f_i \\ \boldsymbol{T}_c \\ \boldsymbol{T}^s_i \end{Bmatrix} = \begin{Bmatrix} \boldsymbol{Q}^f_i \\ \boldsymbol{Q}_c \\ \boldsymbol{Q}^s_i \end{Bmatrix}. \tag{2.37}$$

Although the coupled Equation (2.37) can be solved directly by the SC method, the weak coupling method is also suitable and commonly used for the thermal coupling of the conjugate heat transfer problems.

As mentioned earlier, to solve the conjugate heat transfer problems, we need to exchange heat flux and temperature across the fluid–structure interface (Figure 2.9). Transferring temperature from the higher conductivity side to the lower conductivity side, rather than transferring the heat flux or flow from the lower conductivity side to the higher conductivity side, provides a better convergence in interface load transfers. A detailed explanation about the reason of this is presented in Chapter 3.

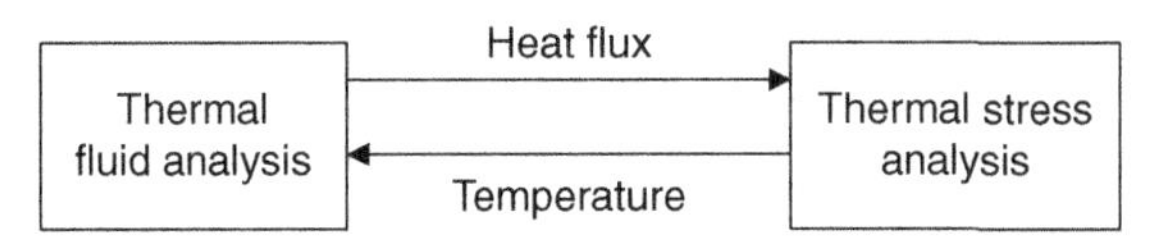

Figure 2.9 The data transfers for conjugate heat transfer problems.

2.9 Acoustic–structure coupling

Similar to the coupling of the NS equation based fluid flow with the structure, the acoustic–structure coupling problems need to satisfy the same interface coupling condition, that is, the continuity condition Equation (2.23) and the equilibrium conditions Equation (2.24), although the acoustic pressure will be the only source of fluid force.

The equilibrium equation of structure element in an element matrix form can be expressed as:

$$\boldsymbol{M}_e^s \ddot{\boldsymbol{u}}_e + \boldsymbol{C}_e^s \dot{\boldsymbol{u}}_e + \boldsymbol{K}_e^s \boldsymbol{u}_e = \boldsymbol{F}_e^s + \boldsymbol{F}_e^{pr}. \tag{2.38}$$

Where, $\boldsymbol{M}_e^s$, element structure mass matrix; $\boldsymbol{C}_e^s$, element structure damping matrix; $\boldsymbol{K}_e^s$, element structure stiffness matrix; $\boldsymbol{F}_e^s$, element structure load vector; and $\boldsymbol{F}_e^{pr}$, element fluid pressure load vector at the interface Γ_0.

The fluid pressure load vector can be obtained by integrating pressure over the area on the interface surface:

$$\boldsymbol{F}_e^{pr} = \boldsymbol{R}_e \boldsymbol{p}_e. \tag{2.39}$$

Where, $\boldsymbol{p}_e$ is the element pressure vector, and, $\boldsymbol{R}_e$, is the transformation matrix, from $\boldsymbol{p}_e$ to $\boldsymbol{F}_e^{pr}$.

$$\boldsymbol{R}_e = \int_{\Gamma_0} \boldsymbol{N}_u \boldsymbol{n} \boldsymbol{N}_p{}^T \mathrm{d}\Gamma, \tag{2.40}$$

with the component,

$$[\boldsymbol{R}_e]_{ai,b} = \int_{\Gamma_0} \boldsymbol{N}_u^a \boldsymbol{n}_i \boldsymbol{N}_p{}^b \mathrm{d}\Gamma.$$

Where, $\boldsymbol{N}_u^a$, is the shape function for displacement at node a of structure element; $\boldsymbol{n}_i$, is unit normal on the coupling interface of the fluid element in i direction; $\boldsymbol{N}_p^b$, is the shape function for acoustic pressure at node b of acoustic fluid element; and $\boldsymbol{N}_p$, is the corresponding matrix of the shape function.

Substituting Equation (2.39) into Equation (2.38), as follows, we get,

$$\boldsymbol{M}_e^s \ddot{\boldsymbol{u}}_e + \boldsymbol{C}_e^s \dot{\boldsymbol{u}}_e + \boldsymbol{K}_e^s \boldsymbol{u}_e - \boldsymbol{R}_e \boldsymbol{p}_e = \boldsymbol{F}_e^s. \tag{2.41}$$

We rewrite the finite element equation for acoustic problem, Equation (1.527), here as reference,

$$\boldsymbol{M}_e^f \ddot{\boldsymbol{p}}_e + \boldsymbol{C}_e^f \dot{\boldsymbol{p}}_e + \boldsymbol{K}_e^f \boldsymbol{p}_e + \rho_0 \boldsymbol{R}_e^{\mathrm{T}} \ddot{\boldsymbol{u}}_e = 0.$$

Combining Equations (1.527) and (2.41), the acoustic-structure coupling Equation (2.42) in assembled matrix form is obtained.

$$\begin{bmatrix} \boldsymbol{M}^s & \boldsymbol{0} \\ \boldsymbol{M}^{fs} & \boldsymbol{M}^s \end{bmatrix} \begin{Bmatrix} \ddot{\boldsymbol{u}} \\ \ddot{\boldsymbol{p}} \end{Bmatrix} + \begin{bmatrix} \boldsymbol{C}^s & \boldsymbol{0} \\ \boldsymbol{0} & \boldsymbol{C}^f \end{bmatrix} \begin{Bmatrix} \dot{\boldsymbol{u}} \\ \dot{\boldsymbol{p}} \end{Bmatrix} + \begin{bmatrix} \boldsymbol{K}^s & \boldsymbol{K}^{sf} \\ \boldsymbol{0} & \boldsymbol{K}^f \end{bmatrix} \begin{Bmatrix} \boldsymbol{u} \\ \boldsymbol{p} \end{Bmatrix} = \begin{Bmatrix} \boldsymbol{F}^s \\ \boldsymbol{0} \end{Bmatrix}. \tag{2.42}$$

There are two coupling terms at the matrix coupling level namely, the coupling mass matrix,

$$\boldsymbol{M}^{fs} = \sum_e \boldsymbol{M}_e^{fs}, \boldsymbol{M}_e^{fs} = \rho_0 \boldsymbol{R}_e^{\mathrm{T}} \tag{2.43}$$

and the coupling stiffness matrix,

$$\boldsymbol{K}^{sf} = \sum_e \boldsymbol{K}_e^{sf}, \boldsymbol{K}_e^{sf} = -\boldsymbol{R}_e. \tag{2.44}$$

The coupling Equation (2.42) can be solved directly by the SC method with acoustic pressure and displacement updated simultaneously. This coupling equation can also be solved in a weak coupling manner, in which the acoustic pressure is transferred to the structure analysis while the displacement and acceleration vectors are transferred for the structure to the acoustic analysis Figure 2.10.

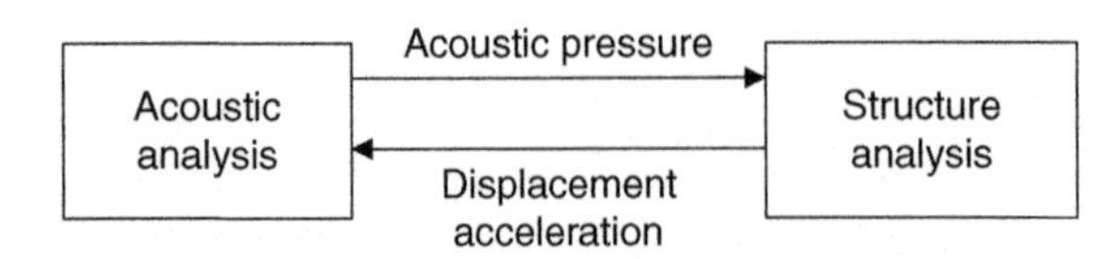

Figure 2.10 The data transfers between acoustic- and structure analysis.

2.10 Piezoelectric analysis

2.10.1 Electromechanical constitutive equations

The electromechanical constitutive equations for linear material behavior are (Allik and Hughes, 1970):

$$\sigma=[D]\varepsilon-[e]E, \tag{2.45}$$

$$D=[e]^T\varepsilon+[\varepsilon]E. \tag{2.46}$$

Where, $\boldsymbol{\sigma}$, is stress vector; $\boldsymbol{D}$, is electric displacement vector; $\boldsymbol{\varepsilon}$, is strain vector; $\boldsymbol{E}$, is electric field vector; $[\boldsymbol{D}]$, is elasticity matrix; $[\boldsymbol{e}]$, is piezoelectric stress matrix; and $[\boldsymbol{\varepsilon}]$, is dielectric matrix.

2.10.2 Formulations in matrix form

Finite element discretization for the displacement $\boldsymbol{u}$ and voltage V,

$$\boldsymbol{u}=\boldsymbol{N}_u{}^T\boldsymbol{u}^e, \tag{2.47}$$

$$V=\boldsymbol{N}_v{}^T\boldsymbol{V}^e. \tag{2.48}$$

Where, $\boldsymbol{N}_u$ and $\boldsymbol{N}_v$ are the shape functions for displacement vector and for voltage, respectively.

Then the strain vector, $\boldsymbol{\varepsilon}$, and electric field, $\boldsymbol{E}$, are related to the displacement and potentials, respectively, represented as follows,

$$\varepsilon=\boldsymbol{B}_u\,\boldsymbol{u}^e, \tag{2.49}$$

$$\boldsymbol{E}=\boldsymbol{B}_V\,\boldsymbol{V}^e. \tag{2.50}$$

Where, $\boldsymbol{B}_u$ is the linear strain matrix of elasticity (Bathe, 2006) and $\boldsymbol{B}_u=\nabla\boldsymbol{N}_u{}^T$. Also, $\boldsymbol{B}_V$ is the gradient of the shape function of voltage, $\boldsymbol{B}_V=\nabla\boldsymbol{N}_V{}^T$.

After the application of variational principle, we can obtain the coupled equation in assembled matrix form,

$$\begin{bmatrix} \boldsymbol{M}^s & \boldsymbol{0} \\ \boldsymbol{0} & \boldsymbol{0} \end{bmatrix}\begin{Bmatrix} \ddot{\boldsymbol{u}} \\ - \end{Bmatrix}+\begin{bmatrix} \boldsymbol{0} & \boldsymbol{0} \\ \boldsymbol{0} & \boldsymbol{0} \end{bmatrix}\begin{Bmatrix} \dot{\boldsymbol{u}} \\ - \end{Bmatrix}+\begin{bmatrix} \boldsymbol{K}^s & \boldsymbol{K}^z \\ \boldsymbol{K}^{zT} & \boldsymbol{C}^{VV}(\boldsymbol{u}) \end{bmatrix}\begin{Bmatrix} \boldsymbol{u} \\ \boldsymbol{V} \end{Bmatrix}=\begin{Bmatrix} \boldsymbol{F}^s \\ \boldsymbol{L}^v \end{Bmatrix}. \tag{2.51}$$

Here, piezoelectric coupling matrix $\boldsymbol{K}^z=\int_\Omega \boldsymbol{B}_u{}^T\boldsymbol{e}\,\boldsymbol{B}_V\,\mathrm{d}\Omega$

Charge vector is $\boldsymbol{L}^v = \boldsymbol{L}^v_{EX} + \boldsymbol{L}^v_C + \boldsymbol{L}^v_{SC}$, where, $\boldsymbol{K}^s$, is the stiffness matrix of the structure; $\boldsymbol{L}^v_{EX}$, is the applied nodal charge vector, $\boldsymbol{L}^v_C$, is the charge density load vector, $\boldsymbol{L}^v_{SC}$, is the surface charge density load vector, and $\boldsymbol{F}^s$, is the external force vector.

The natural boundary force terms in Equation (2.51) can be easily expressed as $\boldsymbol{F}_1 = -\boldsymbol{K}_{12}\boldsymbol{X}_2$ for both structure and piezoelectric fields, so the SC method is easy to implement. Integration domain change also occurs in piezoelectric field due to the structural deformation.

The weak coupling method is also applicable for solving the coupling Equation (2.51) if the coupling terms are moved into the right-hand side of the coupling equations.

2.11 Electrostatic–structure coupling

Combing the governing equations in matrix form for structure and electrostatic analysis, neglecting the coupling term $\boldsymbol{K}^z$, in the coupling Equation (2.52) can be obtained in assembled matrix form:

$$\begin{bmatrix} \boldsymbol{M}^s & \boldsymbol{0} \\ \boldsymbol{0} & \boldsymbol{0} \end{bmatrix} \begin{Bmatrix} \ddot{\boldsymbol{u}} \\ - \end{Bmatrix} + \begin{bmatrix} \boldsymbol{0} & \boldsymbol{0} \\ \boldsymbol{0} & \boldsymbol{0} \end{bmatrix} \begin{Bmatrix} \dot{\boldsymbol{u}} \\ - \end{Bmatrix} + \begin{bmatrix} \boldsymbol{K}^s & \boldsymbol{0} \\ \boldsymbol{0} & \boldsymbol{C}^{VV}(\boldsymbol{u}) \end{bmatrix} \begin{Bmatrix} \boldsymbol{u} \\ \boldsymbol{V} \end{Bmatrix} = \begin{Bmatrix} \boldsymbol{F}^s_{EX} + \boldsymbol{F}^s_{ES} \\ \boldsymbol{L}^v_{EX} + \boldsymbol{L}^v_C + \boldsymbol{L}^v_{SC} \end{Bmatrix}. \tag{2.52}$$

Where, $\boldsymbol{F}^S_{ES}$ is the electrostatic force vector on the structural surface Eq. (1.473) in Section 1.4.5). On the other hand, displacement from structure will affect the analysis domain of the electrostatic field, and this will induce changes in the electrostatic field as well as the electrostatic forces. The load transfers between electrostatic analysis and structure analysis are given in Figure 2.11. Since there is no off-diagonal term in the coupling matrix of Equation (2.52), the weak coupling method can be used directly to solve the electrostatic–structure coupling problem.

2.12 Magneto–structure coupling

As shown in Equations (1.160) and (1.415), there are no common primary variables for static magneto fields and structure fields.

The coupled equations in matrix form for a magneto–structural system is,

$$\begin{bmatrix} \boldsymbol{M}^s & \boldsymbol{0} \\ \boldsymbol{0} & \boldsymbol{0} \end{bmatrix} \begin{Bmatrix} \ddot{\boldsymbol{u}}^s \\ - \end{Bmatrix} + \begin{bmatrix} \boldsymbol{C}^s & \boldsymbol{0} \\ \boldsymbol{0} & - \end{bmatrix} \begin{Bmatrix} \dot{\boldsymbol{u}} \\ \dot{\boldsymbol{A}} \end{Bmatrix} + \begin{bmatrix} \boldsymbol{K}^s & \boldsymbol{0} \\ \boldsymbol{0} & \boldsymbol{K}^{AA}(\boldsymbol{u}) \end{bmatrix} \begin{Bmatrix} \boldsymbol{u} \\ \boldsymbol{A} \end{Bmatrix} = \begin{Bmatrix} \boldsymbol{F}^s_{EX} + \boldsymbol{F}^s_M \\ \boldsymbol{J} \end{Bmatrix}. \tag{2.53}$$

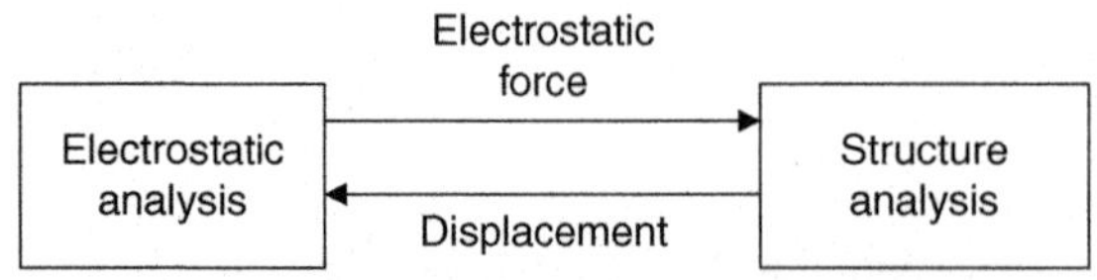

Figure 2.11 The data transfers between electrostatic and structural analysis.

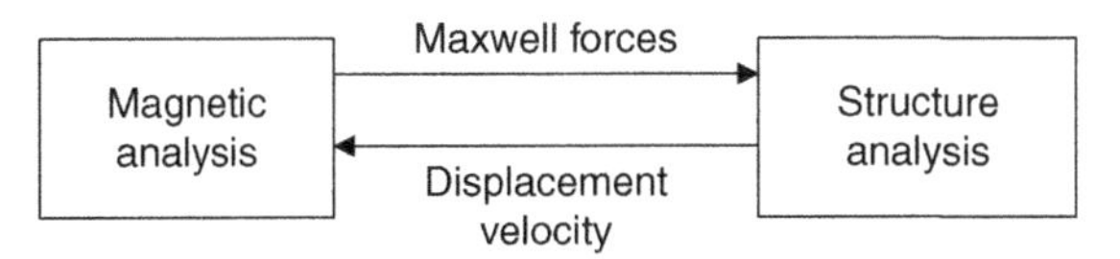

Figure 2.12 The data transfers between magnetic and structural analysis.

Here, $\boldsymbol{F}_M^s = \boldsymbol{F}^{mx}$, is the Maxwell force, as shown in Equation (1.473) derived from the magnetic field; and $\boldsymbol{F}_{EX}^s$, is external load vector for structure except for Maxwell force.

This is a natural boundary coupling to structure solution. The matrix coupling term $\boldsymbol{u}$ in $\boldsymbol{K}^M(\boldsymbol{u})$ stands for the integration domain changes. Since there is no off-diagonal term in the coupled matrix Equation (2.53), it is usually solved directly with the weak coupling method (Figure 2.12).

If the domain changes affect, which include the displacement and velocity, due to negligible structure motion, this bidirectional coupling will switch to a unidirectional coupling problem. As a result, no coupling iteration will be needed in the unidirectional coupling simulation.

$$\begin{bmatrix} \boldsymbol{M}^s & \boldsymbol{0} \\ \boldsymbol{0} & \boldsymbol{0} \end{bmatrix} \begin{Bmatrix} \ddot{\boldsymbol{u}} \\ - \end{Bmatrix} + \begin{bmatrix} \boldsymbol{C}^s & \boldsymbol{0} \\ \boldsymbol{0} & - \end{bmatrix} \begin{Bmatrix} \dot{\boldsymbol{u}} \\ \dot{\boldsymbol{A}} \end{Bmatrix} + \begin{bmatrix} \boldsymbol{K}^s & \boldsymbol{0} \\ \boldsymbol{0} & \boldsymbol{K}^{AA} \end{bmatrix} \begin{Bmatrix} \boldsymbol{u} \\ \boldsymbol{A} \end{Bmatrix} = \begin{Bmatrix} \boldsymbol{F}_{EX}^s + \boldsymbol{F}_M^s \\ \boldsymbol{J} \end{Bmatrix}. \quad (2.54)$$

2.13 Magneto–fluid coupling

The vector coupling term in a magneto-fluid coupled problem, is the Lorentz forces vector $\boldsymbol{F}^{EM}$. $\boldsymbol{F}^{EM}(\boldsymbol{V})$ represents the production-term coupling from primary variables in the flow field (Verardi et al., 1998).

Combining Equations (1.36) and (1.427), the coupled equation for magneto-fluid coupling problem:

$$\begin{bmatrix} \boldsymbol{C}^{AA} & \boldsymbol{K}^{AV} & \boldsymbol{0} & \boldsymbol{0} \\ \boldsymbol{K}^{VA} & \boldsymbol{K}^{VV} & \boldsymbol{0} & \boldsymbol{0} \\ \boldsymbol{0} & \boldsymbol{0} & \boldsymbol{M}^P & \boldsymbol{0} \\ \boldsymbol{0} & \boldsymbol{0} & \boldsymbol{0} & \boldsymbol{M}^V \end{bmatrix} \begin{Bmatrix} \dot{\boldsymbol{A}}^{EM} \\ \boldsymbol{V}^{EM} \\ \dot{\boldsymbol{P}}^f \\ \dot{\boldsymbol{V}}^f \end{Bmatrix} + \begin{bmatrix} \boldsymbol{K}^{AA} + \boldsymbol{K}^{AA}(\upsilon) & \boldsymbol{0} & \boldsymbol{0} & \boldsymbol{0} \\ \boldsymbol{K}^{VA}(\upsilon) & \boldsymbol{0} & \boldsymbol{0} & \boldsymbol{0} \\ \boldsymbol{0} & \boldsymbol{0} & \boldsymbol{\Lambda}^P & \boldsymbol{G}^{\mathrm{T}} \\ \boldsymbol{0} & \boldsymbol{0} & \boldsymbol{G} & \boldsymbol{K}_u + \boldsymbol{\Lambda} \end{bmatrix} \begin{Bmatrix} \boldsymbol{A}^{EM} \\ - \\ \boldsymbol{P}^f \\ \boldsymbol{V}^f \end{Bmatrix} = \begin{Bmatrix} \boldsymbol{J} \\ \boldsymbol{0} \\ \boldsymbol{0} \\ \boldsymbol{F}_{EX}^{\upsilon} + \boldsymbol{F}_{EM}^{\upsilon} \end{Bmatrix}. \quad (2.55)$$

Here, coupling source term from magneto field to the fluid flow is the magnetic force $\boldsymbol{F}_{EM}^{v} = \boldsymbol{F}^{jb} = \boldsymbol{J} \times \boldsymbol{B}$. The terms $\boldsymbol{C}^{AA}$, $\boldsymbol{K}^{AV}$, $\boldsymbol{K}^{VV}$, $\boldsymbol{K}^{AA}$, $\boldsymbol{K}^{AA}(\boldsymbol{v})$, and $\boldsymbol{K}^{VA}(\boldsymbol{v})$, have been illustrated in Section 1.4.3.4.

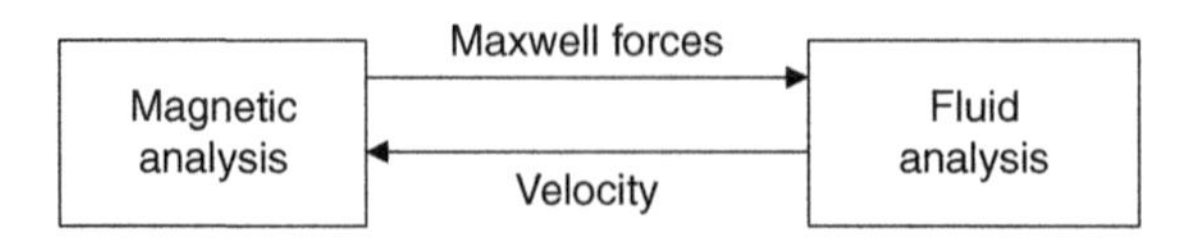

Figure 2.13 The data transfers between magnetic- and fluid analysis.

The coupling Equation (2.55) can be solved by weak coupling, and the volumetric data transfer between magneto and fluid models is demonstrated in Figure 2.13.

2.14 Electrothermal coupling

The coupled thermal electric constitutive equations (Landau et al., 1984) are

$$\boldsymbol{q}=\prod \boldsymbol{J}-\boldsymbol{K}\nabla T, \tag{2.56}$$

$$\boldsymbol{J}=\sigma(\boldsymbol{E}-\alpha\nabla T), \tag{2.57}$$

where, $\prod$, is Peltier coefficient matrix, $\prod = T\boldsymbol{\alpha}$; T, absolute temperature; and

α, is the Seebeck coefficient matrix, $\alpha = \begin{bmatrix} \alpha_{xx} & 0 & 0 \\ 0 & \alpha_{yy} & 0 \\ 0 & 0 & \alpha_{zz} \end{bmatrix}$.

Substituting $\left[\prod\right]$ with $T\boldsymbol{\alpha}$, to further demonstrate the coupling between these two equations,

$$\boldsymbol{q}=T\alpha\boldsymbol{J}-\boldsymbol{K}\nabla T, \tag{2.58}$$

$$\boldsymbol{J}=\sigma(\boldsymbol{E}-\alpha\nabla T). \tag{2.59}$$

Heat generation rate per unit volume is,

$$\boldsymbol{Q}=-\nabla\cdot\boldsymbol{q}+\boldsymbol{J}\cdot\boldsymbol{E}. \tag{2.60}$$

Substituting Equations (2.58) and (2.59) into Equation (2.60), we obtain

$$\boldsymbol{Q}=-\nabla\cdot(k\nabla T)+\boldsymbol{J}\cdot\sigma^{-1}\boldsymbol{J}-T\nabla\cdot(\alpha\boldsymbol{J}). \tag{2.61}$$

The first term on the right-hand side of Equation (2.61) is related to thermal conductivity and the second term is Joule heat. The third term represents the thermal electric effects and can be expanded further into Peltier, Thomson, and Bridgman heat terms (Ney, 1957). If the divergence $\nabla \cdot (\alpha \boldsymbol{J})$ in Equation (2.61) is associated with the temperature of the Seebeck coefficient matrix, then the third term in Equation (2.61) represents the Thomson heat:

$$Q^{Th} = \boldsymbol{J} \cdot \tau \nabla T. \tag{2.62}$$

Where, $\tau = -T\dfrac{d\alpha}{dT}$ is the Thomson heat tensor.

The thermal part gets heat generation $\boldsymbol{Q}^{EM}\,(q^B)$ from electromagnetic solution (natural boundary coupling). As returned, the material properties may be influenced by the temperature from thermal solution (constitutive law coupling).

$$\begin{bmatrix} \boldsymbol{C}^E(\boldsymbol{T}) & \boldsymbol{0} \\ \boldsymbol{0} & \boldsymbol{C}^t \end{bmatrix} \begin{Bmatrix} \overset{*}{\boldsymbol{V}} \\ \overset{*}{\boldsymbol{T}} \end{Bmatrix} + \begin{bmatrix} \boldsymbol{K}^E(\boldsymbol{T}) & \boldsymbol{K}^E_{Vt} \\ \boldsymbol{0} & \boldsymbol{K}^t \end{bmatrix} \begin{Bmatrix} \boldsymbol{V} \\ \boldsymbol{T} \end{Bmatrix} = \begin{Bmatrix} \boldsymbol{I} \\ \boldsymbol{Q} \end{Bmatrix} \tag{2.63}$$

where,

$$\boldsymbol{Q} = \boldsymbol{Q}^{EX} + \boldsymbol{Q}^g + \boldsymbol{Q}^c + \boldsymbol{Q}^j_E + \boldsymbol{Q}^P_E \tag{2.64}$$

$\boldsymbol{K}^E_{Vt}$, is the Seebeck coefficient coupling matrix.

$$\boldsymbol{K}^E_{Vt} = \int_\Omega \left(\nabla \boldsymbol{N}_V{}^T\right)^T \alpha \left(\nabla \boldsymbol{N}_T{}^T\right) \mathrm{d}\Omega, \tag{2.65}$$

where, $\boldsymbol{Q}^{EX}$, is the applied nodal heat flow rate vector; $\boldsymbol{Q}^c$, is the convection surface vector; $\boldsymbol{Q}^g$, is the heat generation rate vector for causes other than Joule heating; $\boldsymbol{Q}^j_E = \boldsymbol{J} \cdot \boldsymbol{E}$, is the heat generation rate for Joule heating; $\boldsymbol{Q}^P_E = -T\nabla \cdot (\alpha \boldsymbol{J})$, is the Peltier heat generation rate; and $\boldsymbol{I}$, is the nodal current vector.

It includes natural boundary condition coupling terms ($\boldsymbol{Q}^j_E, \boldsymbol{Q}^P_E$, and $-\boldsymbol{K}^E_{Vt}\boldsymbol{T}$) and constitutive equation coupling $\left(\boldsymbol{C}^E(\boldsymbol{T}), \boldsymbol{K}^E(\boldsymbol{T})\right)$, in Equation (2.63).

The electro-thermal coupling equations can be solved by the direct SC method. On the other hand, if one solves the electric field and thermal field separately and carries out the load transfers as shown in Figure 2.14, then the weak coupling is used for this system.

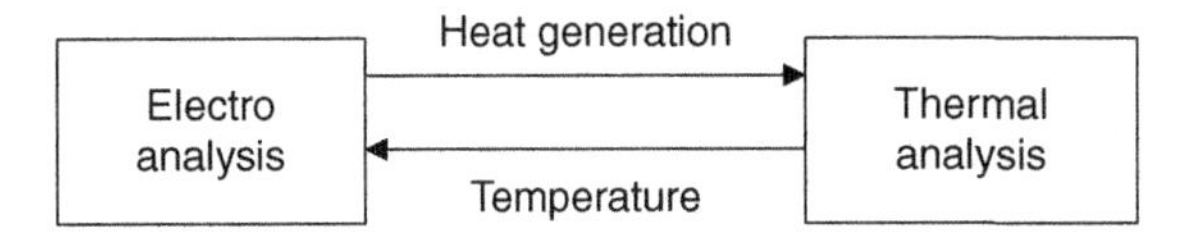

Figure 2.14 The data transfers between electro- and thermal analysis.

2.15 Magnetic–thermal coupling

The coupling matrix equation for magnetic-thermal problems is,

$$\begin{bmatrix} C^{AA}(T) & 0 \\ 0 & C^{t} \end{bmatrix} \begin{Bmatrix} \dot{A} \\ \dot{T} \end{Bmatrix} + \begin{bmatrix} K^{AA}(T) & 0 \\ 0 & K^{T} \end{bmatrix} \begin{Bmatrix} A \\ T \end{Bmatrix} = \begin{Bmatrix} J^{A} \\ Q \end{Bmatrix}. \tag{2.66}$$

Where, $\boldsymbol{C}^{AA}(\boldsymbol{T})$, is eddy current damping matrix, dependent on the current temperature; $\boldsymbol{K}^{AA}(\boldsymbol{T})$, is magnetic reluctivity matrix, dependent on the current temperature; $\boldsymbol{K}^{T}=\boldsymbol{K}^{tb}+\boldsymbol{K}^{tc}$, ($\boldsymbol{K}^{tb}$, thermal conductivity matrix of material, and $\boldsymbol{K}^{tc}$ is thermal conductivity matrix of the convection surface); $\boldsymbol{J}^{A}=\boldsymbol{J}^{nd}+\boldsymbol{J}^{s}+\boldsymbol{J}^{pm}$ ($\boldsymbol{J}^{nd}$, is applied nodal source current vector; $\boldsymbol{J}^{s}$, is source current vector; $\boldsymbol{J}^{pm}$, is coercive force (permanent magnet) vector); $\boldsymbol{Q}=\boldsymbol{Q}^{nd}+\boldsymbol{Q}^{g}+\boldsymbol{Q}^{j}+\boldsymbol{Q}^{c}$ ($\boldsymbol{Q}^{nd}$, is applied nodal heat flow rate vector; $\boldsymbol{Q}^{g}$, is heat generation rate vector for causes instead of Joule heating; $\boldsymbol{Q}^{j}$, is heat generation rate vector for Joule heating; and $\boldsymbol{Q}^{c}$, is convection surface vector).

$\boldsymbol{Q}^{j}$, is the Joule heat-coupling vector in Equation (2.66), and there may be temperature dependency of the magnetic field. Since there is no off-diagonal term in the coupled matrix Equation (2.66), it can be solved directly by the weak coupling way (Figure 2.15).

2.16 Summary of the coupling types

Type of couplings:

1. Fully coupling (essential boundary coupling, $\Gamma_c \neq 0$), $X_1 = X_2$ on Γ_c
2. Natural boundary coupling $F_1(X_2)$ or $F_1 = -K_{12}X_2$,
3. Variables production coupling $K_{11}(X_2)$,
4. Constitutive equation coupling $K_{11}(E(X_2))$,
5. Integration domain coupling, $K_{11}(X_2) = \int_{\Omega(X_2)} f \, d\Omega$

   Mathematically,
   a. Right-hand side load–vector coupling (1 and 2)
   b. Matrix coupling (1, 2, 3, and 4)
   c. Integration domain coupling (5)

   Coupling directions:

1. Unidirectional
2. Bidirectional

The coupling matrix in Table 2.1 demonstrates most of the possible interactions among structural, fluid, thermal, electromagnetic, and acoustic fields. The coupling

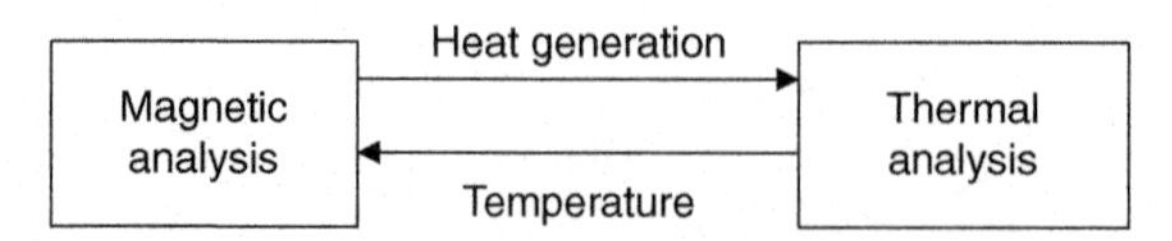

Figure 2.15 The data transfers between magnetic- and thermal analysis.

Table 2.1 The coupling matrices

Thermal	Type: natural boundary coupling (temperature, thermal stress) equation: $F_{th}^{S} = -K_{ut}^{S} \Delta T$, method: SC and WC	Type: constitutive equation coupling, method: W and SC	X	Type: natural boundary coupling, constitutive equation: $F_{th}^{s} = -K_{ut}^{s} \Delta T$ method: SC and WC	Type: constitutive equation (temperature) coupling method: WC	X	Type: constitutive equation (temperature) coupling method: WC	X
Domain coupling Displacement Natural boundary (heat generation term) $Q_{ted}^{T} = -C_{tu}^{T} \dot{u}$, method: SC and WC.	*Solid/structure*	Type: Fully coupled (i.e., displacement and velocity)	Type: domain coupling: displacement, method: WC	X	X	Type: domain coupling: displacement method: WC	X	Type: domain coupling: displacement
Nonlinear natural boundary coupling, method: SC.	Fully coupled (i.e., fluid forces), method: SC and WC.	*Fluid flow*	X	X	X	X	Type: production terms coupling $K^{AA}(v)$ method: WC	Velocity boundary

(Continued)

Table 2.1 The coupling matrices *(cont.)*

X	Natural boundary: electrostatic forces, method: WC.	X	*Electrostatic*	X	X	X	X	X
Domain coupling: displacement natural boundary (heat generation term), $Q^T_{\text{ted}} = -C^s_{tu}\dot{u}$, method: SC and WC.	X	X	X	*Piezoelectric*	X	X	X	X
Natural boundary (Joule heat, Peltier heat) (2.63), method: SC and WC	X	X	X	X	*Electric*	X	X	X

X	Natural boundary: Maxwell forces (2.53), method: WC.	X	X	X	X	*Magneto-static*	X	X
Natural boundary (Joule heat), (2.63), (2.66), method: WC.	X	Natural boundary: Lorentz forces (2.55), method: WC	X	X	X	X	*LF electro-magnetic*	X
X	Natural boundary: acoustic forces.	Acoustic pressure	X	X	X	X	X	*Acoustics*

matrix is asymmetric with cell (*i*, *j*) representing the responses of field *j* induced by field *i*. For instance, cell (2, 8) represents the domain change of electromagnetic field caused by structural motions, and cell (8, 2) stands for the structural response caused by Maxwell or magnetic forces. The diagonal terms, in a coupling sense, are related to domain decomposition for the same physics models. From Table 2.1, we know that all coupling could be bidirectional coupling. If the influences from one field to the other are negligible, a unidirectional coupling approach would be considered, and a lower (or upper) triangular in the coupling matrix will be applied. It is unnecessary to solve unidirectional coupling problems in a coupling manner. The degree of coupling in bidirectional coupling problems can be roughly classified into strongly coupled problems and weakly coupled problems. In the strongly coupled problems, the responses of one physics model tightly depend on the others and vice versa, and the coupled system is highly nonlinear. In the weakly coupled problems, the dependences among physics models are weak, and the nonlinearity of the coupled system is not strong. The details about the coupling method will be given in the next chapter.

The coupling methods

3

Chapter Outline

3.1 Introduction to coupling methods

Coupling methods can be classified into two major types, the strong coupling method (i.e., the direct coupling method), and the weak coupling method (i.e., the iterative coupling method). The strong coupling method solves coupled equations directly and updates all variables of the coupled system simultaneously. On the contrary, the weak coupling method solves each physics model separately, and the coupling conditions are satisfied by transferring data between different physics models. The strong coupling method is usually used for solving strongly coupled problems. As mentioned in the previous chapter, it is unnecessary to use the strong coupling method to solve unidirectional coupled problems. The strong coupling method is inevitable for some strongly coupled problems, and the modified weak coupling methods may be suitable for some levels of strongly coupled problems.

Q. Zhang & S. Cen: Multiphysics Modeling. http://dx.doi.org/10.1016/B978-0-12-407709-6.00003-1

3.2 The strong coupling method

In the strong coupling method, the coupled equations need to be assembled directly, and the coupled system is established through a strong coupling way, in which the coupled equations are solved implicitly. This means that the variable vectors of the coupled system are obtained by solving the linearized coupled equations, the variable vectors of the coupled system are updated simultaneously, and the interface coupling conditions are satisfied implicitly.

Key features of strong coupling method include

- Single, fully coupled monolithic equation set
- Solving coupled physics equations directly
- The coupling conditions are satisfied at the matrix equation level
- Updating all variables of the coupled system simultaneously

The typical simulation procedure for a transient strong coupling problem is shown in Figure 3.1. The coupling time step loops from initial time until the end of simulation. The time step size can vary during the simulation process. However, the time integration scheme and time step size control are unified for all physics models. The nonlinear iteration loop is used to achieve a convergence solution of the coupled system inside the time loop. The convergence indicator includes the residual of each physics model as well as the coupling variables on the interface. The coupled equations are solved at once inside the iteration, and the solution vectors are also updated simultaneously. For nonstructural

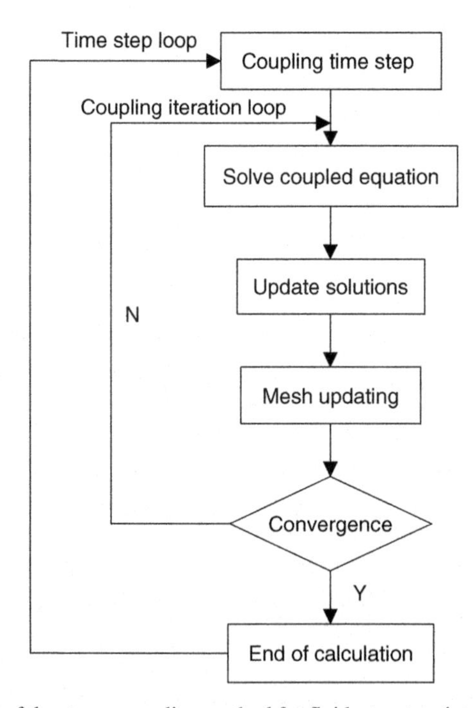

Figure 3.1 Flowchart of the strong coupling method for fluid–structure interaction (FSI) problems.

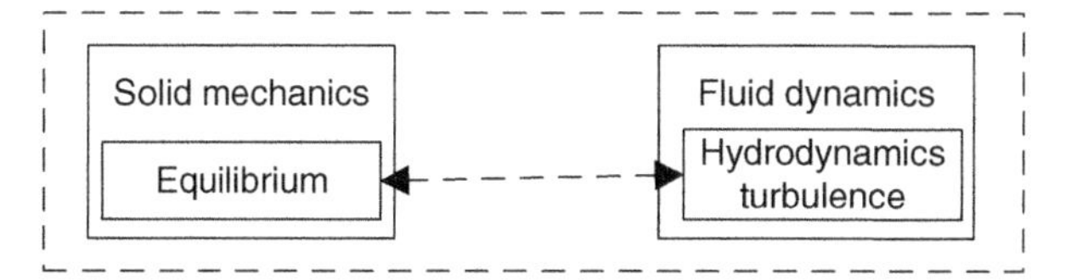

Figure 3.2 Strong coupling system of FSI problem.

physics models with moving boundary, the mesh movement (morphing or remeshing) needs to be calculated every coupling time step or every iteration loop. After the global convergence condition is achieved, simulation enters to next coupling time step.

Figure 3.2 demonstrates the strongly coupled equation system for a fluid–structure interaction (FSI) problem. The monolithic equation set can be obtained by appropriate interface coupling methods, for example, direct matrix assembly (DMA), direct interface coupling (DIC), multipoint constraint (MPC), Lagrange multiplier, or penalty method (PM). The details for these coupling methods will be given in a later section.

3.2.1 Direct matrix assembly strong coupling method

If the physics models involved in the coupling have no common degrees of freedom (DOFs), but the coupling phenomenal can be expressed as matrix form as shown in Equation (3.1); then the DMA method can be applied.

$$\begin{bmatrix} \boldsymbol{K}_{11}^{1}(\boldsymbol{X}_1,\boldsymbol{X}_2) & \boldsymbol{K}_{12}^{12}(\boldsymbol{X}_1,\boldsymbol{X}_2) \\ \boldsymbol{K}_{21}^{21}(\boldsymbol{X}_1,\boldsymbol{X}_2) & \boldsymbol{K}_{22}^{2}(\boldsymbol{X}_1,\boldsymbol{X}_2) \end{bmatrix} \begin{Bmatrix} \boldsymbol{X}_1 \\ \boldsymbol{X}_2 \end{Bmatrix} = \begin{Bmatrix} \boldsymbol{F}_1(\boldsymbol{X}_2) \\ \boldsymbol{F}_2(\boldsymbol{X}_1) \end{Bmatrix}. \tag{3.1}$$

The coupling can be volumetric or on the interaction surface, and the off-diagonal terms in Equation (3.1) can be obtained by volumetric or surface integration.

Types of the coupled problems belonging to DMA class (Chapter 2 for details)

- Thermal–stress (volumetric coupling: temperature with displacement)
- Thermal–fluid (volumetric coupling: temperature with velocity and pressure)
- Piezoelectric (volumetric coupling: voltage with displacements)
- Acoustics–structure (surface coupling: acoustics pressure with displacement)

Remark 3.1 Most of these matrix-coupling problems can also be solved with the weak coupling method by transferring loads across physics models (Section 3.3).

3.2.2 Direct interface coupling (DIC) method with co-related interface DOFs

If the mesh discretizations of the coupled physics models on the interfaces are identical, which means that their interface mesh share the same interface nodes and the underline elements, and the nodes and the underline elements have co-related DOFs, then the DIC can be used to satisfy the interface coupling conditions.

Key features of the DIC method include

- Unified interface meshing on coupling interface across physics models
- Co-related interface DOFs for coupling physics models
- All coupling conditions are satisfied by matrix or vector assembly process without introducing new interface mesh or coupling variables.

3.2.2.1 The DIC method for fluid–structure interaction (Zhang and Hisada, 2001)

In FSI analysis, if the fluid and structure domain share the common interface nodes, as shown in Figure 3.3, then DIC method can be applied. All coupling conditions are achieved simply by the matrix and vector assembly process for the interface mesh.

The corresponding DOFs for this coupling are the displacement vector on the solid side and the mesh displacement vector on the fluid domain. Lagrangian description is used for both the fluid and the solid on the interface. The variable vector of the coupled system can be divided into three parts:

$$\boldsymbol{\Phi}^{fs} = \begin{Bmatrix} \boldsymbol{\Phi}_i^f \\ \boldsymbol{\Phi}_c^{fs} \\ \boldsymbol{\Phi}_i^s \end{Bmatrix}, \ \boldsymbol{d}^s = \begin{Bmatrix} - \\ \boldsymbol{d}_c^{fs} \\ \boldsymbol{d}_i^s \end{Bmatrix}. \tag{3.2}$$

Where, $\boldsymbol{\Phi}_i^f$ ($\boldsymbol{\Phi}_i^f = \{\boldsymbol{P}^f, \boldsymbol{V}_i^f\}^T$) is the pressure and the internal velocity vector of fluid, $\boldsymbol{\Phi}_c^{fs}$ ($\boldsymbol{\Phi}_c^{fs} = \boldsymbol{V}_i^{fs}$) is the coupled velocity vector, $\boldsymbol{\Phi}_i^s$ ($\boldsymbol{\Phi}_i^s = \left\{\boldsymbol{V}_i^s, \overset{*}{\boldsymbol{P}^s}\right\}^T$) is the

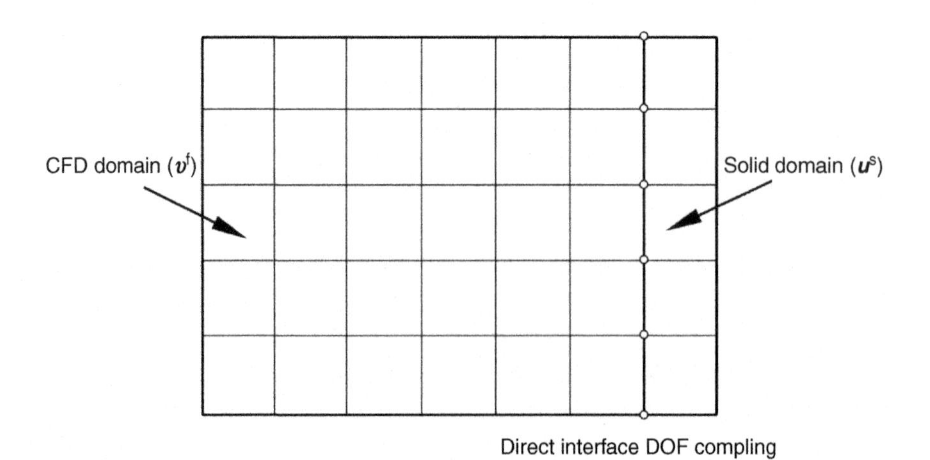

Figure 3.3 Direct interface DOF coupling.

internal velocity vector and the time derivatives of the pressure vector of the solid domain, $\boldsymbol{d}_c^{fs}$ ($\boldsymbol{U}_c^{fs}$) is the coupled displacement vector, and $\boldsymbol{d}_i^s$ ($\boldsymbol{d}_i^s = \left\{\boldsymbol{U}_i^s, \boldsymbol{P}^s\right\}^T$) is the internal displacement vector and pressure vector of solid. Based on the definition of the variable vector of the coupled system Equation (3.2), combining Equation (1.36) and Equation (1.160), the coupled equation can be expressed as follows:

$$\boldsymbol{Q}^{fs} = \boldsymbol{F}^{fs} \tag{3.3}$$

where the internal force vector $\boldsymbol{Q}^{fs}$ includes the inertial force of the coupled system,

$$\boldsymbol{Q}^{fs} = \boldsymbol{M}^{fs}\overset{*}{\boldsymbol{\Phi}}{}^{fs} + \boldsymbol{C}^f\boldsymbol{\Phi}^{fs} + \boldsymbol{Q}^s\left(\boldsymbol{d}^s\right). \tag{3.4}$$

$\boldsymbol{F}^{fs}$ is the external force vector. In Equation (3.4), * denotes the time derivative in ALE coordinate for fluid and time derivative in Lagrangian coordinate for interface and solid, respectively. $\boldsymbol{M}^{fs}$ is the mass matrix composed of those for the fluid and the solid, and $\boldsymbol{C}^f$ consists of the divergence, viscous, and convective terms of fluid. The details of these coupled matrices and vectors can be expressed as follows:

$$\boldsymbol{M}^{fs} = \begin{bmatrix} \boldsymbol{M}^P & 0 & 0 & 0 & 0 \\ 0 & \boldsymbol{M}_{ii}^f & \boldsymbol{M}_{ic}^f & 0 & 0 \\ 0 & \boldsymbol{M}_{ci}^f & \boldsymbol{M}_{cc}^f + \boldsymbol{M}_{cc}^s & \boldsymbol{M}_{ci}^s & 0 \\ 0 & 0 & \boldsymbol{M}_{ic}^s & \boldsymbol{M}_{ii}^s & 0 \\ 0 & 0 & 0 & 0 & 0 \end{bmatrix},$$
$$\boldsymbol{C}^f = \begin{bmatrix} \boldsymbol{\Lambda}^P & \boldsymbol{G}_i^T & \boldsymbol{G}_c^T & 0 & 0 \\ -\boldsymbol{G}_i & \boldsymbol{\Lambda}_{ii} + \boldsymbol{K}_{\mu ii} & \boldsymbol{\Lambda}_{ic} + \boldsymbol{K}_{\mu ic} & 0 & 0 \\ -\boldsymbol{G}_c & \boldsymbol{\Lambda}_{ci} + \boldsymbol{K}_{\mu ci} & \boldsymbol{\Lambda}_{cc} + \boldsymbol{K}_{\mu cc} & 0 & 0 \\ 0 & 0 & 0 & 0 & 0 \\ 0 & 0 & 0 & 0 & 0 \end{bmatrix}, \tag{3.5}$$

$$\boldsymbol{Q}^s = \begin{Bmatrix} - \\ - \\ \boldsymbol{Q}_c^s \\ \boldsymbol{Q}_i^s \\ \boldsymbol{Q}_P^s \end{Bmatrix}, \quad \boldsymbol{F}^s = \begin{Bmatrix} 0 \\ \boldsymbol{F}_i^f \\ \boldsymbol{F}_c^{fs} \\ \boldsymbol{F}_i^s \\ 0 \end{Bmatrix}. \tag{3.6}$$

In the previous equations, the subscripts c and i denote the coupling parts and the internal parts of variables, respectively. The incremental form of the coupled equation (Equation (3.4)) can be obtained by taking linearization.

$$ {}^{t}\boldsymbol{M}^{fs}\Delta\overset{*}{\Phi}{}^{fs} + {}^{t}\boldsymbol{C}^{f}\Delta\Phi^{fs} + {}^{t}\boldsymbol{K}^{s}\Delta\boldsymbol{d}^{s} = {}^{t+\Delta t}\boldsymbol{F}^{fs} - {}^{t}\boldsymbol{Q}^{fs}. \tag{3.7} $$

Here, ${}^{t}\boldsymbol{K}^{s}$ is the tangent stiffness matrix and the coefficient matrices and ${}^{t}\boldsymbol{M}^{fs}$ and ${}^{t}\boldsymbol{C}^{f}$ are secant matrices. The linearization of ${}^{t}\boldsymbol{M}^{fs}$ and ${}^{t}\boldsymbol{C}^{f}$ has been discussed by Zhang (1999). In practice, if an effective time step controlling algorithm (Zhang and Hisada, 2001) is used to control the time step Δt, the secant matrices of ${}^{t}\boldsymbol{M}^{fs}$ and ${}^{t}\boldsymbol{C}^{f}$ frequently work quite well. The algebraic Equation (3.8) can be obtained by using implicit time integration method (Newmark-β) for the linearized coupled Equation (3.7):

$$ \left[{}^{t}\boldsymbol{M}^{fs} + \gamma\Delta t\,{}^{t}\boldsymbol{C}^{f} + \beta\Delta t^{2}\,{}^{t}\boldsymbol{K}^{s} \right]\Delta\overset{*}{\boldsymbol{\Phi}}{}^{fs} = {}^{t+\Delta t}\boldsymbol{F} - {}^{t}\boldsymbol{Q}^{fs}. \tag{3.8} $$

where γ and β are the parameters that can be chosen so as to obtain integral accuracy and stability and $\Delta\overset{*}{\boldsymbol{\Phi}}{}^{fs}$ is the increment of variable vector of the coupled system. Furthermore, Equation (3.8) can be expressed as the simplest form Equation (3.9).

$$ \Lambda^{fs}\Delta\overset{*}{\Phi}{}^{fs} = \boldsymbol{R}^{fs}. \tag{3.9} $$

where $\boldsymbol{\Lambda}^{fs}$ is the effective mass matrix of the coupled system and $\boldsymbol{R}^{fs}$ is the residual vector of the coupled system.

3.2.3 Multipoint constraint equation-based strong coupling method

When the different physics models occupy the same interface geometry (e.g., vertex, edge, face, solid) are discretized by unmatching interface meshes and the nodes and the underline elements have co-related DOFs, the constraint equation-based coupling method can be applied, as shown in Figure 3.4. By using mapping technology, unmatching mesh discretization is allowed to across the coupling interface.

Considering the fluid–structure coupling problem, if the fluid mesh and structure mesh on the coupling interface are incompatible, appropriate searching tools need to be applied to find correlations between the fluid nodes and the structure nodes. The fluid nodes on the interface are treated as slave nodes. In contrast, the structure nodes on the interface are treated as master nodes. The mesh displacement, velocity, and acceleration of the fluid nodes on the interface are forced equal to the corresponding value of the structure. The constraint equations and the fluid force distributions on

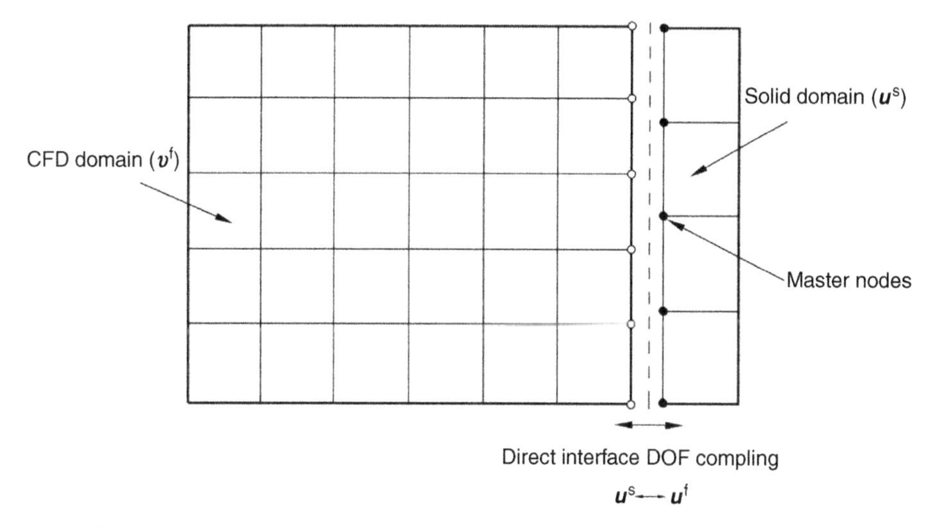

Figure 3.4 Schematic of a mapping technology.

the structure are based on the mapping results. The constraints for velocity vector are given by Equation (3.10).

$$\mathbf{v}^f = \sum_{i=1}^{N} w_i \mathbf{v}_i^s. \tag{3.10}$$

Here, $\mathbf{v}^f$ is the velocity vector of a fluid node (slave node) on the interface, N is the number of related master/structure nodes, w_i is the weight factor on structure at node i, and $\mathbf{v}_i^s$ is the velocity vector of structure at the same master node.

Using the MPC method, one can get similar coupled equations as in the DIC method in Equation (3.3).

Key features of the MPC method include

- Incompatible interface meshes are allowed for different physics models.
- The coupling conditions can be written in an explicit form, that is,

$$\mathbf{v}^f = \mathbf{v}^s \text{ on } \Gamma_c. \tag{3.11}$$

- The interface variables of the master nodes are the only factors that need to be solved in the coupling equations. The interface variables of the slave mesh are calculated directly by the constraint equations.
- Interface meshes and additional interface variables are not needed.
- The sparse property of the coupling matrix on the interface variables may get lost.

Remark 3.2 Typical coupling problems for MPC based on strong coupling include fluid–structure interaction, conjugate heat transfer (thermal–thermal coupling) problems, and so on.

3.2.4 Lagrange multiplier (LM)-based strong coupling method

If the constraint conditions on the coupling interface cannot be expressed explicitly as

$$\boldsymbol{u}_{\mathrm{A}} = \boldsymbol{u}_{\mathrm{B}} \qquad \text{on } \Gamma_c, \tag{3.12}$$

but an implicit function:

$$\boldsymbol{f}(\boldsymbol{u}_{\mathrm{A}}, \boldsymbol{u}_{\mathrm{B}}) = 0 \quad \text{on } \Gamma_c. \tag{3.13}$$

The Lagrange multiplier method is suitable for imposing this type of constraint conditions on the coupling interface.

The Lagrange multiplier $\boldsymbol{\lambda}$ is introduced in this method. The potential formulation (or equivalent formulation for other physics type) of the constraint conditions can be expressed as:

$$W_\lambda = \int_{\Gamma_c} \left(\boldsymbol{f}(\boldsymbol{u}_{\mathrm{A}}, \boldsymbol{u}_{\mathrm{B}}) \cdot \lambda \right) d\Gamma. \tag{3.14}$$

We normally add this formulation to the original potential equation (or equivalent equation for other physics types),

$$W_{\mathrm{L}} = W + W_\lambda, \tag{3.15}$$

and take the variation for $\boldsymbol{u}_{\mathrm{A}}$, $\boldsymbol{u}_{\mathrm{B}}$ and $\boldsymbol{\lambda}$,

$$\delta W + \int_{\Gamma_c} \left(\delta\lambda \cdot \boldsymbol{f}(\boldsymbol{u}_{\mathrm{A}}, \boldsymbol{u}_{\mathrm{B}}) \right) d\Gamma + \int_{\Gamma_c} \left(\left(\frac{\partial \boldsymbol{f}(\boldsymbol{u}_{\mathrm{A}}, \boldsymbol{u}_{\mathrm{B}})}{\partial \boldsymbol{u}_{\mathrm{A}}} \delta \boldsymbol{u}_{\mathrm{A}} + \frac{\partial \boldsymbol{f}(\boldsymbol{u}_{\mathrm{A}}, \boldsymbol{u}_{\mathrm{B}})}{\partial \boldsymbol{u}_{\mathrm{B}}} \delta \boldsymbol{u}_{\mathrm{B}} \right) \cdot \lambda \right) d\Gamma = 0. \tag{3.16}$$

Here, W is the original form of the potential equation.

Key features of LM method include

- Incompatible interface meshes are allowed for different physics models.
- The coupling conditions can be written in either explicit or implicit form, that is,

$$\boldsymbol{f}(\boldsymbol{u}_{\mathrm{A}}, \boldsymbol{u}_{\mathrm{B}}) = 0 \quad \text{on } \Gamma_c.$$

- Additional interface variables for the Lagrange multiplier are needed, and the interface meshes for the interpolation of Lagrange multiplier are needed.
- The interface condition can be satisfied exactly.
- The sparse property of the coupling matrix on the interface variables may be lost.

Remark 3.3 The LM-based strong coupling can be used for electromagnetic coupling interface (section 1.4.6), structural contact (Belytschko et al., 2013), and so on.

3.2.5 Penalty method-based strong coupling method

There is another way to do the strong interface coupling by using penalty method. For a potential-based problem, the control equation is listed as:

$$\min\left(W + \int_{\Gamma_c} \left(\frac{1}{2}\beta\, \boldsymbol{f}(\boldsymbol{u}_A, \boldsymbol{u}_B) \cdot \boldsymbol{f}(\boldsymbol{u}_A, \boldsymbol{u}_B) \right) d\Gamma \right), \tag{3.17}$$

where β is the penalty factor, $\boldsymbol{f}(\boldsymbol{u}_A, \boldsymbol{u}_B)$ is the constraint equation, and W is the original potential equation. The variation form of Equation (3.17) is expressed as:

$$\delta W + \delta \int_{\Gamma_c} \left(\frac{1}{2}\beta\, \boldsymbol{f}(\boldsymbol{u}_A, \boldsymbol{u}_B)\, \boldsymbol{f}(\boldsymbol{u}_A, \boldsymbol{u}_B) \right) d\Gamma = 0, \tag{3.18}$$

$$\delta W + \int_{\Gamma_c} \left(\beta \left(\frac{\partial \boldsymbol{f}(\boldsymbol{u}_A, \boldsymbol{u}_B)}{\partial \boldsymbol{u}_A} \delta \boldsymbol{u}_A + \frac{\partial \boldsymbol{f}(\boldsymbol{u}_A, \boldsymbol{u}_B)}{\partial \boldsymbol{u}_B} \delta \boldsymbol{u}_B \right) \boldsymbol{f}(\boldsymbol{u}_A, \boldsymbol{u}_B) \right) d\Gamma = 0, \tag{3.19}$$

Key features of the PM include

- Incompatible interface meshes are allowed for different physics models.
- The coupling conditions can be written in either explicit or implicit form, that is, $\boldsymbol{f}(\boldsymbol{u}_A, \boldsymbol{u}_B) = 0 \quad \text{on } \Gamma_c$.
- Neither additional interface variables nor interface meshes are needed.
- Depending on the value of factor β, the interface conditions are satisfied approximately, and the quality of the coupling matrix is also impacted by the value of β.
- The sparse property of the coupling matrix on the interface variables may get lost.

Remark 3.4 The PM-based strong coupling is usually for mechanical contact (Section 1.3.5) and thermal contact problems (Section 1.1.8).

3.3 Weak coupling methods

3.3.1 Features and definition of the weak coupling method

The weak coupling method solves each physics model separately, and the coupling conditions are satisfied by transferring data across different physics models.

Features of the weak coupling method include

- Multiple segregated equation sets are assembled and solved iteratively.
- Physics models are coupled by passing loads across coupling interfaces.

- Each physics model is solved in parallel or in serial.
- There is flexibility on solution methods that is optimal for each physics type.

In this section, the following four questions will be answered;

1. When do we need weak coupling?
2. How do we make the weak coupling method robust and efficient?
3. How many ways are there to do weak coupling?
4. What are the key techniques of the weak coupling method?

3.3.2 Solution process and controls for the weak coupling method

3.3.2.1 Solution process

The flowchart in Figure 3.5 shows the procedure of doing the weak coupling method for microelectro-mechanical systems (MEMS) device with Gauss–Seidel iteration. The coupling time loops control the overall time marching and the coupling time step size. Inside the coupling time control is the nonlinear coupling stagger loop. The convergence of the interface loads transfers as well as the convergence of each physics model need to be achieved in the coupling stagger loop.

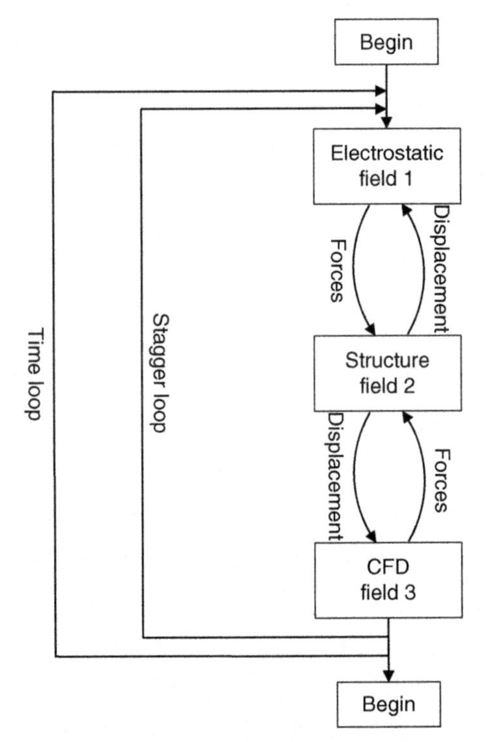

Figure 3.5 Flowchart of weak coupling method by Gauss–Seidel iteration for MEMS device.

One of the major advantages of the weak coupling method over the strong coupling method is the flexibility in time integration method and the usage of optimal time step size for different physics solvers. Autostepping and subcycling (using a smaller time step in the physics field than the coupling time step size) can be used for the physics solver, although the load transfers must happen at each coupling time step, and special treatment may be needed to achieve smoothness and energy conservation in the time domain. The flexibility and efficiency in the application of time integration can be maintained for each physics model in the weak coupling method. It is unnecessary for each physics model to reach a complete converged state during the coupling stagger loop except the very last one.

3.3.2.2 Convergence controls

We check the interface loads convergence as well as the convergence of each physics solver during coupling implicit stagger loop.

The evaluation of interface convergence:

$$\varepsilon^* = \frac{\log(\varepsilon/\text{TOLER})}{\log(10.0/\text{TOLER})}. \tag{3.20}$$

Where ε is the normalized change of the interface loads.

$$\varepsilon = \| \mathbf{\Phi}_{\text{new}} - \mathbf{\Phi}_{\text{pre}} \| / \| \mathbf{\Phi}_{\text{new}} \|,$$

$\mathbf{\Phi}_{\text{new}}$, the current interface load vector; $\mathbf{\Phi}_{\text{pre}}$, the interface load vector at previous coupling iteration; $\| \ \|$, L2 norm of a vector; and TOLER, user input tolerance.

Convergence occurs when $\varepsilon^* \leq 0$, that is, $\varepsilon < \text{TOLER}$.

3.3.2.3 Load transfers among physics models

The flowchart of the MEMS device with the weak coupling method (Figure 3.5) demonstrates the data transfers among physics models at which the electrostatic model is solved first and the resulting electrostatic forces are transferred to the structure model and receive the displacement across the same interface. After receiving the electrostatic forces and the fluid forces, the mechanical behavior of the structure model is solved, and it transfers the displacement to fluid dynamics as mesh displacement. The fluid dynamics model with moving boundary from the fluid–structure interface will be solved at last, and finally by repeating the coupling iteration, all of the interface continuity equations and dynamic equilibrium conditions are satisfied when appropriate convergence criteria are met.

3.3.3 Ways to do weak coupling

There are two ways to do weak coupling. The first is single-code coupling, and the second is multiple-code coupling through the code coupling interface. For single-code coupling, the physics models are implemented in a single code. Since all physics models run sequentially, the Gauss–Seidel iteration must be used. For intersolver coupling, each physics model can be implemented in a different solver code, and the coupling process and data transfer can be controlled through a third-party code coupling interface. There is publicly commercial software available for doing intersolver coupling: MpCCI (Fraunhofer, 2007). Both Gauss–Seidel iteration and Jacobi iteration are available for intersolver coupling, and details about intersolver coupling are given in Section 3.3.5.

3.3.4 The key techniques in weak coupling method

3.3.4.1 The mapping algorithm

The mapping algorithm is used to find the corresponding node or element set on the source side for a given node or element on the target. The mapping algorithm is used for unmatching the strong coupling interface, the load transfer process in the weak coupling problem, and the remeshing procedure for large domain changes of nonstructural physics models. In this section, point cloud mapping, node-to-element mapping, and control-surface mapping (CSM) are discussed.

3.3.4.1.1 Point cloud mapping

For point cloud mapping, the user only needs to provide the coordinates of the points on the source side and the coordinates of the points on the target side separately. The mapping algorithm is done among those points.

One way of doing point cloud mapping is to find the closest point on the source side for the target point in question, as shown in Figure 3.6, and set up the nodal weight factor to 1. Another way to do point cloud mapping is to create tetrahedral elements from the points on the source side in the three-dimensional (3D) case (triangle element in two-dimensional (2D) case, as shown in Figure 3.7), and then do the point to element mapping to find the right element and the nodal weights. The first option lacks accuracy but uses less computational time compared with the second way.

Key features of point cloud mapping include

- Only nodal data are needed on both target mesh and source mesh.
- The elements may be created for more accurate mapping.
- Global search or Oct tree search method can be used.
- Linear interpolation is restricted.

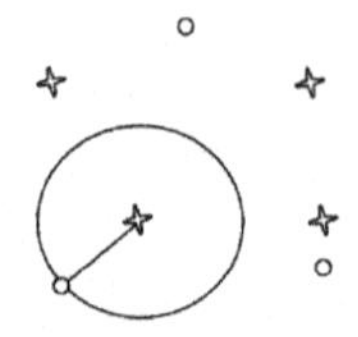

Figure 3.6 Point cloud mapping that finds the closest point on the target side.

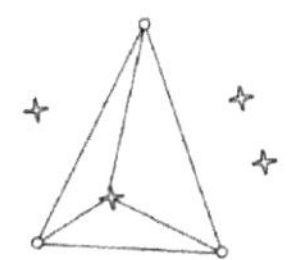

Figure 3.7 Point cloud mapping that creates triangle elements in 2D case.

3.3.4.1.2 Point element mapping by the bucket-search method

For point element mapping, both the nodal and element information on the source side are needed, although only the nodal coordinates information on target side are required. The point-to-element mapping is carried out for each node in question on the target side to find the right element on the source mesh. The local coordinates for the target point in the mapped element and the nodal weights for each node of the mapped element will be calculated. The bucket search method is a useful searching algorithm for point element mapping, at which the global search method will be applied for the node in equation, but the candidate elements are restricted to the selected bucket (Figure 3.9).

Key features of point element mapping include

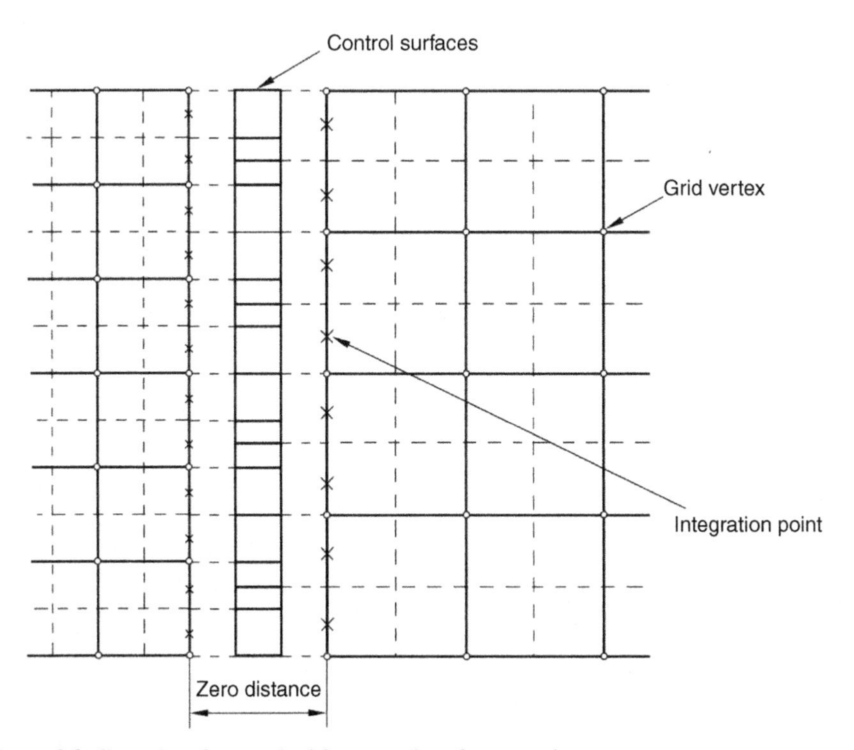

Figure 3.8 Control surface method for control-surface mapping.

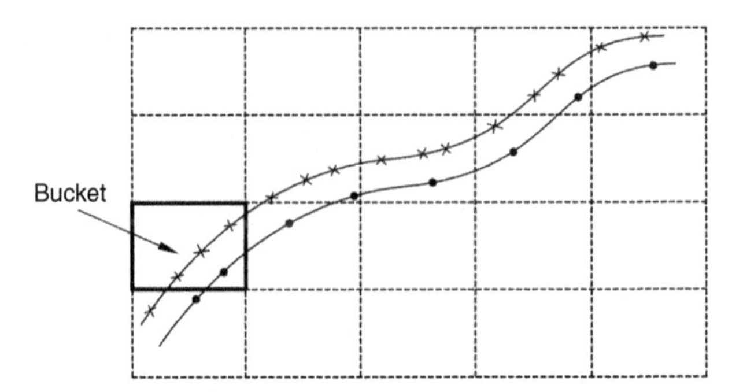

Figure 3.9 Bucket mapping-based surface searching approach.

- Nodal data are needed on the target side and element data on the source side.
- The bucket search method is efficient for point element mapping.
- The applicable interpolation scheme for point-element can be conservative or profile preserving (Section 3.3.4.3).

3.3.4.1.3 Control surface mapping

In the control surface method, meshes with node and element information are needed for both the target and source side. The key characteristic for the CSM is finding the intersection between the target elements and source elements, and this mapping algorithm can be profile preserving as well as global conservative. The intersection virtual mesh is created from interface meshes of the target side and the source side. The CS method is more expensive for memory usage and is computationally time consuming but can maintain the best accuracy for the data interpolation (i.e., can be both profile preserving and global conservative).

Key features of CS method are:

- Need element data on both source mesh and target mesh
- Need to create the intersection mesh from meshes on source side and target side
- Can be both global conservative and profile preserving
- This mapping option is for both nonconservative data (e.g., temperature, displacement) and conservative data (e.g., force and heat flow)
- It is the most computationally expensive compared with other methods.

3.3.4.2 The searching method

3.3.4.2.1 Global search method

The global search method is the simplest searching algorithm but also the most computationally expensive one. It looks through all the nodes or elements on the source side for a node or element in question on the target side.

Key features of the global search method include

- Simple to implement and reliable
- Most expensive one that has a complexity of $O(n \times m)$ where n is the number of nodes or elements to be mapped on the target side and m is total number of nodes or elements on the source side

3.3.4.2.2 Bucket search method

For the bucket search method (Jansen et al., 1992), all nodes or elements on the source side are distributed into buckets (Figure 3.9). The node or element in question on the target side is then located in a bucket. The global search method is applied for the node or element in equation, but the candidate elements or nodes are restricted to the selected bucket.

Key features of the bucket search method include

- For a given node or element, the bucket search method restricts the elements or nodes over which it loops.
- Efficiently compared with the global search method, the bucket search method has a complexity of $O(n \times m')$, where n is the number of nodes or elements to be mapped onto m elements and m' is the number of nodes or elements in each bucket on the source side, and it is usually much smaller than total number m.

3.3.4.2.3 Oct Tree search method

The Oct Tree search method can be used for CSM (Figure 3.10). It is developed on the basis of the well-known Oct Tree data structure, a means of storing spatial occupancy information that is widely used in particulate simulation. It subdivides 3D space into different regions according to different levels. Lower regions are generated from upper regions. Nodes or elements are distributed into corresponding region. For a center node or element, only check targets in its neighborhood.

Key features of Oct Tree search method include

- Easily implemented by recursive algorithm and provide great memory efficiency.
- Has a complexity of $O(n \times \log(m))$ for n nodes or elements mapped onto m elements.

3.3.4.2.4 Mapping diagnostics

The mapping diagnostics process finds and handles the imperfect matching geometric and nonoverlap regions in coupling analysis.

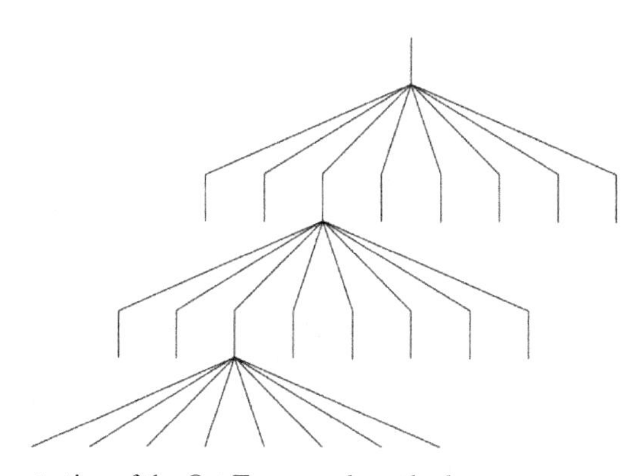

Figure 3.10 Demonstration of the Oct Tree search method.

The normal distance for surface mapping

For surface coupling, the normal distance from the surface mesh on the source to the node in question is one factor to decide whether the node in question is appropriately mapped to this candidate face or not.

If the normal distance exceeds the allowed normal tolerance (Figure 3.11), then consider this node as a mismatched surface node.

Remark 3.5 If there are more than one candidate facets found on the source mesh, then the one with closest distance from the point in question is chosen.

Misaligned surface nodes

For the misaligned node (e.g., the node in question is unable to be projected onto the target mesh for surface mapping; Figure 3.12), if the offset value exceeds the specified tolerance, then we consider this node as a misaligned surface node.

Improperly mapped nodes for volumetric mapping

For volumetric mapping either in 2D or 3D, if the source node in question is out of the range of the target mesh or point cloud with the distance from the region exceeding the specified tolerance, we call them improperly mapped nodes (Figure 3.13).

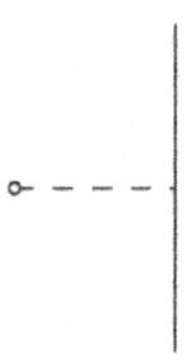

Figure 3.11 The normal distance for the surface mapping exceeds the normal tolerance.

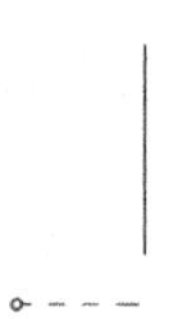

Figure 3.12 The misaligned surface nodes.

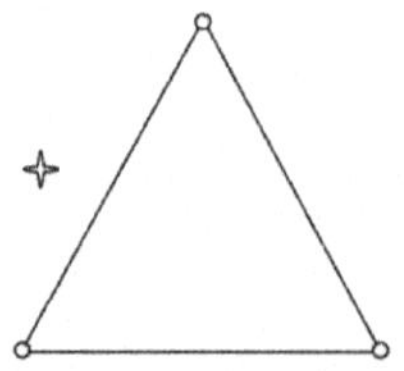

Figure 3.13 Improperly mapped nodes for volumetric mapping.

If the node or element in question on the target side does not pass the mapping diagnostic check, then we consider it as nonoverlap.

3.3.4.3 Interpolation methods

After the mapping is done, the interpolation procedure is carried out for the load transfer in the weak coupling method or to get the new mesh solution in the remeshing process.

3.3.4.3.1 Linear interpolation (element shape function based)

In linear interpolation, the value on the target node or element is calculated by a weighted sum from the contributors on the source mesh.

$$u^{\mathrm{T}} = \sum_{i=1}^{N} w_i u_i^{\mathrm{S}}, \tag{3.21}$$

where u^{T} is the value at node (or element) on the target mesh, w_i is the weight factor at node (or element) i on the source mesh that is in the range of [0, 1], u_i^{S} is the value at the same node (or element), and N is the number of related nodes (or elements) on the source side.

3.3.4.3.2 Control–surface-based interpolation

Compared with linear interpolation, control–surface-based interpolation is more robust and accurate. The control–surface mesh is created during the mapping and interpolation process, and it has the resolution of both meshes on the source side and target side. The intersection information between the target mesh and source mesh is used for the data interpolation in this method, and this makes the method both global conservative and profile preserving.

3.3.4.3.3 Least square-based interpolation

For a global consideration, the calculation of the value on the target side can also be done in the least square method. In this case, mapping is not required; instead, coupling interface integration and solution of algebraic equations are needed.

$$\min\left(\int_{\Gamma_c} \left(u^{\mathrm{T}} - u^{\mathrm{S}}\right)^2 d\Gamma\right). \tag{3.22}$$

This can be done by integrating on the coupling interface of the target mesh. The vector of u^T on the target side can be obtained by solving the resulting linear equations that are derived from Equation (3.22).

3.3.4.3.4 Profile-preserving interpolation for nonconservative data

The profile-preserving interpolation is for nonconservative data (e.g., displacement, temperature, and density):

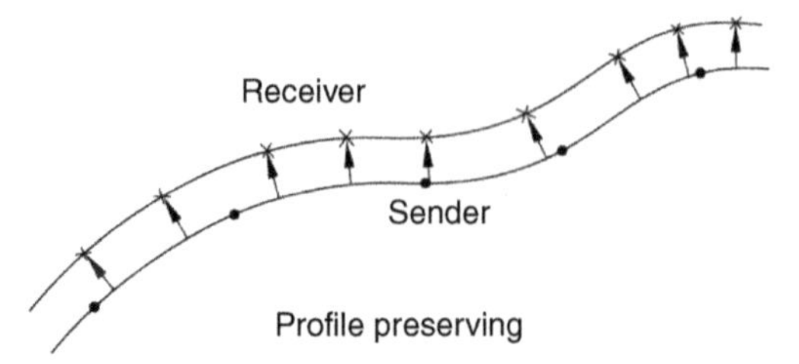

Figure 3.14 Profile-preserving interpolation.

$$u^{\mathrm{T}} = u^{\mathrm{S}} \quad \text{on} \quad \Gamma_c. \tag{3.23}$$

If unmatching mesh is used across the coupling interface and the distribution of the u^{S} is far from uniform (Figure 3.14), then it is very likely,

$$\int_{\Gamma_c^{\mathrm{T}}} u^{\mathrm{T}} dS \neq \int_{\Gamma_c^{\mathrm{S}}} u^{\mathrm{S}} dS. \tag{3.24}$$

This means that the conservation cannot be sustained without a global scaling.

The easiest way to do profile-preserving interpolation in the load transfer is to consider the mesh that receives data as the target mesh and map the node into the sender mesh. Bucket mapping can be used here. Then the linear interpolation can be used to get the value on the receiver nodes (Figure 3.14). This load interpolation method has better accuracy for the case in which the receiver mesh is finer than the sender mesh.

Remark 3.6 Control–surface-based mapping is also a profile-preserving method.

3.3.4.3.5 Global conservative data transfer for conservative data

The global conservative interpolation is for conservative data (e.g., forces, heat flow, heat generation). The data can be distributed density or concentrated nodal value, either surface load or volumetric load. The conservation means that the integrated total load on target mesh is equal to those on the source side, expressed by Equation (3.25),

$$\int_{\Gamma_c^{\mathrm{T}}} f^{\mathrm{T}} d\Gamma = \int_{\Gamma_c^{\mathrm{S}}} f^{\mathrm{S}} d\Gamma, \tag{3.25}$$

or the sum of nodal load on the target side equals to that on the source side, shown in Equation (3.26).

$$\sum_{i=1}^{N^{\mathrm{T}}} F_i^{\mathrm{T}} = \sum_{i=1}^{N^{\mathrm{S}}} F_i^{\mathrm{S}}. \tag{3.26}$$

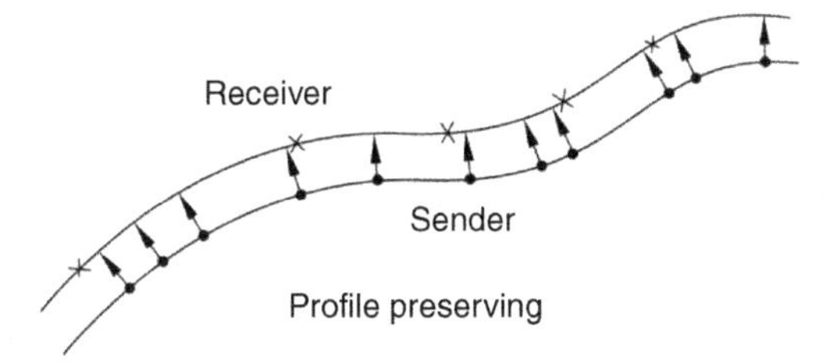

Figure 3.15 Global preservation data transfer.

Where, N^{T} is the number of nodes on the target mesh, N^{S} is the number of nodes on the source side, F_i^T is the concentrated nodal force on the target side at node i, and F_i^S is the concentrated nodal force on the source side at node i.

Ways to do global conservative data transfer in weak coupling:

If the load density is available, the profile-preserving data interpolation will be carried out. If the total value on the receiver side is not equal to the value on the sender side (Equation (3.27)), then a scaling will be applied to achieve global conservation (Equation (3.28)),

$$\text{if} \sum_{i=1}^{N^{\text{Receiver}}} F_i^{\text{Receiver}} \neq \sum_{i=1}^{N^{\text{Sender}}} F_i^{\text{Sender}}, \tag{3.27}$$

$$F_i^{\text{Receiver}} = f_{\text{scale}} * F_i^{\text{Receiver}}, \tag{3.28}$$

where the scaling factor f_{scale}:

$$f_{\text{scale}} = \frac{\sum_{i=1}^{N^{\text{Sender}}} F_i^{\text{Sender}}}{\sum_{i=1}^{N^{\text{Receiver}}} F_i^{\text{Receiver}}} \tag{3.29}$$

Note: If the $\sum_{i=1}^{N^{\text{Receiver}}} F_i^{\text{Receiver}}$ is equal to or close to zero, then a partial interface scaling may be needed.

Control–surface mapping and interpolation is another way to produce global conservative data for load density transfer.

For concentrated nodal load on the sender:

The nodes on the sender are considered as target mesh. Map the nodes into the receiver mesh and distribute the nodal value on the sender into the related nodes on the receiver.

$$F^{\mathrm{S}} = \sum_{i=1}^{N^R} w_i F_i^R. \tag{3.30}$$

Since $\sum_{i=1}^{N^R} w_i = 1$, summation is taken over the mapped nodes on the receiver side. The global conservation equation can be obtained as follows:

$$\sum_{i=1}^{N^R} F_i^R = \sum_{i=1}^{N^S} F_i^S. \tag{3.31}$$

So in this way, global conservation can be maintained.

Note: This method has better interpolation accuracy if the sender mesh is finer than the receiver mesh.

3.3.4.4 The Open MP-based parallel mapping and interpolation method

Due to the independent and simple data structure of the mesh information, the shared memory open MP-based parallel algorithm is easily implemented with excellent scalability for the mapping and interpolation calculation.

3.3.4.5 Ways to make weak coupling robust and efficient

The sources of nonlinearity for a coupled system include the nonlinearity of each physics model that is produced by the improper treatment of the coupling interface conditions for each physics model.

If the separated physics model becomes ill conditioned because of the introduced improper interface conditions that cannot be avoided, then the resulting weak coupling system becomes very unstable (e.g., the flexible structure interacts with enclosed incompressible fluid flow) (Figure 3.20).

If one of the physics models in the coupling system is highly nonlinear and is very sensitive to the coupling interface conditions, such as the FSI problem with structural buckling, or with very soft stiffness but the fluid flow with relative large density and with complex flow field, then the weak coupling method may be invalid because of the convergence issue or the impractical computational expense.

Finding ways to make weak coupling more robust and efficient for coupled problems has been an interesting topic for years. In this section, several major ways of improving the stability and efficiency of weak coupling in strongly coupled problems are presented and discussed.

3.3.4.5.1 Right load transfer direction

The choice of load transfer direction may largely affect the convergence rate in the weak coupling method. In other words, an essential coupling problem, in which the physics model will receive the Dirichlet boundary condition and send the load and in which the physics model will receive natural boundary condition and send the essential boundary value to other physics model, is critical to the coupling convergence. For example, in a partitioned conjugate heat transfer problem, the solid side usually has a higher thermal conductivity than the fluid side. The coupling conditions on the interface are

$$T^s = T^f \quad \text{on } \Gamma_c, \tag{3.32}$$

$$k^s \frac{\partial T^s}{\partial n} = k^f \frac{\partial T^f}{\partial n} \quad \text{on } \Gamma_c. \tag{3.33}$$

In this case, for a good convergence, we need to transfer the temperature from the solid thermal model with higher thermal conductivity to the fluid thermal model with lower thermal conductivity and the heat flux from fluid thermal to solid thermal physics model (Figure 3.16). It is very likely to produce a divergence solution if we do the load transfers in the opposite way.

3.3.4.5.2 Subcycling and nonlinear iteration algorithm

Subcycling algorithm

One of the advantages of the weak coupling method over the strong coupling method is the flexibility in the usage of the time integration scheme and time step size for each physics solver. Subcycling (each physics model can have a smaller time step size than the coupling time step, although the load transfers only happen at the coupling time step level; Figure 3.17) can not only increase the global computational efficiency but

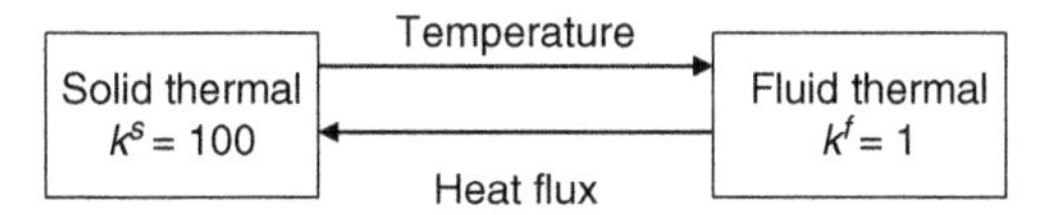

Figure 3.16 Load transfer for a conjugate heat transfer problem.

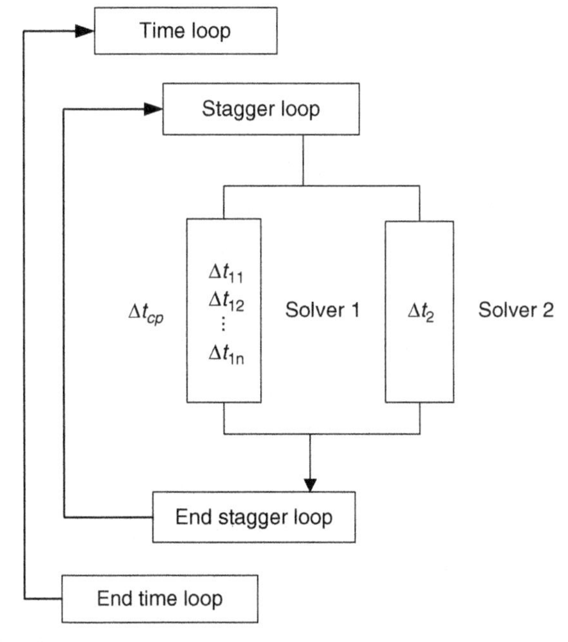

Figure 3.17 Subcycling algorithm.

also improve the convergence property of the coupling system. It allows the usage of the optimal time step size for each individual physics model without scarifying to the smallest one for all physics, as in strong coupling method.

Coupling iteration method

The Gauss–Seidel iteration method and Gauss–Jacobi iteration method can be used for the physics model iteration. As shown in Figure 3.18, the Gauss–Seidel iteration solves physics models sequentially, and each physics solver always gets the most recent interface loads from other physics models. In contrast, the Gauss–Jacobi iteration (Figure 3.19) solves the physics models in parallel, and the coupling conditions are the values from the previous Jacobi iteration. Based on our experience, the Gauss–Seidel iteration method normally has better convergence behavior than the parallel Gauss–Jacobi iteration.

Remark 3.7 In the Gauss–Seidel iteration method, the simulation order may also affect the convergence behavior. An optimal simulation sequence can at least save one coupling iteration for a highly nonlinear problem. It is good to choose the solution-driven physics model and the one with a good initial guess as the starting physics model.

3.3.4.5.3 Interface load underrelaxation algorithm

In the standard weak coupling method, the coupling affect at matrix level is usually neglected, and the coupling conditions are satisfied through load transfers. The neglecting of the coupling matrix usually causes an overshot of the solution (e.g., the neglecting of the added mass matrix and added viscous matrix from the fluid in the structure solution will produce larger interface response). To stabilize the coupling system, the underrelaxation method is normally used:

$$f^{\text{apply}} = f^{\text{previous}} + \alpha \times \left(f^{\text{current}} - f^{\text{previous}} \right). \tag{3.34}$$

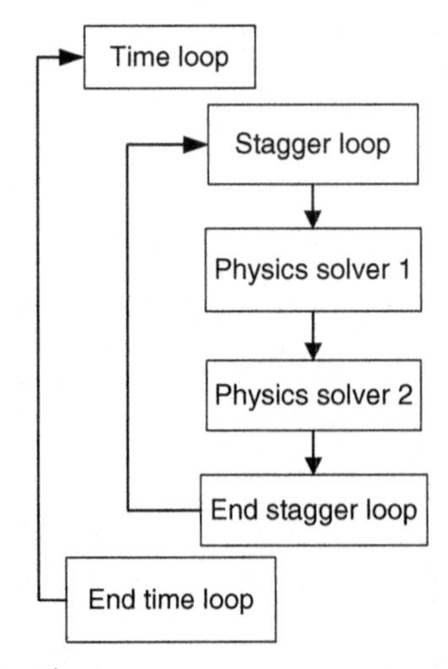

Figure 3.18 Gauss–Seidel iteration.

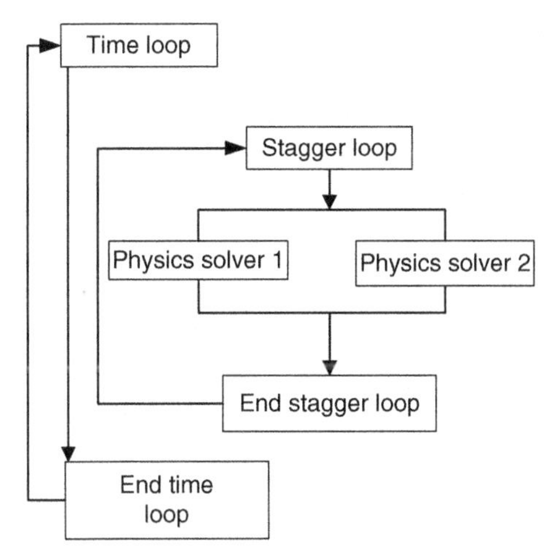

Figure 3.19 Gauss–Jacobi iteration.

The relaxation factor: $\alpha \in [0, 1.0]$.
where f^{previous} is the applied load at previous iteration, f^{current} is the current load directly from corresponding physics models, and f^{apply} is the newly updated value of the applied load. If the relaxation factor $\alpha = 1.0$, the full loads from corresponding physics models are directly applied on the target physics model, which usually produces a divergence solution. The relaxation factor $\alpha = 0.0$ means that the applied load will not be affected or changed by the corresponding physics models. The appropriate value of α depends on the nonlinear characteristics of the coupling system. A smaller value of α will produce more stable consistent convergence but may need more coupling iterations. A larger value of α makes the coupling load from other physics models transferred quickly but may cause a divergence solution. Problems with higher coupling nonlinearity will need a smaller relaxation factor α and a larger iteration number. In contrast, a larger relaxation factor can produce quicker convergence for the low nonlinearity problems, and the default value can be set between 0.5 and 0.75.

3.3.4.5.4 Introduce added matrices (mass and viscous) for CSD side

It is a natural and rational thinking of introducing added matrix in the weak coupling method to overcome the overshooting problem and increase the stability of the coupling system. For example, in the weak coupling FSI analysis, the added interface mass and viscous matrix can be introduced and added into the structure analysis, which can increase the robustness and the convergence rate in the strongly coupled FSI problem (Zhang and Hisada, 2004). We call this method the modified weak coupling method, and the key factor in this method is the way of choosing the added matrix, upon the nonlinearity of the coupling system, and the available information from the other physics models. A consistent or approximated added matrix may be used. A detailed discussion about the added matrix in FSI analysis can be found in Zhang and Hisada (2004).

3.3.4.5.5 Introduce artificial compressibility for the CFD side

As mentioned in Section 3.3.4.5, it is a great challenge for the weak coupling method to deal with the incompressible or slightly compressible fluid flow in an enclosed domain, and it couples with flexible structure (the enclosed water in a piston with flexible membrane in Figure 3.20; and the numerical examples in Sections 8.1.1 and 9.1.1).

There is no problem to solve this type of coupling problems with a strong coupling method because the whole system is well posed and the mass conservation for the incompressible water is guaranteed by the specified movement of the piston and the corresponding deformation of the flexible membrane in the coupling iteration process.

In the standard weak coupling method, the fluid flow and structure model are solved separately. As shown in Figure 3.21, the boundaries of the fluid model consist of the Dirichlet boundary condition (i.e., the velocity is given on fluid-structure interface from structure solution), the forced moving piston boundary, and wall boundaries. The resulting problem in incompressible fluid domain is enclosed by pure Dirichlet boundaries. If the mass conservation condition is destroyed by these boundaries, and this is most likely the case during nonlinear iterations for the coupling problems, then infinite pressure will be caused. After we apply this extremely large fluid force to the flexible structure (Figure 3.22), it will result in huge structure deformation. Transferring

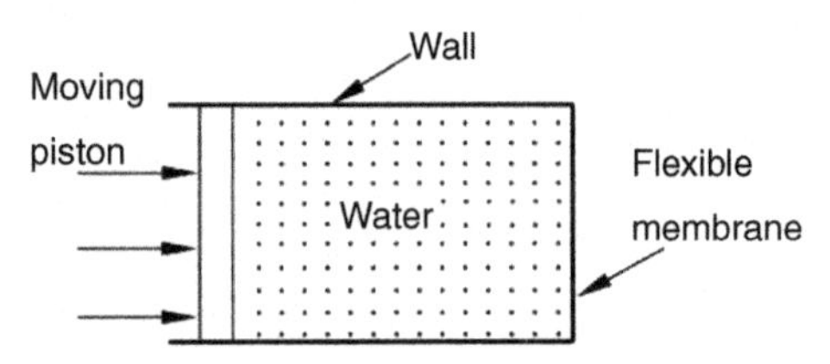

Figure 3.20 Enclosed water in a piston with a flexible membrane.

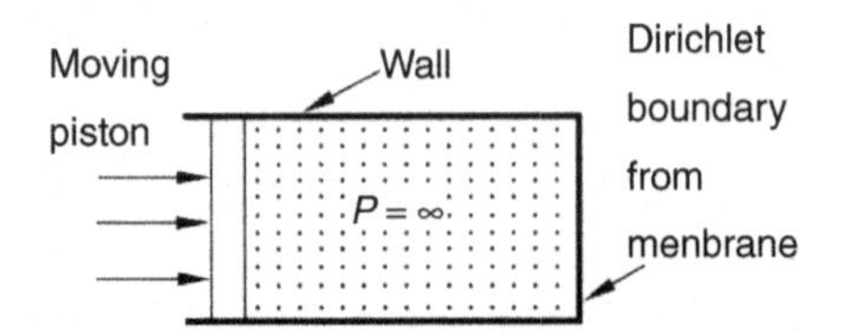

Figure 3.21 Fluid model of the piston problem in weak coupling.

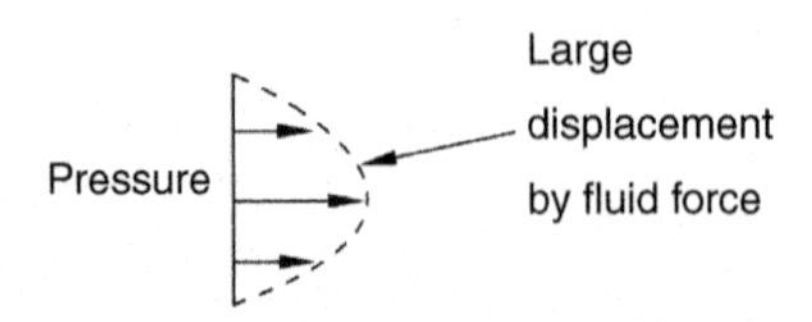

Figure 3.22 Structure model of the piston problem in weak coupling.

the resulting unphysical large structure displacement will cause even worse fluid force. It is easy to conclude that the improper weak coupling treatment causes this unstable weak coupling system.

With a consideration of the membrane's flexibility, the equivalent compressibility of bulk modules of the fluid domain can be approximated. The compressibility is unphysical but can stabilize the nonlinear coupling iteration in the weak coupling fluid physics model. If the error caused by the artificial compressibility is acceptable, then this stabilization method can be used, although we strongly recommend using the strong coupling method for an accurate and stable solution to this type of coupling problems.

3.3.4.5.6 Field solution acceleration algorithm (Aitken algorithm)

The underrelaxation method is easy to implement with consistent performance to stabilize the solution. By using an unique and constant relaxation value for all interface DOFs, but not an optimal and dynamic value, the convergence rate is usually not optimal, and the solution may diverge for certain strongly coupled fluid–structure problem. For some high nonlinear problems, a small relaxation factor with an extremely large number of iteration may be needed. This is impractical for most engineering problems. The modified weak coupling method with added matrix normally has better convergence behavior than the underrelaxation method, although it is more expensive and needs more information from the coupling physics models than the underrelaxation method. The Aitken acceleration algorithm (Küttler and Wall, 2008) uses the relation of the changes of the interface load vector with the changes of interface variables to approximate the added matrix. It is more efficient than the underrelaxation method but needs less matrix level information from the corresponding physics models than the modified matrix-based weak coupling method. Details about Aitken acceleration algorithm it can be found in Küttler and Wall (2008).

3.3.5 Intersolver coupling technique

For the intersolver code coupling technique, each physics model that is involved in the coupling can be run on the same or different physics solver code with different processes. A driver process runs simultaneously with each physics solver code to control the global coupling process (e.g., the coupling time step control, the nonlinear coupling iteration, and convergence controls for load transfers). The mapping and interpolation tools are called directly by each physics solver.

3.3.5.1 Motivation of intersolver coupling technique

A general code coupling module does all the coupling and data transfer controls and minimizes modification of the existing physics solver.

- Allowing multiple code communications among physics solvers
- Supporting surface and volumetric loads and general data or parameter transfers among physics solvers
- Coupling simulation process controls provided by the coupling driver function

- Advanced convergence control, relaxation algorithms, and other coupling methods to handle strongly coupled problems
- Physics solver flexibility: Allows physics solvers to run any way as needed (e.g., simulation type, machine type, parallel algorithms, discretization in space and time)

3.3.5.2 The key components in the intersolver coupling technique

The key components in ;the intersolver coupling technique include

- A driver code that controls global coupling process, including the global coupling controls, data transfer and convergence controls, and so on
- The mapping and interpolation library, directly called by physics solvers
- Communication library for data transfer.

3.3.5.3 The requirements for the physics solver to support the intersolver coupling technique

- Each code needs to support time-repeated iteration.
- Implementing the get and put functions with specified format to get and put coupling control related data as well as coupling interface boundary related data
- Linking with the provided mapping and interpolation library and the communication library
- Inserting the coupling-related data communication calls at each synchronization point

3.3.5.4 Communication scheme in intersolver coupling

The server client communication concept is useful for data transfer among the physics solvers and between the driver code and physics solvers.

The data communication scheme can also use the file-based intersolver data communication technique or socket-based intersolver data communication library.

3.3.5.5 Flowchart for the intersolver coupling technique

In the intersolver coupling technique, as demonstrated in Figure 3.23, the driver manager code and each physics solver runs simultaneously on different computer processes. Each physics solver solves one physics model. The driver manager controls the coupling time step loops, the nonlinear coupling iteration loops, and the convergence controls for load transfers. There are five synchronization points in the coupling simulation process: the code information of each physics solver, the global controls, and the restart information of the simulation are exchanged at synchronization point (SP) 1 between the coupling manager and each physics solver. Based on the data transfer requirements, the mesh data are exchanged between physics solvers at SP2. At SP3, the data of time controls and iteration controls are transferred from the coupling manager to each physics solver. The mapping and interpolation tools are called directly by each physics solver. Most important, SP4 will take care of load and boundary condition transfer across physics solvers. Based on the way of running physics solver in Gauss–Seidel or Gauss–Jacobi iteration, the load transfer can happen at either the beginning or at the end of the solving process. In the meantime, the coupling manager must wait to enter SP5.

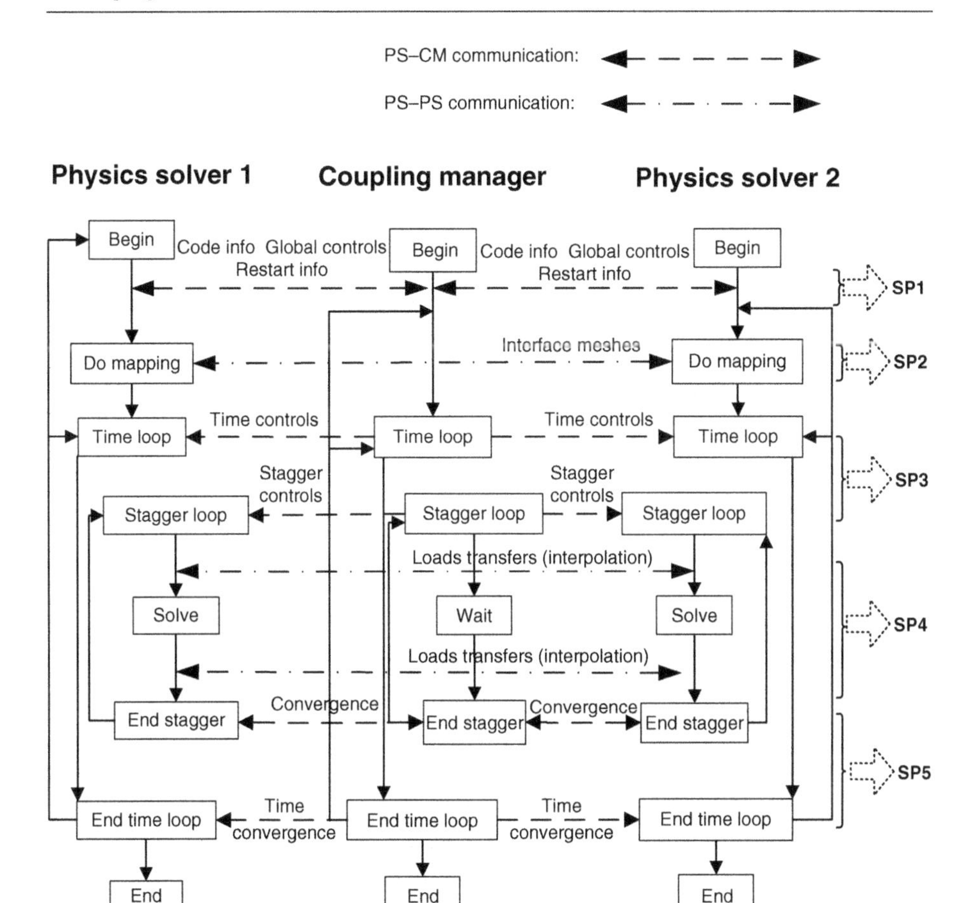

Figure 3.23 Flowchart of the intersolver coupling method.

SP5 exchanges the convergence information of the interface loads and each physics solver, and the physics solver receives time control information from the coupling manager. After the transferred interface loads and the physics model are converged, the coupling simulation goes to the next coupling time step until the end of the simulation time.

Remark 3.8 Although only two physics solvers are involved in the flowchart of the intersolver coupling method, this is applicable to any number of problems involving physics solvers.

3.4 Comparisons of the strong and weak coupling methods

In this section, we will compare the strong and the weak coupling methods in regards to their robustness, efficiency, accuracy, and code implementation.

3.4.1 Strong coupling method

Advantages include

- Solution of a coupled equation system achieved in a single step
- Robust convergence for strongly coupled problems

Disadvantages include

- Requires complete rewriting of solvers for various new coupled physics problems
- The matrix system tends to be very ill conditioned because of the difference in "stiffness" of various physics regions.
- Large problems become computationally expensive in the linear equation solver.
- Less flexibility for different physics models (e.g., mesh discretization, time integration scheme)

3.4.2 Weak coupling method

Advantages include

- Not required to rewrite physics solvers
- Able to leverage main features of each solver
- Different analysis types and unmatching mesh interfaces
- More economical for weakly coupled problems
- Able for code coupling

Disadvantages include

- Load vector coupled analysis requires relaxation techniques
- Slower convergence or divergence for certain classes of strongly coupled problems

3.5 Time integration scheme for transient multiphysics problems

3.5.1 Auto time stepping and bisection scheme

For a time transient and nonlinear problem, it is common that the level of nonlinearity changes during the analysis. This means that the appropriate time step size may vary during the simulation. It is optimal and sometimes necessary to adjust the time step size to achieve a stable and fast solution. The auto time stepping scheme used here is shown in the following equations.

$$\begin{aligned}
&\Delta t^{n+1} = \frac{(\Delta t^{n} \times N_{\text{target}})}{N_{\text{actual}}} \\
&\text{if } \Delta t^{n+1} > \Delta t_{\max} \text{ then } \Delta t^{n+1} = \Delta t_{\max} \\
&\text{if } \Delta t^{n+1} < \Delta t_{\min} \text{ then } \Delta t^{n+1} = \Delta t_{\min}.
\end{aligned} \tag{3.35}$$

Here, Δt is the time step size, N_{target} is the user-specified target number of iterations, and N_{actual} is the actual number of iterations in time step n. If the actual number of iterations is larger than the user-specified target number, then time step size will be reduced for the next time step and vice versa, and the time step size will not exceed the user-specified range.

Besides auto time stepping, we also provide bisection capability to allow redo a diverged solution with half time step size of the original one. If the time step size is larger than the minimum value and the number of iteration also reaches the maximum value, the solver will be automatically back to the previous time step, reduce the time size by half and start a new time step simulation. This bisection operation will also be performed when negative volume of fluid element occurs, or any other divergence is caused due to high nonlinearity.

3.5.2 PMA method (Newmark-β-based method)

The Newmark-β method is now widely used in practical analysis because of its simple and robust characteristics (Newmark, 1959; Bathe and Wilson, 1973). The PMA method is based on the Newmark-β integration method (Zhang, 1999). It belongs to the class of discrete and implicit single-step integration schemes, adopting the same assumptions as in the Newmark-β method. We take structural analysis, for example,

$$\upsilon_{n+1} = \upsilon_n + [(1-\gamma)\boldsymbol{a}_n + \gamma \boldsymbol{a}_{n+1}]\Delta t. \tag{3.36}$$

$$\boldsymbol{d}_{n+1} = \boldsymbol{d}_n + \upsilon_n \Delta t + \left[\left(\frac{1}{2} - \beta\right)\boldsymbol{a}_n + \beta \boldsymbol{a}_{n+1}\right]\Delta t^2 \tag{3.37}$$

At the same time, the acceleration is assumed to be constant and equal to $(1/2)(\boldsymbol{a}_n + \boldsymbol{a}_{n+1})$ at time interval $t \sim t + \Delta t$. By contrast to the Newmark-β method, the PMA method predicts the initial physic quantities at every discrete point in time.

Parameters γ and β in Equations (3.36) and (3.37) are used to control integration accuracy and stability. An unconditionally stable Newmark-β integration scheme requires $\gamma \geq 0.5$ and $\beta \geq 0.25(\gamma + 0.5)^2$. The integration scheme corresponding to $\gamma = 0.5$ and $\beta = 0.25$ (also called the trapezoidal rule) results in a period elongation of about 3% and no amplitude decay, achieving the most desirable accuracy characteristics. Numerical damping is introduced if γ is set to be greater than 0.5, which usually increases the computational stability but sacrifices the computational accuracy. The amount of dissipation, for a fixed time step, is increased by increasing γ.

In every time loop, considering nonlinear iteration, we give predicted solution at time point $t + \Delta t$ as follows:

$$\boldsymbol{a}_{n+1}^{(0)} = 0 \tag{3.38}$$

$$\upsilon_{n+1}^{(0)} = \upsilon_n + \Delta t(1-\gamma)\boldsymbol{a}_n \tag{3.39}$$

$$\boldsymbol{d}_{n+1}^{(0)} = \boldsymbol{d}_n + \Delta t \upsilon_n + \Delta t^2 \left(\frac{1}{2} - \beta \right) \boldsymbol{a}_n \tag{3.40}$$

From Equations (3.36) to (3.40), we could derive relationships of physic quantity increments from two successive iterations:

$$\Delta \boldsymbol{d}_{n+1}^{(\mathrm{i})} = \beta \Delta t^2 \, \Delta \boldsymbol{a}_{n+1}^{(\mathrm{i})}, \tag{3.41}$$

$$\Delta \upsilon_{n+1}^{(\mathrm{i})} = \gamma \Delta t \Delta \boldsymbol{a}_{n+1}^{(\mathrm{i})}. \tag{3.42}$$

Unlike in the central difference method, the solution at time $t + \Delta t$ is based on equilibrium equations at time $t + \Delta t$.

$$\boldsymbol{M} a_{n+1} + \boldsymbol{C} \upsilon_{n+1} + \boldsymbol{Q}(d_{n+1}) = \boldsymbol{F}_{n+1}. \tag{3.43}$$

Substituting Equations (3.41) and (3.42) into Equation (3.43), we get the final equation based on the acceleration increment to be solved:

$$\left(\boldsymbol{M} + \gamma \Delta t \boldsymbol{C} + \beta \Delta t^2 \, \boldsymbol{K}_{n+1}^{i-1} \right) \Delta \boldsymbol{a}_{n+1}^{(\mathrm{i})} = \boldsymbol{F}_{n+1} - \boldsymbol{Q}_{n+1}^{(i-1)} - \boldsymbol{M} \boldsymbol{a}_{n+1}^{(i-1)} - \boldsymbol{C} \upsilon_{n+1}^{i-1}. \tag{3.44}$$

In every time loop, this iteration process is repeated until the required convergence criteria for the solution $\boldsymbol{a}_{n+1}$ as specified by the user are met, in which $\upsilon_{\mathrm{n+1}}$ and $\boldsymbol{d}_{\mathrm{n+1}}$ can also be updated by using Equations (3.41) and (3.42).

3.5.3 α Method

In this section, an outstanding time integration method known as the generalized α method is considered (Chung and Hulbert, 1993). This method also belongs to the class of discrete and implicit single-step integration schemes and owns second-order accuracy and unconditional stability for linear problems. The method was first proposed to calculate structural dynamics problems, possessing a good accuracy combined with user controlled high frequency damping. Then, it is developed and analyzed for linear, first-order systems. Furthermore, it is extended to the filtered Navier-Stokes equations of incompressible flows (Jansen et al., 2000), and works well in both laminar flow and turbulent flow cases, where the spatial domain was not moving. Also, the generalized α method has been extended to FSI analysis, considering the mesh morphing. The results show that α method acquires a better accuracy and stability compared with Newmark-β-based method.

The generalized α method adopts the same assumption as in Newmark (e.g., structure analysis):

$$\boldsymbol{v}_{n+1} = \boldsymbol{v}_n + \left[(1 - \gamma) \boldsymbol{a}_n + \gamma \boldsymbol{a}_{n+1} \right] \Delta t, \tag{3.45}$$

$$\boldsymbol{d}_{n+1} = \boldsymbol{d}_n + \boldsymbol{v}_n \Delta t + \left[\left(\frac{1}{2} - \beta \right) \boldsymbol{a}_n + \beta \boldsymbol{a}_{n+1} \right] \Delta t^2. \tag{3.46}$$

In contrast to the Newmark method, the equilibrium equation is not established at a discrete time; rather, it is established at semidiscrete point:

$$\boldsymbol{M}\boldsymbol{a}_{n+1-\alpha_{\mathrm{m}}} + \boldsymbol{C}\boldsymbol{v}_{n+1-\alpha_{\mathrm{f}}} + \boldsymbol{Q}(\boldsymbol{d}_{n+1-\alpha_{\mathrm{f}}}) = \boldsymbol{F}\left(\boldsymbol{t}_{n+1-\alpha_{\mathrm{f}}} \right). \tag{3.47}$$

where

$$\boldsymbol{d}_{n+1-\boldsymbol{a}_{\mathrm{f}}} = \left(1 - \alpha_{\mathrm{f}}\right) \boldsymbol{d}_{n+1} + a_{\mathrm{f}} \boldsymbol{d}_n \tag{3.48}$$

$$\boldsymbol{v}_{n+1-\alpha_{\mathrm{f}}} = \left(1 - \alpha_{\mathrm{f}}\right) \boldsymbol{v}_{n+1} + \alpha_{\mathrm{f}} \boldsymbol{v}_n \tag{3.49}$$

$$\boldsymbol{a}_{n+1-\alpha_{\mathrm{m}}} = \left(1 - \alpha_{\mathrm{m}}\right) \boldsymbol{a}_{n+1} + \alpha_{\mathrm{m}} \boldsymbol{a}_n \tag{3.50}$$

$$\boldsymbol{t}_{n+1-\alpha_{\mathrm{f}}} = \left(1 - \alpha_{\mathrm{f}}\right) \boldsymbol{t}_{n+1} + \alpha_{\mathrm{f}} \boldsymbol{t}_n \tag{3.51}$$

Parameters α_{f}, α_{m}, β, and γ in Equations (3.45) to (3.51) are used to control integration accuracy and stability and could be specified by the user. For structural problems, the following parameters for the time integration are optimal:

$$\beta = \frac{1}{4}\left(1 - \alpha_{\mathrm{m}} + a_{\mathrm{f}}\right)^2 \tag{3.52}$$

$$\gamma = \frac{1}{2} - \alpha_{\mathrm{m}} + \alpha_{\mathrm{f}} \tag{3.53}$$

$$\alpha_{\mathrm{f}} = \frac{\rho_\infty}{1 + \rho_\infty} \tag{3.54}$$

$$\alpha_{\mathrm{m}} = \frac{2\rho_\infty - 1}{1 + \rho_\infty} \tag{3.55}$$

where ρ_∞ is the spectral radius for an infinite time-step and has to be chosen among $0 \le \rho_\infty \le 1$, $\rho_\infty = 0$, and $\rho_\infty = 1$, corresponding to minimum and maximum levels of damping in generalized α algorithm, respectively.

In every time loop, the semidiscrete solution at the ith iteration process will be expressed as follows:

$$\boldsymbol{d}_{n+1-\alpha_f}^{(i)} = (1-\alpha_f)\boldsymbol{d}_{n+1}^{(i)} + \alpha_f \boldsymbol{d}_n \tag{3.56}$$

$$\boldsymbol{v}_{n+1-\alpha_f}^{(i)} = (1-\alpha_f)\boldsymbol{v}_{n+1}^{(i)} + \alpha_f \boldsymbol{v}_n \tag{3.57}$$

$$\boldsymbol{a}_{n+1-\alpha_m}^{(i)} = (1-\alpha_f)\boldsymbol{a}_{n+1}^{(i)} + \alpha_m \alpha_n \tag{3.58}$$

Combined with relationships Eqs. (3.41) to (3.42), Equation (3.56) to (3.58) could be written as:

$$\Delta\boldsymbol{d}_{n+1-\alpha_f}^{(i)} = (1-\alpha_f)\beta\Delta t^2 \Delta\boldsymbol{a}_{n+1}^{(i)} \tag{3.59}$$

$$\Delta\boldsymbol{v}_{n+1-\alpha_f}^{(i)} = (1-\alpha_f)\gamma\Delta t \Delta\boldsymbol{a}_{n+1}^{(i)} \tag{3.60}$$

$$\Delta\boldsymbol{a}_{n+1-\alpha_f}^{(i)} = (1-\alpha_m)\Delta\boldsymbol{a}_{n+1}^{(i)} \tag{3.61}$$

The final equation to be solved is thus derived based on the acceleration increment:

$$\begin{aligned} &\left[(1-\alpha_m)\boldsymbol{M} + (1-\alpha_f)\gamma\Delta t\boldsymbol{C} + (1-\alpha_f)\beta\Delta t^2 \boldsymbol{K}_{n+1-\alpha_f}^{(i-1)}\right]\Delta\boldsymbol{a}_{n+1}^{(i)} \\ &= \boldsymbol{F}_{n+1-\alpha_f} - \boldsymbol{Q}_{n+1-\alpha_f}^{(i-1)} - \boldsymbol{M}\boldsymbol{a}_{n+1-\alpha_f}^{(i-1)} - \boldsymbol{C}\boldsymbol{v}_{n+1-\alpha_f}^{(i-1)} \end{aligned} \tag{3.62}$$

In every time loop, this iteration process is repeated until the required convergence criteria for the solution $\boldsymbol{a}_{n+1}$ as specified by the user, are met, in which $\boldsymbol{v}_{n+1}$ and $\boldsymbol{d}_{n+1}$ can also be updated by using Equation (3.41) and(3.42). Similar to a structural problem, $\boldsymbol{v}_n$ and $\boldsymbol{a}_n$ are introduced. For the time integration of the fluid dynamic, we also use the generalized α method as presented by Jansen et al. (2000) for the first-order initial value problems. Equation (3.46) is not needed in fluid dynamic cases.

It is noticeable that for the fluid dynamic equations, different optimal parameters are used because the system is of first order:

$$\gamma = \frac{1}{2} - \alpha_m + \alpha_f \tag{3.63}$$

$$\alpha_f = \frac{\rho_\infty}{1+\rho_\infty} \tag{3.64}$$

$$\alpha_m = \frac{1}{2}\frac{2\rho_\infty - 1}{1+\rho_\infty} \tag{3.65}$$

Nonstructural physics with moving boundary

4

Chapter Outline

4.1 The moving domain problem in multiphysics simulation

4.1.1 ALE formulation

The Arbitrary Lagrangian Eulerian (ALE) formulation is used for moving boundary problems of nonstructural physics or structural problems with extremely large deformation, for example, the metal-forming problem. The basic idea of the ALE method is that the mesh movement does not have to be identical with the physical particle movement, as in the Lagrange method. Moreover, unlike the Eulerian method, in which the meshes are fixed in space, the ALE method allows the meshes to move arbitrarily to get a better mesh quality and element accuracy.

Q. Zhang & S. Cen: Multiphysics Modeling. http://dx.doi.org/10.1016/B978-0-12-407709-6.00004-3

$$\begin{cases} \text{Lagrange description:} & \Omega_X, \quad \boldsymbol{X} \\ \text{Euler description:} & \Omega_x, \quad \boldsymbol{x} \\ \text{ALE description:} & \Omega_\chi, \quad \chi \end{cases}$$

The velocity of physical particle in Euler reference frame:

$$v_i = \left.\frac{\partial x_i(\boldsymbol{X},t)}{\partial t}\right|_X. \tag{4.1}$$

The velocity of physical particle in ALE reference frame:

$$w_i = \left.\frac{\partial \chi_i(\boldsymbol{X},t)}{\partial t}\right|_X. \tag{4.2}$$

The velocity of the ALE reference frame in Euler reference frame:

$$\hat{v}_i = \left.\frac{\partial x_i(\chi,t)}{\partial t}\right|_\chi. \tag{4.3}$$

According to the chain law, time derivative of any physics value *f*(*X*,*t*) in Lagrange frame can be expressed as:

$$\left.\frac{\partial f}{\partial t}\right|_X = \left.\frac{\partial f}{\partial t}\right|_\chi + \frac{\partial f}{\partial \chi_i} w_i = \left.\frac{\partial f}{\partial t}\right|_\chi + \frac{\partial f}{\partial x_i}\frac{\partial x_i}{\partial \chi_j} w_j. \tag{4.4}$$

If we set f as x_i in Equation (4.4), one has:

$$\left.\frac{\partial x_i(\boldsymbol{X},t)}{\partial t}\right|_X = \left.\frac{\partial x_i}{\partial t}\right|_\chi + w_j \frac{\partial x_i}{\partial \chi_j}, \tag{4.5}$$

Equation (4.5) can be rewritten as:

$$\upsilon_i = \hat{\upsilon}_i + w_j \frac{\partial x_i}{\partial \chi_j}, \tag{4.6}$$

where $\hat{\upsilon}_i$ is the mesh velocity, and relative velocity of the physical particle with respect to the ALE frame of reference can be expressed as:

$$c_i = \upsilon_i - \hat{\upsilon}_i = w_j \frac{\partial x_i}{\partial \chi_j}. \tag{4.7}$$

Substituting Equation (4.7) into Equation (4.4), we have:

$$\left.\frac{\partial f(\boldsymbol{X},t)}{\partial t}\right|_{X} = \left.\frac{\partial f}{\partial t}\right|_{\chi} + c_i \frac{\partial f}{\partial x_i}. \tag{4.8}$$

For the ALE formation, we describe the governing equations in the ALE frame of reference. Mesh displacement and mesh velocity need to be known in the solution process. The ALE formulations for slightly compressible fluid flow are presented in Chapter 1.

4.1.2 Methods in the academic field to do moving boundary simulation

Besides the ALE method, some other popular methods deal with the moving domain problems for nonstructural problems, especially for multiphysics simulation with moving structural coupling boundaries.

- Deforming-Spatial-Domain/Stabilized Space–Time (DSD/SST) formulation (Bazilevs et al., 2013)
- Immersed Boundary Method (IBM) (Lai and Peskin, 2000; Peskin, 2002; Mittal and Iaccarino, 2005)
- Fictitious Domain Method (Glowinski et al., 1994, 1999; Khadra et al., 2000)
- X-FEM and Level Set Method (Sukumar et al., 2001; Chessa and Belytschko, 2004)
- Smoothed Particle Hydrodynamics (SPH) formulation (Liu and Liu, 2003; Monaghan, 2005)

We will limit our discussion to the ALE method and for details about other ways to do moving boundary problems, please refer the aforementioned references.

4.2 Advanced morphing method

The ALE automatic mesh moving scheme (e.g., Laplace equation–based or elasticity equation–based mesh moving scheme) is used to avoid poorly distorted fluid meshes in fluid–structure interaction (FSI) analysis.

4.2.1 Moving boundary problem in nonstructural physics

For fluid–solid interaction analysis, the boundaries for controlling mesh motions of fluids consist of rigid walls, open boundaries, and interfaces with deformable solid. Under these boundary conditions, the following methods are employed for mesh control in the fluid domain. The mesh is fixed in space for rigid walls and open boundaries, that is,

$$\boldsymbol{u}_m = 0. \tag{4.9}$$

where, $\boldsymbol{u}_m$ is the mesh displacement vector of the fluid.

On the interface with a deformable solid, fluid nodes are attached to the solid, that is, the Lagrangian motion is assumed as,

$$\boldsymbol{u}_m = \boldsymbol{u}^s. \tag{4.10}$$

where $\boldsymbol{u}^s$ is the displacement vector of the structure. For incompatible mesh discretization on the coupling interface, an appropriate mapping and interpolation scheme is needed to get appropriate displacement boundaries for the fluid domain.

4.2.2. Morphing equations for solving the mesh motion of interior nodes

To control the motion of interior nodes of a deformable physics domain, mesh distortion must be effectively avoided. Since the geometrical configuration of the fluid domain can be complex, and the deformation of the fluid domain can be large, a Dirichlet-type boundary value problem is usually solved to determine nodal displacements in each direction as in Equation (4.11):

$$\nabla \cdot (k \nabla u_m) = 0, \qquad \text{in} \quad \Omega_t^f \tag{4.11}$$

where k is the element mesh stiffness. Generally, k is the function of element size, level of element distortion, or distance from the wall. We apply Equation (4.11) to each direction of the simulation domain; it means that the mesh deformations in each direction are uncoupled.

However, while using the automatic mesh-updating algorithm (also called morphing), without changing element topology, it is difficult to handle FSI problems with extremely large boundary movements. In these cases, remeshing is necessary to maintain the element quality and avoid morphing failure.

4.2.3 Some strategies for the best mesh quality controls in the morphing process

- Laplace, elasticity, diffusion equation, or spring system-based mesh morphing;
- Adaptive mesh stiffness, such as, mesh stiffness as a function of wall distance, element size, level of element distortion, etc.
- History-independent mesh deformation for multiple cycle-simulation, incremental form or total value form for mesh displacement.
- Time integration scheme for the problems with mesh velocity is needed.

4.3 Automatic remeshing technology

Although the automatic mesh updating algorithm does not change the element topology, it is difficult to handle the FSI problems with extremely large boundary movements. Therefore, in this case, remeshing (generating a new mesh) is helpful to maintain the element quality and avoid morphing failure.

It is essential for a CFD (Computational Fluid Dynamic) code to have a capability to handle fluid flow with extreme domain changes. INTESIM offers totally automatic mesh-updating and remeshing features for fluid flow with moving boundary and FSI problems. Based on the element information (nodal coordinates, element connectivity, and nodal displacements), the INTESIM mesh generator regenerates a new mesh from a selected element group. INTESIM provides three options to select the candidate elements for remeshing, such as, remesh all fluid elements if the quality of the worst element falls below any specified quality requirement, selectively remesh the fluid elements that have a quality below any specified quality requirement, and remesh a predefined element group if the quality of the worst element in that group falls below any specified quality requirement. It uses three parameters to evaluate the element quality and distortion level; element aspect ratio, changes of element size, and changes of element aspect ratio (Johnson and Tezduyar, 1996). A user can also control remeshing frequency by specifying element quality checkpoints, every time step, every n time steps, or at a single time point. The fluid elements will be automatically remeshed if any quality requirement is not met at a checkpoint.

The INTESIM mesh generator allows a user to control size of the new elements. The remeshing capability is limited to triangle and tetrahedral meshes for 2D and 3D geometries, respectively. It also retains the topology of the element edges/faces on the boundary of a candidate element group.

The element birth and death feature works together with this remeshing capability to cope with fluid–structure problems with extreme domain changes (e.g., pinched flow, or FSI problems with contact of structural domains).

4.3.1 Flowchart for the meshing moving problem with automatic remeshing

The flowchart in Figure 4.1 shows the solution process with remeshing. In Section 4.2, solution and morphing parts have been discussed. Here, we focus on how to measure the mesh quality and employ remeshing controls.

4.3.2 Mesh quality measurement

Three simple ways to measure element qualities have been presented here to decide when remeshing is necessary (Johnson and Tezduyar, 1996).

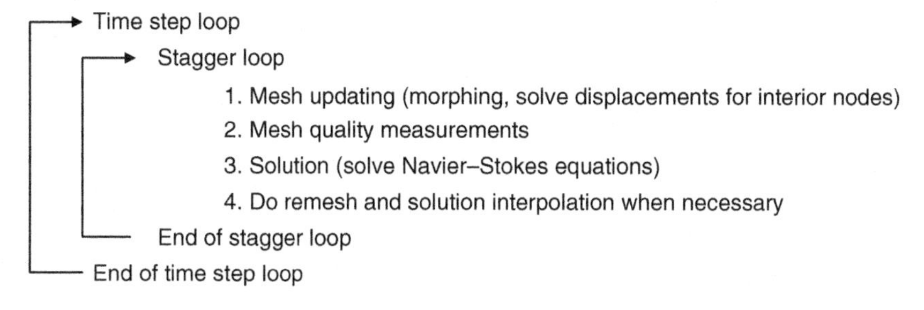

Figure 4.1 Flowchart for meshing moving problem with automatic remeshing.

4.3.2.1 Generalized element aspect ratio AR

For the 2D triangle element,

$$\mathrm{AR} = \frac{\left((l_{\mathrm{avg}})^2/A\right)}{2.3}. \tag{4.12}$$

For the 3D tetrahedron element,

$$\mathrm{AR} = \frac{\left((l_{\mathrm{avg}})^3/V\right)}{8.48}. \tag{4.13}$$

Here, l_{avg} is the average length of the element edges, A is the element area, V is the element volume; 2.3 in Equation (4.12) represents the aspect ratio of an equilateral triangle, and 8.48 in Equation (4.13) is the aspect ratio of a regular tetrahedron element.

4.3.2.2 Changes of element size

For the 2D triangle element,

$$V_{\mathrm{CH}} = \exp\left|\log\left(\frac{A(t)}{A(0)}\right)\right|. \tag{4.14}$$

For the 3D tetrahedron element,

$$V_{\mathrm{CH}} = \exp\left|\log\left(\frac{V(t)}{V(0)}\right)\right|. \tag{4.15}$$

Here, $A(t)$ and $V(t)$ are the element area and volume at time t, respectively, and $A(0)$ and $V(0)$ are the element area and volume at time 0, respectively.

4.3.2.3 Changes of element aspect ratio

$$\mathrm{AR}_{\mathrm{CH}} = \exp\left|\log\left(\frac{\mathrm{AR}(t)}{\mathrm{AR}(0)}\right)\right|. \tag{4.16}$$

If any element quality reaches the specified value, remeshing will be considered for that element. These mesh quality measurements are easy to implement and suitable for evaluating the qualities of triangle mesh and tetrahedral mesh.

4.3.3 Remeshing tool

Based on exterior surface mesh and total displacement at the exterior nodes of a selected element group, we adopt the Delaunay triangular method (Shewchuk, 1998) as the remeshing tool to create a new mesh and then carry out swapping, coarsening, refining, smoothing, and so on to improve the mesh qualities. This can guarantee a new mesh with more desirable element qualities than the one applied with mesh improvement operations only.

4.3.4 Remeshing controls

Based on the exterior surface mesh and the total displacement at exterior nodes of a selected element group, the selected remeshing tool will use the advancing front method to create a new mesh and then carry out swapping, coarsening, refining, smoothing, and so on to improve the mesh qualities. This can guarantee a new mesh with more desirable element qualities than the one that uses mesh improvement operations only (Johnson and Tezduyar, 1996).

The INTESIM automatic remeshing tool also offers flexibility to control the remeshing process.

4.3.4.1 Elements for remeshing

Based on the characteristics of the problem, there are three options to do element selection for remeshing.

Whole-domain remeshing: Remesh all fluid elements if the quality of the worst element falls below any specified quality requirement listed in Section 4.3.2. Whole-domain remeshing is suitable when the moving boundary affects all of the fluid elements because all the fluid elements may become distorted during solution. A cylinder passing through a channel, in Section 4.6.3, is one example that uses whole-domain remeshing.

Selected-domain remeshing: Remesh a predefined element group identified with a specified component name if the quality of the worst element in that group falls below any specified quality requirement. Selected-domain remeshing is suitable when the distorted elements are locally distributed and the location of the distorted elements is known before solution and will not change much during solution (Section 4.6.4: the FSI analysis of a printer paper problem).

Partial-domain remeshing: Remesh the fluid elements that have a quality below any specified quality requirement. Partial-domain remeshing is suitable when the distorted elements are locally distributed and the location of the distorted elements is not known before solution or changes during solution. If you choose to use partial-domain remeshing, some smoothing algorithm may be automatically applied to obtain an element group with smooth boundaries. (The remeshing operation will apply only to the distorted elements when partial-domain remeshing is used for the FSI analysis of a printer paper case in Section 4.6.4).

Besides these three options, you can also use the element exclusion feature to exclude all elements connected to boundary nodes from remeshing. For an FSI analysis, you can also exclude the elements connected to fluid–structure interfaces. This feature is useful to maintain the boundary layer resolution and accuracy during remesh.

4.3.4.2 Remeshing frequency controls

It is possible to provide very flexible ways to control remeshing frequency. You can specify element quality checkpoints in three ways: every time step, every n time steps, or at a single time point to limit the remeshing frequency. The fluid elements will automatically remesh if any quality requirement is not met at a checkpoint. This feature is useful for the problem when the mesh quality is not adequate to decide the remeshing frequency.

4.3.4.3 Element size controls

The default element size for remeshing uses the nearest element size on the boundary of the element domain. You can also size internal elements so that they are approximately the size of the boundary elements times an expansion or contraction factor.

4.3.5 Mapping and interpolation for solutions

Interpolating the nodal or element based values (include boundary conditions) from the old mesh (before remeshing) to the new mesh (after remeshing), is the last step in the remeshing process. We use the bucket search method (Jansen et al., 1992) to search the mapping between an old mesh and a new mesh. It loops over all elements on the old mesh for each new node or new element. Element local coordinate-based linear interpolation will be used to get the nodal solution for the interior nodes. Since there are no topology changes on the boundary nodes of the remeshing domain, the nodal values on the boundary of the new mesh can be obtained directly from the identical node of the old mesh. We also update the element based FSI boundary conditions on the new fluid mesh.

The interpolated solutions on elements or nodes may not satisfy the governing equation, so more iterations may be needed for the coming iteration, especially when the mass conservative law is destroyed by the new interpolated solutions.

4.4 Mesh controls for rotating machinery

There are four ways to deal with the domain change problems for a rotating machinery.

- Morphing and remeshing scheme.
- Sheared layer scheme.
- Sliding interface scheme.
- Multiple frame of reference (MFR) scheme without physical mesh movements

The morphing and remeshing scheme, as shown in Figure 4.2, has the following features:

- Morphing is applied to rotor mesh.
- Remesh should be applied to deformed mesh in rotor region.
- Mapping and interpolation are needed for remeshing domain.

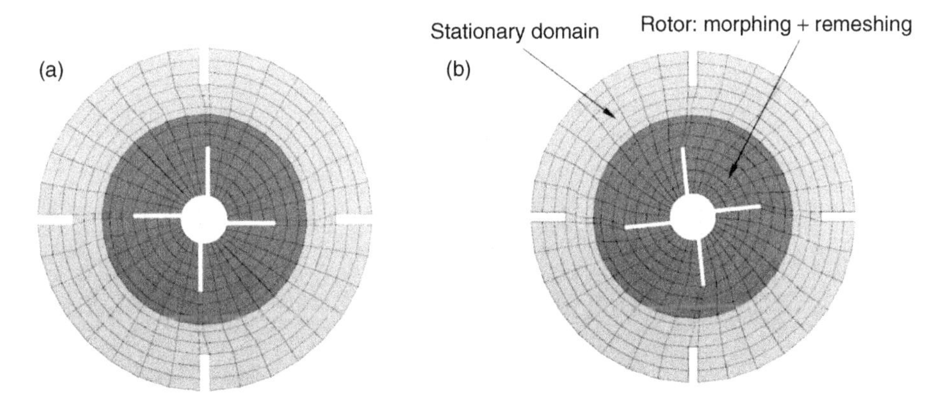

Figure 4.2 Morphing and remeshing scheme. (a) Original model and (b) morphing and remeshing model (rotating 6°).

Sheared layer method is demonstrated in Figure 4.3. Its characteristics are,

- Rigid body morphing is applied to the rotor mesh with one deformable layer excluded.
- Mesh swap is applied to one soft layer between rigid rotor and rigid stator.
- Mapping and interpolation are needed for the remeshed soft layer.

Morphing with a sliding interface (single frame of reference) is shown in Figure 4.4. Features of this method are,

- Rigid body morphing is applied to rotor mesh.
- Sliding mesh boundary is applied to couple the interface.
- 360° is needed.
- All interface conditions are accounted for in the global stationary coordinate system.
- Mapping is needed for coupling the sliding interface for each and every time step.

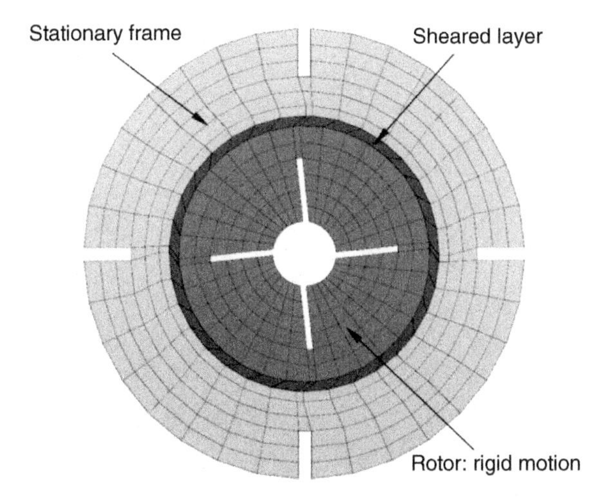

Figure 4.3 Sheared layer method.

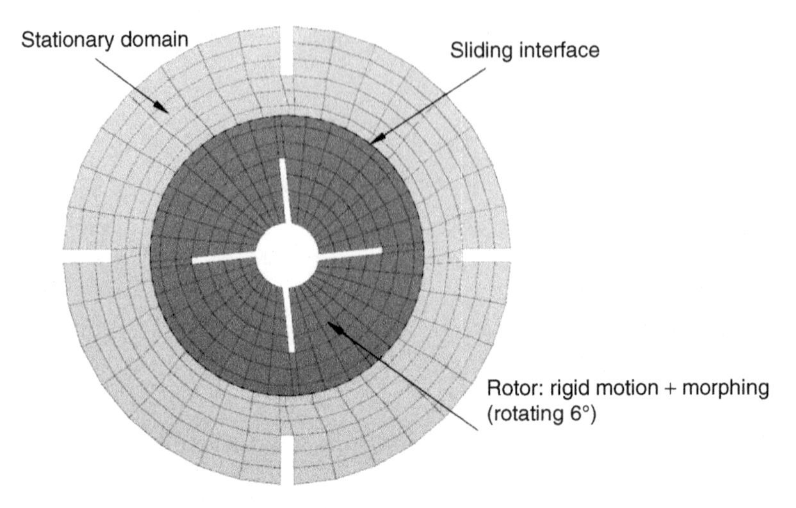

Figure 4.4 Morphing with a sliding interface.

Sliding interface with MFR is shown in Figure 4.5. For this method,

- Mapping and interpolation, in which mesh replication is needed for transient rotor stator.
- Interface coupling conditions need to be satisfied across different frames of reference.
- Transient rotor stator or frozen rotor stator can be employed (ANSYS, 2013).
- No data mapping or interpolation is needed for rotor or stator parts, and only the sliding interface mesh replication is needed for transient rotor stator.

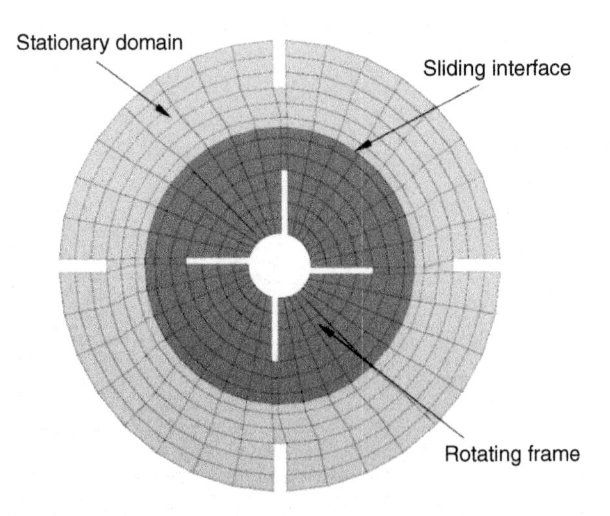

Figure 4.5 Sliding interface with multiple frames of reference.

4.5 Treatment for pinched flow problems

Pinched flow means that the fluid flows in a narrow space that may completely be closed up during the operating process, for example, the valve is completely closed in a water tank. Engineering problems that involve pinched flow include, but are not limited to, piston problems, antibreak systems in automobile engineering, and devices with valves that are opened and closed during operation. There are some numerical challenges for the coupling simulation of pinched flow problems. Discussions have been given here, and some numerical examples have been presented in Section 4.6.

4.5.1 *Element birth and death*

The element birth and death feature is available for fluid elements. It will freeze the node, exclude it from solution, and treat it as Dirichlet-type boundary when the distance from this node to the walls (including FSI boundary) falls below a specified value, and it will freeze an element, if all nodes of that element are frozen. This element birth and death feature, working together with the remeshing capability, can cope with pinched flow and FSI problems with contact of the structural domains. Section 4.6.4 gives an example.

4.5.2 *Transfer of contact information*

The key to the pinched flow problems is that the contact status in structure field needs to conform to element birth and death control of the flow domain, the location between the contactor, and the target of the structure model. One way to do this is to set the same normal tolerance for the contact condition and for the element death and birth conditions before the coupling simulation. A more robust and automatic way to address the pinched flow problem is that in the coupling model, the fluid interface will receive the contact status from the structure physics, and it will turn off the flow, if the corresponding structure is in contact; otherwise, the flow will be activated.

4.6 Examples for mesh control

4.6.1 *Extreme morphing case of 2D artificial heart (Zhang and Hisada, 2001)*

4.6.1.1 *Analytical model*

The 4Q1/P0 (four nodes for bilinear velocity interpolation/constant pressure field) quadrilateral mixed elements are used for 2D analysis of the fluid. The structure is discretized by two-node Bernoulli–Euler beam elements with Hermite interpolation. The 2D model of an artificial heart is shown in Figure 4.6.

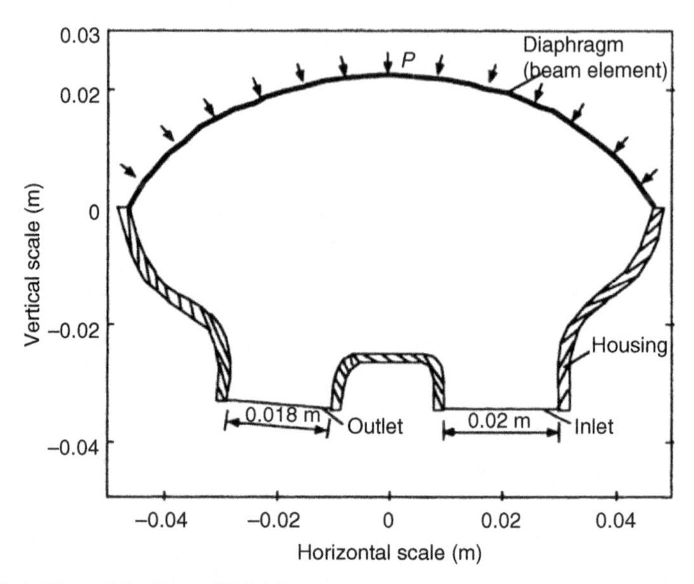

Figure 4.6 A 2D model of an artificial heart.

The material properties of blood are; mass density, $\rho = 1.06 \times 10^3$ kg/m^3; dynamic viscosity, $\mu = 4.71 \times 10^{-3}$ Pa s; bulk modulus, $B = 2.06 \times 109$ Pa. The material properties of the diaphragm are; Young's modulus, $E = 2.0 \times 10^7$ Pa; Poisson's ratio, $\nu = 0.45$; mass density, $\rho = 1.13 \times 10^3$ kg/m^3; and thickness, $t = 0.35$ mm. Initial conditions are; velocity, $v = 0$; and pressure, $p = 81\,\text{mmHg}$.

4.6.1.2 Pulsation procedures

The pulsation procedure (one cardiac cycle) of the artificial heart is divided into three phases. The pressure histories applied to the diaphragm, outlet, and inlet are shown for one cardiac cycle in Figure 4.7a–c, respectively.

1. Contraction phase (t = 0.0–0.25 s): the inlet is closed and the outlet is open. Pressure is applied to the diaphragm in the sequence 81 → 125 → 103 mmHg and to the outlet in the sequence 81 → 125 → 103 mmHg.
2. Isovolumetric relaxation phase (t = 0.25–0.30 s): both outlet and inlet are closed. Pressure is applied to the diaphragm in the sequence 103 → 11.9 mmHg.
3. Relaxation phase (t = 0.30–0.75 s): the inlet is open, and the outlet is closed. Pressure applied to the diaphragm is 11.9 mmHg (constant), and that to the inlet is 11.9 → 12.2 → 13.0 mmHg.

4.6.1.3 Numerical results

The deformations and velocity distributions of blood domain and corresponding mesh motions are shown in Figures 4.8a–f and 4.9a–f, respectively. It is seen from Figure 4.8a–f that vortices appear in the isovolumetric relaxation and the relaxation phases but not in the contraction phase. As seen in Figure 4.9c, the analysis domain shrinks at the end of the contraction phase (t = 0.25 s), especially near the outlet where

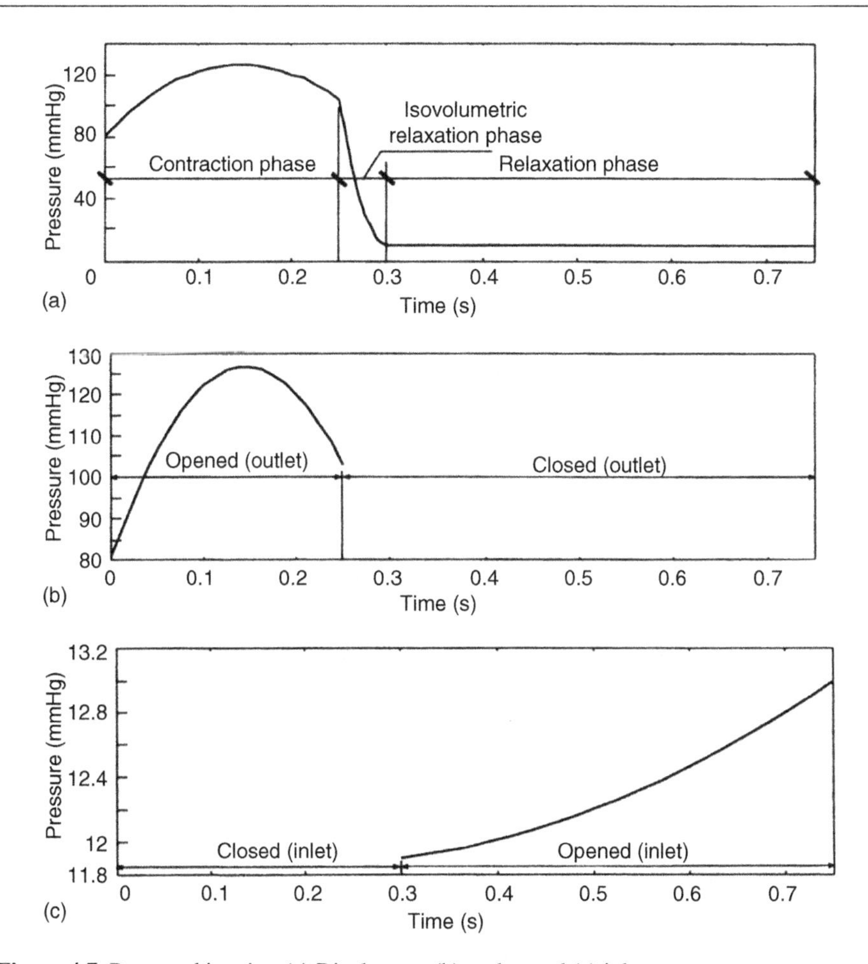

Figure 4.7 Pressure histories. (a) Diaphragm, (b) outlet, and (c) inlet.

the mesh becomes dense. However, the determinant of the Jacobian operator of the isoparametric finite element keeps positive at all of the quadrature points, and the maximum element aspect ratio and element skew remain smaller than 37.0 and 15.0, respectively. As a result, reasonable velocity distributions are still obtained at $t = 0.25$ s, and convergence is obtained without causing numerical difficulties. In the relaxation phase, the mesh distortion recovers gradually, and at $t = 0.62$ s, it almost returns to the initial configuration (Figure 4.9f).

4.6.2 Rotating problem with a sliding interface

An application of a sliding interface to an impeller–baffle interaction problem has been shown here. This case mainly shows the mesh control technology of the MRF method, for a transient fluid rotation problem. The geometry is a section of a 3D channel

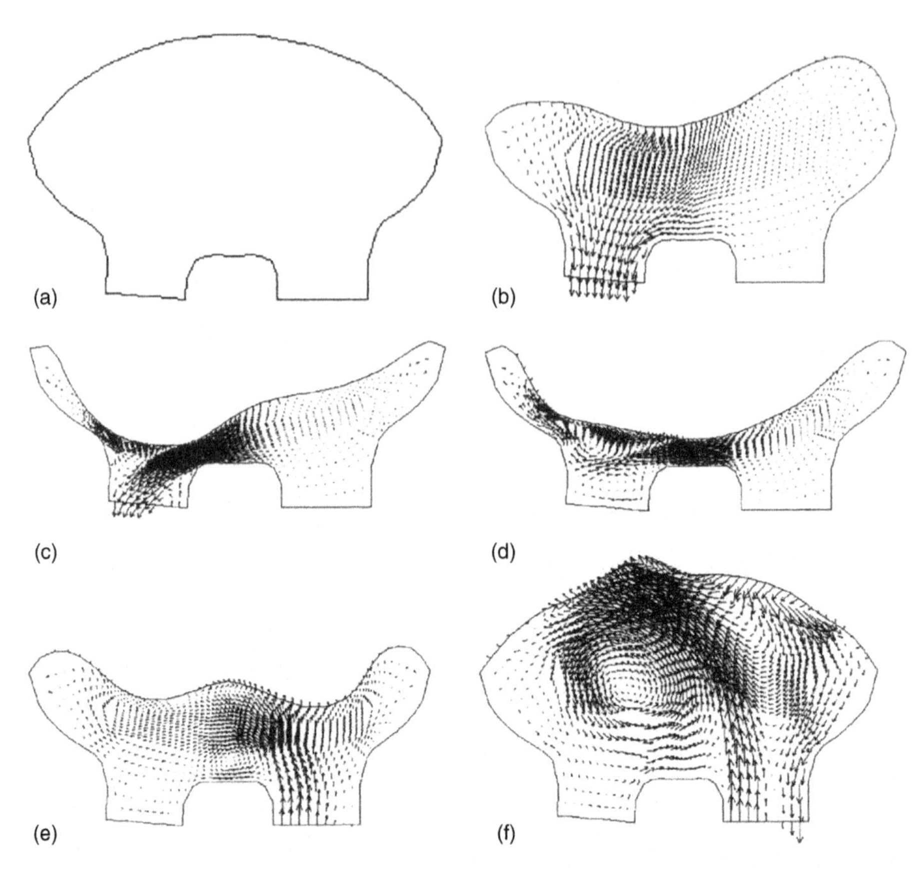

Figure 4.8 Deformation and velocity distributions. (a) Time: $t = 0.0$ s, (b) time: $t = 0.08$ s, (c) time: $t = 0.25$ s, (d) time: $t = 0.30$ s, (e) time: $t = 0.50$ s, and (f) time: $t = 0.62$ s.

with a four-blade impeller at the center and four baffles at the wall, which is shown in the Figure 4.10. The flow is driven by a rotating impeller, and the flow domain is divided into two regions, the central region with a rotating mesh and the closed wall region with a stationary mesh. A sliding interface is applied at the interface of the two regions. As far as the central rotating part is concerned, the analysis is conducted in a rotating frame, the rotational speed is 1 rad/s, and the rotation axis is Z. The symmetry boundary is applied on the top and bottom surfaces, and the inner wall is considered to be a no-slip wall. The stationary part is conducted in the stationary frame, the symmetry boundary is applied on the top and bottom surfaces, and the outer wall with four baffles is considered a no-slip wall. The transient time step size is 0.05 s and the total analysis time is 1 ss.

The solution of flow field is depicted by the mesh motion in Figure 4.11, and the velocity vector plot and the pressure plot are shown in Figures 4.12–4.13, respectively.

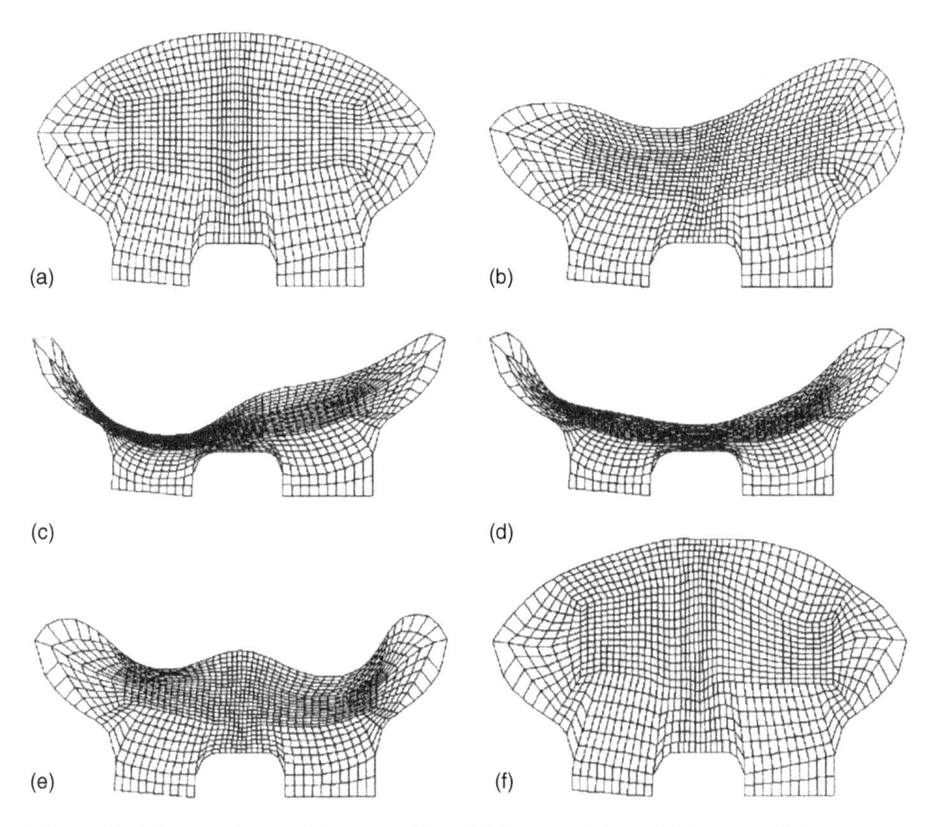

Figure 4.9 Mesh motions. (a) Time: $t = 0.0$ s, (b) time: $t = 0.08$ s, (c) time: $t = 0.25$ s, (d) time: $t = 0.30$ s, (e) time: $t = 0.50$ s, and (f) time: $t = 0.62$ s.

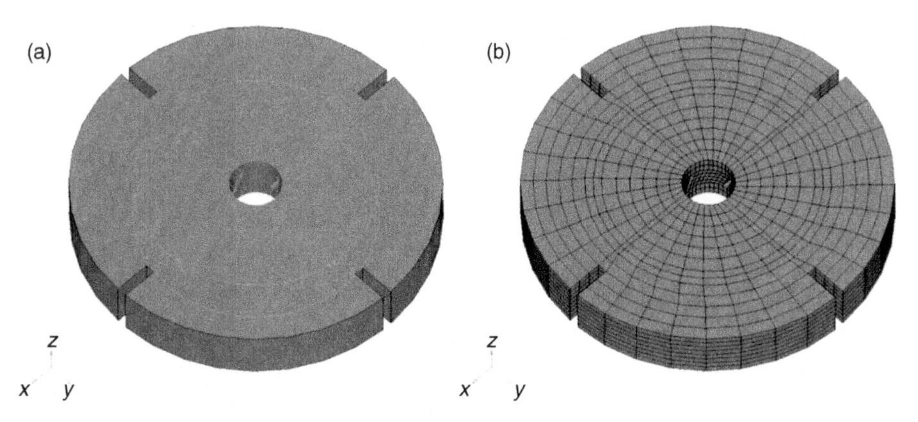

Figure 4.10 A model of an impeller–baffle interaction problem. (a) The geometry model and (b) the mesh model.

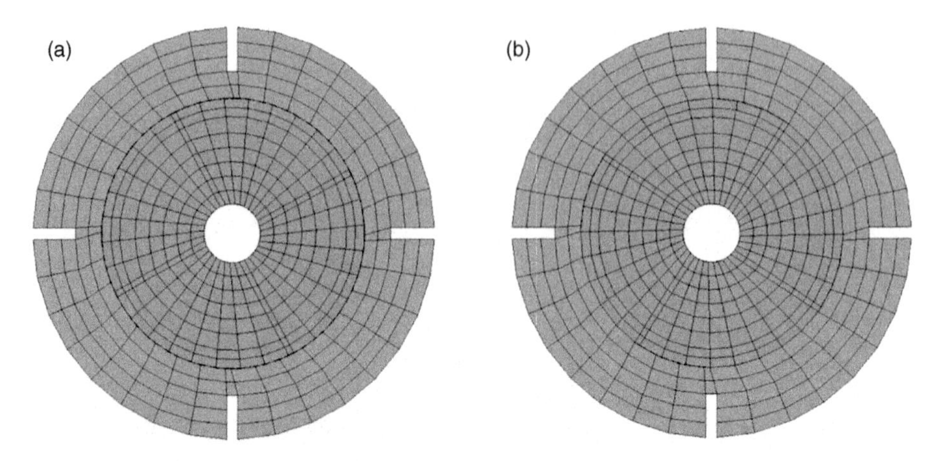

Figure 4.11 The mesh motion at two different times. (a) $t = 0.5$ s and (b) $t = 1$ s.

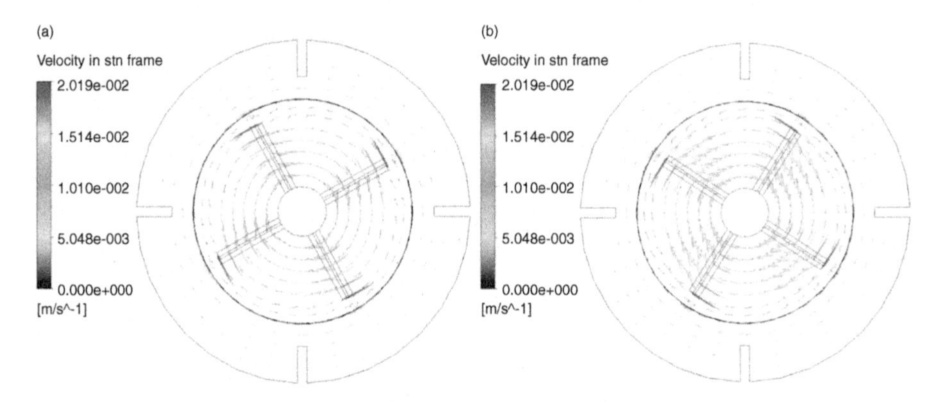

Figure 4.12 Velocity vector plot at two different times. (a) $t = 0.5$ s and (b) $t = 1$ s.

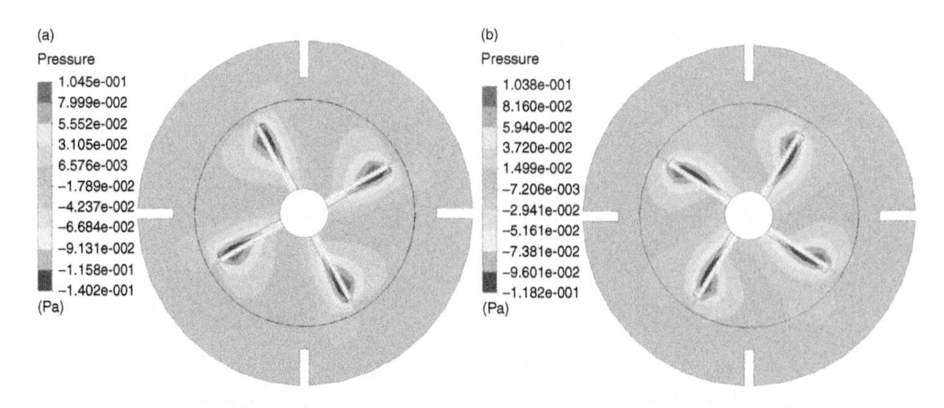

Figure 4.13 Pressure plot at two different times. (a) $t = 0.5$ s and (b) $t = 1$ s.

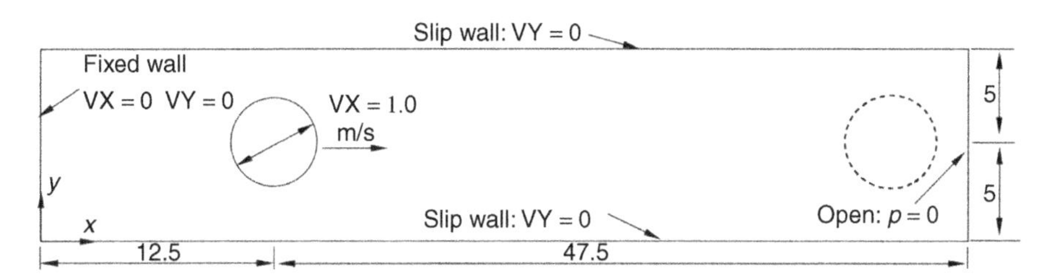

Figure 4.14 Cylinder passing through a channel. Fixed mesh motions applied on all exterior boundaries.

Figure 4.11 shows the mesh motion at two different times. At $t = 0.5$ s, the central rotating mesh rotates 28.66°, and at $t = 1$ s, the central rotating mesh rotates at 57.33°. Figure 4.12 shows the velocity vector, which is driven by the rotating impeller at two different times, and Figure 4.13 shows the pressure plot at the cutting plane of whole flow domain at location Z = 0.

4.6.3 Cylinder passing through a channel

This problem involves a cylinder passing through a channel with a constant velocity VX = 1.0 m/s. The 2D triangle elements model the air at 25°C (Figure 4.14). The simulation uses ALE formulation, and the whole-domain remeshing algorithm is employed. The specified element quality requirements are aspect ratio = 5.0, change of element size = 3.0, and change of aspect ratio = 3.0. The transient time step size is 1.0 s and the total analysis time is 44 s.

The remeshing was done nine times during the solution. Figure 4.15 shows the mesh deformation at four different times. These figures show that the remeshing capability successfully handles fluid flow with large moving boundary walls. The mesh keeps good element qualities in the whole solution process. Element quality criteria can be increased appropriately to reduce the total number of remeshing.

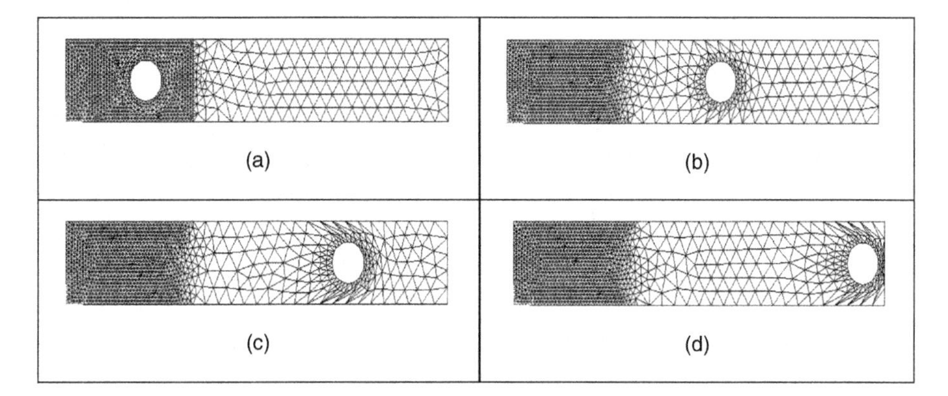

Figure 4.15 Mesh deformations of cylinder passing through a channel.

4.6.4 FSI analysis of a printer paper problem

Figure 4.16 demonstrates an FSI problem of sucking a piece of printer paper onto a plate. The movement of the flexible paper is driven by the pressure difference between inlet and outlet. We simplify this problem as a 2D model. As the paper is moving down, fluid elements around the tip of the paper will be distorted quickly, but the rest of the fluid elements will still keep good shape. So we use the *selected-domain remeshing* scheme in this analysis. Figure 4.17a shows the preselected element group for remeshing. The mesh motions of the fluid domain are fixed on all exterior boundaries in all directions except for the displacements in the vertical direction along the right inlet boundary, which are allowed. Contact elements are used for dealing with the contact between beam element and bottom of the upper chamber. Element birth and death criterion, 0.00001, is set up for the fluid elements that are located between the paper and target surface. Figure 4.17b,c shows the mesh deformation during solution; it is obvious that contact occurs and element death happens for the fluid elements between the paper and target surfaces in the late stage of solution. Figure 4.18 gives the velocity distributions of the fluid domain at $t = 0.06$ s, while the displacement history of the tip node of the paper is shown in Figure 4.19. The FSI solution will diverge at a very early time ($t = 0.24$ s) without remeshing.

4.6.5 Conclusions and future works

ALE mesh morphing and remeshing work well for pure fluid flow and FSI problems with large domain changes. It is suitable for interior element failures but not suitable for surface element failure because there are no topology changes for the boundary

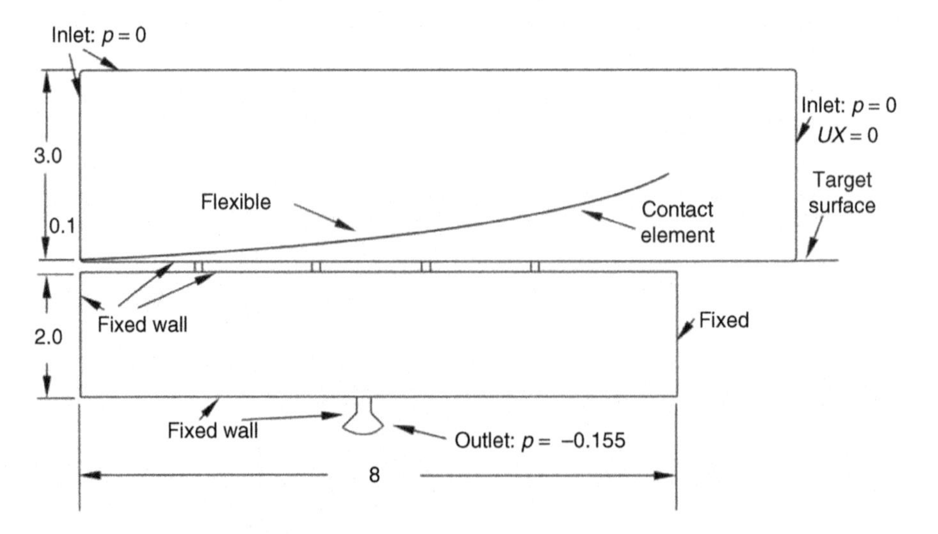

Figure 4.16 FSI analysis of paper problem (units: in., lb, s).

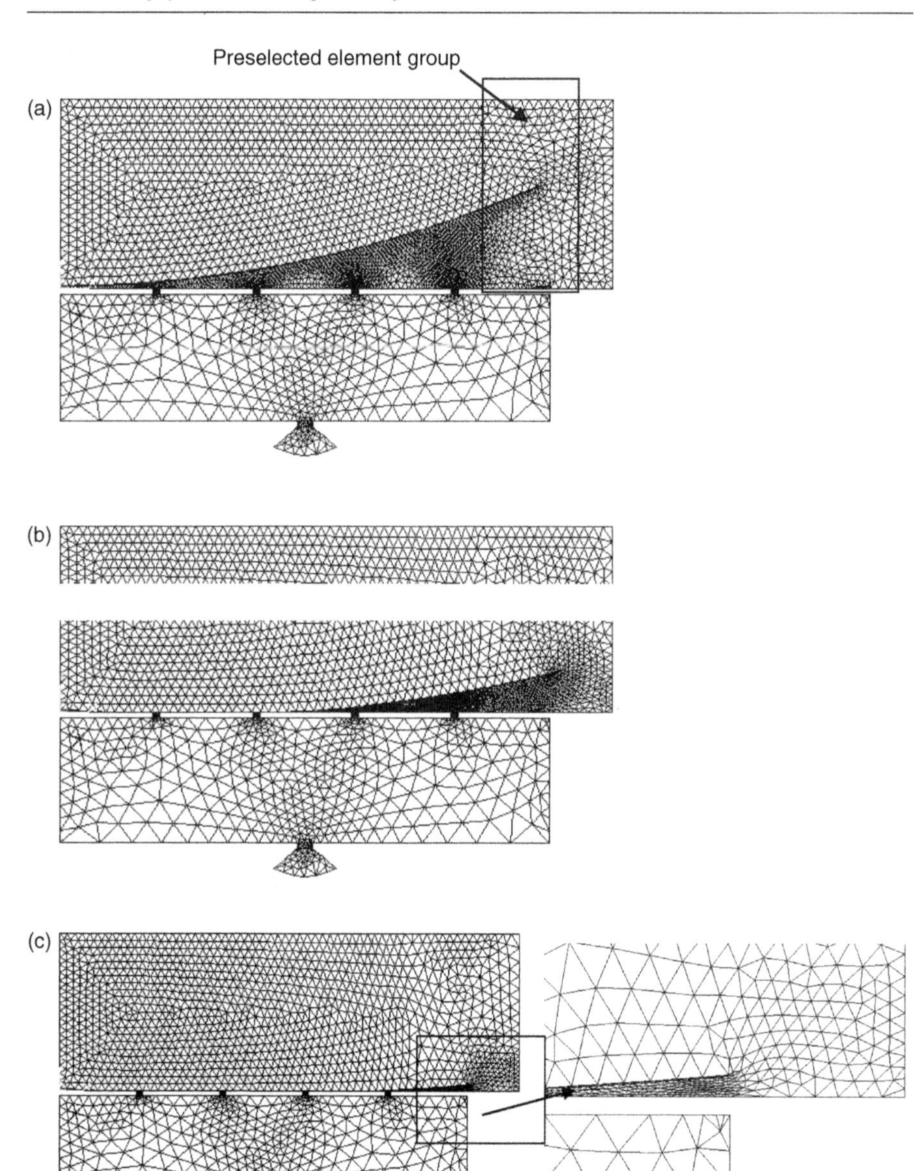

Figure 4.17 (a) Original mesh and predefined element group for remeshing, (b) mesh deformation at $t = 0.06$ s, and (c) mesh deformation at $t = 0.08$ s.

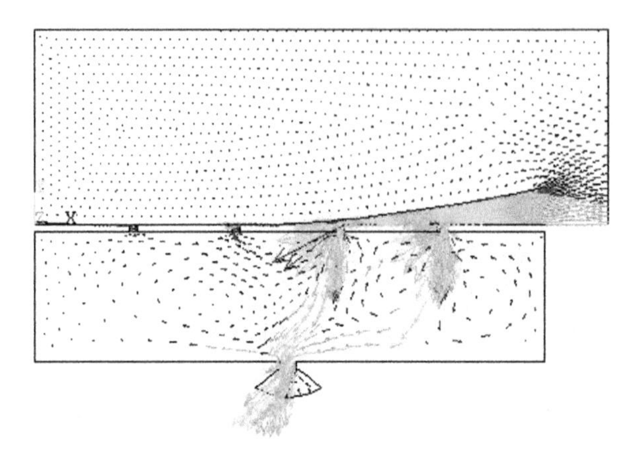

Figure 4.18 Velocity distributions at $t = 0.06$ s.

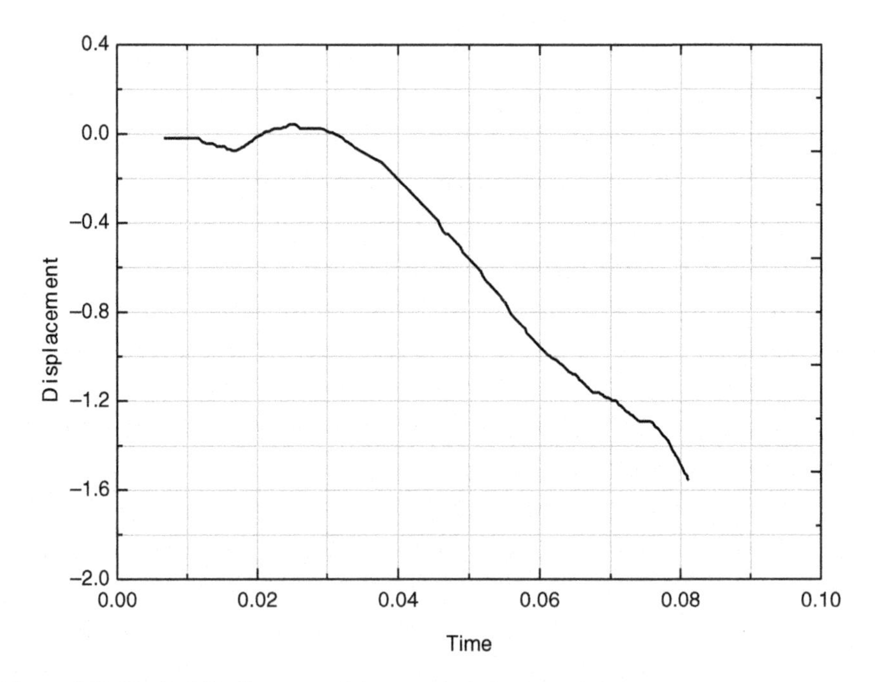

Figure 4.19 Vertical displacement of the paper's tip (units: s, in.).

nodes. It is quite effective and convenient to select remeshing elements, control remeshing frequency, and specify element sizes.

Automatic remeshing features for other nonstructural elements, such as electrostatic element, electric element, electromagnetic element, will be supported in the future INTESIM release. Remeshing capability in INTESIM-CFD will also be supported for FSI analysis in INTESIM coupling.

Stabilization schemes for highly nonlinear problems

Chapter Outline

5.1 An overview of stabilization methods

For a highly nonlinear coupled system, numerical stabilization schemes are needed to improve the computational stability and efficiency during the process of time integration and nonlinear iteration. An overview of stabilization methods in space and time that may be used for multiphysics simulation is given in this chapter, which includes the stabilization method in space and time, as well as during the nonlinear iteration process. The chapter discusses problems such as convection-dominated transport problems, stabilization methods for incompressible analysis, gauge methods for edge-based electromagnetic formulation, stabilization methods in time integration, and so on. The purpose is to present an overview and give readers a whole picture of the numerical stabilization methods in multiphysics simulation. The chapter also provides some evidence that demonstrates the validity of these stabilization methods on stability and accuracy. For the technical details of each method, readers should refer to the provided reference books or papers.

Q. Zhang & S. Cen: Multiphysics Modeling. http://dx.doi.org/10.1016/B978-0-12-407709-6.00005-5

5.2 Stabilization methods in spatial domain

5.2.1 Spatial stabilization methods in the advection-diffusion equation

For a fluid flow equation with high convection term, the Streamline Upwind Petrov/Galerkin (SUPG) method (Brooks and Hughes, 1982) is proposed to stabilize the velocity field in finite element formulation. Although the same shaped function and interpolation are used for velocity and pressure degree of freedoms in the elements presented in Chapter 1, the Pressure-Stabilizing/Petrov–Galerkin (PSPG) stabilization method (Tezduyar and Sathe, 2003) is used to avoid the pressure oscillation.

5.2.1.1 Advection-diffusion equation

The scalar, time-dependent form of the advection-diffusion equation has the following form:

$$\rho\frac{\partial\phi}{\partial t}+\rho v\cdot\nabla\phi-\mu\nabla(\cdot\nabla\phi)=f. \tag{5.1}$$

Here, ϕ is the primary scalar variable, v is the component of known velocity vector, ρ is the density, f is the source term, and μ is the dynamic viscosity.

Dividing Equation (5.1) by ρ, we obtain the classical form of the scalar advection-diffusion equation:

$$\frac{\partial\phi}{\partial t}+v\cdot\nabla\phi-v\nabla(\cdot\nabla\phi)=f/\rho \tag{5.2}$$

where, $v=\mu/\rho$ is the kinematic viscosity. Equation (5.2) is often used to describe the transport problem of species, and it is also often used as a starting point for the development of numerical formulations in CFD because of its simplicity.

5.2.1.2 Definition of the Peclet number

The definition of Peclet number for the advection-diffusion equation:

$$\mathrm{Pe}=\frac{\rho}{\mu}vh_e. \tag{5.3}$$

Where, h_e is the element size and v is the length of the velocity vector. It is noted that Pe represents the significance of advection relative to diffusion. With large Pe for Equation (5.2), the advection will dominate, which will lead to numerical difficulty because of the nonsymmetry in the finite element coefficient matrix arising from the discretization of strong advection term. Specifically, the nonphysical oscillation of the solution given by central differencing scheme was found with Pe $\geq$ 2.0. To improve the quality matrix and obtain a stable solution, the spatial stabilization method for convection dominated transport problems may be needed. The SUPG method is presented in the next section.

5.2.1.3 Properties of discretization schemes

When solving the advection-diffusion equation, proper discretization schemes are needed to discretize the advection term as well as the diffusion term. Otherwise, they will result in unsuitable solution (e.g., central differencing scheme may result in non-physical oscillation of the solution when it is used to discretize the advection term with a high Peclet number). The most important properties of a discretization scheme for advection-diffusion equations are summarized as follows:

- Conservativeness

  Conservativeness means that the flux of ϕ leaving a control volume across a certain face must be equal to the flux of ϕ entering the adjacent control volume through the face, which is an important characteristic when constructing a discretization scheme for convection–diffusion problems. A lack of conservativeness in a discretization scheme may give rise to an unsuitable solution that does not satisfy global conservation.
- Boundedness

  Boundedness means that in the absence of source terms in the internal nodes, the value of ϕ should be bounded by the boundary value (Versteeg and Malalasekera, 1995), Equation (5.4).:

$$\phi_b^{\min} \leq \phi_i \leq \phi_b^{\max} \tag{5.4}$$

  The discretized equations at each node represent a set of algebraic equations that need to be solved. Sufficient condition for a convergent iteration method can be expressed in terms of the values of coefficient of the discretized equation:

$$\left.\frac{\sum |a_{nb}|}{|a'_P|}\right|_i = \begin{cases} \leq 1 & \text{at all nodes} \\ < 1 & \text{at one node at least} \end{cases}. \tag{5.5}$$

  Here, a'_P is the net coefficient of the central node P and a_{nb} is the coefficient of the neighboring nodes of P. The summation in Equation (5.5) is taken over for all the neighboring nodes of node P. If the differencing scheme produces coefficients that satisfy the previously mentioned criterion, the resulting matrix is diagonally dominant, which is a desirable feature for the boundedness criterion. Another essential requirement for boundedness is that all coefficients of the discretized equations should have the same sign. Physically, this implies that an incensement of ϕ at one node will result in an incense of ϕ at neighboring nodes. If the discretization scheme does not satisfy the boundedness condition, it is very likely to produce divergence, or undershoots and overshoots solution (Figure 5.1). The numerical stabilization schemes given in this chapter mathematically are to improve the matrix quality and make the resulting matrix satisfy the boundedness conditions.
- Transportiveness

  Transportiveness is another important property of the discretization scheme for convection-dominant problems. The influence factor from the upstream point W to the condition of point P is larger than that from the downstream point E (Figure 5.2). In the case of

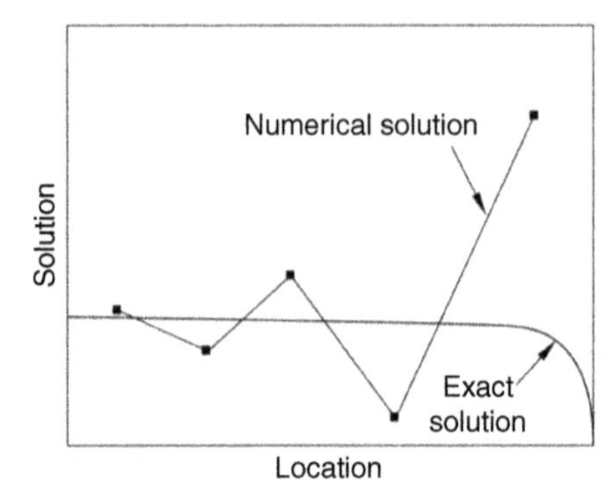

Figure 5.1 Under- and overshoots solution.

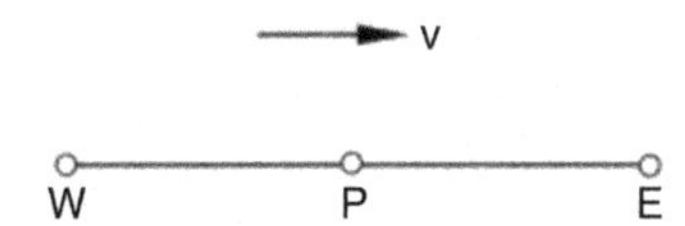

Figure 5.2 Discretization points with flow convection.

pure convection, Pe→∞, the conditions at point *P* are completely affected by the upstream source but unaffected by the downstream source. In contrast, in the case of Pe→0, the influences to the conditions at point *P* are equal.

It is very important that the relationship between the directionality of the influencing, the flow direction, and the magnitude of the Peclet number, known as the transportiveness, should be borne out in the discretization scheme (Versteeg and Malalasekera, 1995). The upwind scheme, SUPG scheme, and total variation diminishing (TVD) scheme have the properties of transportiveness.

5.2.1.4 Stabilization methods for advection-diffusion equation

5.2.1.4.1 Central differencing approach

The central differencing approach has second-order accuracy, often used to discretize the diffusion term. However, it may produce numerical oscillation when it is applied to discretize the advection term with the Peclet number greater than 2. It can be represented as follows:

$$\left.\frac{\partial \phi}{\partial x}\right|_i = \frac{\phi_{i+1} - \phi_{i-1}}{2h_e}. \tag{5.6}$$

Where, ϕ_i denotes the primary scalar variable at node i and h_e is the element size (Figure 5.3).

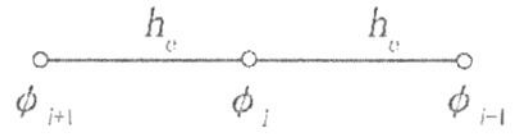

$$\overset{h_e}{\circ\!\!-\!\!-\!\!-\!\!\circ}\overset{h_e}{\!\!-\!\!-\!\!-\!\!\circ}$$
ϕ_{i+1} ϕ_i ϕ_{i-1}

Figure 5.3 One-dimensional demonstration of central differencing approach.

5.2.1.4.2 Upwind method for convection-dominant transport equations (first order accuracy)

The upwind differencing scheme is the most stable and unconditionally bounded scheme when it is used to discretize the advection terms, but it has only first-order accuracy because it is established on the backward differencing formula introducing a high level of false diffusion. The upwind differencing scheme takes account of the flow direction when determining the value at a cell face: the convective value of ϕ at the cell face is taken to be equal to the value at the upstream node, having the following form:

$$\left.\frac{\partial\phi}{\partial x}\right|_i = \left\{\begin{array}{ll} \dfrac{\phi_{i+1}-\phi_i}{h_e} & \text{flow direction: } i+1 \rightarrow i \\ \dfrac{\phi_i-\phi_{i-1}}{h_e} & \text{flow direction: } i-1 \rightarrow i \end{array}\right\} \tag{5.7}$$

5.2.1.4.3 The Streamline Upwind Petrov/Galerkin method

The SUPG method is the most commonly used stabilization method in convection-dominant fluid flow problems. It enhances the stability of the Gakerkin's method for the advection-diffusion equations without compromising the accuracy. The weak form of SUPG formulation for the transport equation is given in Equation (5.8). For details about the implementation of the stabilization method for transport, please refer to Bazilevs et al. (2013).

Find pressure $\phi^h \in \mathcal{S}^h$, such that $\forall w^h \in \mathcal{W}^h$:

$$\begin{aligned} &\int_\Omega w^h \cdot \rho\left(\frac{\partial\phi^h}{\partial t} + v^h \cdot \nabla\phi^h - f\right)\mathrm{d}\Omega + \int_\Omega \nabla w^h \cdot \nabla\phi^h\,\mathrm{d}\Omega - \int_{\Gamma_h} w^h h^h\,\mathrm{d}\Gamma, \\ &+\sum_{e=1}^{n_{el}}\int_{\Omega_e}\tau_{\mathrm{SUPG}}\left(v^h\cdot\nabla\phi^h\right)\cdot r_e\left(\phi^h\right)\mathrm{d}\Omega = 0 \end{aligned} \tag{5.8}$$

Here n_{el} is the number of elements and r_e is the residual vector of the momentum equation:

$$r_e\left(\phi^h\right) = \rho\left(\frac{\partial\phi^h}{\partial t} + v^h \cdot \nabla\phi^h - f^h\right) - \nabla\cdot\left(\nu\nabla\phi^h\right). \tag{5.9}$$

$\mathcal{S}^h$ denotes the set of finite-dimensional trail functions for the scalar ϕ; $\mathcal{W}^h$ is the set of test functions (weighting functions) for the advection-diffusion equation.

The commonly used definition of τ_{SUPG} (Bazilevs et al., 2013) is given by:

$$\tau_{\text{SUPG}} = \left(\frac{1}{\tau^2_{\text{SUPG}_1}} + \frac{1}{\tau^2_{\text{SUPG}_2}} + \frac{1}{\tau^2_{\text{SUPG}_3}} \right)^{-\frac{1}{2}}. \tag{5.10}$$

Where,

$$\tau_{\text{SUPG}_1} = \left(\sum_{i=1}^{n_{\text{en}}} \left| \boldsymbol{v}^h \cdot \nabla N_i \right| \right)^{-1}, \tag{5.11}$$

$$\tau_{\text{SUPG}_2} = \frac{\Delta t}{2}, \tag{5.12}$$

$$\tau_{\text{SUPG}_3} = \frac{h_r^2}{4\nu}, \tag{5.13}$$

here, h_r is the generalized element size,

$$h_r = 2.0 \left(\sum_{i=1}^{n_{\text{en}}} \left| \boldsymbol{j} \cdot \nabla \boldsymbol{N}_i \right| \right)^{-1}, \tag{5.14}$$

and $\boldsymbol{j}$ is the unit velocity of the gradient of the norm of the convective velocity,

$$\boldsymbol{j} = \frac{\nabla \phi^h}{\left\| \nabla \phi^h \right\|}. \tag{5.15}$$

n_{en} is the number of nodes per element, and N_i is the element shape function at node i.

5.2.1.5 Discontinuity capturing operator (DCO)

The SUPG method is not always monotonicity preserving for highly convection-dominated problems, and it may result in numerical oscillation and overshoot problem of the solution. The idea of the discontinuity capturing method (for compressible flow with shocks and for incompressible problems with large solution gradients) is to introduce directional viscosity as a limiter to bound and smooth the solution without losing accuracy, which could be used for shock capturing in compressible problems and avoiding the overshoot problem in incompressible problems. The additional term of discontinuity capturing method to Equation (5.8) is:

$$S_{\mathrm{DCDD}} = \sum_{e=1}^{n_{\mathrm{el}}} \int_{\Omega_e} \rho \nabla w^h : \left(\nu_{\mathrm{DCDD}} \boldsymbol{\kappa} \cdot \nabla \phi \right) \mathrm{d}\Omega. \tag{5.16}$$

Where,

$$\boldsymbol{\kappa} = \boldsymbol{I}. \tag{5.17}$$

The Y, Z, β method (Tezduyar and Senga, 2007),

$$\nu_{\mathrm{DCDD}} = \left| Y^{-1} Z \right| \left(\sum_{i=1}^{n_{sd}} \left| Y^{-1} \frac{\partial \phi}{\partial x_i} \right|^2 \right)^{\frac{\beta}{2}-1} \left(\frac{h_r}{2} \right)^{\beta}, \tag{5.18}$$

where

$$Y = \phi_{\mathrm{ref}}, \tag{5.19}$$

$$Z = \boldsymbol{v}^h \cdot \nabla \phi^h, \tag{5.20}$$

$$\beta = 1 \text{ or } 2. \tag{5.21}$$

Remark 5.1: The choice of β value is the key to the discontinuity capturing method to avoid the nonphysical oscillation without seriously losing accuracy.

Remark 5.2: From Equation (5.16), we know that the discontinuity capturing operator nothing else but with the directional viscosity added to the system, and the isotropic viscosity is defined if $\boldsymbol{\kappa} = \boldsymbol{I}$ is adopted.

5.2.1.6 Total variation diminishing method for finite volume method

The total variation (TV) for coupled systems is defined as:

$$TV(\phi) = \sum_{i=1}^{N} \left| \phi_{i+1} - \phi_i \right|. \tag{5.22}$$

In order to satisfy the monotonicity condition, TV must not increase with iteration or time, that is,

$$TV\left({}^{n+1}\phi\right) < TV\left({}^{n}\phi\right). \tag{5.23}$$

Equation (5.23) shows the TVD condition, where n and $n + 1$ refer to consecutive time steps or iterations.

The upwind differencing scheme that satisfies the TVD condition is the most stable and unconditionally bounded scheme and does not give any "wiggles," but it introduces a high level of false diffusion because of its low order of accuracy. The central differencing scheme has higher second-order accuracy but may result in spurious oscillation or "wiggles" when the Peclet number exceeds. In contrast, TVD schemes are designed for finite volume and the finite difference method to address this unrealistic oscillation behavior with high-order accuracy. In TVD schemes, the tendency toward oscillation is counteracted by adding an artificial diffusion fragment or a weighting toward upstream contribution. Several differencing schemes that satisfy the TVD conditions are presented by Versteeg and Malalasekera (1995). For the finite volume method, we only present the basic concept in the book; for details, please refer to Versteeg and Malalasekera (1995).

5.2.1.7 TVD method for finite element method

The algebraic-based TVD method was introduced to finite element method (FEM) by Kuzmin and Turek (2004), which improves the matrix properties and makes it satisfy the TVD condition during the matrix assembly process. Although the operation is at the stage of the matrix assembly and directly works on the modification of the value of the matrix term, it is easy to implement in the FEM procedure. However, in our experience, it excessively smears the solution, and the accuracy of the solution may lose dramatically (see the example given later).

5.2.1.8 Example of convection dominated heat transfer problem

A discussion about convection-dominated heat transfer problems is developed by a tube flow case in this section. Figure 5.4 shows the mesh of the tube, which has an inlet temperature of 300°C, an inlet velocity 0.04 m/s, and zero outlet pressure. The no-slip wall has a heat transfer coefficient of 10 W/(m^{2}K), and the environment temperature is 20°C. The element size is 1.25e-3m. The material properties of the model are as listed follows: mass density $\rho = 0.615$ kg/m^3, dynamic viscosity $\mu = 2.97 \times 10^{-5}$ Pa s, bulk modulus $B = 1.0 \times 10^5$ Pa, thermal conductivity $k = 0.0261$ W/m°C, and specific heat capacity $C = 1004.4$ J/(kg°C). Here, two examples with different Peclet numbers are conducted to demonstrate the validity of the stabilization method in the convection-dominated heat transfer problem. Figure 5.5 shows the temperature distributions with different Peclet numbers when different advection schemes are used.

Table 5.1 shows the results with different Peclet numbers when different advection schemes are used. Through a comparison with CFX (FVM with TVD), it is known that a variety of the FEM-based advection schemes such as CD, TVD,SUPG, and SUPG with DCO work well for heat transfer problems with Peclet numbers less than 2.0. Although with large Peclet numbers, CD method and SUPG method have an overshoot problem, FEM with the TVD method has the problem that the lower bound of temperature is too high because of the high viscosity. In contrast, the spatial stabilization SUPG with discontinuity capturing method for FEM and TVD stabilization for FVM could give a stable solution with acceptable accuracy.

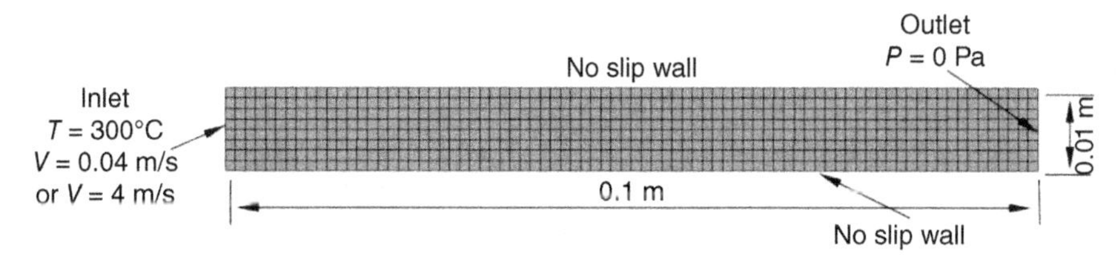

Figure 5.4 Model of a tube flow case.

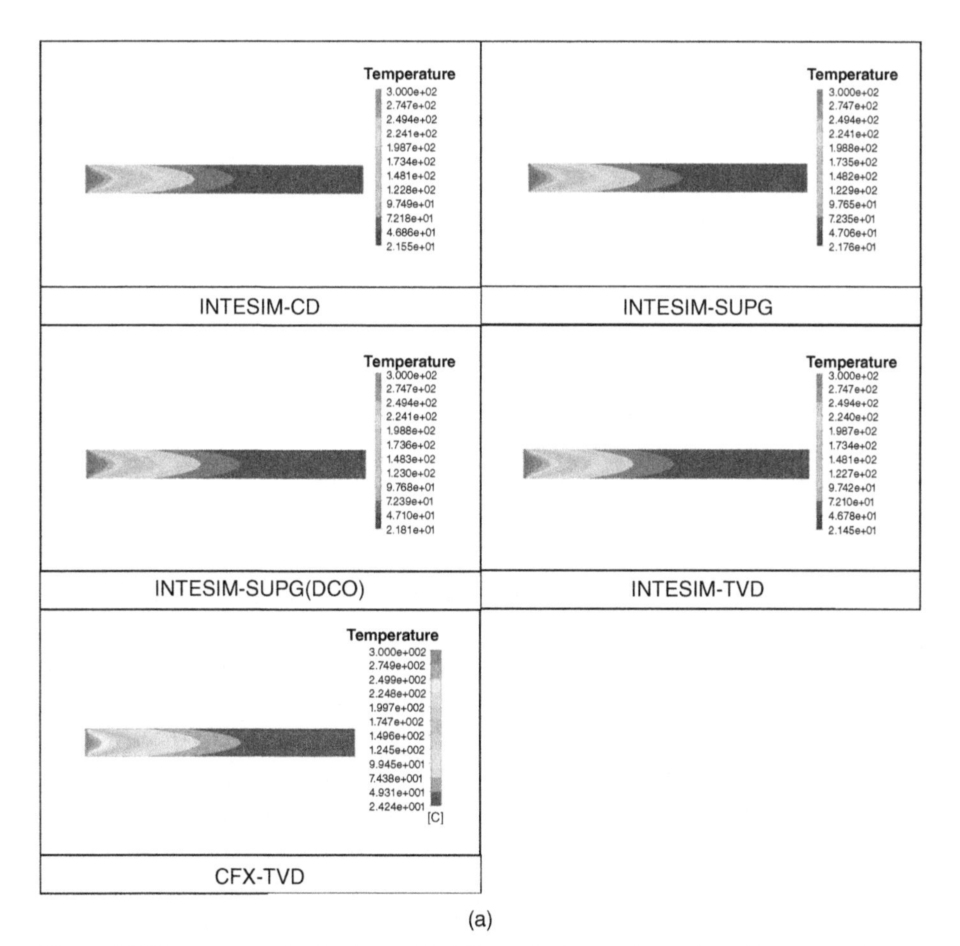

Figure 5.5 Temperature distributions with different Peclet numbers when different methods are used. (a) Pe = 1.03 and (b) Pe = 103.

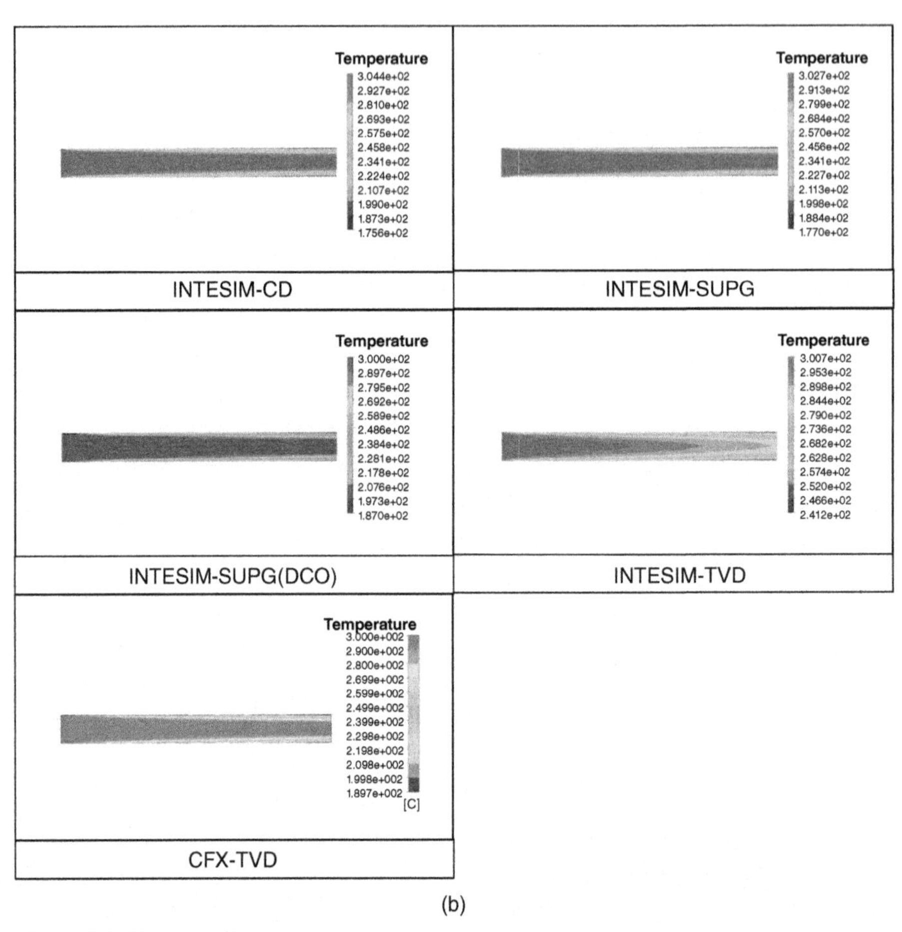

Figure 5.5 *(Continued)*

Table 5.1 Temperature distribution comparison with different Peclet number when different advection schemes are used

Advection scheme	V = 0.04 m/s, Pe = 1.03 (°C)	V = 4 m/s, Pe = 103 (°C)
INTESIM-CD	21.55-300	175.58-304.38
INTESIM-SUPG	21.76-300	176.97-302.73
INTESIM-SUPG with DCO	21.81-300	187.00-300
INTESIM-TVD	21.45-300	241.18-300.66
CFX-TVD	24.2-300	189.7-300

5.2.2 Stabilization methods for incompressible problems

5.2.2.1 Inf-sub conditions for incompressible problems

The mixed u/p formulation is widely used for the incompressible and almost incompressible problems in solid mechanics (e.g., hyperelasticity) and fluid dynamics.

The adoption of mixed u/p formulation is to avoid the instability of the pressure field, which relates to the satisfaction of the inf-sup condition (Bathe, 2006). To satisfy the inf-sup condition, the element must be volume locking free. Generally, the smaller of the pressure space $\mathcal{S}_p^h$ than the velocity or displacement space $\mathcal{S}_v^h$, the easier it is to satisfy the inf-sup condition. For the mixed u/p formulation, the weakened inf-sup condition can be written in the following form (Bathe, 2006):

$$\inf_{q_h \in P_h(D_h)} \sup_{v_h \in V_h} \frac{q_h (\nabla v_h) \mathrm{d}\Omega}{\|q_h\| \, \|v_h\|} \geq \beta > 0. \tag{5.24}$$

Where, β is a constant independent of h. The projection operator P_h is defined by:

$$\int_{\Omega} \left[P_h (\nabla \cdot v_h) - \nabla \cdot v_h \right] q_h \, \mathrm{d}\Omega = 0 \, \text{for all} \, q_h \in \mathcal{S}_p^h. \tag{5.25}$$

$\mathcal{S}_p^h$ is a pressure space to be chosen, and $\mathcal{S}_p^h$ is always contains $P_h(D_h)$, but $\mathcal{S}_p^h$ is sometimes larger than $P_h(D_h)$.

Bathe (2006) gives a detailed discussion about the inf-sup condition and presents a list of mixed u/p elements that satisfy the inf-sup condition. Also, some elements that do not satisfy the inf-sup condition but can still be used reliably for many practical problems are also recommended. In this book, we use mixed u/p elements that satisfy the inf-sup condition for incompressible hyperelasticity. The PSPG stabilization method is used for the incompressible fluid element with equal-order shape function for velocity and pressure.

5.2.2.2 PSPG method for fluid flow with equal-order shape function for pressure and velocity

It is convenient to have equal-order shape function for the pressure and velocity, but this may result in a spurious pressure mode for incompressible or almost incompressible fluid flow because the inf-sup condition is not satisfied. One way to stabilize the pressure field is to add an additional term to the weak form of the NS equations, called the PSPG method. The formulation of PSPG method has been given in Sections 1.2.7.1.

5.2.3 Gauge method for* un-unique solution in space: *electromagnetic analysis with edge element

For the edge-based electromagnetic element in terms of magnetic vector potential (Chapter 1), the solution is not unique, so the Gauge method is used to unify the solution. As discussed in Chapter 1, spanning tree gauge, iterative solvers, and Lagrange multiplier gauge can be used to regularize the solution and obtain a well-posed equation system with a unique solution. In this section, we demonstrate the validity of the proposed methods by the given example.

This numerical example has been used to investigate the performance of different numerical methods (Preis et al., 1991). In this numerical example, an iron cube with

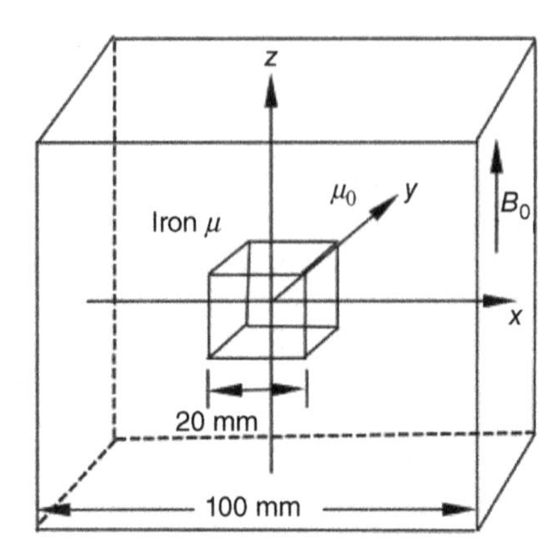

Figure 5.6 Iron cube in a homogeneous magnetic field.

Figure 5.7 The analyzed region, the finite element model.

relative permeability equaling $1000(\mu = 1000\mu_0)$ surrounded by air is situated in a homogeneous magnetic field ($B_0 = (200\text{ T})\boldsymbol{e}_z$). We will verify the distribution of magnetic induction by using different methods.

As shown in Figure 5.6, the normal component of the magnetic flux density on the planes $x = 0$ mm and $y = 0$ mm is equal to zero (i.e., Γ_B type boundaries). The tangential component of the magnetic field intensity on the plane $z = 0$ mm is equal to zero (i.e., Γ_H type boundary). The homogeneous field in the z direction is simulated by artificial boundaries conditions defined by potentials on the planes $x = \pm 50$ mm, $y = \pm 50$ mm. The artificial far boundaries placed at $z = \pm 50$ mm are Γ_H type. Considering the symmetry characters of the structure and boundary conditions mentioned earlier, the analysis of the whole problem is simplified into an eighth of it $x \geq 0$ mm, $y \geq 0$ mm, $z \geq 0$ mm as shown in Figure 5.7, and corresponding boundary conditions are listed in Table 5.2.

In the absence of iron, the distribution of magnetic flux density is homogenous and along z direction (the value is 200 T) in Figure 5.6 by specifying boundary conditions defined by vector potential. In our study, the iron is set in the center of air region. We

Table 5.2 **Boundary conditions in an eighth of the model**

Plane	$x = 0$ mm	$x = 50$ mm	$y = 0$ mm	$y = 50$ mm
Along x-axis	$Az = 0$	$Az = 0$	$Az = 0$	$Az = 0$
Along y-axis	$Az = 0$	$Az = 10$	$Az = 0$	$Az = 0$

will verify and compare the distribution of magnetic flux density approximated by different formulations or discretization methods.

The magnetic flux density approximated by different formulations or discretization methods is similar to each other: large (condensed) in the iron region, especially in the corns, and small (sparse) in the air region. However, a more detailed comparison shows that difference exists in different formulations. Considering the change of magnetic flux density along the line $y = 0$ mm, $z = 0$ mm, and line $y = 2.5$ mm, $z = 0$ mm, and line $y = 5$ mm, $z = 0$ mm and line $y = 7.5$ mm, $z = 0$ mm, we find that the solution by mixed FEM method and the solution by tree gauge are consistent with the reference results, achieving a more accurate solution than the node-based FEM results (Figure 5.8).

The inaccurate distribution approximated by Coulomb–Gauged node element especially near the iron–air interface has been found (Preis et al., 1991). To overcome this, a CG/ICCG iterative solver without imposing Coulomb gauge is needed to solve this singular system. Unfortunately, this method leads to a large number of iterations. As an alternative method to the nodal element method, the edge element method, combined with spanning tree gauge or Lagrange multiplier gauge, can avoid the mentioned problems.

5.2.4 *Turbulence model for CFD analysis: unresolution in space*

In cases in which the mesh resolution is not enough for the fine-scale response (e.g., turbulence flow with high Reynolds number), the turbulence model may be needed to model the effects of fine-scale characteristics on the mesh scale resolution.

Although we have presented k-epsilon model and k-omega model as well as the LES model in Chapter 1, we will skip the details here. We know that without using the turbulence model for a high Reynolds number problem, it usually causes the problems of solution divergence, unphysical oscillation, inaccurate result, and so on. The direct contribution from the turbulence model to the laminar NS equations is the turbulence viscosity, which not only improves the accuracy of the solution but also enhances the numerical stability on the mesh scale resolution. So we can consider the turbulence model as a generalized numerical stabilization method in the sense of multiscale phenomena.

5.2.5 *Regularization for discontinuity of viscous model for sliding viscous contact problems*

The Coulomb viscous model for contact problem is highly nonlinear, as discussed in Chapter 1. The viscous force is discontinuous on sliding contact. As the relative velocity

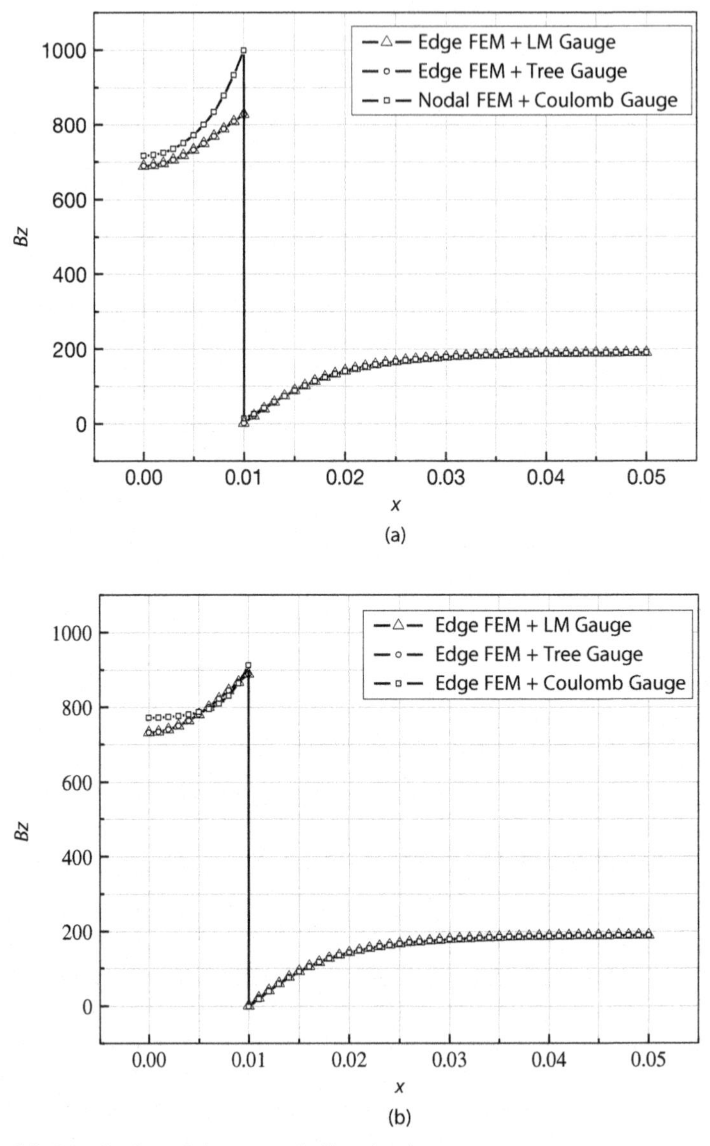

Figure 5.8 Distribution of the magnetic flux density.

$\boldsymbol{v}_T$ reverses the direction, $\boldsymbol{F}_T$ reverses the direction, correspondingly (Figure 5.9). This may result in divergence of the contact simulation. The solution to this is to regularize the frictional viscous model by the equation:

$$\boldsymbol{F}_T = -\mu |F_N| \frac{2}{\pi} \arctan\left(\frac{V_T}{c}\right) \boldsymbol{V}_T / \|\boldsymbol{V}_T\|. \tag{5.26}$$

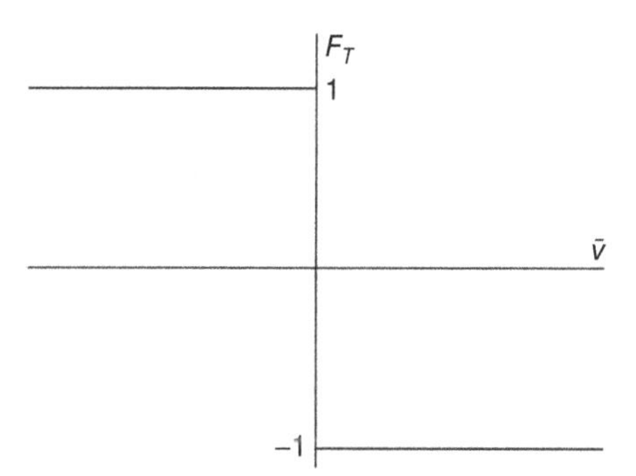

Figure 5.9 Coulomb viscous model ($F_T = 1$).

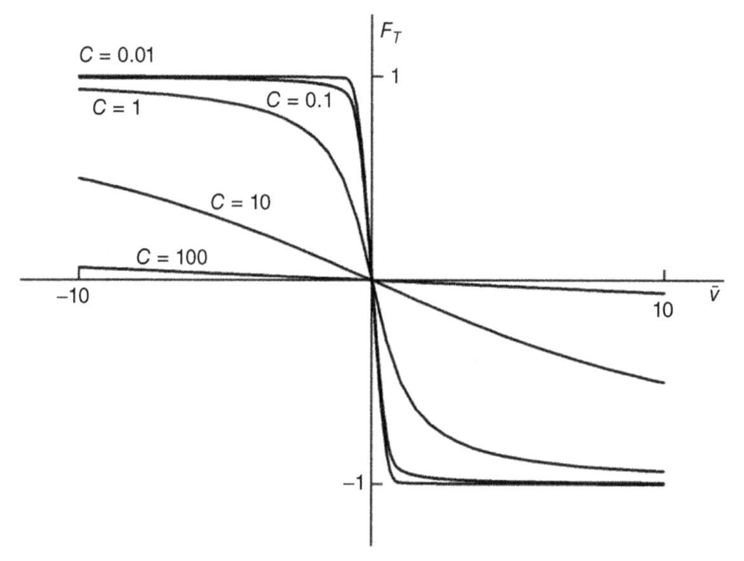

Figure 5.10 Viscous model after regularization.

Where, c is the regularization factor and $\boldsymbol{F}_T$ and F_N are the tangent and normal contact force, respectively. The smaller c is, the steeper the viscous model curve becomes (Figure 5.10). The regularization for the discontinuity of viscous model in contact analysis is one of the effective ways to smooth out the discontinuity and makes it more stable and easier to solve than the original formulation.

5.3 Stabilization in the time integration scheme

When using FEM to discretize the spatial domain, the spatial resolution of these high-frequency modes typically is poor, and controllable numerical dissipation in the higher frequency modes is desirable, so that spurious high-frequency responses could be damped out. On the other hand, high-frequency numerical dissipation could improve the convergence of the nonlinear iterative process when solving highly nonlinear problems. In this section, stabilization methods in terms of time discretization by introducing numerical dissipation for implicit schemes are discussed, including adjusting integration parameters in the Newmark-β method, using the α method, and using the modified matrix method for thin structure with rotational DOFs. Numerical examples are given to demonstrate the validity of stabilization methods used in time integration.

5.3.1 Direct time integration scheme for time discretization

5.3.1.1 Central differencing method

Second-order accuracy but conditionally stable

For the acceleration:

$$ {}^{t}\ddot{u} = \frac{1}{\Delta t^2}\left({}^{t-\Delta t}\boldsymbol{u} - 2\,{}^{t}\boldsymbol{u} + {}^{t+\Delta t}\boldsymbol{u}\right). \tag{5.27} $$

For the velocity:

$$ {}^{t}\dot{\boldsymbol{u}} = \frac{1}{2\Delta t}\left({}^{t}\,{}^{t+\Delta t}\boldsymbol{u} - {}^{t-\Delta t}\boldsymbol{u}\right). \tag{5.28} $$

An important consideration for the central differencing method is the request for a small time step size, which cannot exceed the critical value:

$$ \Delta t < \Delta t_{cr} = \frac{T_n}{\pi}. \tag{5.29} $$

Where T_n is the smallest time period of the FE assemblage with n DOFs.

5.3.1.2 Backward Euler method

First-order accuracy, unconditionally stable

$$ {}^{t+\Delta t}\dot{\boldsymbol{u}} = \frac{1}{\Delta t}\left({}^{t+\Delta t}\boldsymbol{u} - {}^{t}\boldsymbol{u}\right). \tag{5.30} $$

The backward Euler integration is first-order accuracy and unconditionally stable method.

5.3.1.3 Newmark-β method

Second-order accuracy, unconditionally stable with suitable parameters

Combining the advantages of the central difference method and the backward Euler method, the Newmark-β method is a kind of unconditionally stable method with second-order accuracy if only the integration parameters satisfy certain conditions. It could be considered as an extension of the linear acceleration method, and the following assumptions are used (Bathe, 2006):

$$ {}^{t+\Delta t}\dot{u} = {}^{t}\dot{u} + \left[(1-\delta)\,{}^{t}\ddot{u} + \delta\,{}^{t+\Delta t}\ddot{u}\right]\Delta t, \tag{5.31} $$

$$ {}^{t+\Delta t}u = {}^{t}u + {}^{t}\dot{u}\Delta t + \left[\left(\frac{1}{2}-\alpha\right){}^{t}\ddot{u} + \alpha\,{}^{t+\Delta t}\ddot{u}\right]\Delta t^{2}. \tag{5.32} $$

Where α and δ are parameters that should be determined by:

$$ \delta \geq 0.5, \alpha \geq \frac{1}{4}\left(\gamma + \frac{1}{2}\right)^{2} \tag{5.33} $$

to satisfy the unconditionally stable property.

When $\delta = 1/2$ and $\alpha = 1/6$, Equations (5.31) and (5.32) correspond to the linear acceleration method. When $\delta = 1/2$ and $\alpha = 1/4$, Equations (5.31) and (5.32) correspond to the constant–average–acceleration method (also called the trapezoidal rule), which is widely used in practical engineering because it is unconditionally stable and second-order accurate and, shows no amplitude decay, with acceptable period elongation.

5.3.2 Ways to stabilize the solution in time domain

5.3.2.1 *Artificial damping for the Newmark time integration method*

The Newmark-β method is an unconditionally stable method with suitable integration parameters when a linear analysis is applied, which is now widely used in practical analyses. Specifically, the trapezoidal rule, one of the Newmark family, is recognized to be most effective: unconditionally stable, second-order accurate showing no amplitude decay, with acceptable period elongation (Bathe, 2006). However, the trapezoidal rule may become unstable and even fail to converge during the nonlinear iteration process when nonlinear analysis is pursued (e.g., large deformation dynamic analyses). As discussed previously, the introduction of numerical dissipation into time integration methods can improve the convergence.

For the Newmark-β method, the artificial damping can be introduced by increasing the time integration parameter in Equation (5.33).

With a larger δ and α value, numerical dissipation could be introduced to damp out the high-frequency response, and the solution becomes stable.

5.3.2.2 Using the α method

As discussed in the Section 5.3.2.1, to improve the convergence of nonlinear iteration process with Newmark-β method, numerical dissipation should be introduced by adjusting the integration parameters. However, as we adjust the integration parameters δ and α to damp out the high-frequency response, the lower frequency response is also largely affected, leading to inaccuracy. To overcome these drawbacks, the α-method was proposed. It is unconditionally stable for linear analysis. In particular, it possesses a good accuracy combined with user-controlled high-frequency damping, showing good convergence during nonlinear iteration process. The example used to demonstrate the validity of α method is given in Section 5.3.4.

5.3.3 Modified matrix method: stabilization method for thin structure with rotational DOFs

For the extreme thin shell structure, the rotational degree of freedom-related Eigenvalue becomes very high. Then the criteria of time–step–size will be very small and impractical. The modified mass matrix for the rotational degree of freedom can resolve this issue (Hughes et al., 1977, 1978; Zhang, 1999). This modified mass is able to lower the maximum frequency and will weaken the higher order vibration model to stabilize the dynamic solution of the thin shell structure. The consistent mass matrix for mixed interpolation tensorial components (MITC) shell element is:

$$\boldsymbol{M}_u = \rho_0 a \int_A N_i N_j \,\mathrm{d}A, \tag{5.34}$$

$$\boldsymbol{M}_\alpha = \rho_0 \frac{a^3}{12} \int_A N_i N_j \,\mathrm{d}A \left({}^t\psi^i\right)^T {}^t\psi^j, \tag{5.35}$$

where a is the thickness of the shell element and A is the area of the shell element.

$${}^t\psi^i = \left\{ -{}^tV_2^i \quad {}^tV_1^i \right\}. \tag{5.36}$$

For four-node shell element, Equation (5.37) is used to modify the mass matrix of rotational DOFs.

$$\boldsymbol{M}_\alpha = \rho_0 \frac{a\gamma}{8} \int_A N_i N_j \,\mathrm{d}A \left({}^t\psi^i\right)^T {}^t\psi^j, \tag{5.37}$$

where $\gamma = A$.

This modified mass matrix will not affect the solution much but dramatically stabilizes the dynamic solution.

5.3.4 Example demonstrating stabilization in time integration

A transient analysis of structural mechanics and fluid dynamics is mainly carried out using direct time integration of the equations of motion. In the case of a linear dynamic system, the main topic is usually computational accuracy, and the criterion for stability is easy to satisfy relatively. On the other hand, in the case of a nonlinear dynamic system, the main interest may be focused on the numerical stability of the algorithms. For reliable solutions in a nonlinear dynamic system, much research effort has been directed to improve the stability of integration schemes for dynamic analysis (Kuhl and Crisfield, 1999; Bathe and Baig, 2005). The α-method (Chung and Hulbert, 1993), as discussed in Chapter 2, is a kind of outstanding time integration method that is widely used in dynamic analysis, having superior numerical performance of computational stability. In this section, we will demonstrate it by comparing it with the Newmark-β method.

Here we consider an example (Bathe and Baig, 2005) shown in Figure 5.11, modeled with 20×5 mesh of a four-node MITC shell elements and subjected to pressure loading and with dimension of 0.4 m $\times$ 0.1 m $\times$ 0.001 m. The density $\rho = 2700$ kg/m^3, the Young's modulus $E = 7 \times 10^{10}$ N/m^2, and the Poisson's ratio $\nu = 0.33$. The total analysis time is 2 s, and the time interval $\Delta t = 0.002$ s. The large deformation of structure is considered.

We present numerical results by using Newmark-β time integration scheme with $\delta = 0.5$ and $\alpha = 0.25$ and the Newmark-β time integration scheme with $\delta = 0.6$ and $\alpha = 0.3025$, and α method with $\rho = 0.8$.

We first solve the problem by using the Newmark-β time integration scheme with $\delta = 0.5$ and $\alpha = 0.25$. The displacement, velocity, and acceleration of the free end of the shell model along the z-direction are plotted in Figure 5.12. Newmark-β method with $\delta = 0.5$, $\alpha = 0.25$ performs well for about four time periods, but after that, the responses deteriorate noticeably. Eventually, the solution is diverged as shown in Figures 5.12–5.14.

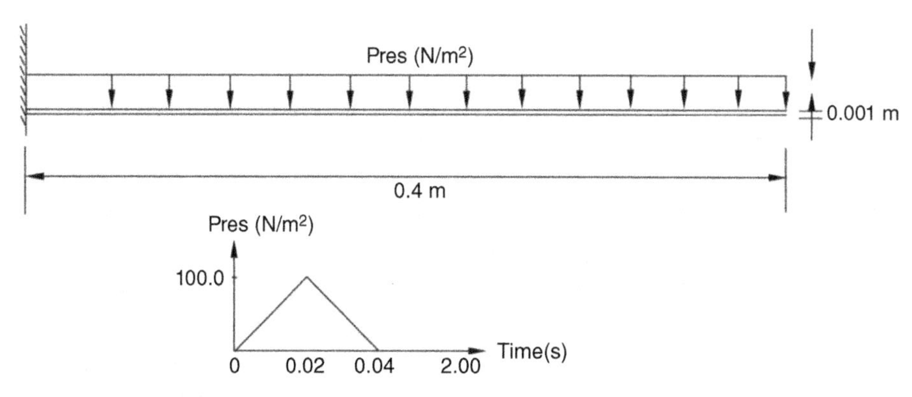

Figure 5.11 Shell model with large deformation subjected to pressure loading.

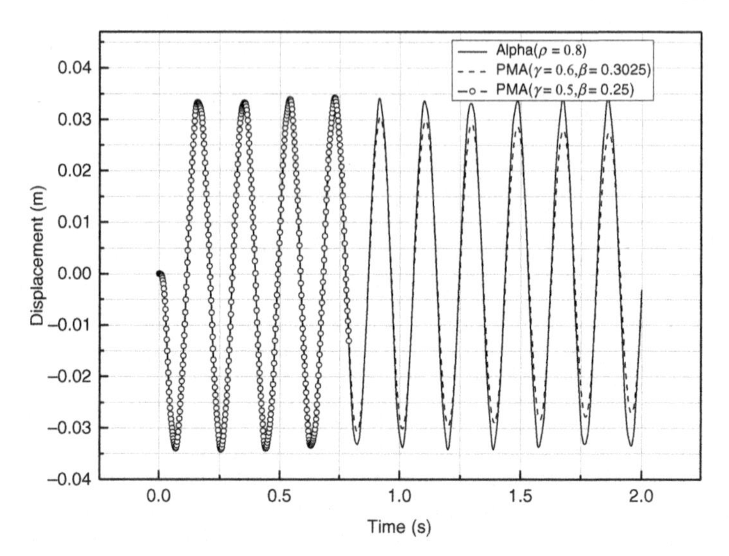

Figure 5.12 Displacement response at the free end using different methods.

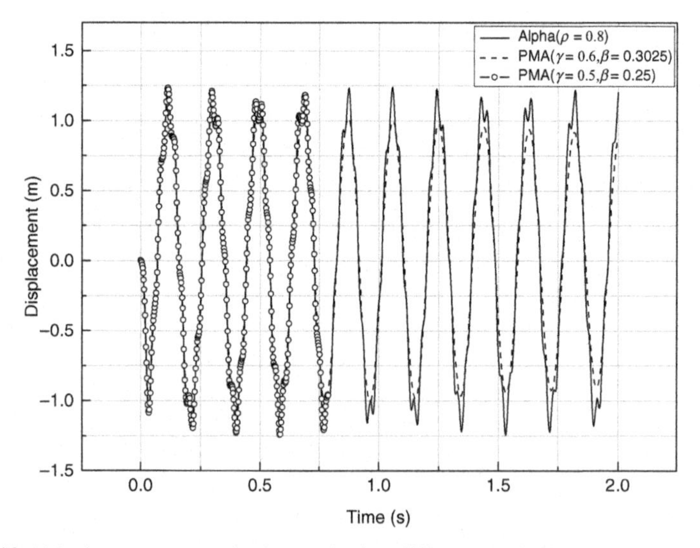

Figure 5.13 Velocity response at the free end using different methods.

The problem is next solved using the Newmark-β time integration scheme with $\delta = 0.6$ and $\alpha = 0.3025$. It has better stability but worse accuracy because great numerical damping reduces the accuracy of the solution, as shown in Figures 5.12–5.14.

Next we solve the problem by using the α method (with $\rho = 0.8$). The solution obtained is stable and accurate, as shown in Figures 5.12–5.14, and in our experience, the α method remains stable and accurate even if a larger time step size is used.

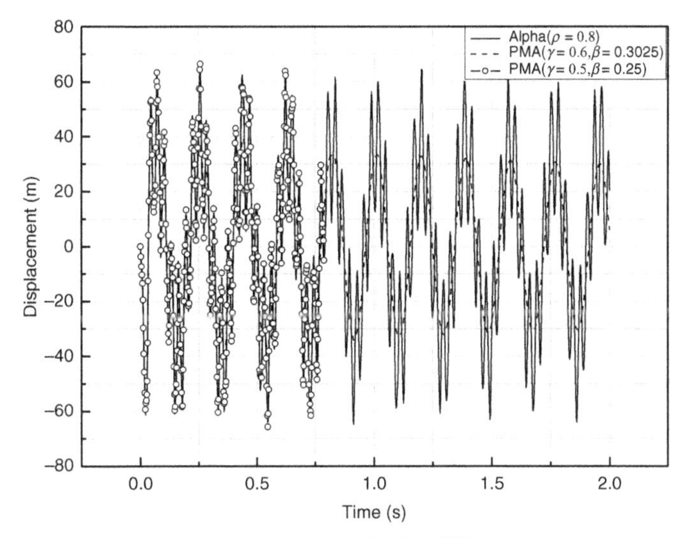

Figure 5.14 Acceleration response at the free end using different methods.

5.4 Underrelaxation of the solution vector

5.4.1 Direct underrelaxation of the solution and load vector

The relaxation method is another well-known approach to stabilizing a nonlinear simulation procedure. Relaxing the solution may slow down the convergence rate but will make the nonlinear iteration more stable. It is helpful for the highly nonlinear-coupled physics problems. Equation (5.38) is used to relax the changes of solution vector in this book.

$$\phi_{i+1} = \phi_i + \alpha \Delta\phi, 0 < \alpha \le 1.0 \tag{5.38}$$

where ϕ_{i+1} is the solution vector at iteration $i + 1$, ϕ_i is the solution vector at iteration i, $\Delta\phi$ is the incremental of the solution vector from the coupled incremental equation solver, and α is the relaxation factor with $\alpha = 1.0$ representing that no relaxation is applied. By reducing the α value from 1 to 0, it will make the nonlinear iteration stable but slow down the convergence rate.

Relaxation can be applied to all solution variables in physics solvers, the interface loads for weak coupling load transfers, and any other system nonlinear properties, which could always slow down the convergence rate and stabilize the nonlinear iteration process. The optimal choice of the relaxation factor is the maximum value that achieves consistent convergence behavior of the nonlinear system, although it is not easy to obtain in practice.

5.4.2 Inertial relaxation for the steady problem

The importance of the quality of diagonally dominant matrix has been discussed in a previous section. The inertial-relaxation for steady state makes an equation system more diagonally dominant. It is commonly used in solving the compressible pressure equations and turbulence equations.

The original algebraic equation (e.g., finite element equations) for node i can be representatively written as:

$$a_{ii}\phi_i + \sum_{j\neq i} a_{ij}\phi_j = f_i, \tag{5.39}$$

With the inertial relaxation, this equation changes to:

$$\left(a_{ii} + A_{ii}^d\right)\phi_i + \sum_{j\neq i} a_{ij}\phi_j = f_i + A_{ii}^d\phi_i^{\text{old}}, \tag{5.40}$$

where

$$A_{ii}^d = \frac{\int_{\Omega_i} \rho w_i \, d\Omega}{\alpha}$$

α is the inertial relaxation factor.

At convergence, $\phi_i^{\text{old}} = \phi_i$, the inertial relaxation terms in Equation (5.40) cancel out. Then Equation (5.40) with inertial relaxation backs to the original form in Equation (5.39).

5.5 Capping for the solution

For some problems without internal sources, if we know the solution boundaries prior to solving the problem, we can set limiters to cap the solution with the expected ranges. Due to numerical reasons, it is possible for part of the solutions to exceed the physically allowed boundaries (undershoots or overshoots) during the iteration process. As long as this happens, we cut the solution off for any value out of the specified ranges. This will avoid unreasonable solutions and thus stabilize the nonlinear iteration process.

Capping can be employed to the relative value of pressure and the absolute value of temperature. You need to cap the total temperature for compressible thermal analysis, and it will help ensure negative properties that do not enter the calculation or the overshoot or undershoot in convective heat transfer problems.

5.6 Trade off the stability, accuracy, and efficiency

For a numerical scheme, the stability and accuracy are always similar to conflict partners. This means that the desire for a more stable numerical scheme usually causes less accuracy with the numerical scheme. The key factor in engineering simulations is to have an acceptable accuracy with practical computational requests for both the memory usage and the computing time. Moreover, the solution process should be stable and reliable. So how to choose appropriate methods (i.e., stabilization methods in numerical simulation) to satisfy the requests mentioned previously is critical, usually determining the success of a simulation. For a highly nonlinear multiphysics problem, we may choose the best combination of the computational stability that guarantees a converged solution and acceptable accuracy. Zhang and Hisada (2001) show that the SUPG and PSPG methods for incompressible fluid flow, modified mass matrix for thin shell structure, with automatic time stepping and artificial damping, make it possible to solve the totally embedded artificial heart problems with extremely large domain changes and structural buckling.

Coupling simulation for rotating machines

6

Chapter Outline

6.1 Reference frames

6.1.1 Global stationary Cartesian frame

The most commonly used coordinate system is the global stationary Cartesian frame, as demonstrated in Figure 6.1. It follows the right-hand law, with origin O and coordinate axes X, Y, and Z fixed in space and time.

Q. Zhang & S. Cen: Multiphysics Modeling. http://dx.doi.org/10.1016/B978-0-12-407709-6.00006-7

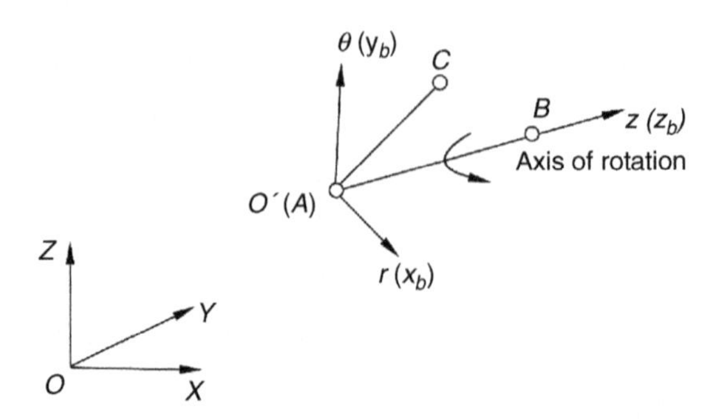

Figure 6.1 Coordinate systems.

6.1.2 Body-attached rotating reference frame

6.1.2.1 Definition of a rotating reference frame

The axis of rotation for a rotational frame can be defined by point $\boldsymbol{A}$ and point $\boldsymbol{B}$ in the global stationary frame as shown in Figure 6.1. The local rotational coordinate system (O', x_b, y_b, z_b) at the initial state can be set up by a given point $\boldsymbol{C}$ that is located in the rotating body but not on the rotational axis, and the following vector product calculations (Equations (6.1)–(6.3)) can be used to define the axes:

The rotational axis:

$$\boldsymbol{e}_z = \frac{\boldsymbol{AB}}{\|\boldsymbol{AB}\|}, \tag{6.1}$$

The coordinate axis in the rotational direction:

$$\boldsymbol{e}_\theta = \boldsymbol{e}_z \times \frac{\boldsymbol{AC}}{\|\boldsymbol{AC}\|}, \tag{6.2}$$

The coordinate axis in the radial direction:

$$\boldsymbol{e}_r = \boldsymbol{e}_\theta \times \boldsymbol{e}_z. \tag{6.3}$$

6.1.2.2 Matrix and vector transformation

The rotational transformation matrix $\boldsymbol{R}_{xb\_X}$ in Equation (6.4) can be used for a vector transformation calculation from the global stationary Cartesian coordinate to the rotational body-attached frame, as shown in Equation (6.5).

$$\boldsymbol{R}_{xb\_X} = \begin{bmatrix} \boldsymbol{e}_r^{\mathrm{T}} \\ \boldsymbol{e}_\theta^{\mathrm{T}} \\ \boldsymbol{e}_z^{\mathrm{T}} \end{bmatrix} = \begin{bmatrix} \boldsymbol{e}_{rx} & \boldsymbol{e}_{ry} & \boldsymbol{e}_{rz} \\ \boldsymbol{e}_{\theta x} & \boldsymbol{e}_{\theta y} & \boldsymbol{e}_{\theta z} \\ \boldsymbol{e}_{zx} & \boldsymbol{e}_{zy} & \boldsymbol{e}_{zz} \end{bmatrix}, \tag{6.4}$$

$$\boldsymbol{v}_{xb} = \boldsymbol{R}_{xb\_X}\boldsymbol{v}_X. \tag{6.5}$$

Where; $\boldsymbol{v}_{xb}$, the vector in the body-attached frame $(O',\ x_b, y_b, z_b)$; and $\boldsymbol{v}_X$, the vector in the global stationary frame $(O,\ X, Y, Z)$.

6.2 General coupling boundary conditions

6.2.1 The internal coupling for physics models

The internal coupling in a single physics is the coupling due to the domain partitions where the continuity conditions as well as the corresponding conservative conditions are needed to be satisfied across the partition–internal coupling interfaces (Figure 6.2).

6.2.1.1 Structure mechanics

Primary variables for structure mechanics: displacement vector, $\boldsymbol{u}$; rotation vector, β; and pressure (scalar value, only for mixed formulation), P.

Continuity conditions:

$$\boldsymbol{u}_A = \boldsymbol{u}_B \quad \text{on} \quad \Gamma_c, \tag{6.6}$$

$$\beta_A = \beta_B \quad \text{on} \quad \Gamma_c. \tag{6.7}$$

Corresponding conservative/balance properties: forces and moment;

$$\boldsymbol{F}_A = -\boldsymbol{F}_B \quad \text{on} \quad \Gamma_c \tag{6.8}$$

$$\boldsymbol{M}_A = -\boldsymbol{M}_B \quad \text{on} \quad \Gamma_c. \tag{6.9}$$

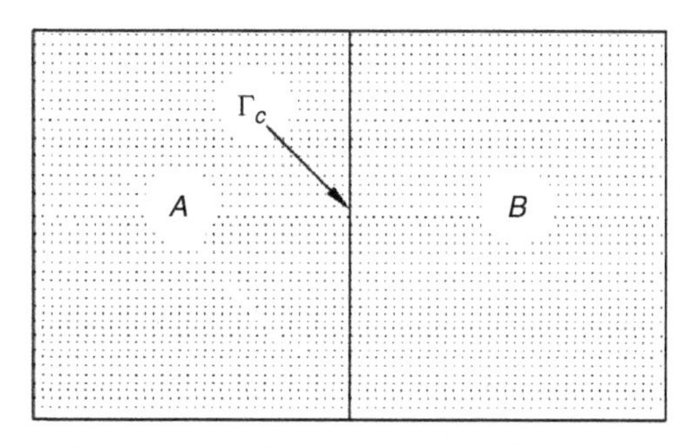

Figure 6.2 Internal coupling interface Γ_c between parts A and B.

6.2.1.2 Fluid mechanics

Primary variables: velocity vector, $\boldsymbol{v}$; pressure (scalar value), P.

Continuity conditions: velocity and pressure;

$$\boldsymbol{v}_A = \boldsymbol{v}_B \quad \text{on} \quad \Gamma_c \tag{6.10}$$

$$P_A = P_B \quad \text{on} \quad \Gamma_c. \tag{6.11}$$

Corresponding conservative/balance properties: forces, continuity term;

$$\boldsymbol{F}_A = -\boldsymbol{F}_B \quad \text{on} \quad \Gamma_c. \tag{6.12}$$

6.2.1.3 Thermal analysis

Primary variables: temperature, T.

Continuity conditions: temperature;

$$T_A = T_B \quad \text{on} \quad \Gamma_c \tag{6.13}$$

Corresponding conservative/balance properties: heat flow;

$$\boldsymbol{Q}_A = -\boldsymbol{Q}_B \quad \text{on} \quad \Gamma_c. \tag{6.14}$$

Note: The view factor must be updated as the mesh rotates.

6.2.1.4 Electromagnetic analysis

Primary variables: electric scalar potential, V; and magnetic vector potential, $\boldsymbol{A}$ ($\boldsymbol{A}_z$ is a directional value along each edge for edge based element Figure 6.3, the constant tangential component of $\boldsymbol{A}$ along edge).

Continuity conditions: the tangential components of the magnetic vector potential, $\boldsymbol{A}$ must be continuous across the internal coupling interface:

$$\boldsymbol{A}_\tau^A = \boldsymbol{A}_\tau^B \quad \text{on} \quad \Gamma_c \tag{6.15}$$

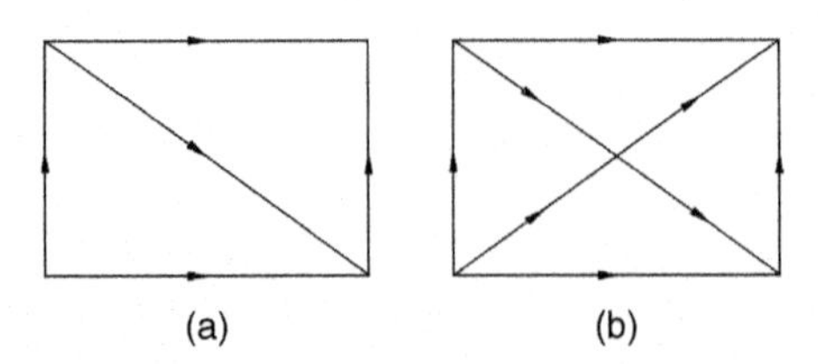

Figure 6.3 Interface mesh and the edge based magnetic vector potential degree of freedoms. (a) Side A and (b) side B.

where,

$$A_{\tau}^{A} = A^{A} \wedge n^{A} = A^{A} - \left(A^{A} n^{A}\right) n^{A}, \tag{6.16}$$

where Γ_c, the coupling interface: control surface meshes on the coupling interface, surface meshes on side A, or surface meshes on side B; n^A, the surface normal vector on the side A; and n^B the surface normal vector on the side B.

Using the least square method in tangential direction (or penalty method),

$$\min \int_{\Gamma_c} \alpha\left(A^{A} \wedge n^{A} - A^{B} \wedge n^{B}\right)^{\mathrm{T}} \left(A^{A} \wedge n^{A} - A^{B} \wedge n^{B}\right) \mathrm{d}S, \tag{6.17}$$

where α is the weight factor and can be unity across Γ_c.

Lagrange multiplier method (or Mortar element method):

$$\int_{\Gamma_c} \left(A^{A} \wedge n^{A} - A^{B} \wedge n^{B}\right) \phi \,\mathrm{d}S \tag{6.18}$$

where, ϕ is the Lagrange multiplier.

From Equations. (6.17) and (6.18), it is obvious that the explicit mapping matrix M and interpolation relations of Equation (6.19) for the transformation of vector potential A from side A to side B is not easy to obtain. This results in the difficulty of using the multipoint constraint method, although the Lagrange multiplier method is applicable for this type of problem. The details about Lagrange multiplier formulation have been presented in Section 1.4.7.

$$A^{A} = MA^{B} \tag{6.19}$$

In contrast, for the electric scalar potential V, the standard linear interpolation and multipoint constraint method can be used to satisfy the continuity and conservative/balance conditions across the internal coupling interface.

6.2.2 Periodic boundary conditions

6.2.2.1 Translational periodic boundary (Figure 6.4)

Mapping for translational periodic boundary: due to the translational distance between the two periodic interfaces, a forced translational motion is applied onto the interface mesh on side A to make it match the corresponding mesh on side B (Equation (6.20)), and a standard mapping process can be carried out:

$$X_{A} = X_{A} + \left(X_{B}^{C} - X_{A}^{C}\right), \tag{6.20}$$

Figure 6.4 Translational periodic boundary.

where $\boldsymbol{X}_A^C$, geometry center of mesh on side A and; $\boldsymbol{X}_B^C$, geometry center of mesh on side B.

Coupling condition on the periodic boundary:

$$\boldsymbol{u}_A = \boldsymbol{u}_B$$

The internal force $\boldsymbol{Q}_i$ for an internal node i is shown in Equations. (6.21)–(6.23) for vector value and scalar value, respectively.

For vector value:

$$\boldsymbol{Q}_i = \boldsymbol{K}_{iA}\boldsymbol{u}_A = \boldsymbol{K}_{iA}\left(\sum_{j=1}^{n}\alpha_j \boldsymbol{u}_B^j\right) = \sum_{j=1}^{n}\left(\boldsymbol{K}_{iA\_jB}\boldsymbol{u}_B^j\right), \tag{6.21}$$

where

$$\boldsymbol{K}_{iA\_jB} = \alpha_j \boldsymbol{K}_{iA}, \tag{6.22}$$

α_j is the weight function from the mapping and $\sum_{j=1}^{n}\alpha_j = 1$.

For scalar value:

$$\boldsymbol{Q}_i = \boldsymbol{K}_{iA}\phi_A = \boldsymbol{K}_{iA}\left(\sum_{j=1}^{n}\alpha_j \phi_B^j\right) = \sum_{j=1}^{n}\left(\boldsymbol{K}_{iA\_jB}\phi_B^j\right) \tag{6.23}$$

Similarly:

$$K_{iA\_jB} = \alpha_j K_{iA}. \tag{6.24}$$

Note: For the translational periodic condition, we can specify this type of internal coupling for each component of velocity vector v_x, v_y, and v_z separately.

Pressure drop and mass flow: it is often useful to model periodic flow by allowing a pressure drop between the two sides of the interface.

If the pressure drop is specified, the solver will calculate the flow field, including the mass flow rate. Alternatively, if the mass flow rate through the interface is known, the solver will calculate the flow field and the pressure drop.

$$\boldsymbol{K}_{AA}\boldsymbol{u}_A + \boldsymbol{K}_{AB}\boldsymbol{u}_B = \boldsymbol{F}_A, \tag{6.25}$$

$$\boldsymbol{K}_{BA}\boldsymbol{u}_A + \boldsymbol{K}_{BB}\boldsymbol{u}_B = \boldsymbol{F}_B. \tag{6.26}$$

Subtracting Equation (6.26) from Equation (6.25), replacing $\boldsymbol{u}_A$ by the production of the transformation matrix $\boldsymbol{T}_{AB}$ and vector $\boldsymbol{u}_B$, one has:

$$\sum\left(\boldsymbol{K}_{BA}\boldsymbol{T}_{AB}\boldsymbol{u}_B\right)-\sum\left(\boldsymbol{K}_{AA}\boldsymbol{T}_{AB}\boldsymbol{u}_B\right)+\sum\left(\boldsymbol{K}_{BB}\boldsymbol{u}_B\right)-\sum\left(\boldsymbol{K}_{AB}\boldsymbol{u}_B\right)=\sum\left(\boldsymbol{F}_B\right)-\sum\left(\boldsymbol{F}_A\right). \tag{6.27}$$

The right-side term represents the pressure drop, and the terms on the left side represents the flow field and flow rate.

6.2.2.2 Rotational periodic boundary

Mapping for rotational periodic boundary:

To carry out the mapping for rotational periodic boundary, we first need to convert the periodic interface meshes on both sides onto (r, z) coordinate in the cylindrical coordinate representation and then perform the mapping on (r, z) coordinate in a standard way.

Periodic condition (*type* 1): the rotational periodic boundary as demonstrated in Figure 6.5 can be expressed as:

$$u_r^A = u_r^B, \quad u_\theta^A = u_\theta^B, \quad u_z^A = u_z^B \text{ on interfaces } A \text{ and } B, \tag{6.28}$$

In Cartesian coordinates, this condition can be expressed as:

$$\boldsymbol{R}_A\boldsymbol{u}^A = \boldsymbol{R}_B\boldsymbol{u}^B, \tag{6.29}$$

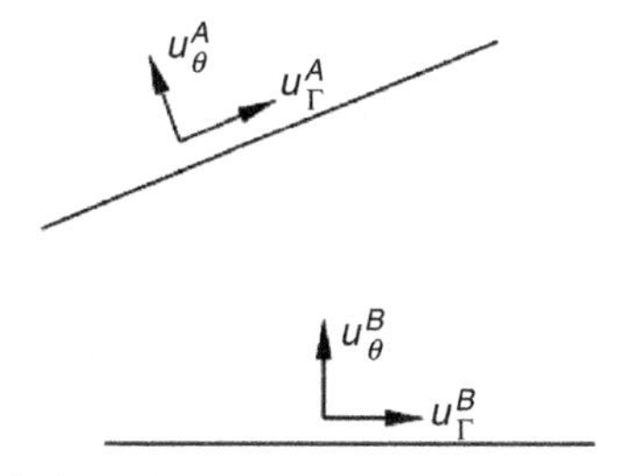

Figure 6.5 Rotational periodic boundary (type 1).

$\boldsymbol{R}_A$, $\boldsymbol{R}_B$ transformation matrix from global Cartesian coordinates to cylindrical coordinates, and then we have:

$$\boldsymbol{u}^A = \boldsymbol{R}_A^{\mathrm{T}} \boldsymbol{R}_B \boldsymbol{u}^B. \tag{6.30}$$

The calculation of $\boldsymbol{R}$ can be found in Section 6.1.2.2.

The internal force $\boldsymbol{Q}_i$ for an internal node i is shown in Equation (6.31) for vector value.

For vector value:

$$\boldsymbol{Q}_i = \boldsymbol{K}_{iA} \boldsymbol{u}_A = \boldsymbol{K}_{iA} \left(\sum_{j=1}^{n} \alpha_j \boldsymbol{R}_A^{\mathrm{T}} \boldsymbol{R}_B \boldsymbol{u}_B^j \right) = \sum_{j=1}^{n} \left(\boldsymbol{K}_{iA\_jB} \boldsymbol{u}_B^j \right) \tag{6.31}$$

where

$$\boldsymbol{K}_{iA\_jB} = \alpha_j \boldsymbol{K}_{iA} \boldsymbol{R}_A^{\mathrm{T}} \boldsymbol{R}_B, \tag{6.32}$$

α_j is the weight function from the mapping, and $\sum_{j=1}^{n} \alpha_j = 1$.

For scalar value, the equations for the translational periodic boundary (Equations (6.23) and (6.24)) are also applicable for the rotational periodic case.

Periodic condition (*type* 2): the second type of the rotational periodic boundary as shown in Figure 6.6 can be expressed as:

$$u_{\mathrm{r}}^A = u_{\mathrm{r}}^B, \quad u_\theta^A = -u_\theta^B, \quad u_{\mathrm{z}}^A = u_{\mathrm{z}}^B \text{ on interfaces } A \text{ and } B, \tag{6.33}$$

For the rotational matrix, $\boldsymbol{R}_A^{\mathrm{T}}$ can be obtained by substituting the negative value of $\boldsymbol{e}_\theta$ into Equation (6.4).

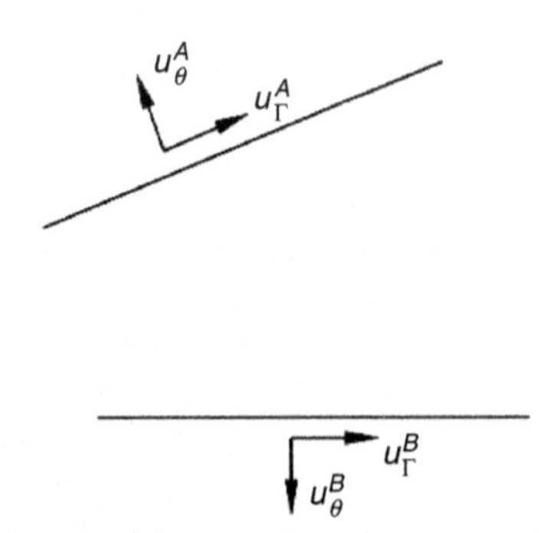

Figure 6.6 Rotational periodic boundary (type 2).

6.3 Governing equations in body-attached rotating frame

6.3.1 Basic formulation in rotational body-attached frame of reference

The absolute motion of a point in the body-attached frame can be expressed as (Figure 6.1):

$$V(C) = V(A) + \omega \times AC + v(C) \tag{6.34}$$

where

$V(C)$, velocity vector of point C in global stationary coordinate system (O, X, Y, Z); $V(A)$, velocity vector of point A in global stationary coordinate system (O, X, Y, Z);

ω, rotation vector of body-attached coordinate system (O', x_b, y_b, z_b), defined in Section 6.1.2.1 corresponding to the stationary coordinate;

AC, vector from A to an arbitrary point C; and $v(C)$, relative velocity vector of point C in body-attached coordinate (O', x_b, y_b, z_b).

From Equation (6.34), we can get the relation between the accelerations in two different frames.

$$A(C) = A(A) + A_r + \dot{\omega} \times r' + \omega \times (\omega \times r') + 2\omega \times v_r. \tag{6.35}$$

Here, $A(A)$, acceleration of point A in the stationary coordinate system; A_r, relative acceleration in the body-attached frame; r', coordinate of node C in the body-attached frame; $\dot{\omega} \times r'$, rotational acceleration term; $\omega \times (\omega \times r')$, centrifugal term of acceleration; and $2\omega \times v_r$, Coriolis term of acceleration.

6.3.2 Governing equations for fluid flow in rotational body-attached frame

Substituting the definition of Equation (6.35) into the equilibrium equation of fluid flow in global Cartesian system in Equation (1.17), one can obtain the corresponding formulation expressed in the rotating frame of reference Equation (6.36).

$$\rho \frac{\partial v}{\partial t} + \rho c \cdot (\nabla v) = \nabla \cdot \sigma + \rho g \underline{- \rho r \dot{\omega} e_\theta + \rho r \omega^2 e_r - 2\rho \omega \times v} \quad \text{in } \Omega_t^f \tag{6.36}$$

The underline parts indicate additional source terms in the body-attached frame, with the origin assumed to be fixed in space, that is, $A(A) = 0$. The details for each term are listed as $-\rho r \dot{\omega} e_\theta$, rotational acceleration force, $\rho r \omega^2 e_r$, centrifugal force; and $-2\rho \omega \times v$, Coriolis force.

6.3.3 Governing equations for mechanics systems in a rotational body-attached frame

The additional source terms for the computational solid mechanics in a body-attached rotating frame are the same as those of the underline parts of fluid flow Equation (6.36), and it can be expressed as:

$$\rho \frac{\mathrm{d}^2 \boldsymbol{u}}{\mathrm{d}^2 t} = \nabla \cdot \sigma + \rho g \underline{- \rho r \dot{\omega} \boldsymbol{e}_\theta + \rho r \omega^2 \boldsymbol{e}_r - 2\rho\omega \times v} \quad \text{in } \Omega_t^S. \tag{6.37}$$

The meaning of the underline terms is the same as those in Equation (6.36).

6.3.4 For thermal systems

There is not any extra term for heat-transfer equation when expressing in the body-attached frame, but it is necessary to update the radiation view factor for surface radiation caused by the rotational motion of the body. For the conservation of the total energy of compressible flow, the advection of the total enthalpy needs to be replaced by the advection of rotational stagnation enthalpy (ANSYS CFX-Solver Theory Guide, 2013).

6.3.5 Governing equations for electromagnetic systems in a rotational body-attached frame

The additional term from rotation for Maxwell equations (Equation (1.365)) in the global stationary frame is:

$$-\nabla \cdot \left([\sigma] (\boldsymbol{v} \times \boldsymbol{B}) \right) = -\nabla \cdot \left([\sigma] (\boldsymbol{\omega} \times \boldsymbol{r} \times \boldsymbol{B}) \right), \tag{6.38}$$

where

$\boldsymbol{v}$ is the absolute velocity vector caused by rigid body motion in a global stationary frame.

$\boldsymbol{B}$ is the absolute magnetic field flux density vector in a global stationary frame.

$\boldsymbol{\omega}$ is the absolute rotation vector of the body frame in a global stationary frame.

$\boldsymbol{r}$ is the position vector of node C from point A in a global stationary frame.

This rotational term of Equation (6.38) in the body-attached frame will be eliminated from the Maxwell equation.

6.4 Multiple frames of references for rotating problems

For a coupled system, typically for rotating machinery with multiple rotors and stators, whenever the stationary frame is used for the stator and rotating frames with different rotating speeds adopted for different parts of rotors for a convenient mathematical formation, we call this methodology a multiple frame of reference. The appropriate coupling technique is needed to deal with the cross-frame coupling.

6.4.1 Choosing the simulation domain

For simulation of a rotating machine, you can solve the entire 360° of the model without using the periodic characteristics. But the computational resources needed are usually expensive and very costly. An alternative way is to pick one sector of rotor and stator, that is, θ_{pitch} in Figure 6.7a. Make sure the pitch ratio of the rotor and stator is close to 1. In practice, components of unequal pitch can be treated by solving N passages on one side and M passages on the other side, with N and M determined such that the net pitch change across the interface is close to unity. Pitch change needs to be scaled properly for conservative value (e.g., force, flow, heat). For example in Figure 6.7a, one passage in the rotor side and two passages in the stator side are chosen as the simulation domain, which makes the pitch ratio of the rotor and stator equal to 1.

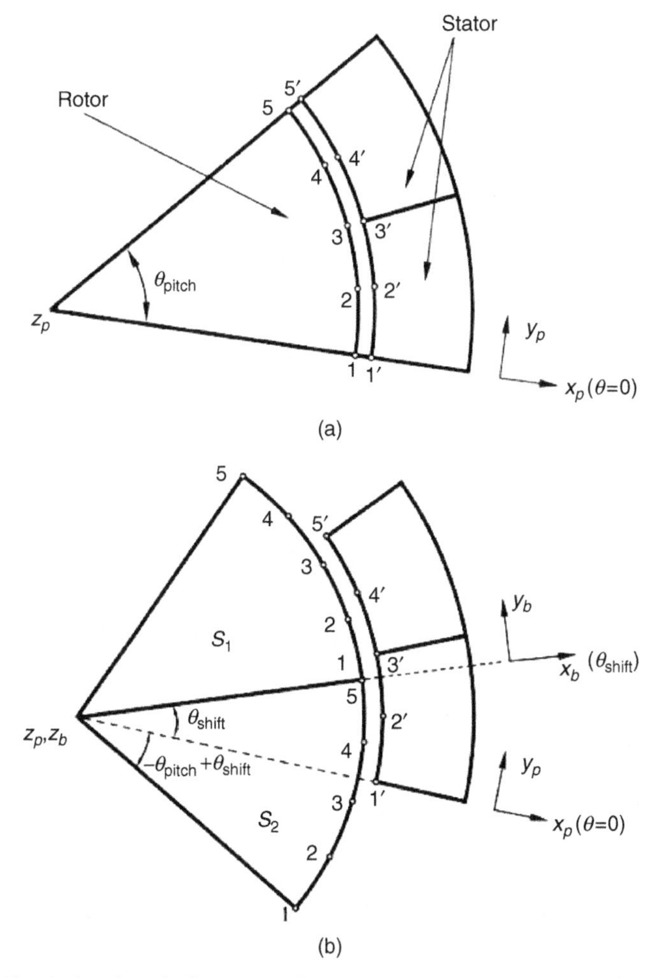

Figure 6.7 Simulation domain for rotor and stator.

In Figure 6.7, the reference frame (O', x_p, y_p, z_p) is the initial state of the body-attached frame (O', x_b, y_b, z_b) and is a Cartesian frame fixed in space and time. This reference frame is used as an intermediate coordinate system between the global stationary frame and the body-attached frame.

6.4.2 The mapping process for rotating interface

On the basis of the value of θ_{shift}, the interface meshes on the rotor side are replicated to make sure the replicated rotor interfaces overlap all the interface meshes on stator side (Figure 6.7b). Again on the basis of the θ_{shift} and θ_{pitch}, you will be able to find $\theta(t)$ for all the interface nodes on rotor side that overlap with the original stator interface mesh. Then you will find the right mapping for all the nodes on the stator interface (1′,2′,…,5′) to the interface nodes on the rotor side (1,2,…,5).

Step 1: Calculate θ_{shift}: input the actual rotating angle of ωt from original configuration at current time t and evaluate the minimum positive θ_{shift}:

$$\theta_{\text{shift}} = \omega t + n\theta_{\text{pitch}}. \tag{6.39}$$

Choose n (…, 2, 1, 0, −1, −2, …) to make sure

$$0 < \theta_{\text{shift}} \leq \theta_{\text{pitch}}. \tag{6.40}$$

Step 2: Interface mesh replication: replicate interface mesh based on (O', x_p, y_p, z_p) coordinates.

For sector 1:

$$\theta_1(t) = \theta(0) + \theta_{\text{shift}}. \tag{6.41}$$

Where, $\theta(0)$ is the initial value of θ for any point in sector 1 of the rotor and $\theta_1(t)$ is the corresponding value at time t.

For sector 2:

$$\theta_2(t) = \theta(0) + \theta_{\text{shift}} - \theta_{\text{pitch}}. \tag{6.42}$$

Basing on $\theta_1(t)$ and $\theta_2(t)$, we can prepare interface meshes on (O', x_p, y_p, z_p) frame for mapping tools.

$$\boldsymbol{x}_p = \begin{Bmatrix} x_p \\ y_p \\ z_p \end{Bmatrix} = \begin{Bmatrix} \cos(\theta(t))r_b \\ \sin(\theta(t))r_b \\ zb \end{Bmatrix} = \boldsymbol{R}(\theta_{\text{shift}})\boldsymbol{x}_b \tag{6.43}$$

where $r_b = \sqrt{x_b \bullet x_b + y_b \bullet y_b}$ and x_b, y_b, z_b are coordinates at the body-attached frame.

Note: We will solve the rotor in the body-attached system (O', x_b, y_b, z_b). But for the interfacial contribution in the global Cartesian frame, the matrices and vectors need to be evaluated in the stationary (O', x_p, y_p, z_p) frame. All of the vectors, scalars, and coordinates need to be evaluated in this system. The force contribution is needed to be converged to global (O, X, Y, Z) before merging. Another option is to perform element calculation directly in (O, X, Y, Z) frame, but all of the scalars, vectors, and coordinates need to be evaluated in this global stationary frame.

Step 3: Do the mapping.

Note: Node id shift may be needed for sector 2 before calling the mapping tools.

Mapping result: the resulting mapping coefficient matrix can be expressed as:

$$V_S = MV_M. \tag{6.44}$$

Residual merge for master nodes

$$\delta V_S^{\mathrm{T}} F_S = \delta V_M^{\mathrm{T}} F_M = \delta V_M^{\mathrm{T}} M^{\mathrm{T}} F_S, \tag{6.45}$$

$$F_M = M^{\mathrm{T}} F_S, \tag{6.46}$$

$$\delta V_S^{\mathrm{T}} C_{ss} V_S = \delta V_M^{\mathrm{T}} C_{MM} V_{MS} = \delta V_M^{\mathrm{T}} M^{\mathrm{T}} C_{ss} M V_M. \tag{6.47}$$

$$C_{MM} = M^{\mathrm{T}} C_{ss} M. \tag{6.48}$$

Where, in Equations (6.44)–(6.48), the subscripts M and S represent the value on the master side and on the slave side, respectively.

6.4.3 Transformation matrices for velocity vector

The three basic transformation matrixes

1. From (O, X, Y, Z) to (O', x_p, y_p, z_p): R_{xp_X}. Please see Section 6.1.2.2. for the definition of R_{xp_X}

$$\begin{Bmatrix} v_{xp} \\ v_{yp} \\ v_{zp} \end{Bmatrix} = R_{xp\_X} \begin{Bmatrix} v_X \\ v_Y \\ v_Z \end{Bmatrix}. \tag{6.49}$$

2. From $\left(O', x_p, y_p, z_p\right)$ frame to local coordinate ($\theta = \theta(t)$), one has:

$$\boldsymbol{R}(\theta(t)) = \begin{bmatrix} \cos(\theta(t) & \sin(\theta(t)) & 0 \\ -\sin(\theta(t)) & \cos(\theta(t)) & 0 \\ 0 & 0 & 1 \end{bmatrix}, \tag{6.50}$$

$$\begin{Bmatrix} v_r \\ v_\theta \\ v_{zp} \end{Bmatrix} = \boldsymbol{R}(\theta(t)) \begin{Bmatrix} v_{xp} \\ v_{yp} \\ v_{zp} \end{Bmatrix}. \tag{6.51}$$

3. From the Cartesian to the cylindrical system in the body-attached local system,

$$\boldsymbol{R}(\theta(0)) = \begin{bmatrix} \cos(\theta(0) & \sin(\theta(0)) & 0 \\ -\sin(\theta(0)) & \cos(\theta(0)) & 0 \\ 0 & 0 & 1 \end{bmatrix}. \tag{6.52}$$

For an any given vector,

$$\begin{Bmatrix} v_r \\ v_\theta \\ v_{zb} \end{Bmatrix} = \boldsymbol{R}(\theta(0)) \begin{Bmatrix} v_{xb} \\ v_{yb} \\ v_{zb} \end{Bmatrix}. \tag{6.53}$$

From Equations (6.49), (6.51), and (6.53) and since the equivalence between Equations (6.51) and (6.53) caused by the periodicity, we have

$$\boldsymbol{R}(\theta(0)) \begin{Bmatrix} v_{xb} \\ v_{yb} \\ v_{zb} \end{Bmatrix} = \boldsymbol{R}(\theta(t)) \boldsymbol{R}_{xp\_X} \begin{Bmatrix} v_X \\ v_Y \\ v_Z \end{Bmatrix}. \tag{6.54}$$

As a result:

$$\begin{Bmatrix} v_X \\ v_Y \\ v_Z \end{Bmatrix} = \boldsymbol{R}_{xp\_X}{}^{\mathrm{T}} \boldsymbol{R}\left(\theta(t)\right)^{\mathrm{T}} \boldsymbol{R}(\theta(0)) \begin{Bmatrix} v_{xb} \\ v_{yb} \\ v_{zb} \end{Bmatrix}. \tag{6.55}$$

and

$$\boldsymbol{R}_{X\_xb} = \boldsymbol{R}_{xp\_X}{}^{\mathrm{T}} \boldsymbol{R}(\theta(t))^{\mathrm{T}} \boldsymbol{R}(\theta(0)) = \boldsymbol{R}_{xp\_X}{}^{\mathrm{T}} \boldsymbol{R}(\theta_{\mathrm{shift}})^{\mathrm{T}}. \tag{6.56}$$

So we have

$$V_X = R_{X\_xb} v_{xb}. \tag{6.57}$$

From Equation (6.56), we have

$$R_{X\_xb} = R(\theta_{\text{shift}}) R_{xp\_X}. \tag{6.58}$$

6.4.4 Vector transformation equations

Then velocity vector for an arbitrary point C in the global stationary frame (O, X, Y, Z) is (Figure 6.1):

$$V(C) = V(A) + \omega \times AC + v(C), \tag{6.59}$$

$$AC = X(C) - X(A), \tag{6.60}$$

$$X(C) = X(A) + R_{xp\_X}{}^{\mathrm{T}} x_p. \tag{6.61}$$

From Equations (6.60) and (6.61), we have

$$AC = R_{xp\_X}{}^{\mathrm{T}} x_p \tag{6.62}$$

$$V(C) = V(A) + \omega \times R_{xp\_X}{}^{\mathrm{T}} x_p + R_{X\_x_b} v_{xb}. \tag{6.63}$$

Equation (6.63) is in global (O, X, Y, Z) system, and from this equation, we have:

$$v_{xb} = R_{X\_x_b}{}^{\mathrm{T}} \left(V(C) - V(A) - \omega \times R_{xp\_X}{}^{\mathrm{T}} x_p \right). \tag{6.64}$$

From Equation (6.64), one can have the vector evaluated in (O', x_p, y_p, z_p) frame:

$$V(C)_{xp} = \omega_{xp} \times x_p + R_{xp\_xb} v_{xb}, \tag{6.65}$$

$$R_{xp\_xb} = R(\theta_{\text{shift}})^{\mathrm{T}}. \tag{6.66}$$

From Equation (6.65), we have

$$v_{xb} = R_{xp\_xb}{}^{\mathrm{T}} \left(V(C)_{xp} - \omega_{xp} \times x_p \right). \tag{6.67}$$

Equations (6.65) and (6.67) in the stationary (O', x_p, y_p, z_p) frame can be used for the forces and flux calculation, and they need to be converted to (O, X, Y, Z) frame before merged into the interfacial equations.

The force vector conversion from (O', x_p, y_p, z_p) to (O, X, Y, Z) is shown as follows:

$$F_X = R_{xp\_X}{}^{\mathrm{T}} F_{xp}. \tag{6.68}$$

Equation (6.63) gives the transformation of a vector value from local body-attached frame to the global stationary Cartesian frame. It will be used for constraint condition for the interface nodes on the rotor side.

6.4.5 Interface constraints conditions for interface nodes on the rotor side

For the interface coupling conditions, if the multiple point constraint method is used as the interface coupling method and the stator is treated as the master component, then it is necessary to constraint the velocity, as well as the pressure of interface nodes on the rotor side, onto appropriate interface nodes on the stator side through the mapping tool. The velocity on both sides must be converted into the same coordinate system (e.g., global stationary Cartesian).

6.4.6 Interface conservative/balance equations

The balance or equilibrium conditions on the interface will be carried over for the interface nodes on the stator side. The contribution from the rotor side must be evaluated under the correct region (the overlap region with the stator interface) with appropriate $\theta(t)$ for both coordinate locations and the nodal solutions. And all values must be converted into the global stationary coordinate system and consistent with that used in the stator frame. An interface node on the rotor side needs to accumulate the contributions from all the elements attached to it, and those elements must be in the same sector (1 or 2, Section 6.4.1) as this node.

Step1: Loop over all elements.

If all element nodes are interior nodes, then do the assembly in local (O', x_b, y_b, z_b) frame.

$$\delta v_{xb}^{i\ \mathrm{T}} C_{ij}^{xb,xb} v_{xb}^{j}. \tag{6.69}$$

The corresponding residual vector of Equation (6.69) can be expressed as

$$F_{xb}^{i} = -C_{ij}^{xb,xb} v_{xb}^{j}. \tag{6.70}$$

If an element has interface nodes but all the nodes of this element are on the same sector (1 or 2), then assembly needs to be done in (O', x_p, y_p, z_p) frame.

For the interface node i:

$$\delta \boldsymbol{v}_{xp}^{i\ \mathrm{T}} \boldsymbol{C}_{ij}^{xp,xp} \boldsymbol{v}_{xp}^{j}, \tag{6.71}$$

$$\delta \boldsymbol{v}_{X}^{i\ \mathrm{T}} \boldsymbol{R}_{xp\_X}^{i\ \ \mathrm{T}} \boldsymbol{C}_{ij}^{xp,xp} \boldsymbol{v}_{xp}^{j}, \tag{6.72}$$

$$\boldsymbol{F}_{X}^{i} = -\boldsymbol{R}_{xp\_X}^{i\ \ \mathrm{T}} \boldsymbol{C}_{ij}^{xp,xp} \boldsymbol{R}_{xp\_X}^{j} \boldsymbol{v}_{X}^{j}, \tag{6.73}$$

$$\boldsymbol{F}_{X}^{i} = -\boldsymbol{R}_{xp\_X}^{i\ \ \mathrm{T}} \boldsymbol{C}_{ij}^{xp,xp} \boldsymbol{R}_{xp\_X}^{j} \left(\boldsymbol{v}_{j}^{X}(A) + \boldsymbol{\omega} \times \boldsymbol{R}_{xp\_X}{}^{\mathrm{T}} \boldsymbol{x}_{p} + \boldsymbol{R}_{X\_xb} \boldsymbol{v}_{xb}^{j} \right), \tag{6.74}$$

$$\boldsymbol{F}_{X}^{i} = -\boldsymbol{C}_{ij}^{X,X} \boldsymbol{v}_{j}^{X}(C). \tag{6.75}$$

For the interior node, one has

$$\delta \boldsymbol{v}_{xb}^{i\ \mathrm{T}} \boldsymbol{C}_{ij}^{xb,xb} \boldsymbol{v}_{xb}^{j}. \tag{6.76}$$

But here $\boldsymbol{v}_{xb}^{j}$ is from Equation (6.64) or (6.67)

From Equations (6.64) and (6.70), we have

$$\boldsymbol{F}_{xb}^{i} = -\boldsymbol{C}_{ij}^{xb,xb} \boldsymbol{R}_{X\_xb}{}^{\mathrm{T}} \left(\boldsymbol{v}_{j}^{X}(C) - \boldsymbol{v}_{j}^{X}(A) - \boldsymbol{\omega} \times \boldsymbol{R}_{xp\_X}{}^{\mathrm{T}} \boldsymbol{xp} \right), \tag{6.77}$$

$$\boldsymbol{F}_{xb}^{i} = -\boldsymbol{C}_{ij}^{xb,xb} \boldsymbol{v}_{xb}^{j}. \tag{6.78}$$

For an element that has interface nodes on different sectors, you have to evaluate Equation (6.73) in those sectors that belongs to.

Note: From Equations (6.73), (6.75), and (6.76), we know that the element on the interface needs to be evaluated in (O', x_p, y_p, z_p) frame (may be only needed in one sector or needed in two sectors) and body frame (O', x_b, y_b, z_b).

6.4.7 General matrix transformation for matrix assembly

The matrix on the left side of the balance equation can be obtained by taking derivative corresponding to the interface variable $\boldsymbol{v}_{X}^{j}$ in (O, X, Y, Z) frame and interior variable $\boldsymbol{v}_{xb}^{j}$.

Take derivative of Equation (6.70) with respect to $\boldsymbol{v}_{xb}^{j}$, and we have matrix

$$\langle \delta \boldsymbol{v}_{xb}^{i\mathrm{T}} \rangle \boldsymbol{C}_{ij}^{xb,xb} \langle \boldsymbol{v}_{xb}^{j} \rangle. \tag{6.79}$$

Take derivative of Equation (6.73) with respect to $\boldsymbol{v}_X^j$, and we have

$$\left\langle \delta \boldsymbol{v}_X^{i\ \mathrm{T}} \right\rangle \boldsymbol{R}_{xp\_X}^{j\ \mathrm{T}} \boldsymbol{C}_{ij}^{xp,xp} \boldsymbol{R}_{xp\_X}^{j} \left\langle \boldsymbol{v}_X^j \right\rangle, \tag{6.80}$$

$$\left\langle \delta \boldsymbol{v}_X^{i\ \mathrm{T}} \right\rangle \boldsymbol{C}_{ij}^{X,X} \left\langle \boldsymbol{v}_X^j \right\rangle. \tag{6.81}$$

Take derivative of Equation (6.74) with respect to $\boldsymbol{v}_{xb}^j$, and we have

$$\left\langle \delta \boldsymbol{v}_X^{i\ \mathrm{T}} \right\rangle \boldsymbol{R}_{xp\_X}^{j\ \mathrm{T}} \boldsymbol{C}_{ij}^{xp,xp} \boldsymbol{R}_{xp\_X}^{j} \boldsymbol{R}_{X\_xb} \left\langle \boldsymbol{v}_{xb}^j \right\rangle, \tag{6.82}$$

$$\left\langle \delta \boldsymbol{v}_X^{i\ \mathrm{T}} \right\rangle \boldsymbol{C}_{ij}^{X,X} \boldsymbol{R}_{X\_xb} \left\langle \boldsymbol{v}_{xb}^j \right\rangle. \tag{6.83}$$

Take derivative of Equation (6.77) with respect to $\boldsymbol{v}_X^j$, and we have

$$\left\langle \delta \boldsymbol{v}_{xb}^{i\ \mathrm{T}} \right\rangle \boldsymbol{C}_{ij}^{xb,xb} \boldsymbol{R}_{X\_xb}^{\mathrm{T}} \left\langle \boldsymbol{v}_X^j \right\rangle. \tag{6.84}$$

Where i, j represent node id and $\boldsymbol{v}$ represents a vector data at a node.

Note: The coupling term for $\boldsymbol{v}_{xb}^j$ and ω can be obtained by taking a derivative of Equations (6.74) and (6.77) with respect to ω.

The matrix assembly for element 1 in Figure 6.9 is given by Equation (6.85).

$$\left\{ \delta v_{xb}^{1\ \mathrm{T}}, \delta v_{X\_s2}^{2\ \mathrm{T}}, \delta v_{X\_s1}^{3\ \mathrm{T}}, \delta v_{xb}^{4\ \mathrm{T}} \right\} \begin{bmatrix} \boldsymbol{k}_{11}^{xb}, & \boldsymbol{k}_{12}^{xb} \boldsymbol{R}_{X\_xb}, & \boldsymbol{k}_{13}^{xb} \boldsymbol{R}_{xb,X}, & \boldsymbol{k}_{14}^{xb} \\ \boldsymbol{k}_{21}^{s2} \boldsymbol{R}_{X\_xb}, & \boldsymbol{k}_{22}^{s2}, & \boldsymbol{k}_{23}^{s2s1}, & \boldsymbol{k}_{24}^{s2} \boldsymbol{R}_{X\_xb} \\ \boldsymbol{k}_{31}^{s1} \boldsymbol{R}_{X\_xb}, & \boldsymbol{k}_{32}^{s1s2}, & \boldsymbol{k}_{33}^{s1}, & \boldsymbol{k}_{34}^{s1} \boldsymbol{R}_{X\_xb} \\ \boldsymbol{k}_{41}^{xb}, & \boldsymbol{k}_{42}^{xb} \boldsymbol{R}_{xb\_X}, & \boldsymbol{k}_{43}^{xb} \boldsymbol{R}_{xb\,X}, & \boldsymbol{k}_{44}^{xb} \end{bmatrix} \begin{Bmatrix} v_{xb}^1 \\ v_{X\_s2}^2 \\ v_{X\_s1}^3 \\ v_{xb}^4 \end{Bmatrix}, \tag{6.85}$$

where in Equation (6.85),

$$\begin{aligned} \delta \boldsymbol{v}_{X\_s2}^{2\ \mathrm{T}} k_{23}^{s2s1} \boldsymbol{v}_{X\_s1}^{3\ \mathrm{T}} = \delta \boldsymbol{v}_{X\_s2}^{2\ \mathrm{T}} k_{23}^{s2} \boldsymbol{v}_{X\_s2}^{3} = \delta \boldsymbol{v}_{X\_s2}^{2\ \mathrm{T}} k_{23}^{s2} \boldsymbol{R}\left(\theta_{\mathrm{pitch}}\right) \boldsymbol{v}_{X\_s1}^{3} \\ k_{23}^{s2s1} = k_{23}^{s2} \boldsymbol{R}\left(\theta_{\mathrm{pitch}}\right), \end{aligned} \tag{6.86}$$

and

$$\begin{aligned} \delta \boldsymbol{v}_{X\_s1}^{3\ \mathrm{T}} \boldsymbol{k}_{32}^{s1s2} \boldsymbol{v}_{X\_s2}^{2} = \delta \boldsymbol{v}_{X\_s1}^{3\ \mathrm{T}} \boldsymbol{k}_{32}^{s1} \boldsymbol{v}_{X\_s1}^{2} = \delta \boldsymbol{v}_{X\_s1}^{2\ \mathrm{T}} \boldsymbol{k}_{32}^{s1} \boldsymbol{R}\left(-\theta_{\mathrm{pitch}}\right) \boldsymbol{v}_{X\_s2}^{2} \\ \boldsymbol{k}_{32}^{s1s2} = \boldsymbol{k}_{32}^{s1} \boldsymbol{R}\left(-\theta_{\mathrm{pitch}}\right). \end{aligned} \tag{6.87}$$

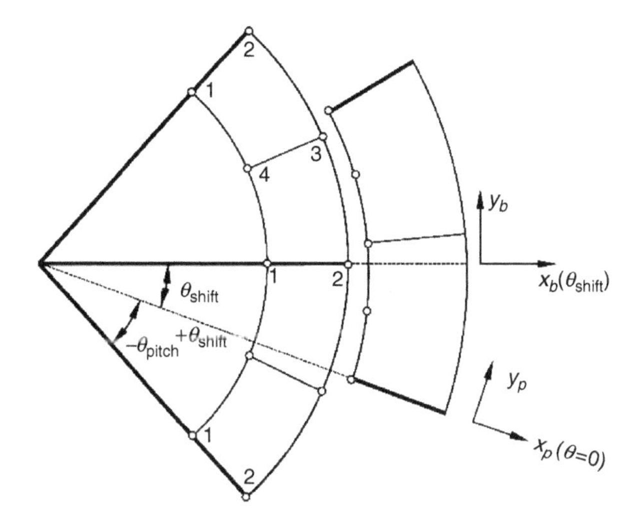

Figure 6.8 Element matrix assembles for rotor in sector 1 (s1) and sector 2 (s2).

In Equations (6.85)–(6.87), the superscripts $s1$ and $s2$ indicate the sector, subscript X and xb represents the reference frame, and the integer number stands for node id of element 1. (Figure 6.8)

Note: For the frozen rotor stator case (ANSYS CFX-Solver Theory Guide, 2013), the interface meshes of sector 1 ($s1$) and sector 2 ($s2$) have no changes and stay coincide with the interface meshes of the stator.

6.5 Morphing technology for rotating problems

In this section, the morphing with the remeshing method or sliding interface is used for the rotational coupling interface. Different from the multiple frame of reference, the rotor is formulated in the global stationary frame of reference in this methodology.

6.5.1 Morphing and remeshing for rotating problems

From specified or calculated rotational DOF (θ) or by solving moving boundary problem with theta DOF only ($u_r = 0$), the rotational mesh motion can be obtained for the rotating problem by solving the mesh morphing equation. The morphing and remeshing process for the rotating problem is given as follows:

- Rigid body morphing is applied for rotor mesh.
- Remesh or mesh swap needs to be applied to one soft layer between the rigid rotor and rigid stator.
- Mapping and interpolation are needed for the remeshing domain.

6.5.2 *Morphing with a sliding interface*

An alternative way of handling the mesh updates for rotational problem is the rigid body motion for the rotor with a sliding interface. It consists of the following procedures:

- Rigid body morphing is applied for rotor mesh.
- Sliding mesh boundary is applied to couple the interface.
- 360° is needed.
- All interface conditions are accounted for in the global stationary coordinate system.
- Mapping is needed for coupling the sliding interface at each time step.

Note: for more details about the morphing technology for rotating machinery, refer Chapter 4.

6.6 Multiphysics simulation for rotating machines

There are quite a number of categories of rotating machineries that involve multiphysics coupling (e.g., turbine machinery, propellers, electric motors). The gas turbine is one of the typical rotating machines with multiphysics coupling in which the complex fluid flow interacts with the rotating and stationary structures and the fluid–thermal–structure coupling takes place in the operating process.

6.6.1 *Multiple frame of reference approach*

It is convenient to use the multiple frame of reference formulation for the fluid–structure interaction modeling of a turbine machine. In this case, the rotor blades and the rotating region of the fluid domain are formulated in a consistent rotating frame of reference, so the direct interface coupling method or direct load transfer-based weak coupling method (Chapter 3) can be used without the complex frame updating for the fluid–structure interface. Although the stator blade as well as the "stationary" region of fluid flow are modeled in the stationary frame of reference, the frame change coupling in the fluid–fluid coupling interface can be treated as transient rotor-stator coupling that accounts for all of the transient coupling effects. The multiple point constraint–based coupling equations have been given in Section 6.4.

To demonstrate the concept of coupling physics problems of turbine machinery, Figure 6.9 gives a simplified model with a rotor blade, stator, and fluid flow. This model is solved by the multiple frame of reference in this section and by the sliding interface method in the Section 6.6.2.

6.6.1.1 Fluid–structure interaction model for the multiple frame of reference approach

An application of multiple frame of reference and sliding interface to a coupling physics problem of the turbine machinery is shown in this section. The solid model is a

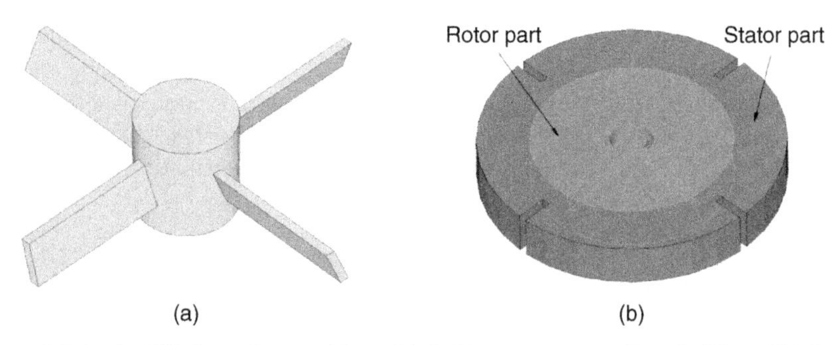

Figure 6.9 A simplified rotating machine with fluid–structure coupling. (a) The solid domain and (b) the fluid domain.

four-blade impeller, and the fluid model is a section of a three-dimensional channel with a four-blade impeller at the center and four baffles at the wall, which are shown in Figure 6.9; the mesh models are shown in Figure 6.10.

The material properties of fluid model are as follows: mass density $\rho = 1.0 \times 10^3$ kg/m^3, dynamic viscosity $\mu = 1.0 \times 10^{-3}$ Pa s, and bulk modulus $B = 2.0 \times 10^8$ Pa. The material properties of the solid model are Young's modulus $E = 2.0 \times 10^{11}$ Pa, Poisson's ratio $\nu = 0.3$, and mass density $\rho = 7.8 \times 10^3$ kg/m^3.

A transient analysis is conducted; the transient time step size is 0.05 s, and the total analysis time is 1 s. As for the solid model, the rotor blades are formulated in a consistent rotating frame of reference. The rotational speed is 1 rad/s, and the central nodes of the shaft are fixed. The fluid region consists of rotor and stator part. As for the rotor part, it is also formulated in a consistent rotating frame of reference with the rotational speed 1 rad/s, and the symmetry boundary is applied on the top and bottom surface and the inner wall is considered as fluid–structure coupling interface. The direct load transfer-based weak coupling method is used for the fluid–structure interface. The stator part is indicated as nonrotation region. The symmetry boundary is applied on the top and bottom surface, and the outer wall is considered a no-slip wall. The fluid rotor–stator coupling interface is treated as a sliding interface.

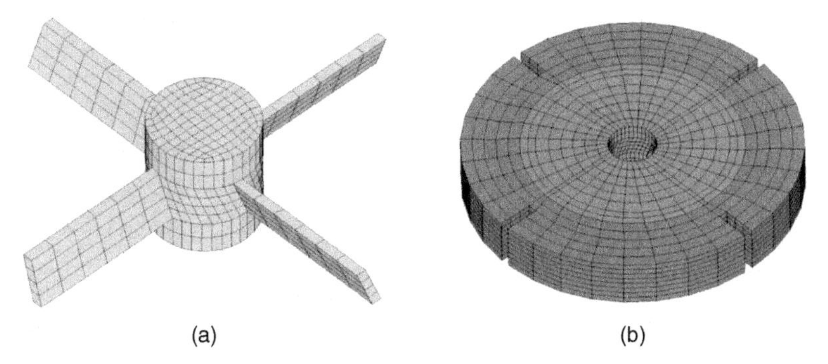

Figure 6.10 Mesh models of the rotating machine. (a) The solid domain and (b) the fluid domain.

The Von Mises stress distributions of the solid blade produced by the centrifugal force and fluid force at two different times are shown in Figure 6.11a,b and the fluid mesh motion is shown in Figure 6.12a,b observed in the global coordinate system. At $t = 0.05$ s, the central rotating mesh rotates 2.865°, and at $t = 1$ s, the central rotating mesh rotates 57.325°. The fluid velocity and pressure distributions are shown in Figure 6.13a,b and Figure 6.14a,b.

6.6.2 Sliding interface approach

If all physics that involved in the multiphysics coupling of the rotation machinery are solved in the global stationary frame, the sliding interface mesh treatment may be needed for the coupling between the rotor and stator interface. In this case, the fluid domain is separated by a sliding interface between the rotor part and stator part, with the rotor part doing rigid body mesh motions along with the rotor blade, although the mesh of the stator part stays "still" with the stator. In this case, the entire 360° model

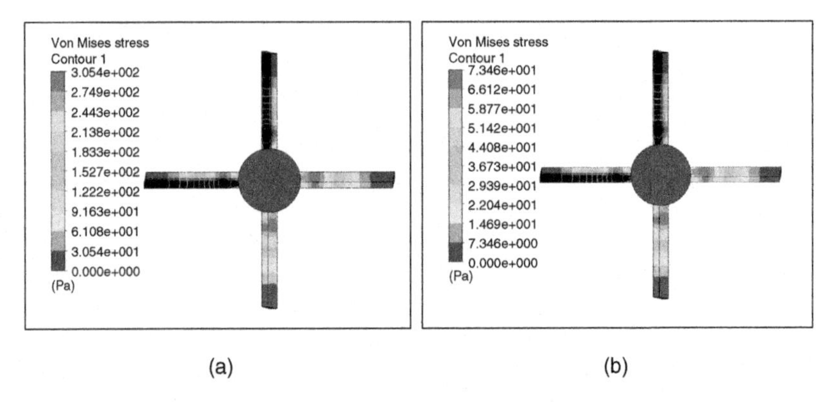

Figure 6.11 The Von Mises stress distributions of the solid blade. (a) $t = 0.05$ s and (b) $t = 1$ s.

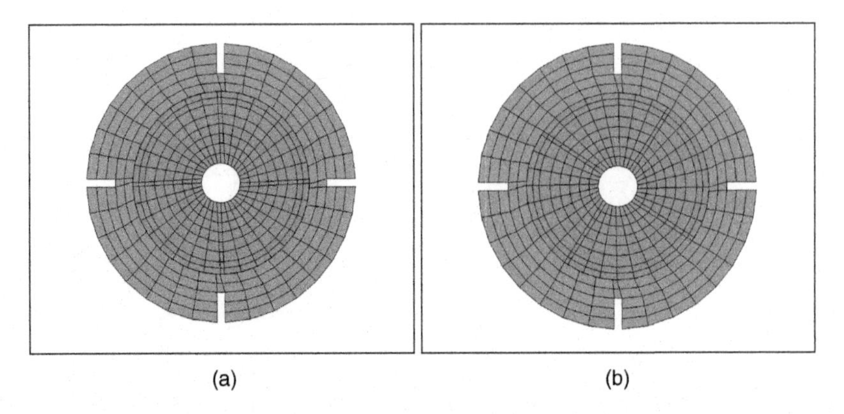

Figure 6.12 The mesh motion result of the fluid domain. (a) $t = 0.05$ s and (b) $t = 1$ s

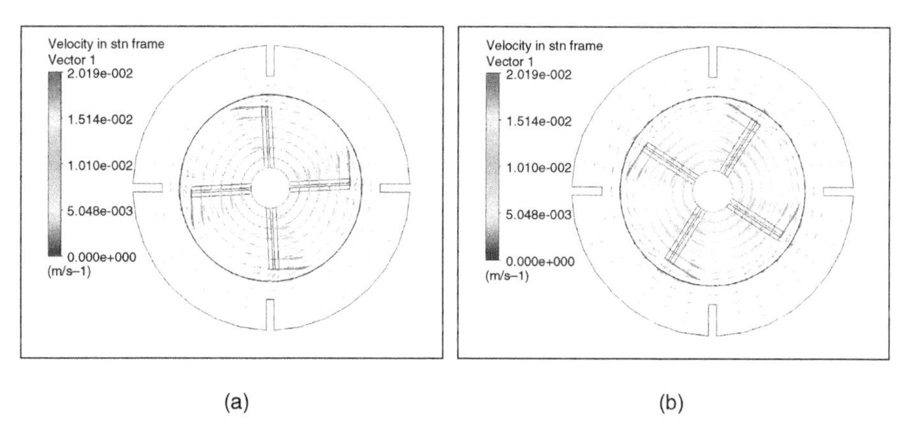

Figure 6.13 The velocity distributions of the fluid domain. (a) $t = 0.05$ s and (b) $t = 1$ s.

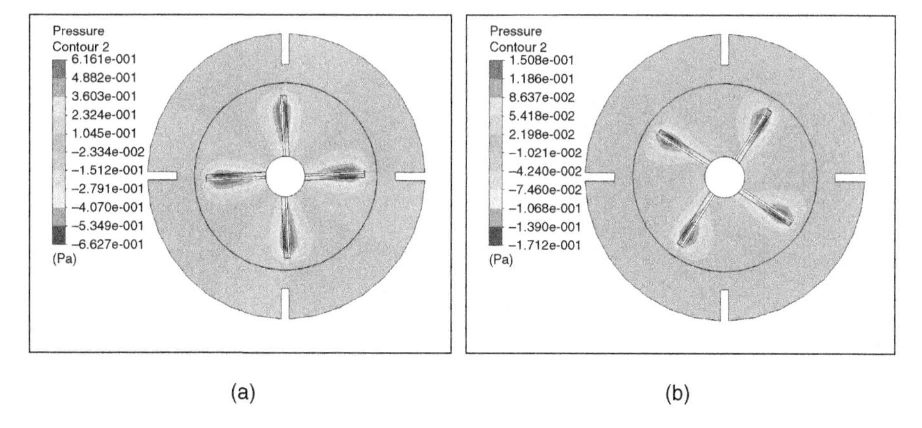

Figure 6.14 The pressure distributions of the fluid domain. (a) $t = 0.05$ s and (b) $t = 1$ s

is needed in the numerical simulation. Morphing with a sliding interface presented in Section 6.5.2 will be used for the interface coupling.

6.6.2.1 Fluid–structure interaction model for morphing with the sliding interface approach

An application of morphing and the sliding interface to a coupling physics problem of the turbine machinery is shown in this section. The analysis model and the material properties are the same as the model in Section 6.6.1.1.

A transient analysis is conducted. The transient time step size is 0.05 s, and the total analysis time is 1 s. As for solid model, the rotor blades are formulated in a rotating frame of reference. The rotational speed is 1 rad/s, and the central nodes of

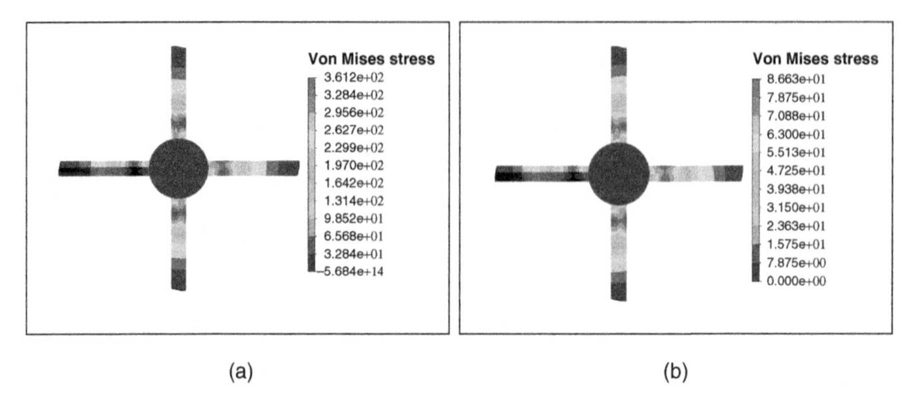

Figure 6.15 The Von Mises stress distributions of the solid blade at two different times. (a) $t = 0.05$ s and (b) $t = 1$ s.

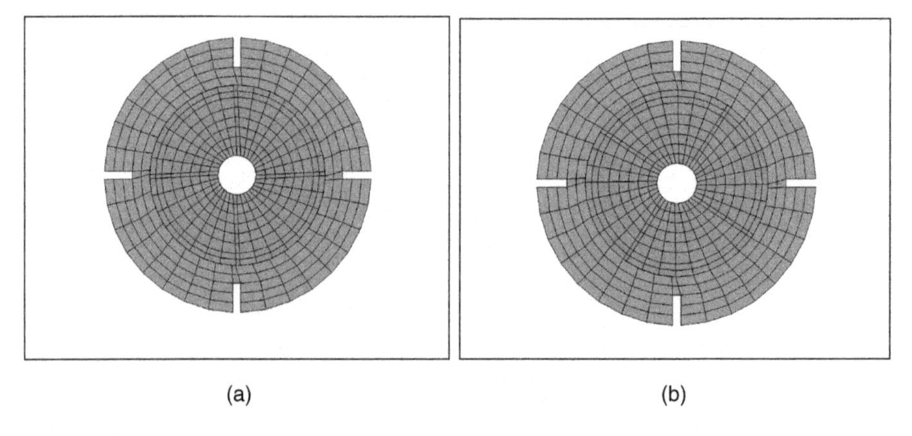

Figure 6.16 The fluid mesh motion results at two different times. (a) $t = 0.05$ s and (b) $t = 1$ s.

the shaft are fixed. The fluid region consists of the rotor and stator part; both the rotor and stator parts are formulated in the stationary frame. As for the rotor part, the rigid body morphing condition is applied at the fluid rotor–stator coupling interface, and the symmetry boundary applied on the top and bottom surfaces and inner wall is considered a fluid–structure coupling interface. The weak coupling method can be used for the fluid–structure interface with the consideration of frame change. As for the stator part, the symmetry boundary is applied on the top and bottom surfaces, and the outer wall is considered a no-slip wall. The fluid rotor–stator coupling interface is treated as a sliding interface.

The Von Mises stress distributions of the solid blade produced by the centrifugal force and fluid force at two different times are shown in Figure 6.15a,b and the fluid mesh motion results at two different times are shown in Figure 6.16 a,b. At $t = 0.05$ s,

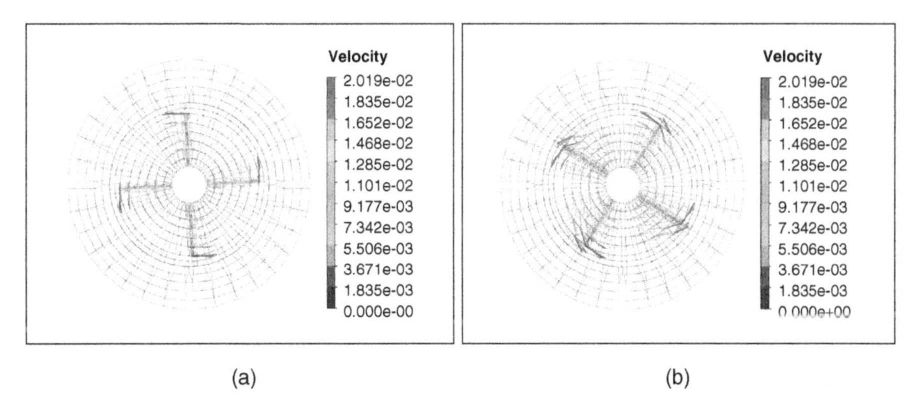

Figure 6.17 The velocity distributions of the fluid domain at two different times. (a) $t = 0.05$ s and (b) $t = 1$ s.

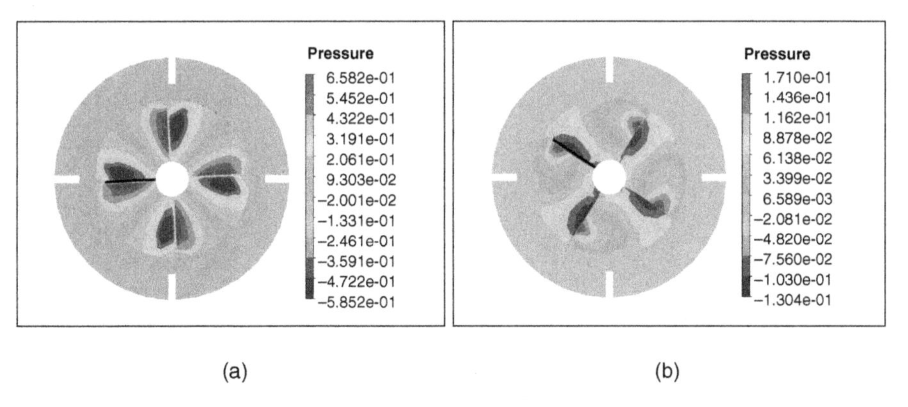

Figure 6.18 The pressure distributions of the fluid domain at two different times. (a) $t = 0.05$ s and (b) $t = 1$ s.

the central rotating mesh rotates 2.865°, and At $t = 1$ s, the central rotating mesh rotates 57.325°. The fluid velocity and pressure distributions at two different times are shown in Figure 6.17 a,b and Figure 6.18a,b.

Compared with results in Section 6.6.1.1, it is known that the Von Mises stress distributions of the solid blade and fluid velocity distributions are approximate in previous two methods; pressure distributions of fluid domain at $t = 1$ s have a difference in the fluid rotor part, which mainly comes from a calculation error with different software.

High-performance computing for multiphysics problems

7

Chapter Outline

7.1 The challenges in large-scale multiphysics simulation

As discussed in Chapter 2, in a multiphysics system, at least two different types of physics models are involved in the simulation, and the different physics models usually have different physical characteristics, resulting in much more different numerical properties of the matrix and vector. The two or more physics models can be solved by a strong coupling method, which will generate a coupling equation system with the consideration of coupling boundary conditions by the multipoint constraint method, Lagrange multiplier method, penalty method, and so on. The strong coupling equations, a large-scale system, usually lose the sparseness on the interface, and the quality of the resulting matrix may get worse than those of each individual physics. The standard iterative linear equation solver may not work for most of these cases, just like the weak coupling method does not work for the strongly coupled problem. The strong coupling method is inevitable for some problems, but the scale and the characteristic of the resulting matrix make the parallel algorithm very challenging. Therefore, this chapter discusses a hybrid method-based parallel algorithm for the strong coupling method.

As presented in Chapter 3, the weak coupling method maintains much independence among physics models. It is not just the time integration method for transient analysis, time step size, and varying differencing schemes for each physics model, but it also allows each physics model in the simulation to use different and appropriate parallel schemes.

Q. Zhang & S. Cen: Multiphysics Modeling. http://dx.doi.org/10.1016/B978-0-12-407709-6.00007-9

7.2 Parallel algorithm for the strong coupling method

It is necessary to use an efficient parallel computing for large-scale multiphysics problems. In this section, a hybrid parallel coupling technique, that is the shared memory OpenMP for the element calculation and MPI-based distributed memory for linear equation solver is presented.

7.2.1 OpenMP-based element matrix and vector calculations

For element matrix and vector calculation of the physics model, we use the OpenMP-based shared memory parallel scheme to improve performance. It has almost linear scalability and is very easy to implement into a computer program. It is applicable to most computer platforms, especially to multicore machines.

The basic idea of the OpenMP-based parallel is to partition the analysis domain into element groups, so each thread will do the element calculation for one assigned group. Except for global vector assembly that needs to be locked from parallel, all other element matrix and vector calculation can be carried out in a shared memory parallel fashion. So this makes the OpenMP-based parallel method have great scalability. Figure 7.1 demonstrates how the OpenMP method works for the element calculation of a multiphysics system.

7.2.2 MPI-based distributed memory direct sparse solver

Since the coupled equations of the multiphysics system may be ill conditioned (e.g., the thin shell structure coupling with water), it is difficult to get a converged solution by iterative solver, or at least a considerable number of iterations are needed to get a

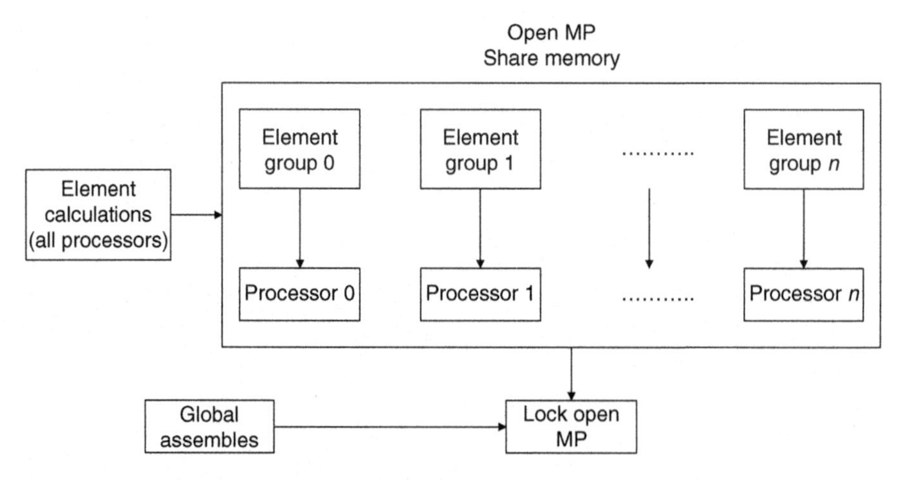

Figure 7.1 OpenMP-based element calculation.

converged solution. Instead of using a parallel iterative solver, the distributed parallel direct sparse solver is more robust and efficient. In this section, an MPI-based multifrontal parallel direct sparse solver is recommended for solving the coupled fluid–structure equations.

7.2.3 The flowchart of hybrid parallel algorithm for the strong coupling problem

The flowchart of the hybrid parallel algorithm for strong coupling problem is shown in Figure 7.2. In this algorithm, the initialization, finalization, and data I/O tasks are carried out in the master thread without parallelization. The OpenMP-based shared memory parallel method is applied for the element calculations. Although the assembly of matrices and vectors is carried in the critical region of OpenMP or on an atomic way, the assembled global linear equations can be solved by the MPI-based parallel direct or iterative solver.

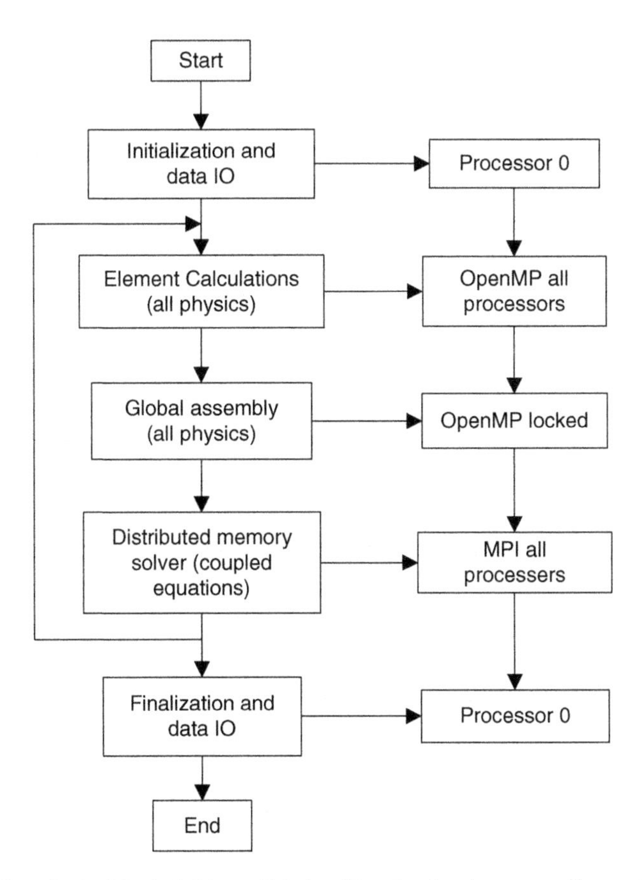

Figure 7.2 Flowchart of the hybrid parallel algorithm for the strong coupling problem.

7.2.4 The limitation of direct solver–based strong coupling method

Although the intensive memory usage and extreme computational request, the direct sparse solver–based hybrid parallel algorithm limits its applications to multiphysics problems of small to medium scale. As soon as the meshes reach millions of nodes, this method may become impractical. So we suggest that as long as the coupled matrix quality is not too bad for an iterative parallel solver to solve, the users should use the iterative parallel solver to solve the coupled matrix.

7.3 Parallel scheme for weak coupling methods

When a strong coupling method is unnecessary, we suggest using a weak coupling method, in which independent time integration scheme, varying time step size, appropriate spatial differencing scheme, and optimal parallel technique can be used for each physics solver. This guarantees an optimal numerical combination for each physics solver with the consideration of efficiency, accuracy, and robustness. The only extra thing that needs to be considered is the choice of appropriate weak coupling strategies discussed in Chapter 3.

The key factors for the parallel scheme of the weak coupling methods are:

- Appropriate parallel solver technology for different physics solver
- Communication overhead
- Load balances across different physics solvers

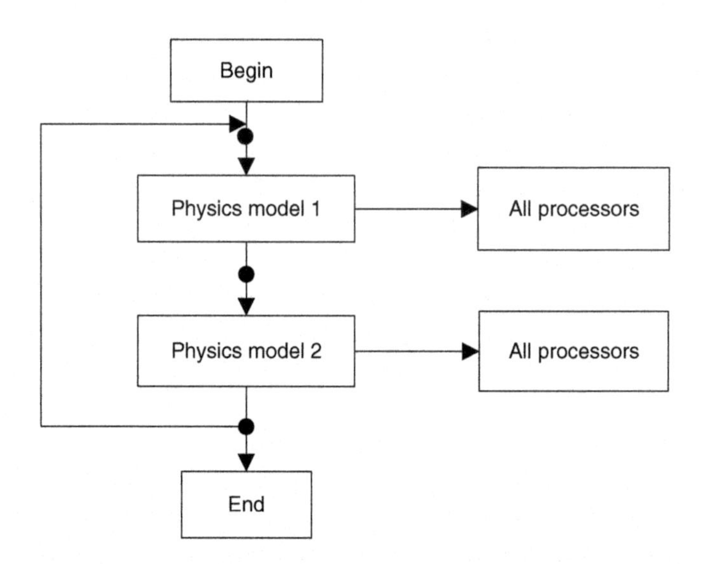

Figure 7.3 Parallel Gauss–Seidel method.

7.3.1 Gauss–Seidel versus Jacobi iteration methods

For the staggered Gauss–Seidel method, each physics model is solved sequentially. This means that each physics solver will have all of the computing resources when it is actively being solved while the others hang and wait for finishing and vice versa. This means that each physics solver should use as many processors as possible when it is active and to stay with minimum memory and processor usage when it is idle. When the Gauss–Seidel iteration method is used, no waiting time is needed. Each physics solver can maintain good scalability and efficiency if it uses the maximum available computing processors when active. Figure 7.3 shows the processors' usage and the data communication points (the dots) for the weak coupling Gauss–Seidel iteration method.

On the other hand, the Jacobi iteration method allows the physics models involved in the coupling to be solved simultaneously (in parallel). Figure 7.3 demonstrates the processors' usages and the synchronization points for the Jacobi iteration method. In this algorithm, the load balance needs to be considered before starting the solving process; otherwise, a longer wait time will be needed at the synchronization load transfer points. Because different physics models have different computational complexity and matrix quality, balancing the loads to make all physics solvers to finish one coupling iteration with closest time amount becomes a challenging work for parallel computing. The Gauss–Seidel method is the easiest to use when each physics solver still has good scalability to use the entire available processors (Figure 7.4).

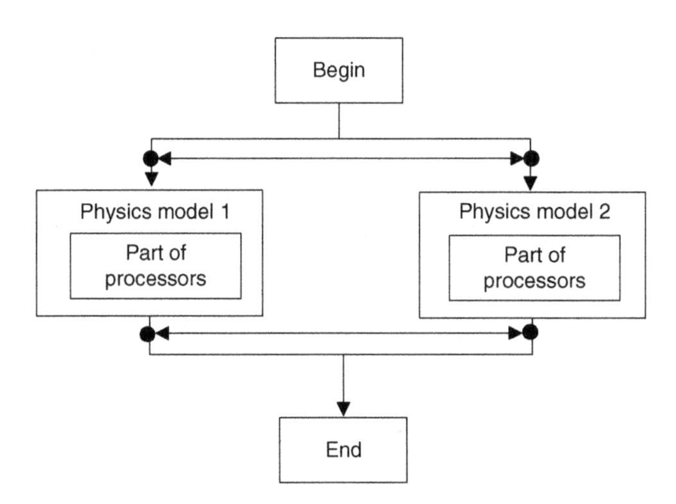

Figure 7.4 Parallel Jacobi iteration method.

General multiphysics study cases

Chapter Outline

8.1 Efficiency studies of strong and weak coupling methods for simple case (Zhang and Toshiaki, 2004)

The key issues differentiating strong coupling method and weak coupling method in fluid–structure interaction (FSI) analysis are the limitations of their application, the computational efficiency, and the requirements of modification of the existing programming codes. The limitations in their application relate to their convergence properties (convergence or divergence), computer-storage limitations, the changes of the interface area (sloshing problem in a flexible tank, FSI with contact), and so forth. Computational efficiency is evaluated with reference to the CPU time for one FSI iteration, the flexibility between the fluid solver and solid solver in time integration and mesh discretization, and the convergence rate. Then a simple case of cylinder vessel is used to make a comparison of the convergence property and the computational efficiency of the strong and weak coupling methods.

8.1.1 Analytical model

As an example of a simple FSI problem, the geometry of the cylinder vessel and the boundary conditions are shown in Figure 8.1. The uniform velocity is prescribed on the left open boundary, from 0 s to 1 s, and the specified velocity is increased from 0 m/s to 0.05 m/s smoothly. The right end of the cylinder vessel is enclosed by a flexible membrane. There is no outflow boundary in this model. The inflow is allowed by

Q. Zhang & S. Cen: Multiphysics Modeling. http://dx.doi.org/10.1016/B978-0-12-407709-6.00008-0

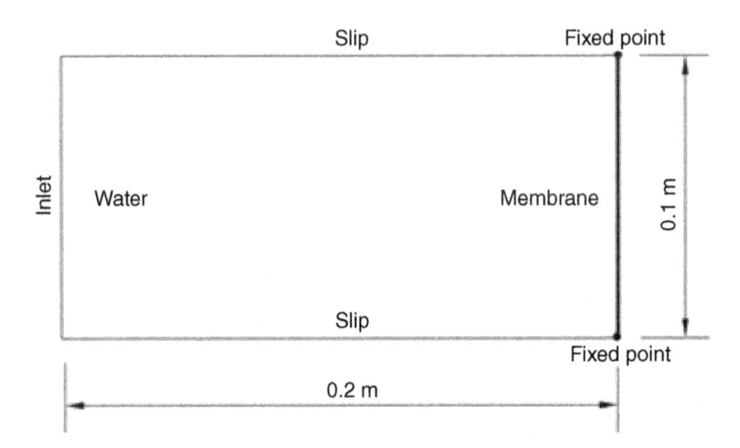

Figure 8.1 Analytical model for a simple interaction problem.

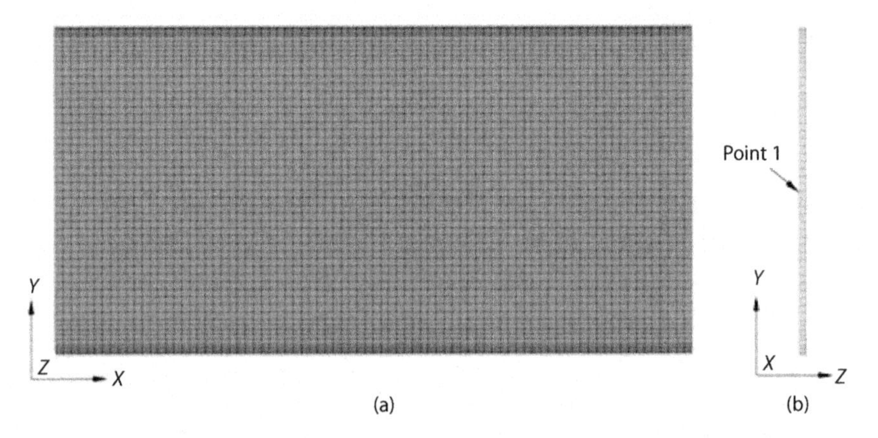

Figure 8.2 Mesh discretization of simplified cylinder vessel.

the deformation of the flexible membrane; as shown here, this causes the numerical difficulty of the weak coupling method. The material properties of the membrane are Young's modulus, $E = 2.0 \times 10^7$ Pa, mass density $\rho = 1.0 \times 10^3$ kg/m^3, and thickness $t = 0.2$ mm. The fluid is water, bulk modulus $B = 2.0 \times 10^9$ Pa, mass density $\rho = 1.0 \times 10^3$ kg/m^3, and dynamic viscosity $\mu = 1.0 \times 10^{-3}$ Pa·s.

The mesh discretization of simplified cylinder vessel for different parts is demonstrated in Figure 8.2, in which 320 fluid elements and 54 shell elementss are generated. The total number of nodes is 9020, which includes 110 interface nodes. The eight-node hexahedral element is used for the fluid domain, and the four-node MITC shell element is used for membrane discretization. The center of the shell structure of point 1 in Figure 8.2 is used to monitor the deformation of the membrane.

Table 8.1 Comparisons of the weak coupling and strong coupling methods

Item	Strong coupling	Weak coupling	Weak coupling
Δt(s)	0.0125	0.0125	1.25e-05
n Iterations	3	Divergence	9
Time (s) (computational time for one time step/one coupling iteration)	9/3		81/9

8.1.2 Results and discussion

The convergence properties (time step size: Δt, number of coupling iterations: nIters) and computational efficiency (time: computational time for one time step and one coupling iteration) at $t = 0.025$ s are listed in Table 8.1. It can be seen from this table that using the strong coupling method, the solution converges by three iterations in 9 s. Table 8.1 also tells us that the weak coupling method causes solution divergence with the same time step size of 0.0125 s as the strong coupling method. As the time step size is reduced to 1.25 e-05 s, the weak coupling method produces converged solution after nine iterations in 81 s.

The vector plots of velocity distribution and the contour plot of pressure under the deformed configuration at 0.025 s are illustrated in Figures 8.3 and 8.4, respectively. Figures 8.5 and 8.6 demonstrate the same information for the simulation results at 0.5 s.

The time history of the displacement at the center of the structure in X-direction is presented in Figure 8.7. This curve reveals that the displacement of the monitored points increases monolithic with time.

In conclusion, for the FSI problems in which the incompressible fluid is enclosed by rigid walls and a flexible structure, the strong coupling method that solves the coupled system simultaneously works properly with a smaller number of iterations. However, the weak coupling method produces a diverged solution when using the

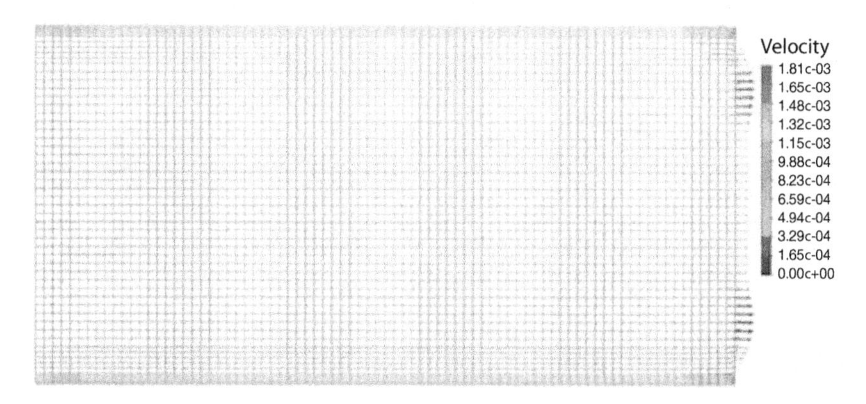

Figure 8.3 Vector plot of velocity in fluid at $t = 0.025$ s.

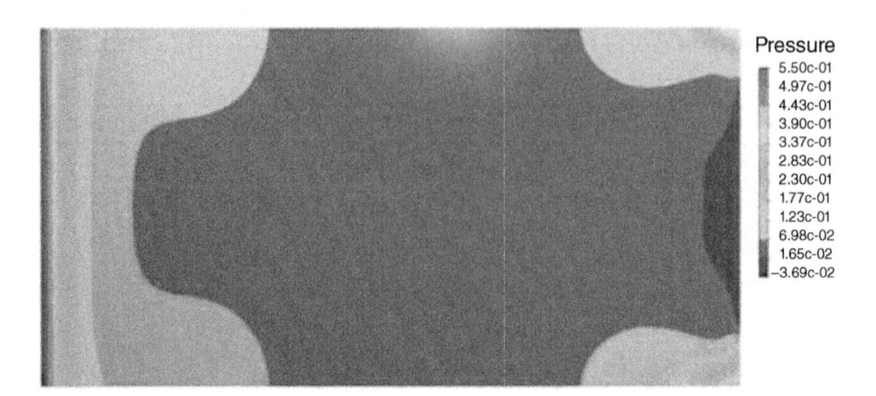

Figure 8.4 Contour plot of pressure of fluid at $t = 0.025$ s.

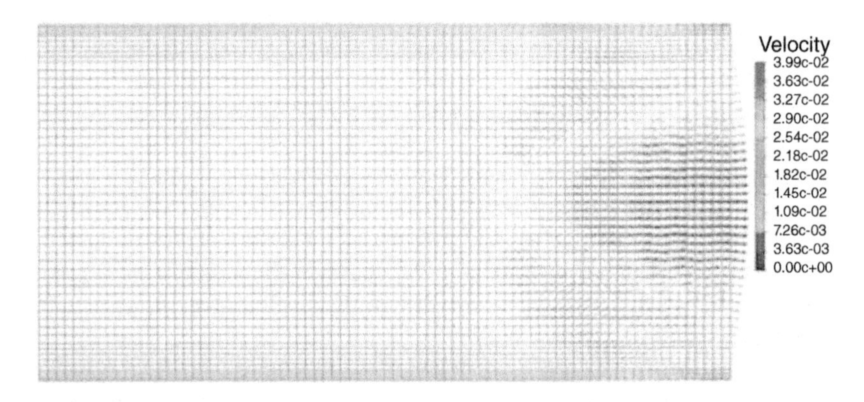

Figure 8.5 Vector plot of velocity of fluid at $t = 0.5$ s.

Figure 8.6 Contour plot of pressure of fluid at $t = 0.5$ s.

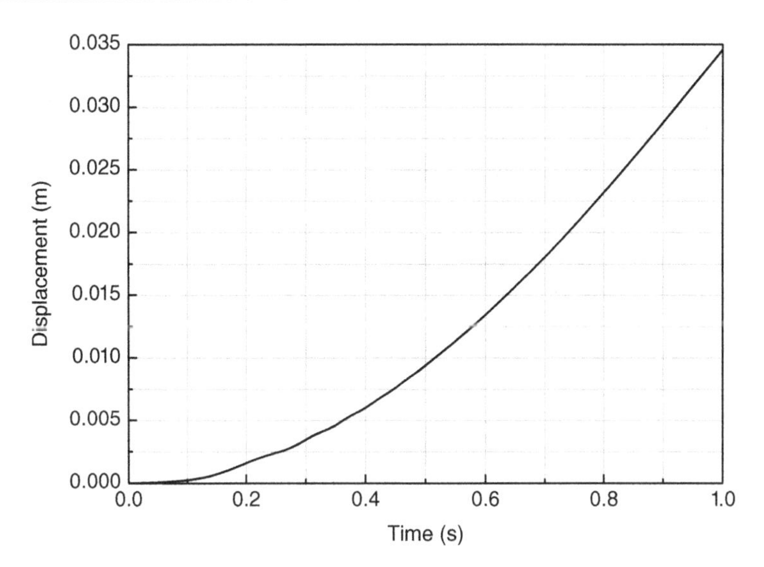

Figure 8.7 The record of displacement with time at point 1.

same time step size as strong coupling method. An impractical smaller time step size may help in the convergence, but very large number of iterations may needed, resulting in an unacceptable amount of computation time.

8.2 Fluid–structure interaction simulation of flow around a cylinder with a flexible flag attached

The problem of flow around a circular cylinder is a classic problem in the fluid dynamics, which has a deep engineering background, such as bridge building, sea poling, and so on. Fluid with different Reynolds numbers could cause different flow field and vortex, and the flexible cantilever beam that is attached at the downstream point of cylinder may produce self-oscillation because of the unbalanced force. This problem demonstrates the bidirectional coupling phenomena of the classic fluid flow around a circular cylinder with a flexible membrane attached. It provides some insights for this type of FSI problem and gives more evidence for the necessity of the strong coupling method in solving the FSI problems with complex flow around a thin and flexible structure.

8.2.1 The physics model

This case simulates the interactions between the fluid flow around a circular cylinder and the flexible structure. The analysis model is shown in Figure 8.8. The membrane structure in Figure 8.8 has very thin thickness, so the shell element is used to simulate the interaction between the structure and the fluid in the analysis. The material density of the structure is $\rho = 180$ kg/m^3; Young's modulus, $E = 3.2e7$ Pa; the Poisson's ratio,

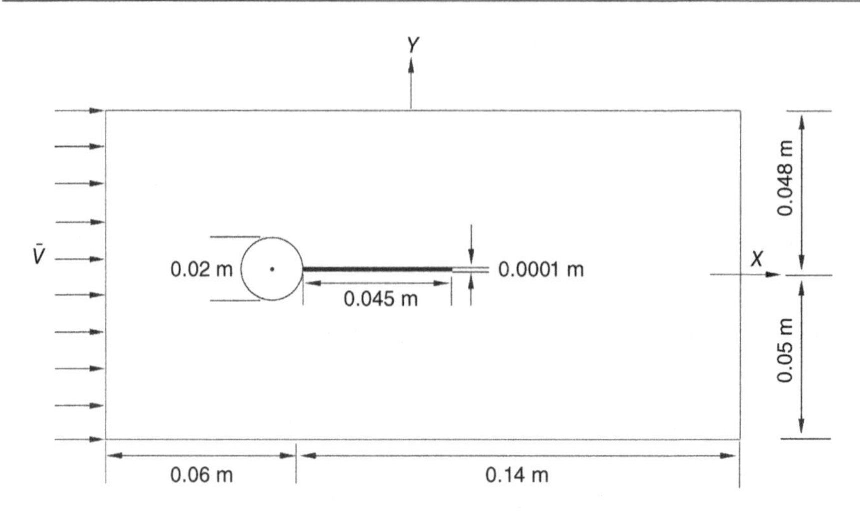

Figure 8.8 The diagram of the model.

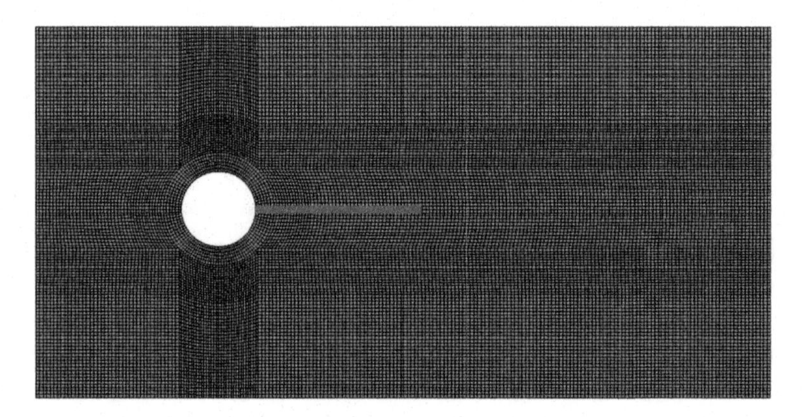

Figure 8.9 The mesh of the analysis model.

$\nu = 0.25$; the density of the fluid, $\rho = 1000$ kg/m^3; the dynamic viscosity, $\mu = 0.001$ kg/m·s; and the bulk modulus, $B = 2\text{e}9$ Pa.

Considering the efficiency and accuracy of the simulation, the appropriate mesh discretization is shown in Figure 8.9. Since the three-dimensional (3D) program INTESIM is used to solve this two-dimensional problem, the mesh of the fluid domain is discretized by hexahedral elements. The total number of elements is 23,410; the mesh size in *Z*-direction size, which is perpendicular to the paper, is 0.001 m; the mesh contains only one layer in *Z*-direction. The MITC quadrilateral shell elements for structure as the thick line are shown in Figure 8.9. The total number of meshes is 50; the mesh size in *Z*-direction is also 0.001 m. Figure 8.10 presents the enlarged view of the mesh around the coupling interface. It is known that at the coupling interface, the matching mesh discretizations are used for fluid and structure.

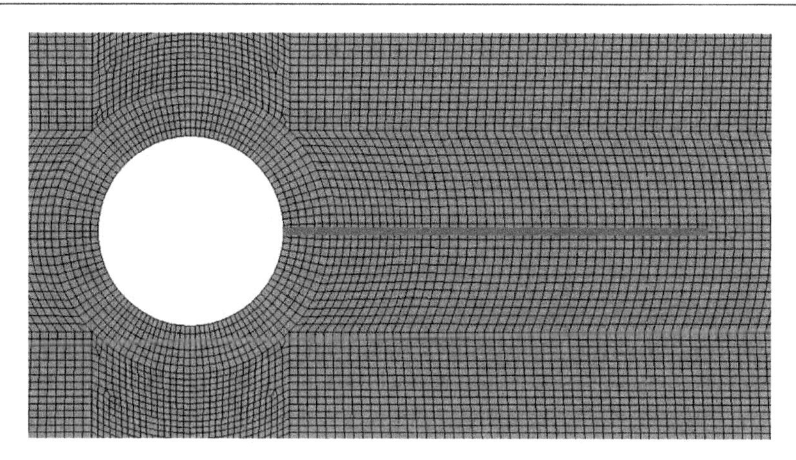

Figure 8.10 Enlarged view of the mesh around coupling interface.

8.2.2 Analysis conditions

In this example, Re = 400 is flow condition to study the interactions of the fluid flow around a circular cylinder with attached flexible membrane structure. The total analysis time is 60 s; in 0−4 s, the time step size is set to 0.025 s; in 4−60 s, the time step is set to 0.05 s. The fluid velocity at the inlet is shown as Equation (8.1); in 0−4 s, the flow velocity increases gradually with time; in 4−60 s, the flow velocity keeps constant value at 0.02 m/s. The outlet is set to opening boundary with pressure $p = 0$ Pa; the upward and downward surfaces are indicated to be symmetry boundaries; other boundaries are indicated as solid wall boundary conditions, shown in Figure 8.11.

$$\overline{V} = \begin{Bmatrix} 0.01\sin\left(\dfrac{\pi}{4}(t-2)\right)+0.01, t \le 4\text{ s} \\ 0.02, 4\text{ s} \le t \le 60\text{ s} \end{Bmatrix} \tag{8.1}$$

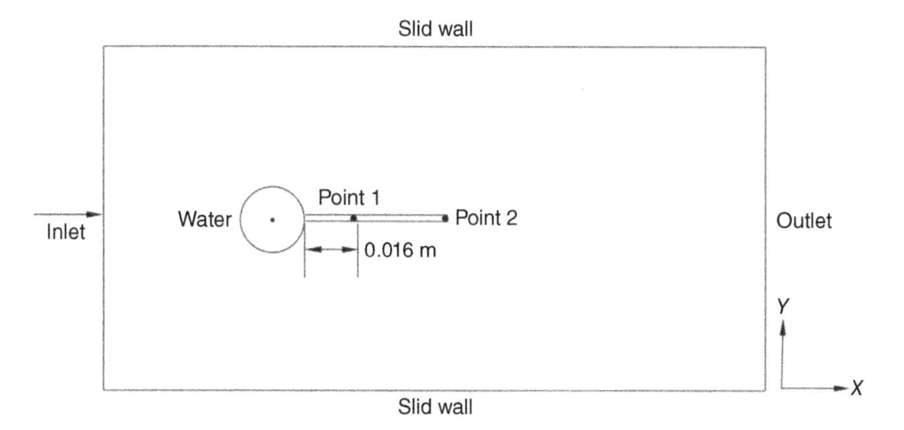

Figure 8.11 Boundary condition settings.

For the structural model, the left end of the shell is attached at the cylinder, and the movements of all nodes in Z-direction are restricted. Also, the displacements in Y-direction of points 1 and 2 (Figure 8.11) are monitored. The strong coupling boundary is set on the structural- and fluid-coupling interface using the multipoint constraint (MPC) method, discussed in Section 3.2.3.

In addition, the INTESIM software is used for the strong coupling analysis in this case. The geometrical nonlinearity is considered for structure analysis, and the fluid model is solved by SUPG/PSPG stabilization method (Section 5.2.1). To calculate the pressure fluctuation more accurately, the quadratic time integration method (alpha method, Section 3.5.3) is used to solve the transient strong coupling system.

8.2.3 Simulation results

The time histories of the displacement in Y-direction at points 1 and 2 are presented in Figure 8.12. This chart indicates that the displacement of the monitored points vibrate periodically. Thus, when the fluid analysis reaches periodically stable state, the shell structure vibrates periodically under the force of vortex shedding in the fluid domain, the period of this case is $T \approx 6.5$ s.

Taking $t = 19.3-25.8$ s in Figure 8.13 as one cycle, four time points are chosen in one cycle, $T/4$, $T/2$, $(3/4)$T, and T; check the results of the fluid velocity distribution, mesh deformation and pressure distribution, etc.

The streamline of the fluid velocity in one cycle is presented in Figure 8.13. As it is shown, at the points of $T/4$ and $(3/4)T$, the fluid-forming vortex behind the cylinder is positioned in the opposite direction of the movement of the membrane. This shows the

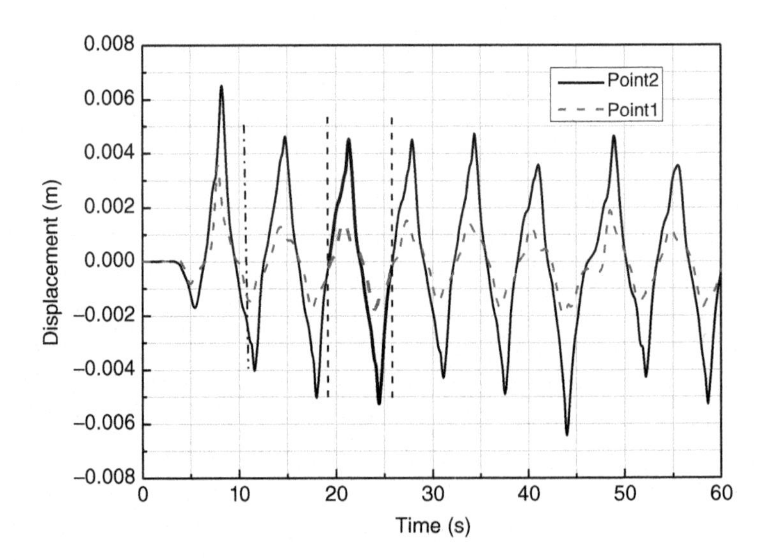

Figure 8.12 The time history of displacement at monitored points.

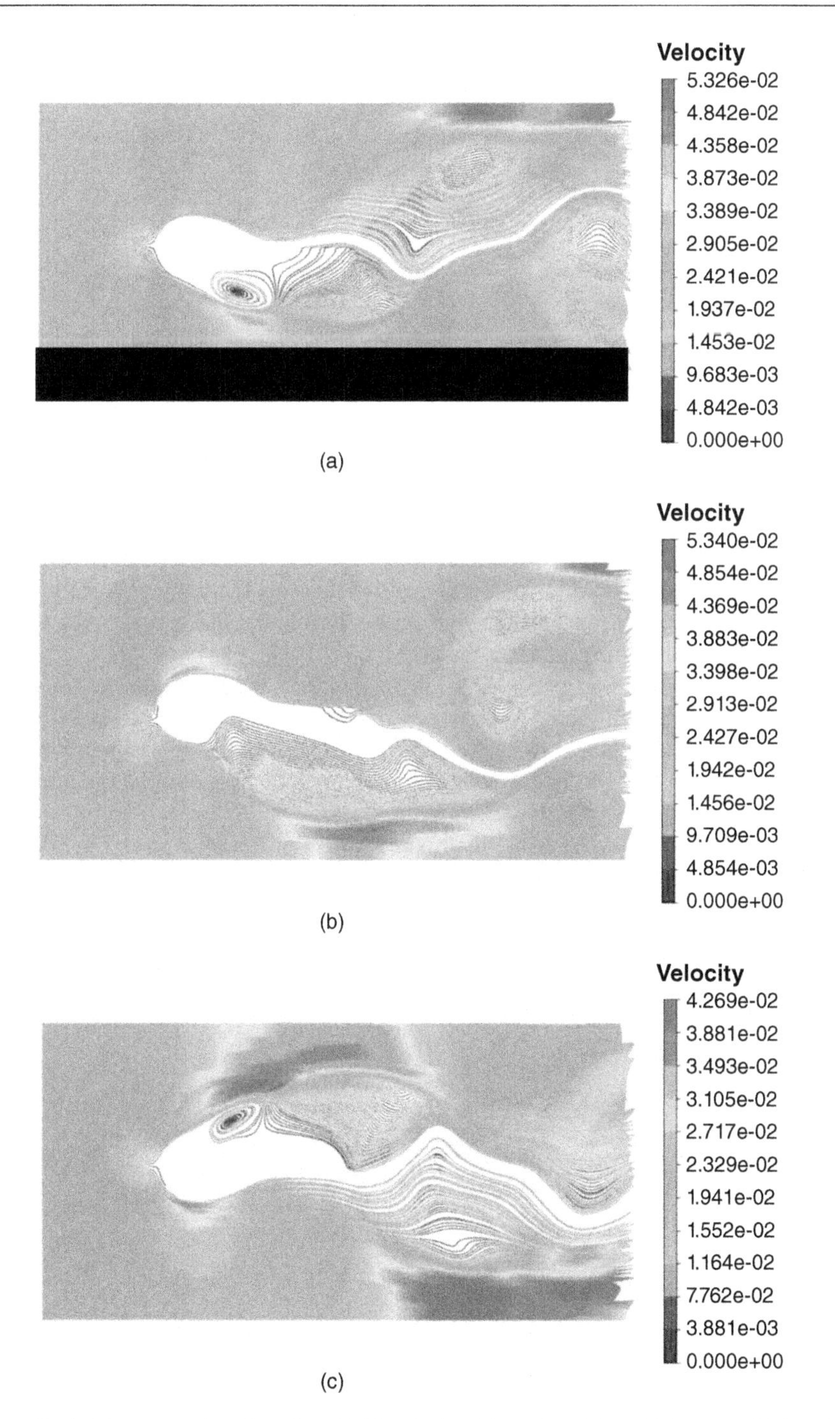

Figure 8.13 The streamline plot of the fluid velocity. (a) $t = T/4$, (b) $t = T/2$, (c) $t = 3/4T$, and (d) $t = T$.

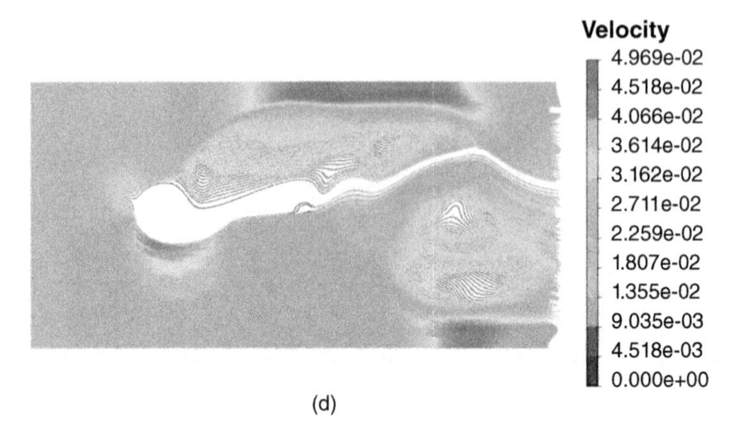

Figure 8.13 *(Cont.)*

effects of the structure vibration on the formation of vortex shedding. As for the points at *T*/2 and *T*, because of the smaller structural deformation, the vortex of the fluid is not observable enough behind the cylinder.

The mesh deformations at *T*/4, *T*/2, (3/4)*T*, and *T* in the fluid domain are shown as Figure 8.14. The highlighted thick lines represent the deformed shell structure. As shown in Figure 8.14, the Arbitrary Lagrangian Eulerian (ALE)–based mesh-morphing technology used by INTESIM works properly to maintain the mesh quality of the fluid domain under large structure deformation.

The contour plot of the pressure distribution at *T*/4, *T*/2, (3/4)*T*, and *T* are demonstrated in Figure 8.15. These figures show that the larger structure vibration produces larger variations of the pressure distributions.

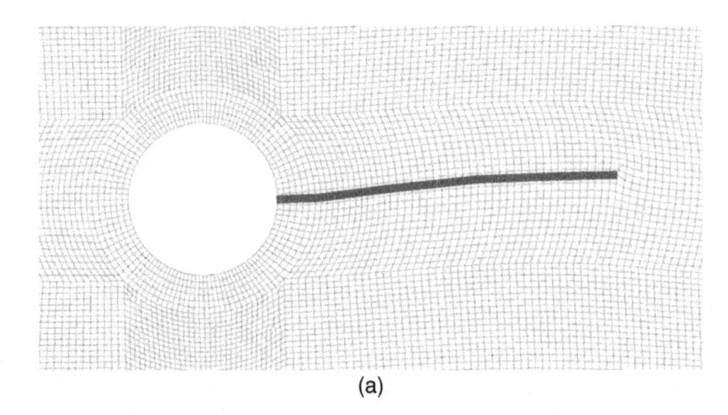

Figure 8.14 The deformation results of the local position. (a) $t = T/4$, (b) $t = T/2$, (c) $t = 3/4T$, and (d) $t = T$.

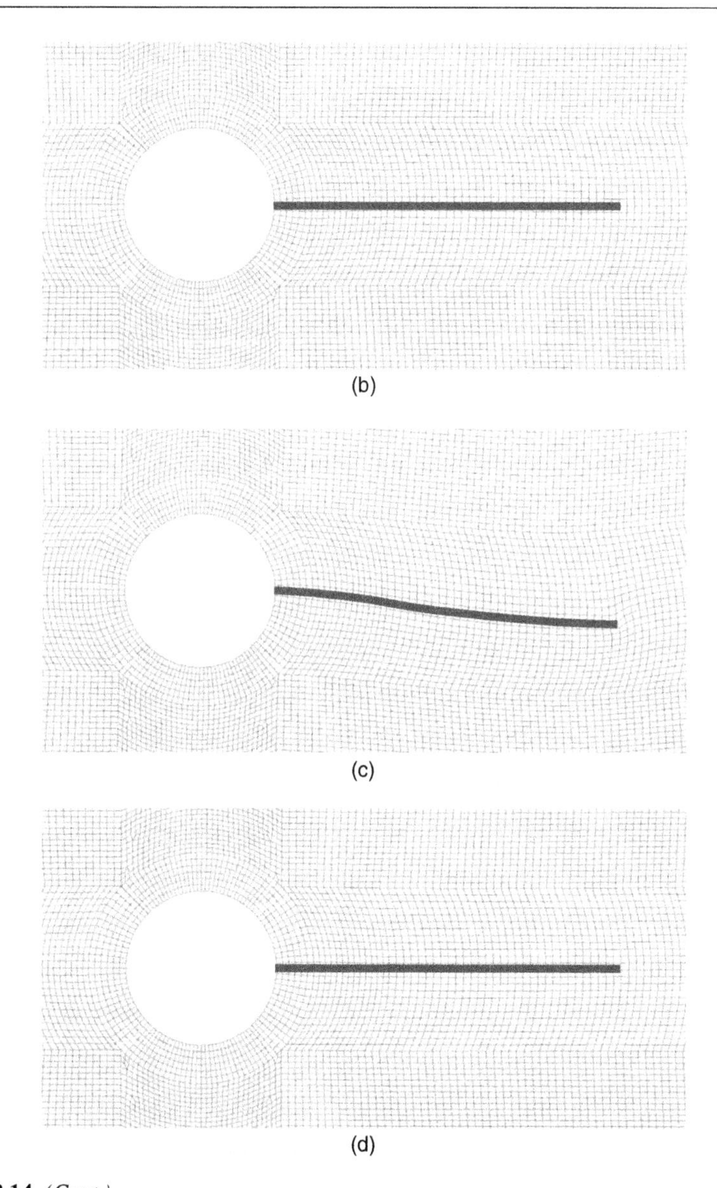

Figure 8.14 *(Cont.)*

8.2.4 Conclusions

The MPC-based strong coupling method is used for the coupling simulation of flow around a cylinder with a flexible structure attached. When the coupling simulation reaches a periodic stable state, under the effect of the fluid force, the structure demonstrates a periodic vibration behavior. This example provides the basis for this kind of

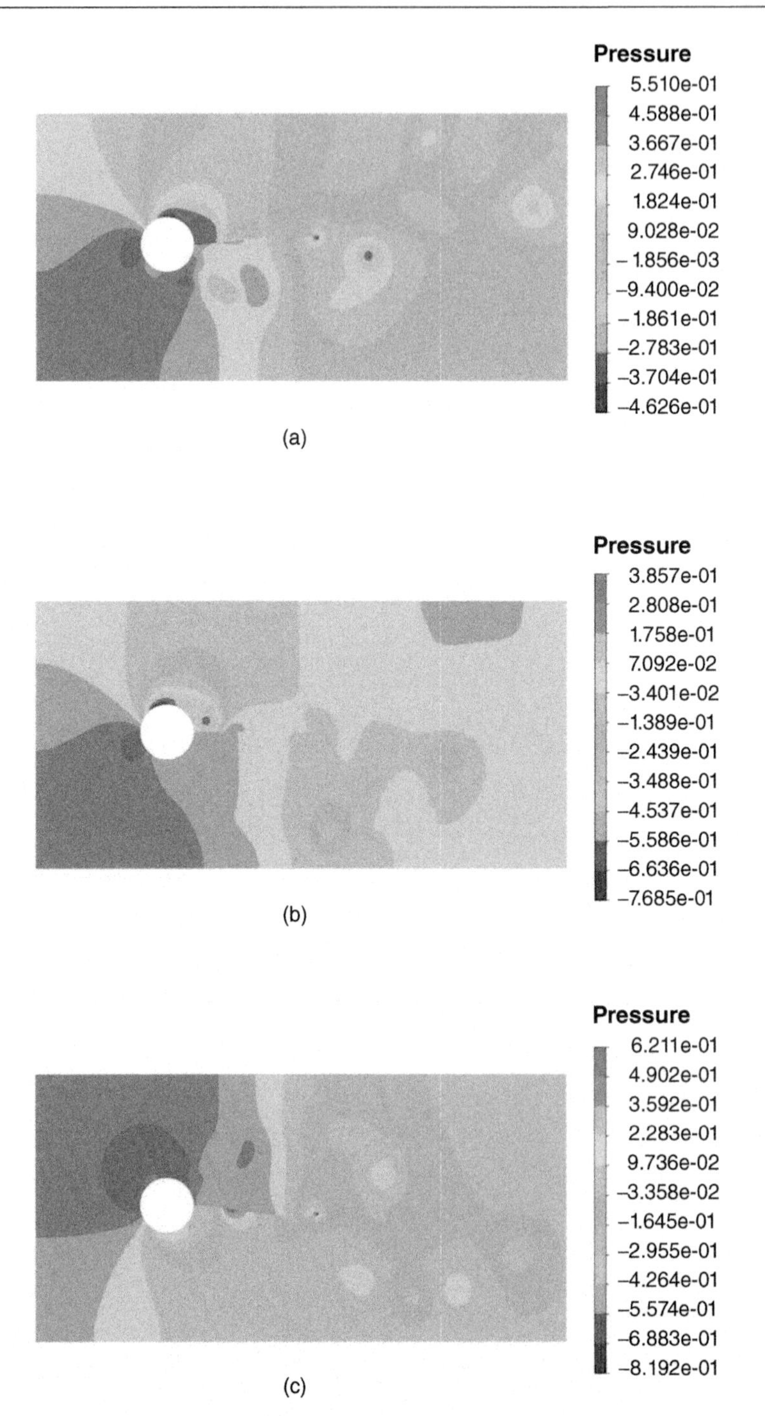

Figure 8.15 Contour plots of pressure at different times. (a) $t = T/4$, (b) $t = T/2$, (c) $t = 3/4T$, and (d) $t = T$.

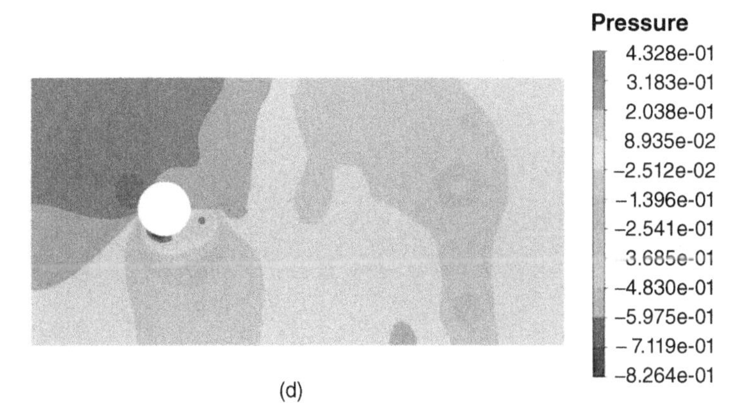

Figure 8.15 *(Cont.)*

fluid–structure strong coupling analysis. Furthermore, we also tried a weak coupling to analyze this problem. However, due to the strong nonlinearity caused by the thin shell structure and the vortex shedding in fluid flow, the convergence in coupling simulation cannot be achieved by the weak coupling method. It indicates that the application scope and major solvable problems for the strong coupling and weak coupling analysis methods are different. Regarding strong nonlinearity and tight interactions between physical fields, the strong coupling method could be the only option to choose.

8.3 Fluid–structure simulation of a flapping wing structure in a water channel (Zhang and Zhu, 2012)

8.3.1 Model of a flapping wing in a water channel

This is an example of a 3D wing structure that flaps in a water channel. The dimensions of the model are shown in Figure 8.16 (front and top views). For the fluid domain, the inlet velocity of 1.0 m/s is specified, an opening boundary with zero pressure is applied on the outlet, and fixed wall boundaries are applied on all of the wall boundaries. The motions of the wing structure in one flapping cycle are dramatically demonstrated in Figure 8.17. In this simulation, a forced motion is applied on the surrounding nodes of the rotating axis (dark gray nodes [red-colored nodes in the web version]) through the rotating axis in Figure 8.18. The motions of the wing in one flapping cycle consist of clockwise rotation from initial position with zero pitching angle to a position with pitching angle of 30° at 0.375 s. In the second phase from 0.375 s to 1.0 s, the wing translates vertically in the channel. Then from 1.0 s to 1.625 s, the wing structure does a counterclockwise rotation that makes it back to a zero pitching angle and at the top

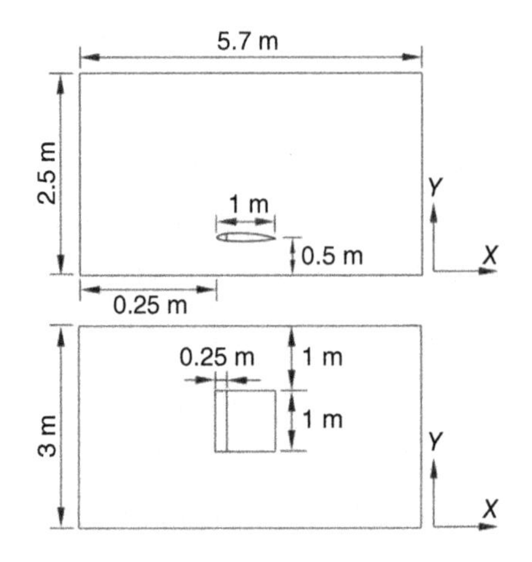

Figure 8.16 Dimensions of the model.

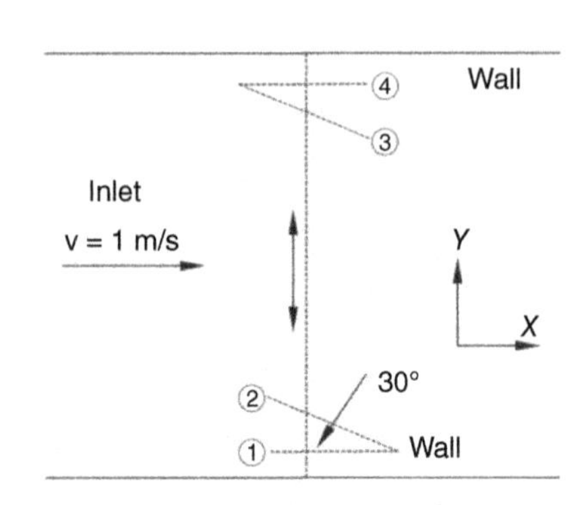

Figure 8.17 Motion of the wing in a flapping cycle.

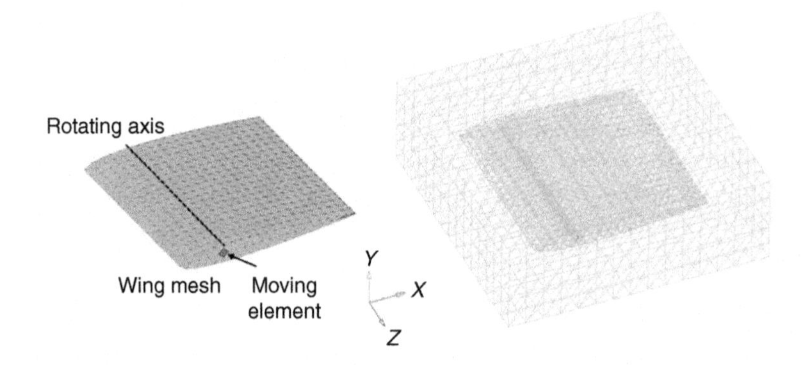

Figure 8.18 The wing mesh and nearby CFD meshes with higher element morphing stiffness (1000 times higher) and excluded from remeshing.

position of the channel. By contrast, in the second half of one flapping cycle, the wing will do a clockwise rotation, vertical translational motion, and counterclockwise rotation back to its original position sequentially.

Due to the large motion of the wing structure, to avoid the mesh distortion, an advanced morphing scheme with automatic partial domain remeshing technique is needed. In the morphing controls, sliding mesh moving conditions are applied on the inlet, outlet and all wall boundaries. To maintain the mesh size and to minimize the mesh deformation in the near wall region, we use varying mesh morphing stiffness for different analysis domain (i.e., a 1000 times larger mesh morphing stiffness is specified in the near-wall region; Figure 8.18) than the far wall regions from the wing in Figure 8.19 (the fluid elements excluding the elements in the near wall box); we

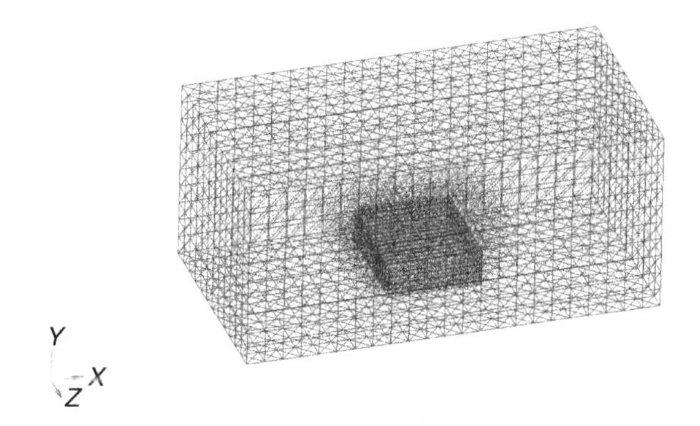

Figure 8.19 CFD region excluding the near wall elements for remeshing.

also exclude those near-wall elements from the remeshing process. Due to the high Reynolds number flow, the LES Smagrinsky model (Ghosal, 1996) is applied in this study case to model the turbulence behavior of the fluid flow.

The material properties of fluid are as follows: mass density $\rho = 1000$ kg/m^3, dynamic viscosity $\mu = 0.001$ Pa s, and bulk modulus $B = 1.0 \times 10^6$ Pa. The material properties of the flexible wing structure in the FSI simulation are Young's modulus $E = 7.0 \times 10^{10}$ Pa, Poisson's ratio $\nu = 0.3$, and mass density $\rho = 2.7 \times 10^3$ kg/m^3.

8.3.2 Simulation results

The time histories of the integrated fluid force on the coupled fluid–wing interface in FSI analysis are plotted in Figure 8.20a. By contrast, Figure 8.20b provides the time history of the fluid forces on the moving boundary from a pure CFD moving boundary simulation, and it assumes a rigid body motion of the wing structure. It shows from Figure 8.20 that because of the flexibility of the wing structure in the coupled FSI, the maximum values of the fluid forces in the *X*- and *Y*-directions are different and slightly larger than those of the pure CFD moving boundary problem. Also because of the high stiffness of the wing structure, the patterns of the time history curves from the two simulations are quite similar.

The motion and velocity distribution of the flow field at different time points in one flapping cycle are demonstrated in Figure 8.21. The deformed meshes after the morphing process and the meshes after remeshing at time 1.375 s are given in Figure 8.22. This example shows that the advanced morphing and partial domain remeshing scheme proposed in this research can handle the periodic extremely large domain changes well. More investigations about the mesh resolution on the boundary layer, detailed discussions about the accuracy of the simulation results, and comparisons with experimental results exceed the scope of this chapter, and we will provide more information on this subject in coming studies.

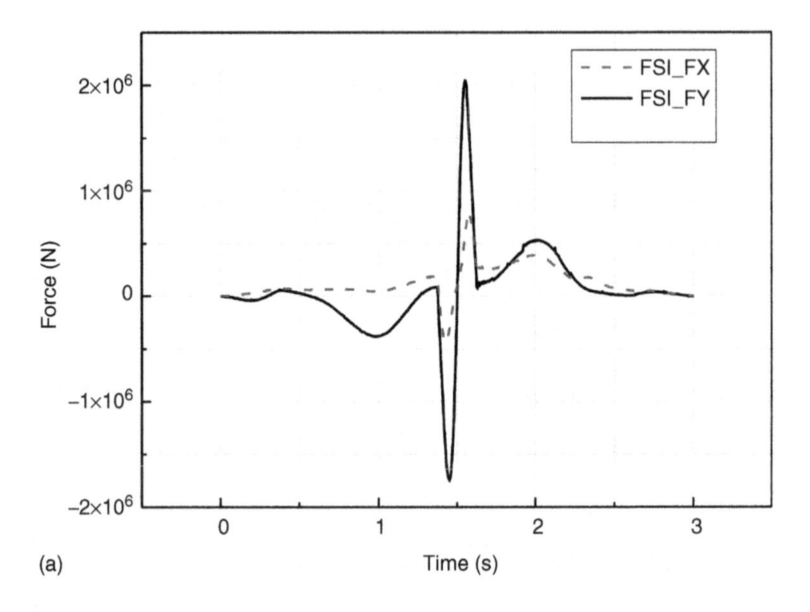

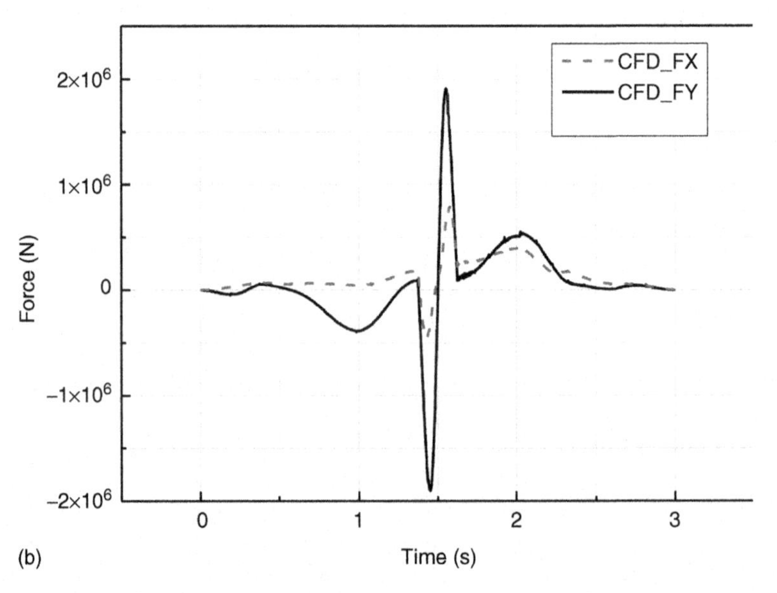

Figure 8.20 Time histories of the fluid forces on the fluid–wing interface. (a) Fluid–structure coupling calculation and (b) pure CFD moving boundary calculation.

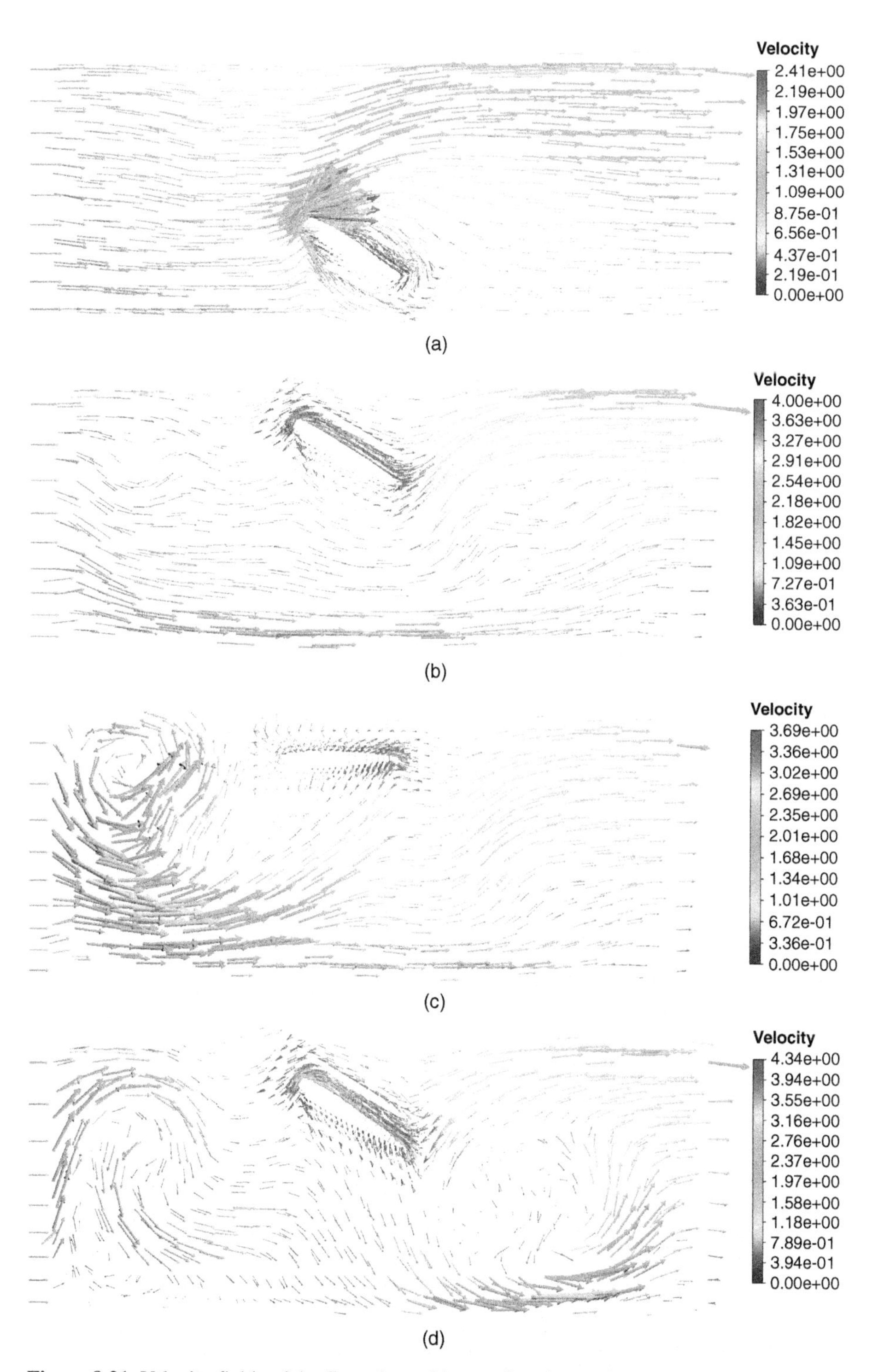

Figure 8.21 Velocity fields of the flow channel in one flapping cycle. (a) time = 0.375 s, (b) time = 1.375 s, (c) time = 1.5 s, (d) time = 1.625 s, (e) time = 2.625 s, and (e) time = 3.0 s.

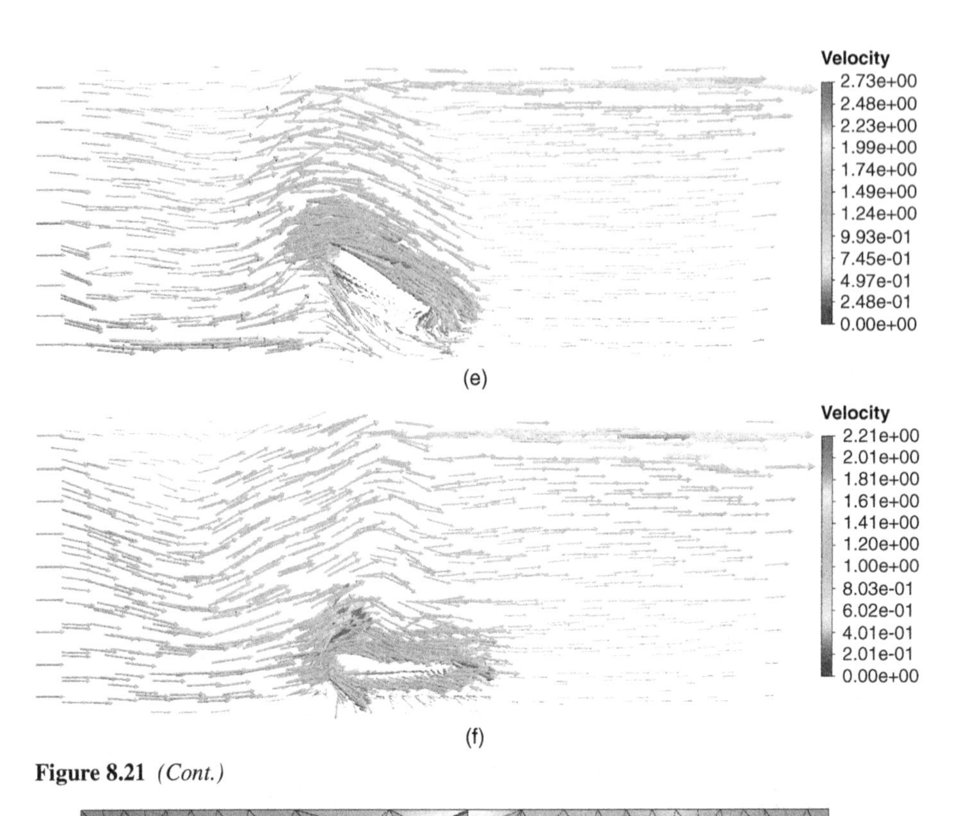

Figure 8.21 *(Cont.)*

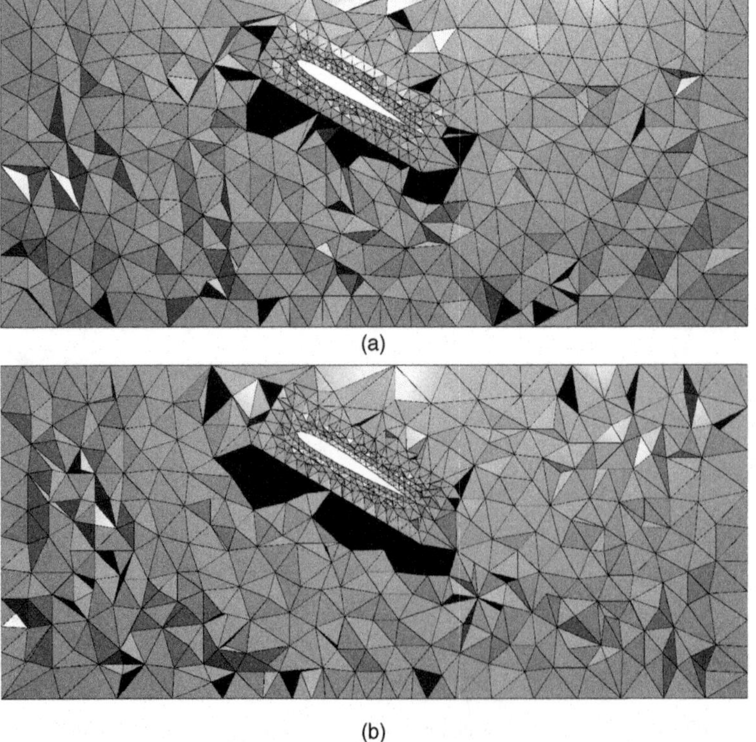

Figure 8.22 The morphed meshes and the meshes after remeshing at time = 1.375 s. (a) The morphed mesh and (b) the remeshed mesh.

Multiphysics applications in automotive engineering

Chapter Outline

9.1 The study of dynamic characteristics of hydraulic engine mounts by strong coupling finite element model

The hydraulic engine mount (HEM) has been developed and used for isolating vibration and noise from the engine to the body of the vehicle over the past 40 years. The mechanical behavior of the HEM under the excitation of high frequency oscillating forces, includes complex fluid–solid interactions (FSI). There are a lot of challenges in numerical simulations: the large domain changes, the strong coupling between the hyperelastic rubber, and the enclosed incompressible viscous fluid flow, the high frequency vibration, and the large scale problem. The numerical simulations will be very useful for clarifying the mechanical characteristics, and doing optimal design of HEM. This is especially true for the cases in which experimental studies are impossible, for example, the vibration of the HEM with very small amplitude at high frequency, and the dynamic behavior of rotating vibrations. However, these numerical difficulties had made it unfeasible to simulate the dynamic responses of the HEM in three dimensions, and in a fully coupling way.

The remarkable progress of the FSI analysis (Bathe et al., 1999; Stein et al., 2000; Taylor et al., 1998; Zhang and Hisada, 2001; Zhang and Zhu, 2012) and the incremental increases in computer performance in recent years make it possible to simulate the HEM in three dimensions. In this section, we introduce a fluid-structure interaction

Q. Zhang & S. Cen: Multiphysics Modeling. http://dx.doi.org/10.1016/B978-0-12-407709-6.00009-2

framework that includes the multipoint constraint equation-based strong coupling method, which is necessary for the strong coupled fluid-solid problems with incompressible liquid enclosed in the solid container, and that allows the nonconforming mesh discretization between the fluid and solid domain. We also use an advanced morphing technique to handle the large computational fluid dynamics (CFD) domain change with minimum mesh distortions. These capabilities are suitable for the fully coupled fluid-solid simulation of the HEM, and have been developed and implemented in the general purposed multiphysics simulation software program INTESIM. The proposed coupling technique will be used to simulate the 3D dynamic behavior of a HEM, and to get the time history of the reflection force on the base of the mount. Then, the dynamic stiffness and delay angle of the HEM are obtained by using those history curves with corresponding displacement excitation, through Fourier transformation. The comparisons of the numerical results with the experimental results are demonstrated in order to show the reliability and the accuracy of the proposed methods.

The arbitrary Lagrangian-Eulerian formulation for fluid flow and the finite element (FE) equation for nonlinear hyper-elasticity materials that will be used to model the rubber material of the engine mount, the strong coupling method, and the advanced mesh morphing method for handling CFD domain changes, are all implemented in INTESIM, and are presented in the previous chapters. The definitions of the dynamic characteristics for HEMs are given in this section.

A real 3D HEM model is presented, and the comparison of simulation results with experimental results will be provided. Finally, the concluding remarks will be given.

9.1.1 Definition of the dynamic characteristics for HEM

The key dynamic characteristics of the HEM are the dynamic stiffness and the delay angle. The definition of the dynamic stiffness and the delay angle are shown as follows. First, we get the complex stiffness by Equation (9.1),

$$K(j\omega_0) = \mathrm{F}(\mathrm{F}(t))/\mathrm{F}(\mathrm{X}(t))\,|_{\omega=\omega_0} = K_1 + jK_2 \tag{9.1}$$

Where $\mathrm{F}(t)$ is the time history of reflection force; $\mathrm{X}(t)$ is the time history of displacement excitation; ω_0 is the base angular frequency of the forced vibration; function $\mathrm{F}(\mathrm{X}(t))$ is the Fourier transformation for time history $\mathrm{X}(t)$. Then, the dynamic stiffness K_d and the delay angle can be defined by Equations (9.2) and (9.3), respectively.

$$\text{Dynamic stiffness: } K_\mathrm{d} = \sqrt{K_1^2 + K_2^2} \tag{9.2}$$

$$\text{Delay angle: } \phi = \tan^{-1}\left(K_2/K_1\right) \tag{9.3}$$

The dynamic stiffness indicates the value of the amplitude of the time history curve of the reflection force divided by the amplitude of the time history curve of the displacement excitation, and represents the stiffness of the system or the device under dynamic load. The delay angle represents the delay of the displacement excitation history curve versus the reflection force time history curve; it reveals the damping effects of the system or the device.

9.1.2 Physical model of HEM

The HEM is a typical engine mount, which is used in the field of automobiles (Zhang and Zhu, 2012; Cao et al., 2012). Its major parts consist of a rubber spring, two fluid chambers, an inertial track, and an uncoupler, a stirrer, a sealed bowl, an inner core, an outer covering, bases, etc. Figure 9.1 shows the outside view of an HEM. Figure 9.2 gives the cross section of an HEM. The inertial track is shown in Figure 9.3. The mesh discretization of 3D, cross section, and fluid domain are demonstrated in Figure 9.4, Figure 9.5, and Figure 9.6, respectively. The rubber spring acts as the main support

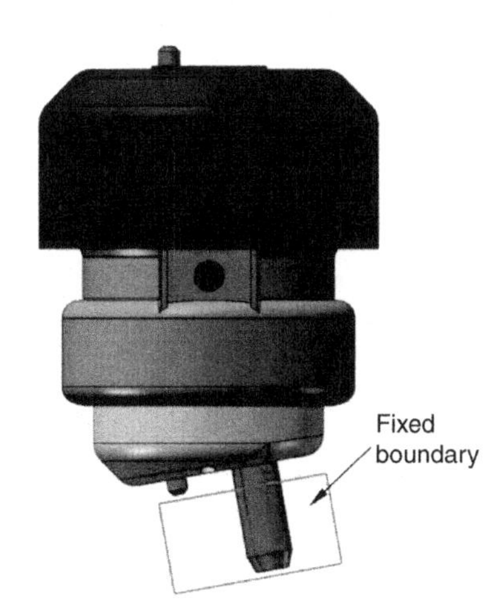

Figure 9.1 3D Geometry of the HEM model.

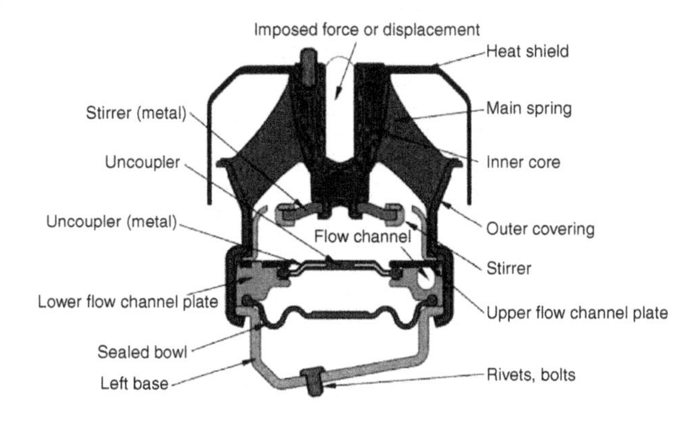

Figure 9.2 Cross-section of the HEM model.

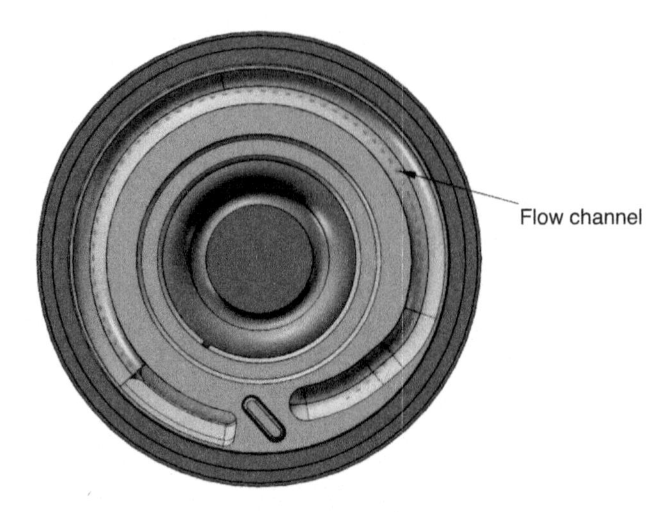

Figure 9.3 Cross-section of the flow channel.

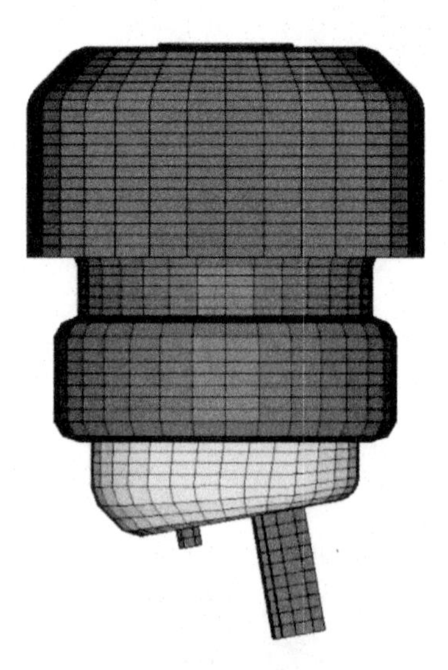

Figure 9.4 Mesh discretization of 3D.

spring. The inertial track plays the role of a tuned isolator damper in order to provide a higher damping in the low frequency range; this is the major contributor for the damping, and also the main reason of the delay angle of the HEM. The uncoupler connecting the upper and lower chamber can provide a lower stiffness to isolate the higher frequency vibration, when the inertial track is clogged.

The material information and mesh discretization for each part are listed in Table 9.1.

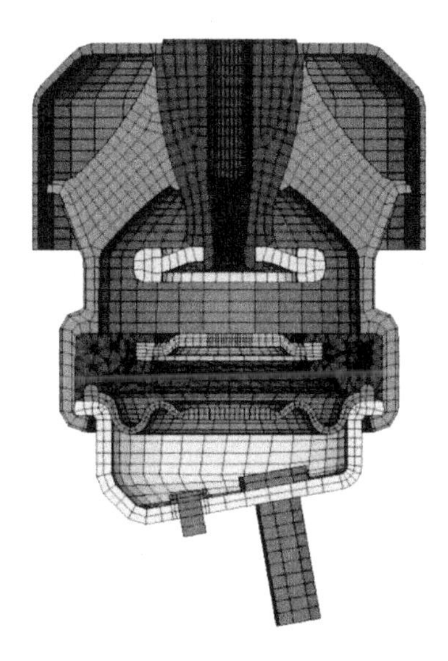

Figure 9.5 Mesh discretization of cross-section.

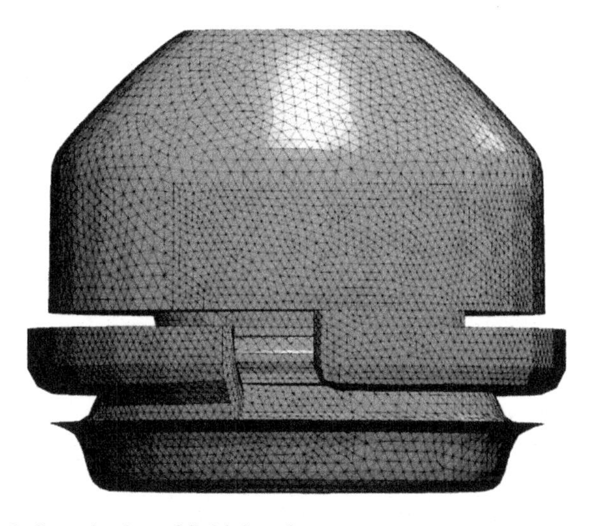

Figure 9.6 Mesh discretization of fluid domain.

The material properties of glycol are as follows: mass density $\rho = 1.038 \times 10^3 \text{kg/m}^3$, dynamic viscosity $\mu = 0.056 \, Pa \cdot s$; and bulk modulus $B = 2.0 \times 10^7$ Pa. The rubbers are assumed to be hyperelastic materials, and the coefficients of the high Mooney–Rivlin model are listed in Table 9.2. The material properties of the steel (engine connector) are: Young's modulus $E = 2.0 \times 10^{11}$ Pa, Poisson's ratio $\nu = 0.266$, and the mass density $\rho = 7.85 \times 10^3 \text{kg/m}^3$. For aluminum, Young's modulus $E = 7.1 \times 10^{10}$ Pa, Poisson's ratio $\nu = 0.33$, and the mass density $\rho = 2.7 \times 10^3 \text{kg/m}^3$.

Table 9.1 Material and element information of the HEM

	Name	Element type	Number of elements	Material name
Solid	Heat shield	Hex	896	Rubber
	Main spring	Hex	4096	Rubber
	Outer covering	Hex	832	Steel
	Inner core	Hex	3968	Steel
	Stirrer	Hex	224	Rubber
	Stirrer (metal)	Hex	224	Steel
	Upper flow channel plate	Tet	6429	
Fluid	Lower flow channel plate			
	Uncoupler	Hex	1504	Rubber
	Uncoupler (metal)	Hex	352	Steel
	Sealed bowl	Hex	2016	Rubber
	Left base	Hex	1210	Steel
	Rivets, bolts	Hex	436	Steel
	Glycol	Tet	133122	Glycol

Table 9.2 Material properties of the rubbers

Locations	C01 (MPa)	C10 (MPa)	Density: ρ (kg/m^3)	Poisson
Upper spring	0.2969	0.0548	1.1	0.49
Uncoupler	0.2969	0.0548	1.1	0.49
Lower diaphragm	0.2969	0.0548	1.1	0.49

9.1.3 Analysis conditions

The analysis procedure consists of a static load phase, and a high frequency forced vibration phase. Firstly, the initial load F_0 is applied slowly (in 1000 s) to the upper steel boundary, and the vibration displacements are applied to the upper steel

Table 9.3 **Simulation conditions**

Initial load	Amplitude of displacement vibration	Range of frequency
$F_0 = -1200$ N	$A = \pm 1$ mm	$f = 1{\sim}50$ Hz

Table 9.4 **Frequency points and corresponding dynamic stiffness and delay angle**

Frequency (Hz)	Dynamic stiffness (N/mm)	Delay angle (°)
2.5	173.25	3.02
5	162.63	9.45
8	157.97	26.75
10	184.32	42.51
12.5	272.94	50.49
15	383.18	44.64
17.5	447.86	34.46
20	478.91	24.81
22.5	478.48	18.07
25	467.23	12.70
30	441.01	5.73
35	411.09	4.61
40	392.67	3.65
50	365.65	3.10

boundary: $d = A \sin(2\pi \bullet ft)$. Here, A is the amplitude of the vibration, f is the frequency, and t indicates the time. Table 9.3 lists the value for F_0, A, and f. There are 14 frequency points investigated (Table 9.4). A true 3D strong coupling fluid-structure interaction simulation is carried out for each point in order to obtain the dynamic stiffness and delay angle curve of the HEM. The fixed boundaries are applied at the bottom of the base of the HEM (Figure 9.1).The initial load condition ($F_0 = -1200$ N), and subsequent displacement excitation conditions, are distributed and applied on the inner face of the inner core, as shown in Figure 9.2.

9.1.4 Results and discussions

The deformations and the von Mises stress distribution of solid parts of the HEM in one vibration cycle are shown in Figure 9.7a–d ($f = 20$ Hz), and the velocity distributions of the fluid domain are shown in Figure 9.8 a–d ($f = 20$ Hz). Although there is a limitation of demonstrating the behavior of the HEM by those figures, we know from the animations at different frequencies that, for the high frequency forced

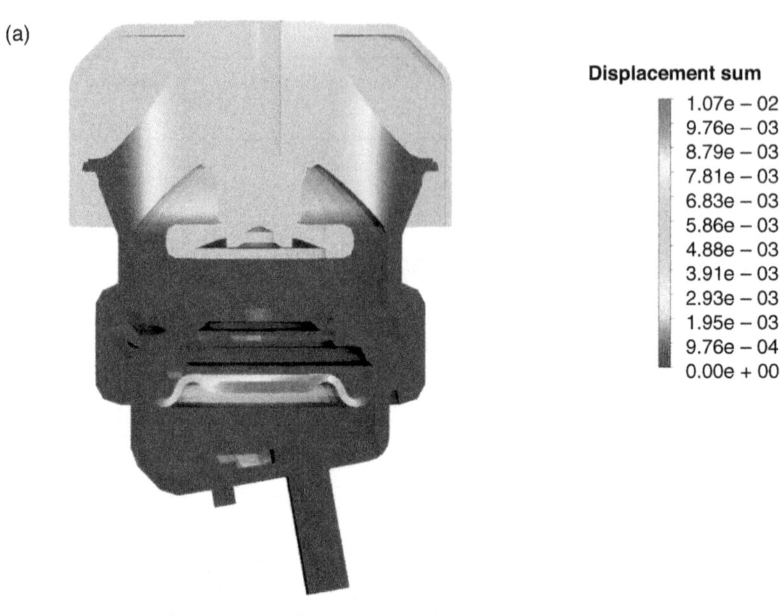

Contour plot of the length of the displacement vector

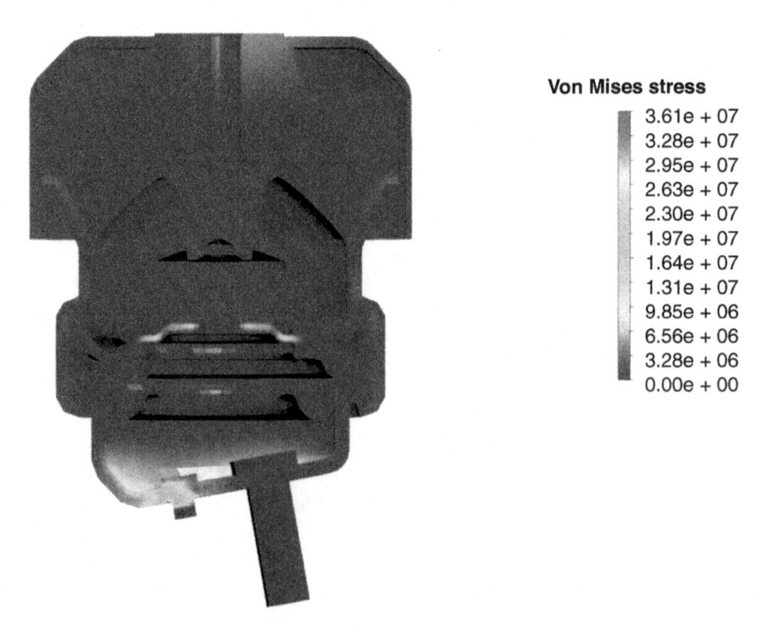

Distribution of the von Mises stress

(a) $t = 0.0$

Figure 9.7 Deformations and von Mises stress distributions of the solid parts ($f = 20$ Hz).

(b)

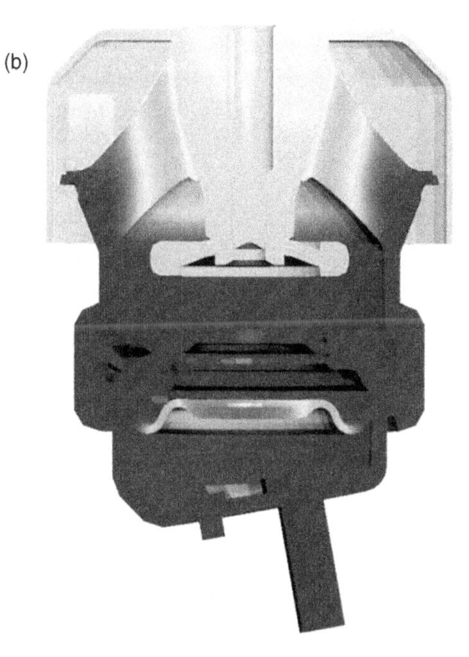

Contour plot of the length of the displacement vector

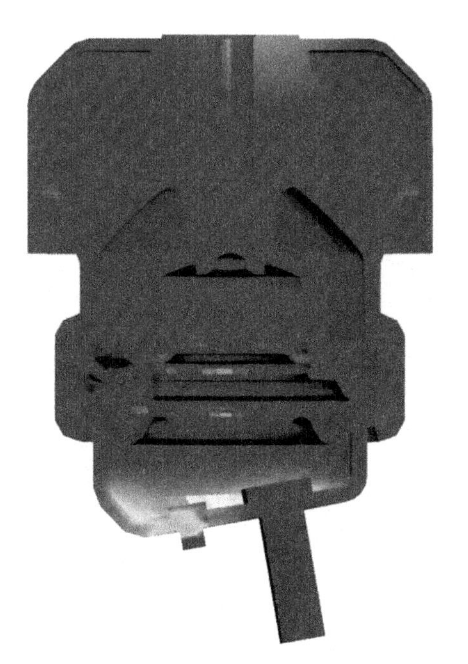

Distribution of the von Mises stress

(b) $t = 1/4T$

Figure 9.7 (*Continued*)

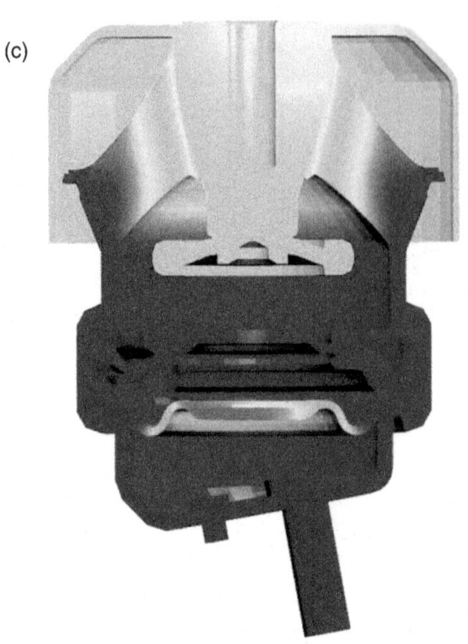

Contour plot of the length of the displacement vector

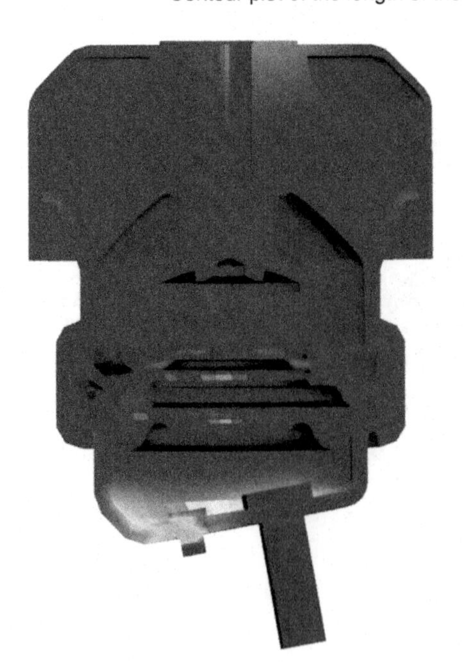

Distribution of the von Mises stress

(c) $t = 1/2T$

Figure 9.7 (*Continued*)

(d)

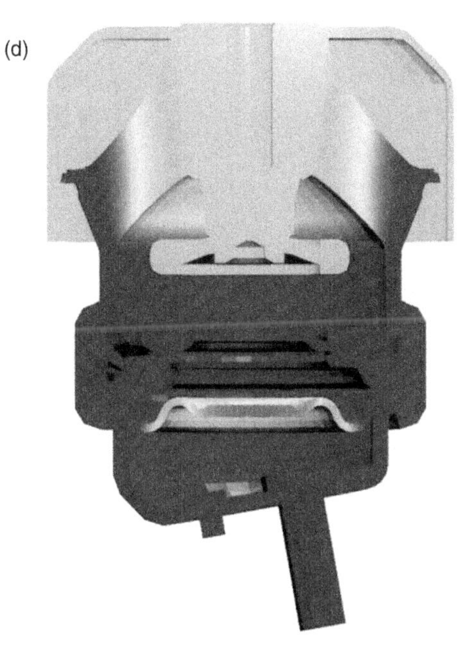

Contour plot of the length of the displacement vector

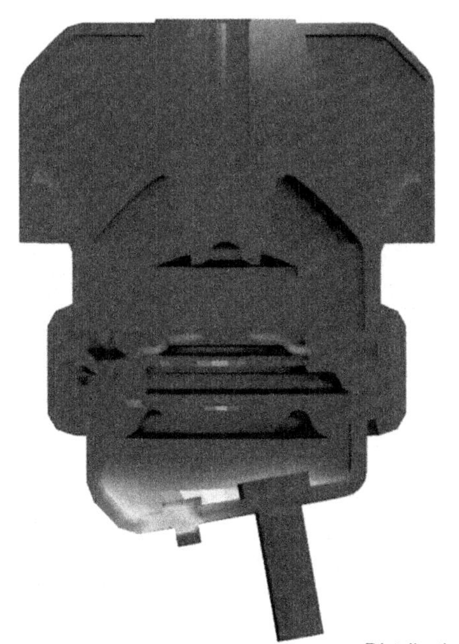

Distribution of the von Mises stress

(d) $t = 3/4T$

Figure 9.7 (*Continued*)

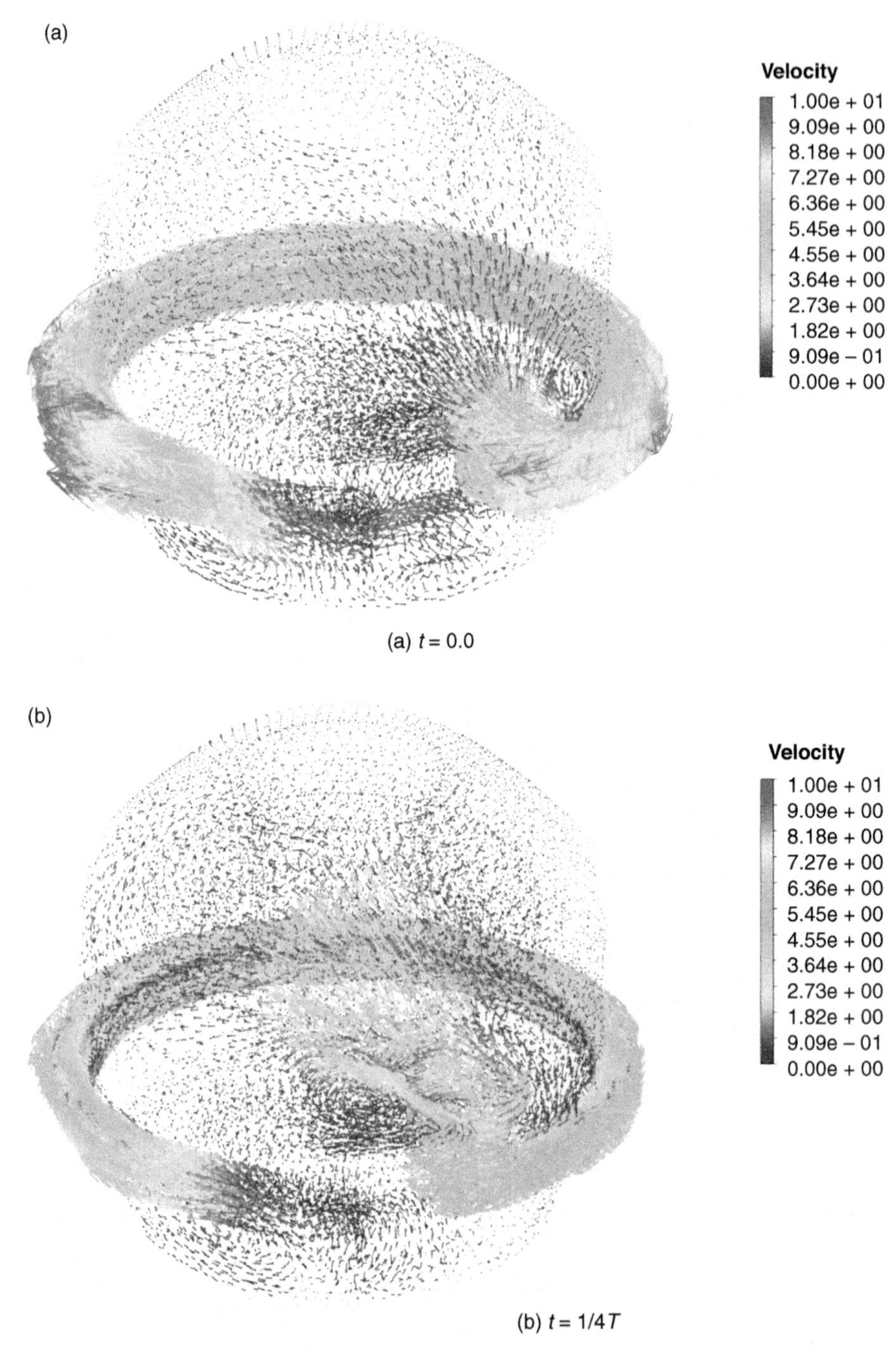

Figure 9.8 Velocity distributions of the fluid domain (f = 20 Hz).

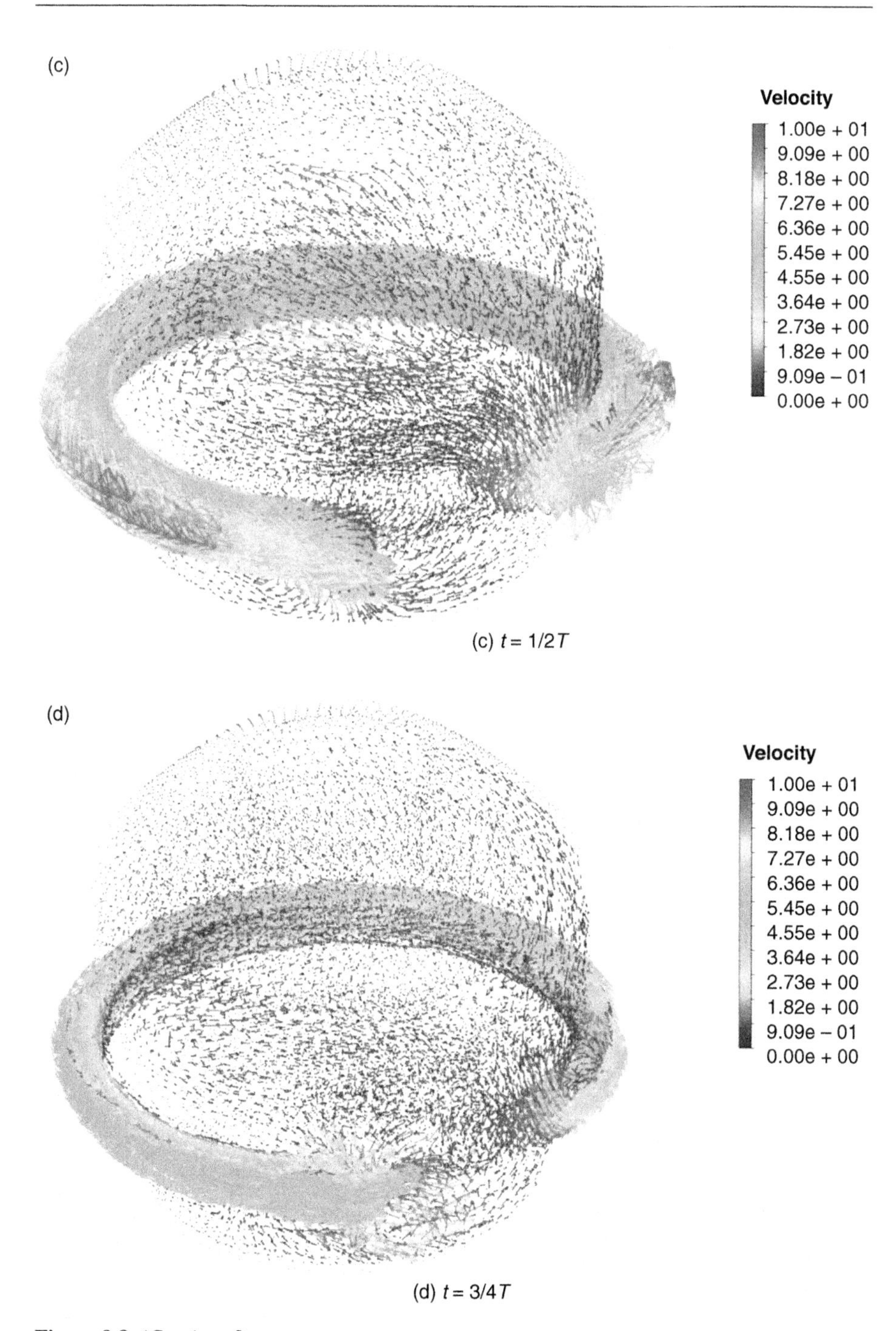

Figure 9.8 *(Continued)*

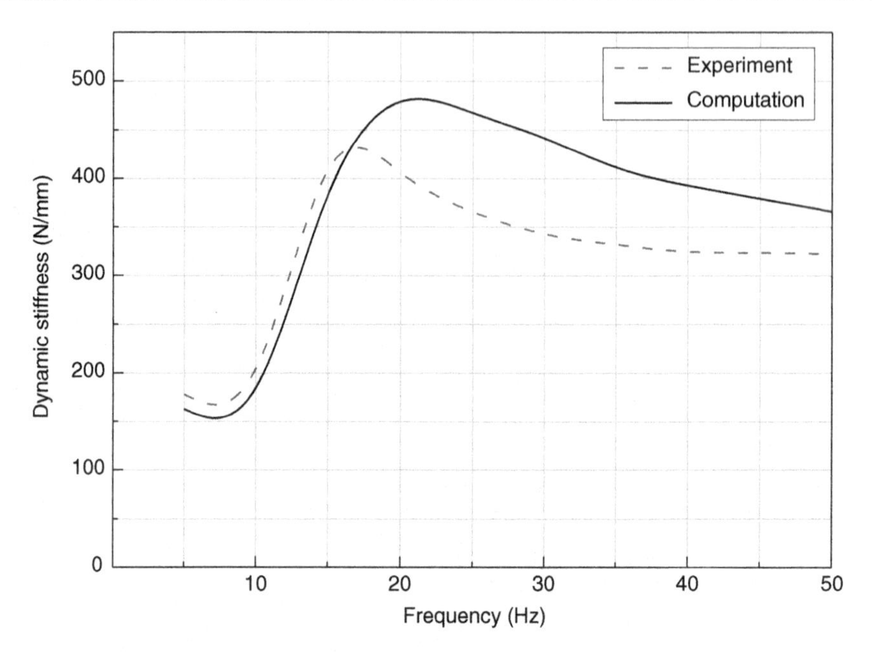

Figure 9.9 Comparison of dynamic stiffness of the HEM of the computational results with experimental results.

vibration, the uncoupler works well. On the other hand, the fluid flow in the inertial track provides more damping for the low frequency vibration. The time history curves of the reflection force on the base of HEM are plotted for all 14 frequency points (Table 9.4), combining with the corresponding displacement excitation curve. Using a Fourier transformation, we have obtained the dynamic stiffness curve, and the delay angle curve. The comparisons with the experimental results done by the R&D Center of China FAW Co., Ltd. are shown in Figures 9.9 and 9.10, respectively. Both figures show a good consistency between the coupled simulation results and the experimental results. The error between simulation results and the experimental results is less than 15%, for the dynamic stiffness and the delay angle. It demonstrates that the numerical capability proposed in this research, and implemented in INTESIM software, is applicable for the study of the dynamic characteristics of HEM in a fully coupled way.

9.1.5 Mesh refinement analysis

We carried out the mesh refinement study in order to investigate the dependence of the mechanical responses and dynamic properties of the HEM on the mesh size. Considering the computational efficiency and the results' accuracy, the mesh of the fluid domain is increased from 133,122 elements to 223,846 elements, while other simulation conditions remain the same as Sections 9.1.2 and 9.1.3.

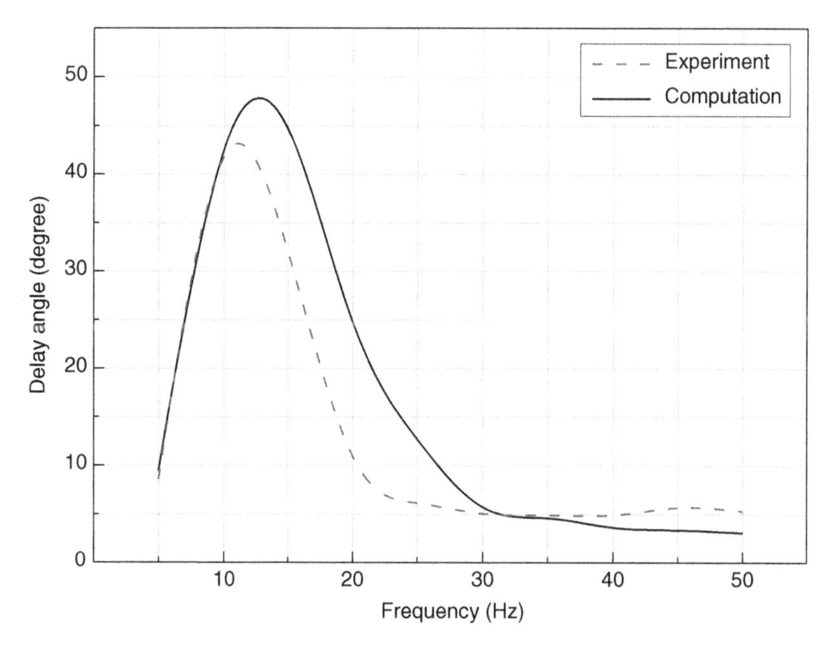

Figure 9.10 Comparison of the delay angle of the HEM of the computational results with experimental results.

Table 9.5 Comparison of results between fine and coarse mesh

	Cases	$t = 0$	$t = T/4$	$t = T/2$	$t = 3T/4$
Maximum stress (MPa)	Coarse mesh	36.1	66.2	78.8	47.3
	Fine mesh	44.6	79.5	91.1	55.2
Maximum displacement of solid (mm)	Coarse mesh	10.7	9.03	10.1	11.7
	Fine mesh	11.3	9.31	11.0	11.9

Table 9.5 shows the comparison of the maximum von Mises stress and the maximum displacement by these two meshes. The patterns of spatial distribution and variation tendencies, with time of the results by the finer mesh and coarser mesh, are very close. But, as can be seen from this table, the maximum difference of the von Mises stresses occurred at time point $t = T/4$. The dynamic stiffness and delay angle simulated by the refined mesh are closer to the experimental values than those by the coarse mesh, especially in the range of high frequency (Figure 9.11 and Figure 9.12). Although we see less sensitivity of the dynamic properties, than the maximum stress on the mesh discretization, this suggests to us to do more investigations on the mesh size of the HEM model, in order to produce mesh independent results in further studies.

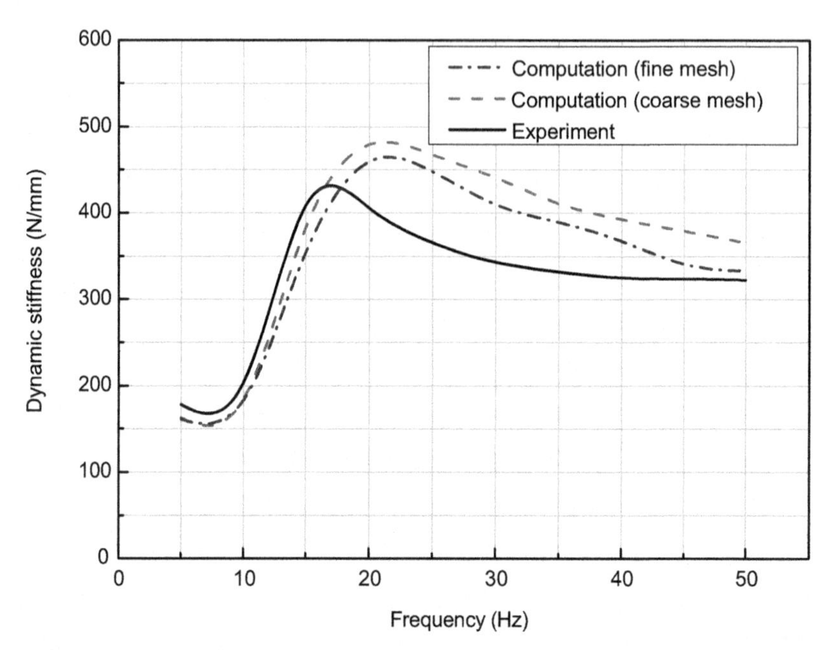

Figure 9.11 Comparison of the dynamic stiffness of the HEM of the computational results with experimental results.

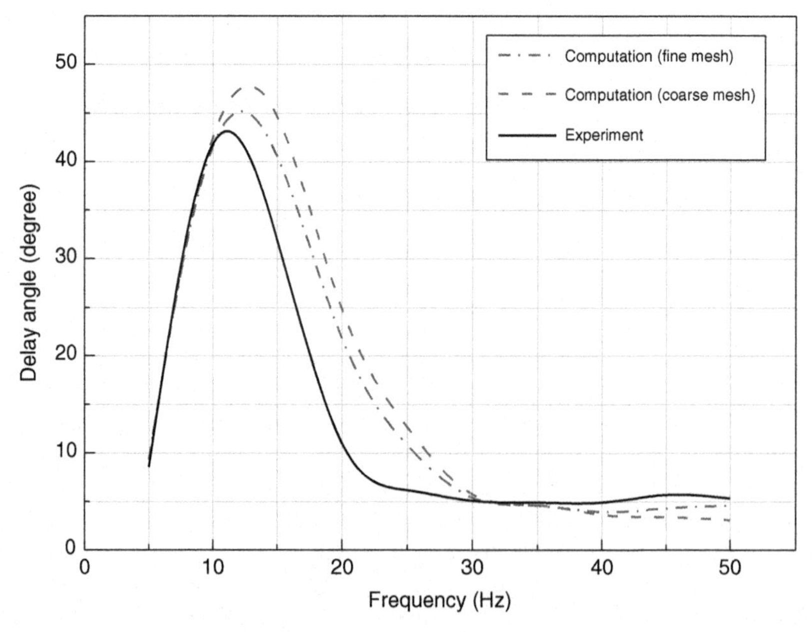

Figure 9.12 Comparison of the delay angle of the HEM of the computational results with experimental results.

9.2 Weak coupling fluid–solid–thermal analysis of exhaust manifold

9.2.1 Physical modeling

Exhaust manifold connecting to the engine cylinder collects the exhaust gas from each cylinder into the exhaust collector. Since the exhaust manifold is the first component through which the high temperature exhaust passes, in the exhaust system of the engine, it works under the harsh condition of an alternating state between high temperature and normal temperature. Problems, such as thermal fatigue cracking, fracture, and leakage may occur due to an extreme stress alternation caused by the cross-shock heating and cooling process.

The working process of the exhaust manifold is a complex conjugate heat transfer problem of fluid–solid coupling (Zhang et al., 2013). Traditionally, the unidirectional weak coupling method is used to analyze the fluid-solid thermal coupled fields of the exhaust manifold. The thermal fluid field enclosed in the manifold is first analyzed in order to acquire the approximate heat transfer coefficients and temperature conditions of the near-wall gas. Then, the approximate conditions are mapped into the finite element model that is used to calculate the stress and deformation of the thermal structure. Since the thermal equilibrium between the thermal structure and hot fluid cannot be perfectly fulfilled, the results predicted by the unidirectional weak coupling method often have large errors. In this case, in order to predict the temperature and stress distribution of the fluid–solid fields accurately, the bidirectional weak coupling method that ensures the thermal equilibrium between the thermal structure and the hot fluid is used to simulate the conjugate heat transfer problem of the exhaust manifold.

The load transfer conditions of the bidirectional weak coupling of the exhaust manifold are given in Figure 9.13. In the exhaust manifold, the steady state of the thermal fluid flow and the steady-state of the thermal stress analysis models are set up separately. Then, the contact surface between the fluid and structure models is set as the coupling interface. When the coupling analysis is carried out, the thermal–fluid analysis is started first. After the thermal–fluid analysis is completed, the fluid loads of the coupling wall and the heat flux are passed to thermal–structure analysis, as load conditions. Then, the thermal–structure analysis calculation is started. After the

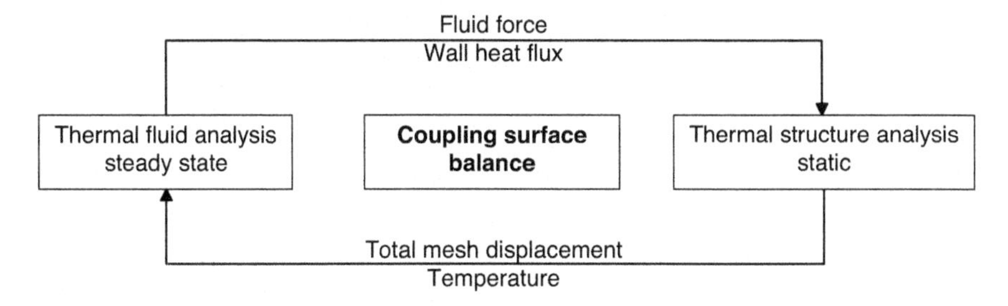

Figure 9.13 Analysis process of the exhaust manifold fluid-solid-thermal bidirectional weak coupling.

thermal–stress analysis is completed, the displacement and temperature on the coupling surface are passed back as a boundary condition to the thermal fluid. Then, the thermal fluid calculation is conducted again. This cycle continues until both the thermal–fluid and the thermal–structure analysis are in stable convergence, and the error between the load transfer and the boundary conditions in the coupling surface is less than the setting precision value. After this, the thermal–fluid and thermal–stress bidirectional weak coupling analysis exit the coupling iteration cycle. Finally, the steady working state is achieved under the heat balance on the coupling interface in the exhaust manifold.

The thermal structure mesh model of an exhaust manifold is shown in Figure 9.14. A hexahedron mesh is employed with 11,402 elements and 17,559 nodes. The inner thermal fluid mesh model is shown in Figure 9.15, and a structure hexahedral mesh

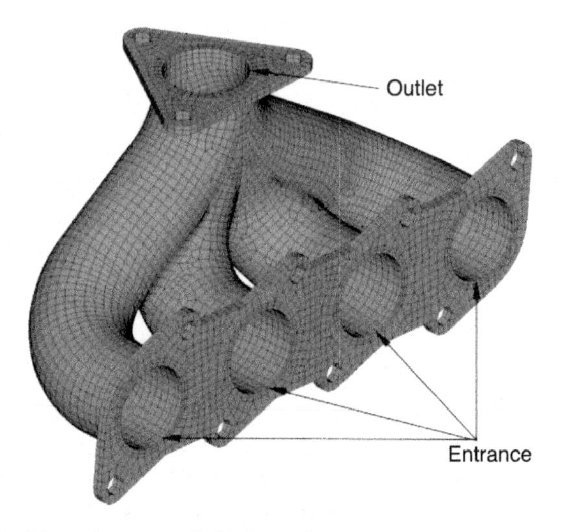

Figure 9.14 Mesh of the exhaust manifold thermal structure.

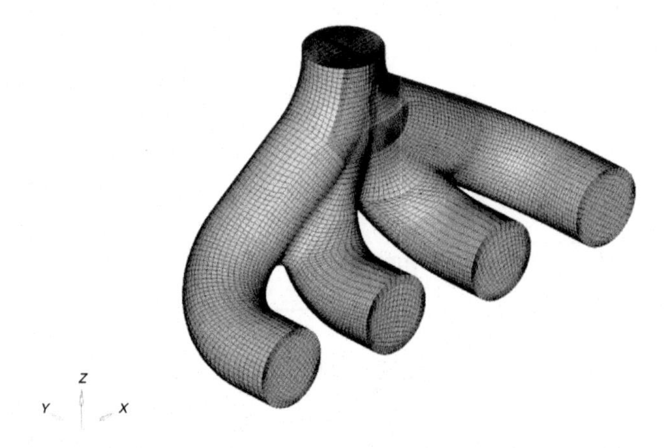

Figure 9.15 Mesh of the thermal fluid exhaust manifold model.

with 99,976 elements and 107,728 nodes is employed. The thermal fluid enters the exhaust manifold through four entrances, flows along the exhaust manifold, and then exhausts from one outlet.

9.2.2 Analysis conditions

Under the assumption of steady state, the exhaust manifold working conditions can be simplified as follows:

1. Because the frequency of exhausting is high, the transient exhaust effect in one cycle is ignored. Only the stable state condition is simulated.
2. An assumption is made that the average flow velocity at four inlets of thermal–fluid is 2 m/s, the gas temperature is 900°C, and the average static pressure at the outlet is zero.
3. Another assumption is made for the outer surface of the thermal–structure analysis model, namely that the external environment convective heat transfer coefficient is 55 W/($m^2 \cdot$ K), and the ambient temperature is 30°C.
4. The thermal–structure analysis model excludes extra parts, such as the cylinder head at the fluid inlet and the bolt connected to it, as well as the turbocharger at the fluid outlet and the bolt connected to it. Therefore, the contact force from the bolts' pretension is ignored. Only the displacement of the bolthole, as well as the normal displacement of the cylinder head and the turbocharger connected plane is constrained. Specified load boundary conditions are shown in Figure 9.16.
5. The exhaust manifold is made of ordinary steel material. The thermal and mechanical properties are temperature-dependent. However, it is assumed that the material deformation of the exhaust manifold is within the elastic range, and plastic deformation is not considered. The specific material parameters are shown in Table 9.6.

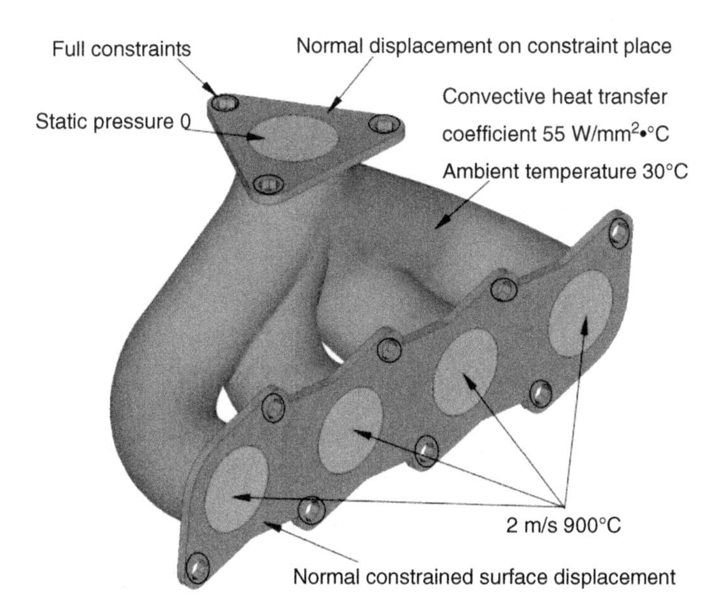

Figure 9.16 Load boundary conditions of the exhaust manifold for fluid–solid–thermal coupling analysis.

Table 9.6 **Exhaust manifold temperature dependence of material properties**

Temperature (°C)	Density	Poisson's ratio	Young's modulus (GPa)	Thermal conductivity (W/m · K)	Thermal expansion
20	7850 kg/m^3	0.28	202	51.08	12.67 e − 6 K^{-1}
100			200	50.04	12.67 e − 6 K^{-1}
200			196	48.99	12.67 e − 6 K^{-1}
300			189	45.85	13.07 e − 6 K^{-1}
400			167	42.71	13.47 e − 6 K^{-1}
600			149	35.59	14.41 e − 6K^{-1}
800			100	25.96	12.64 e − 6 K^{-1}

Table 9.7 **Thermal fluid material parameters of ideal gas (25°C, 1 atm)**

Viscosity (Pa · s)	1.831 e − 05
Density (kg/m^3)	1.185
Thermal expansion (K^{-1})	0.003356
Thermal conductivity (W/m · K)	0.0261
Specific heat capacity (J/kg · K)	1004.4

6. The exhaust manifold fluid adopts the model of the ideal gas, and gas density changing with temperature and pressure is considered. The material parameters at 25°C, 1 atm is shown in Table 9.7. The compressible NS equation is employed in thermal–fluid analysis, standard $k - \varepsilon$ model is employed in turbulence model, and the standard wall function is employed near the wall.
7. The outer surface of thermal fluid and the inner surface of thermal structure are the coupling interfaces. The under-relaxation factor for interface load transfer is set to 0.75. Once the thermal–fluid analysis and the thermal–structure analysis reach the convergence state, as well as the load transfers for thermal and mechanical coupling conditions become stable, the convergence conditions of the coupling system are satisfied.

9.2.3 Results and discussions

The velocity distribution of the inner fluid is shown in Figure 9.17. Due to the dramatic collection of airflow from four entrances, gas flow velocity is significantly increased at the outlet, and the maximum gas flow velocity in the exhaust manifold is about 8 m/s.

The gas pressure distribution along the inner surface is shown in Figure 9.18. It can be seen that the effects of overall gas pressure on the coupling wall is relatively low and, thus, the fluid force mainly caused by gas pressure is also very small and negligible.

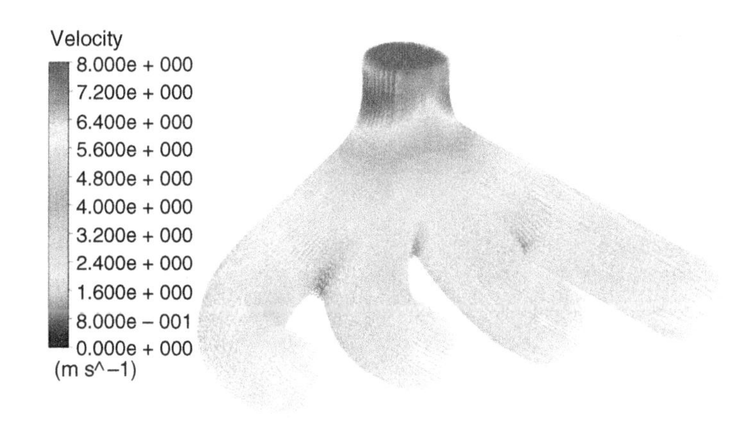

Figure 9.17 Gas flow velocity in the exhaust manifold.

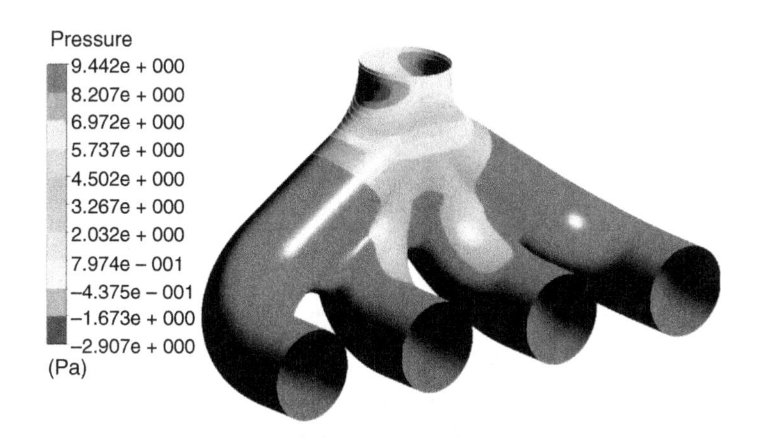

Figure 9.18 Inner surface gas pressure in the exhaust manifold.

Gas temperature distribution in manifold exhaust is shown in Figure 9.19. It can be seen that the fluid temperature changes sharply near the coupling surface, and that there is little difference between the average temperature at entrance and outlet. This is due to the high flow velocity and the short path of the exhaust manifold.

The temperature distribution of the thermal structure is shown in Figure 9.20. It can be seen that temperature distributes show uniform transition at the inner surface of the exhaust manifold. The temperature at the outlet of the exhaust manifold is higher, up to 106°C, and the temperature at the entrance is lower, down to about 55°C. By checking the thermal fluid temperature and thermal structure, respectively, it is found that their coupling wall temperature is consistent, a fact that also means that the thermal–fluid and thermal–structure bidirectional weak coupling analysis is in a stable equilibrium state.

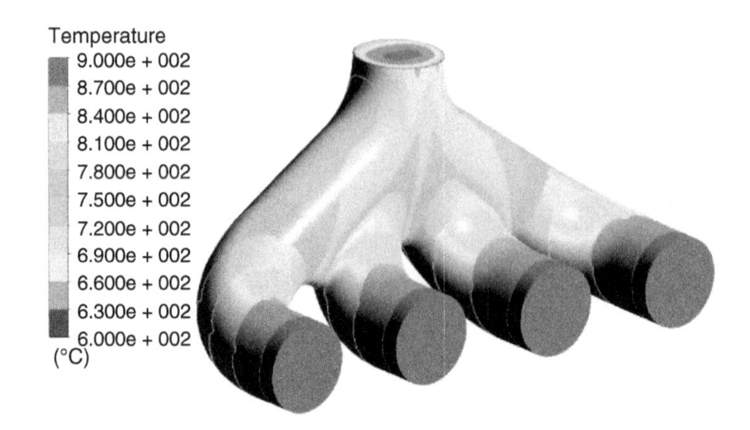

Figure 9.19 Gas temperature in the exhaust manifold.

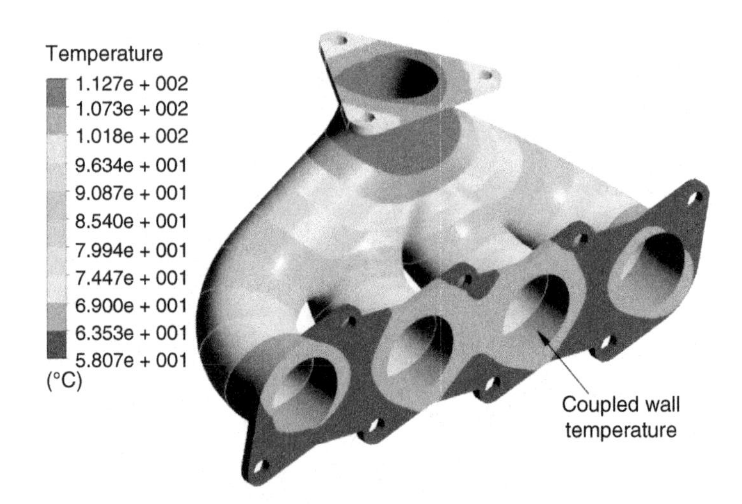

Figure 9.20 Exhaust manifold thermal structure temperature distribution.

Stress distribution of the exhaust manifold thermal structure is shown in Figure 9.21. It can be seen that the stress distribution is not similar to the temperature distribution in Figure 9.20. On the contrary, there is a greater stress at the exhaust manifold exit, the boltholes and other nearby locations. This is due to the combined effect of temperature load and structure constraints. Therefore, the interaction of temperature and structure loads and constraints must be considered to acquire a reasonable analysis result.

Through the fluid–solid–thermal bidirectional coupling analysis, this case simulates the three-field coupling effect of the engine exhaust manifold, under a steady work state. The velocity and temperature distribution of the fluid, as well as the temperature and stress distribution of the solid structure are obtained.

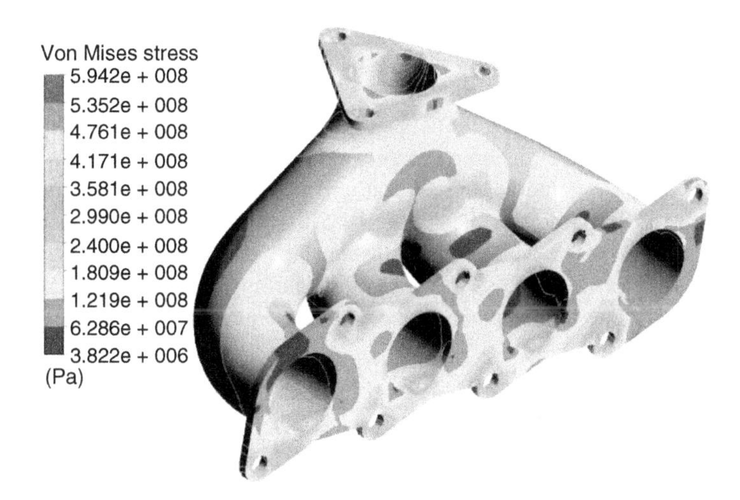

Figure 9.21 Stress distribution of the exhaust manifold.

The results provide a reference for the reliability and safety design of exhaust manifolds, as well as for multiphysics investigations in automotive engineering. This case demonstrates that for fluid–solid–thermal problems, the bidirectional weak coupling method is simple and effective. The proposed algorithm can be easily extended to the fatigue analysis of the exhaust manifold, by considering real transient coupling conditions.

9.3 Coupling analysis of permanent magnet synchronous motor

9.3.1 Cooling simulation of PMSM by multifield coupling technology

The new generation of hybrid/electric power vehicles have become the focus of the world because of the energy crisis, but also because of environmental issues. The electric motor device is the core part of new generation vehicles. The permanent magnet synchronous motor (PMSM) for electric vehicles has high efficiency, high power density, and good control performance. It is widely used in the high performance driving field. The temperature has important influence on the demagnetization of permanent magnetic materials. As a result, the key to the design of the PMSM depends on high precision of the temperature control (Zeng et al., 2014).

The heat generated from operating a PMSM is mainly from copper loss, iron loss, and eddy current loss. It is a complex process, including electromagnetic heating, heat conduction in the solid, and convection in the fluid. Although the studies of the cooling of PMSM are extensive, most of them use the simplified decoupling method. The main procedure of the decoupling method is as follows: the temperature at the fluid–coupling interface is first assumed to start the calculation. The heat transfer coefficient and temperature distribution on the interface obtained from the fluid analysis are then treated as the boundary conditions in the motor thermal analysis. At the same time, the losses obtained from the electromagnetic field analysis are applied to the thermal stability analysis in order to get the temperature distribution of the motor. As such, it is impossible to guarantee the temperature continuity and conservation of the energy flux across the coupling interface, issue that will affect the analysis' accuracy.

For the purpose of predicting the temperature distribution of PMSM accurately, the INTESIM Company has developed a set of interface software, INTESIM-ETFMotor that can perform electromagnetic–thermal–fluid coupling analysis for motors. The details of the coupling control analysis process are introduced in Section 9.3.1.1.

9.3.1.1 Coupling analysis process

INTESIM-ETFMotor is a coupling interface specially developed to manage and control the coupling process for electromagnetic–thermal–fluid coupling simulation of motors. It implements the unidirectional weak coupling between the electromagnetic field analysis software JMAG and the temperature–stress analysis software ABAQUS, as well as the bidirectional coupling between ABAQUS and the CFD software, FLUENT. INTESIM-ETFMotor provides a convenient way for electromagnetic–thermal–fluid coupling analysis. It provides the features of mapping and interpolation for unmatching interface, file-based data communication among physics solvers, the process, and convergence control. The analysis process is listed as Figure 9.22.

1. Conduct the electromagnetic loss analysis in the electromagnetic software, acquiring the copper loss, iron loss in coils, and the eddy current loss in permanent magnetic materials; transfer the average loss as the heat source to the thermal analysis model by mapping and interpolating.
2. Start thermal analysis by script control; output the coupling interface temperature in thermal analysis; take the calculated coupling interface temperature value as the wall boundary condition, and transfer it into the fluid model.
3. Start fluid analysis by script control; output the heat flux value on the fluid–coupling interface; transfer the flux into the thermal model.
4. The steps above are repeated until Δ flux $<$ tolerance; at this point, the coupling balance is reached and the calculation ends.

9.3.1.2 Theoretical foundation

The related theory in electromagnetic analysis, thermal analysis, and fluid analysis has been described in detail in Chapter 1. As such, in this section, the theory involved will be described briefly.

9.3.1.2.1 Electromagnetic loss analysis

For general electromagnetic materials, when the magnetic flux density amplitude is fixed, the iron loss that consists of hysteresis loss and eddy current loss mainly depends on the frequency. The Apply loop provided by JMAG (JMAG-Designer User's Manual, 2010) and the Fast Fourier Transform method are usually adopted in

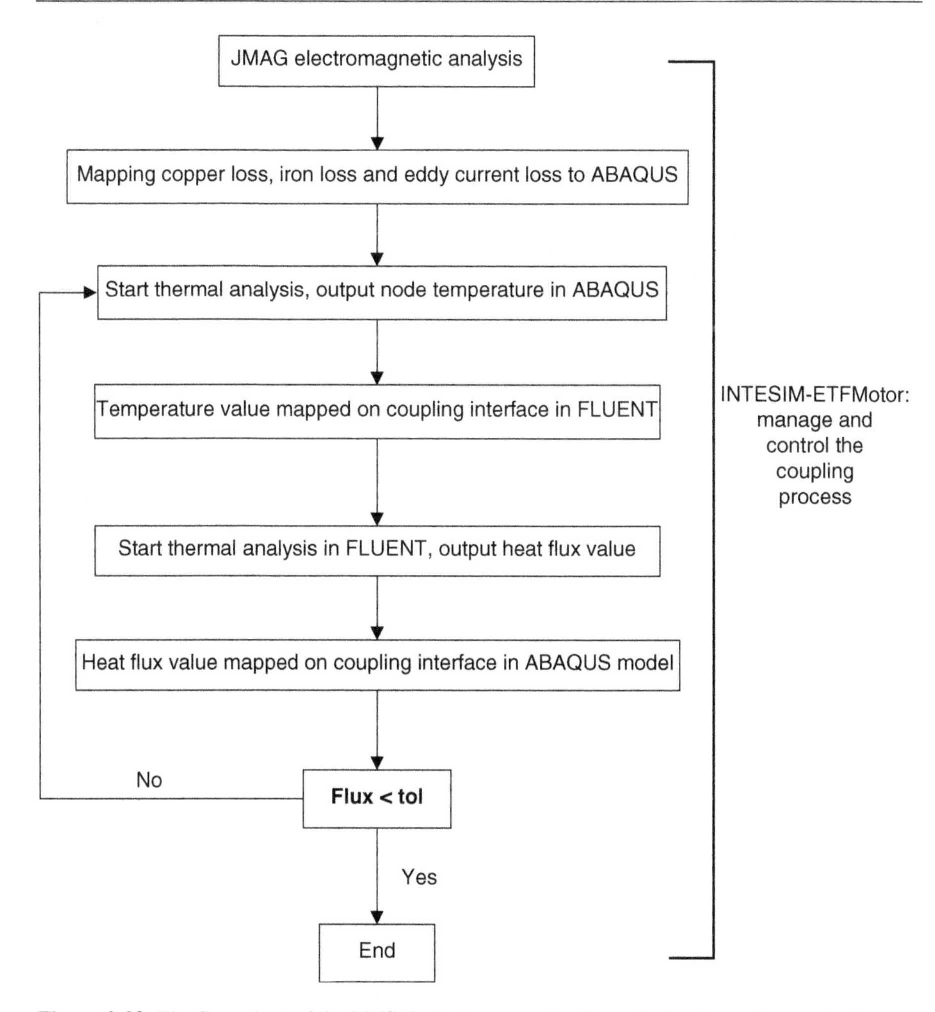

Figure 9.22 The flow chart of the PMSM electromagnetic–thermal–fluid coupling analysis.

order to calculate the hysteresis loss and eddy current loss. Detailed formulations are as follows:

$$W_H = \sum_{e=1}^{n} f \sum_{k=1}^{m} a(B_k) V_e \tag{9.4}$$

$$W_E = \sum_{e=1}^{n} \sum_{j=1}^{N} b(|B_j|, f_j) f_j^2 V_e \tag{9.5}$$

In the equations, B_k is the magnetic flux density amplitude of the k-th loop; $a(B_k)$ is the coefficient of magnetic flux density determined by the frequency separation method; f is the basic frequency; m is the number of loops; f_j is the frequency at the frequency order j; $b\left(\left|B_j\right|, f_j\right)$ is the coefficient at frequency f_j under the magnetic flux density $\left|B_j\right|$ determined by the frequency separation method; N is the maximum frequency order; V_e is the volume of each element; n is the number of elements.

Using Equations (9.4) and (9.5) in JMAG, the iron loss in an entire loop can be calculated. The average loss density in the iron core can also be calculated, which can be used as one of the thermal sources in motor heat analysis.

The copper loss density in the internal coils of the motor, and the eddy current loss density on the permanent magnet, can be calculated as follows:

$$p_C = I^2 R, \tag{9.6}$$

$$p_E = \frac{1}{\sigma} J^2. \tag{9.7}$$

In the equations, I is the current through the coils; R is the resistance of the coils; J is the equivalent current density; σ is the electrical conductivity. Through the eddy current loss analysis in a complete time circle, the average copper loss density and eddy current loss density can be obtained, which can be used as the heat source in thermal analysis.

9.3.1.2.2 Thermal analysis

The temperature of the motor increases due to the heat from the iron loss and copper. The temperature changes inside the motor affect the properties of the electromagnetic and permanent materials; this, in turn, affects the performance of the motor. As a consequence, it is necessary to conduct the numerical simulations for motor design in advance.

There are three kinds of heat transfer mechanisms involved within the motor, namely: heat conduction, convection, and radiation. The heat conduction is caused by the interaction of the internal particles in materials. The temperature difference is the essential condition for heat conduction. Heat convection is the result of heat conducting between fluid and solid, and is closely related to the flow of the fluid and condition of the interface. Radiation is a phenomenon of atomic and molecular electronic structure changes in the heat transfer material, depending on the electromagnetic wave to transmit energy. At present, the thermal analysis model of a motor is built mostly based on the equivalent circuit method, where the heat resistance or heat transfer coefficient is acquired by the heat transfer theory, or experience. With regard to the actual work characteristics of the PMSM, this section treats the thermal contact between different components with different ways in order to complete the construction of the equivalent circuit. As for the heat transfer between the stator and rotor, considering the flow effect of the fluid between the air gaps, the equivalent heat transfer coefficient of the stator and rotor is calculated by the following equation:

$$h_1 = \frac{6.6}{10^5} \frac{v^{0.67}}{d^{0.3}}. \tag{9.8}$$

Where v is the velocity of the rotor; d is the air-gap separation between stator and rotor. It is obvious that the equivalent heat transfer coefficient is proportional to the velocity of the rotor.

The heat transfer coefficient between rotor and coil, or between stator and permanent magnet, can be calculated by the following equation:

$$h_2 = \frac{k}{d}, \tag{9.9}$$

Where k is the thermal conductivity of the intermediate medium; d is the gap between two components.

As for other contact regions in the motor model, a bounded constraint could be used in order to achieve the zero loss of heat transfer; while for the positions in contact with air, the heat dissipation by air convection can be used to build a complete thermal circuit analysis model.

9.3.1.2.3 Fluid analysis

The fluid flow is controlled by the conservation laws of physics, including mass conservation, momentum conservation, and energy conservation. The fluid flow analysis of the cooling liquid outside the water jacket of the motor also follows these three laws. As a result of turbulent flow, when building analysis models, the standard $k - \varepsilon$ model and the standard wall function are adopted. The set of the boundary condition, with the exception of the coupling interface, is consistent with the general fluid analysis.

9.3.1.2.4 The coupling analysis

As shown in Figure 9.23, the whole three-field coupling analysis of the motor is divided into two parts: the electromagnetic–thermal unidirectional coupling module, and the thermal–fluid bidirectional interface coupling module.

1. Electromagnetic-thermal unidirectional coupling module: After completing the electromagnetic field loss analysis, INTESIM-ETFMotor is launched in order to get the average values of copper loss, iron loss, and the eddy current loss on the permanent magnet. The coupling interface maps and interpolates the generated heat into the thermal-stress field of ABAQUS mesh, and uses them as the input heat sources in the thermal–stress analysis.
2. Bidirectional interface coupling module: The INTESIM-ETF software can be used to carry out the bidirectional mechanical and thermal coupling analysis between ABAQUS and FLUENT. Through real time data transfer across the coupling interface, the continuity condition of temperature and the conservative condition for heat flow can be achieved, that is, Equations (9.10) and (9.11):

$$T_w\big|_{\text{fluid}} = T_w\big|_{\text{solid}}, \tag{9.10}$$

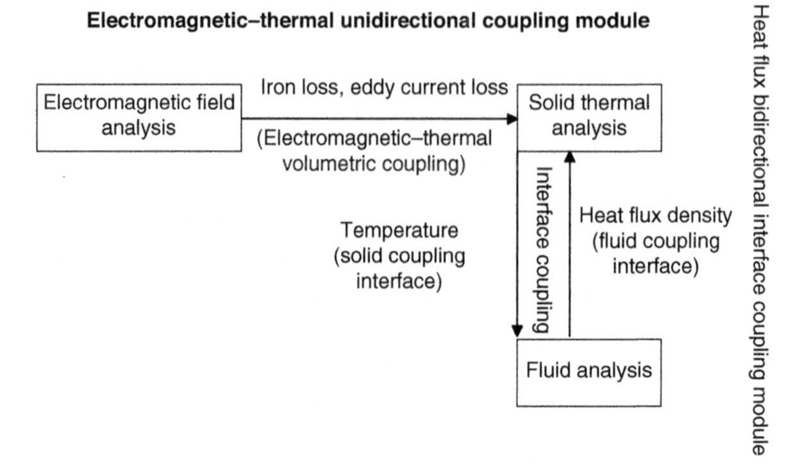

Figure 9.23 Data transfer between two modules.

$$k_{\text{solid}} \left.\frac{\partial T}{\partial n}\right|_{\text{solid}} = k_{\text{fluid}} \left.\frac{\partial T}{\partial n}\right|_{\text{fluid}}. \tag{9.11}$$

where $T_w|_{\text{fluid}}$ is the temperature on the fluid coupling interface; $T_w|_{\text{solid}}$ is the temperature on the solid coupling interface; $\boldsymbol{k}$ is the heat transfer coefficient; $k_{\text{solid}} \left.\frac{\partial T}{\partial n}\right|_{\text{solid}}$ is the heat flux on the solid coupling interface; $k_{\text{fluid}} \left.\frac{\partial T}{\partial n}\right|_{\text{fluid}}$ is heat flux density on the fluid coupling interface.

The electromagnetic–thermal–fluid three-field coupling in the PMSM could be achieved by using the two coupling modules above.

9.3.1.3 Three-field coupling analysis of motor

9.3.1.3.1 Analysis model

The analysis model of the motor is presented in Figure 9.24. This model is the three-phase octupole permanent synchronous alternating current motor, including stator, rotor, coils, permanent magnet, shaft, watercase, and part of the shell. The analysis of this permanent synchronous motor involves multiple physical fields: electromagnetic, thermal, and fluid. The detailed material properties for each physical model are given in the following sections.

1. Electromagnetic field analysis
   The electromagnetic analysis model in the JAMG software is shown in Figure 9.25, and it consists of four parts: stator, rotor, permanent magnet, and coils. Among them, the stator and rotor employ the soft magnetic material, Nippon steel 35H300. The direction of the lamination is parallel to the axis direction, and the coefficient of the lamination is 95%. The permanent magnet uses the material, Hitachi metals (SSMC) NEOMAX-42VH. The direction

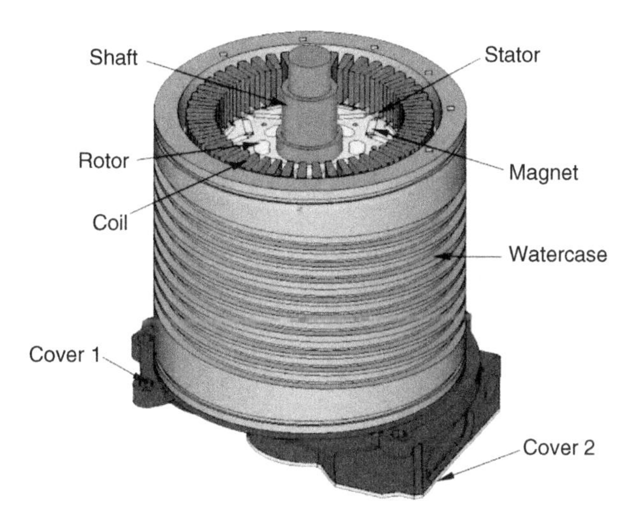

Figure 9.24 Model of the PMSM.

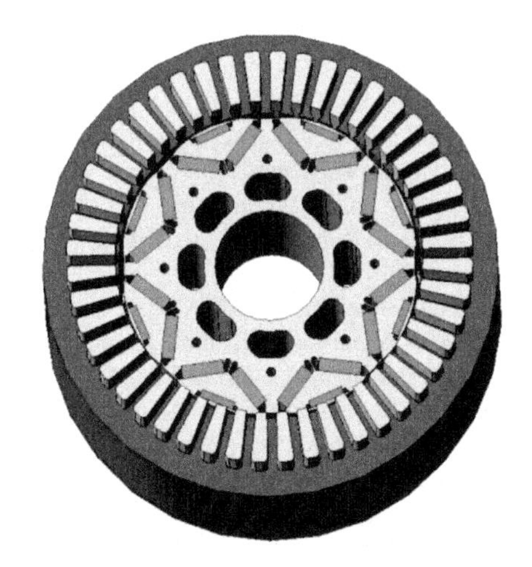

Figure 9.25 The analysis model of electromagnetic field.

of magnetization in the permanent magnet is shown in Figure 9.26. The coils are made of copper, the resistance of which is $1.673\ \mathrm{e}-8\ \Omega \cdot \mathrm{m}$.

Electromagnetic analysis is carried out for two electric angle periods, that is, 0.0204 s. The rotating speed of the motor is 1469 rpm. The setting of the three-phase current is shown below:

$$I_{\mathrm{U}} = A \cdot sin\left(2\pi ft + \theta\pi / 180\right) \tag{9.12}$$

$$I_{\mathrm{V}} = A \cdot sin\left(2\pi ft + (\theta - 120)\pi / 180\right) \tag{9.13}$$

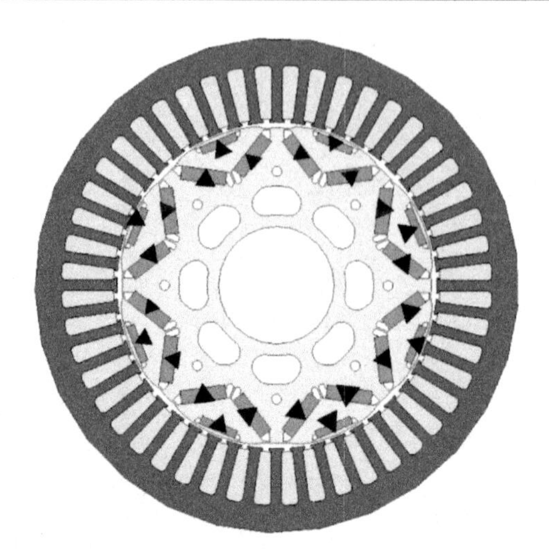

Figure 9.26 Magnetization direction on the permanent magnet.

$$I_W = A \cdot sin\left(2\pi ft + (\theta + 120)\pi / 180\right) \tag{9.14}$$

Where ***A*** is current amplitude, 574.17; ***A***, ***f*** refers to frequency, 97.93 Hz; θ is the initial phase angle, −1.92°. The air-gap between stator and rotor is set to be a sliding boundary on the top, and symmetric boundary on the bottom. Iron loss of the stator and rotor, and the eddy current generated by the permanent magnet, is taken into consideration.

2. Thermal field analysis
The steady thermal analysis model built in ABAQUS is presented in Figure 9.24. This model includes stator, rotor, coils, permanent magnet, shaft, watercase, and part of the shell. The material information of each component is listed in Table 9.8. In order to simulate the heat transfer in the solid model, the thermal contact coefficients between stator and rotor, between stator and coils, and between rotor and permanent magnet are calculated by using Equations (9.8) and (9.9). The thermal contact conditions are listed in Table 9.9. The energy-conservative heat transfer can be achieved by the bounded constraints at rest of the contact boundaries. And the thermal convection conditions are applied to the boundaries opening to the air. Eventually, the complete heat circuit analysis model can be established.

Table 9.8 Material of thermal analysis

Material	Component	Density (kg/m^3)	Heat conductivity (J/K · m · s)	Specific heat (J/K · kg)
Steel	Stator rotor	7850	23	460
Copper	Coil	8950	380	380
Magnet	Magnet	7500	20	460
Shaft	Shaft	7800	80	440
Aluminum	Water_case (cover 1, cover 2)	2719	202.4	871

Table 9.9 **The setting of heat transfer coefficient**

Location	**Heat transfer coefficient (J/s · m^2 · K)**
Stator_to_rotor	160.7
Stator_to_coil	1000
Rotor_to_magnet	540
Stator_to_watercase	100,000
Rotor_shaft	100,000
Watercase_to_cover	100,000
Cover 1_to_cover 2	100,000
Motor_air	20, Ambient temperature, 298 K

3. Fluid analysis

In the FLUENT software, the analysis model is shown in Figure 9.27. The specified material is water at normal temperature; at the inlet, the mass flow of the inlet is 10 L/min; temperature is 298 K; static pressure of the outlet is set to 0. In addition, the specified temperature is assigned at the coupling interface to ensure the continuity conditions.

The result file acquired by the magnetic field analysis, the structure thermal analysis file, and the fluid file can be prepared as the input file of each physical field in the three-field coupling analysis, and the entire process can be controlled by INTESIM-ETFMotor.

9.3.1.3.2 The result of coupling analysis

After the electromagnetic field analysis is finished, the contour plot of Joule loss density and the iron loss density are presented in Figures 9.28 and 9.29, respectively. The Joule loss density includes the copper loss density in the coils, and the eddy current loss density in the permanent magnet. For the iron loss, as shown in Figure 9.29, the density in the magnetic bridge of the rotor is high, the rest of it is almost zero; whereas the iron loss density in the stator structure is uniformly distributed.

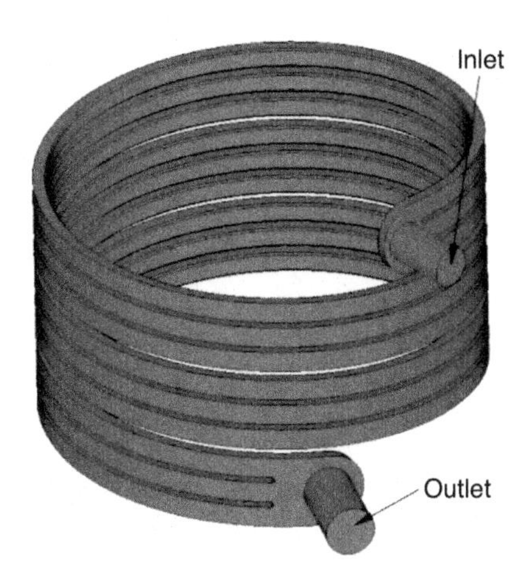

Figure 9.27 Model of fluid analysis.

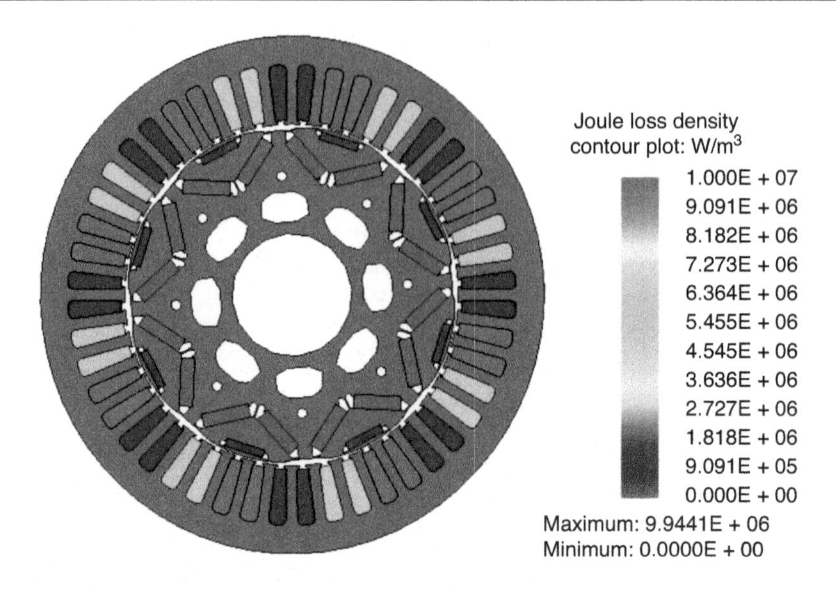

Figure 9.28 Contour plot of Joule loss density.

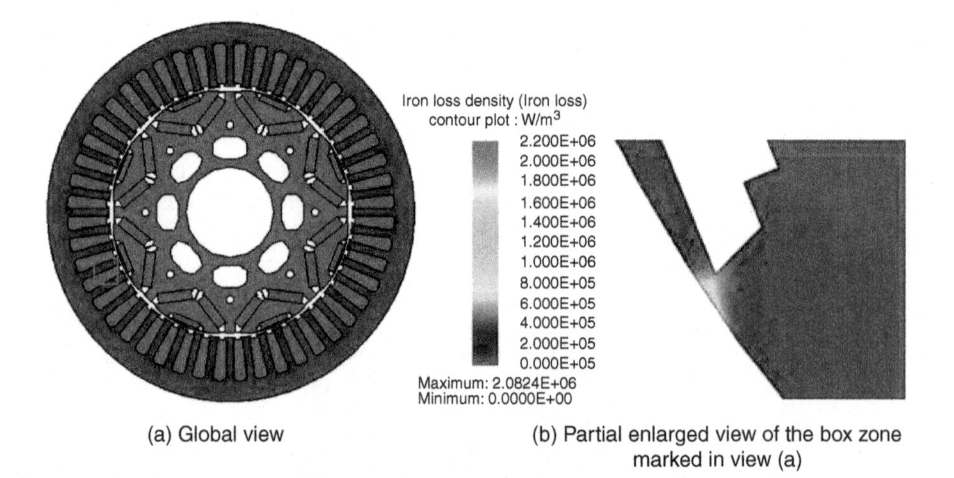

Figure 9.29 Contour plot of iron loss density in the stator and rotor.

The temperature distribution in the thermal field and fluid model, after reaching the equilibrium state, are shown in Figures 9.30 and 9.31, respectively. In Figure 9.30, the highest temperature (~417.2 K) mainly occurs in the coil, and the lowest temperature (~301.8 K) occurs near the waterside of the shell. In Figure 9.31, the highest temperature (~321 K) of the water model occurs at the heat transfer coupling interface, and the lowest temperature (~298 K) occurs at the inlet of the fluid field.

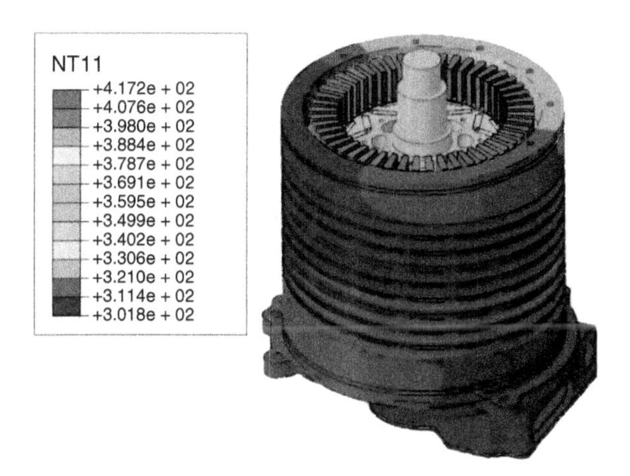

Figure 9.30 Temperature distribution contour plot of the motor.

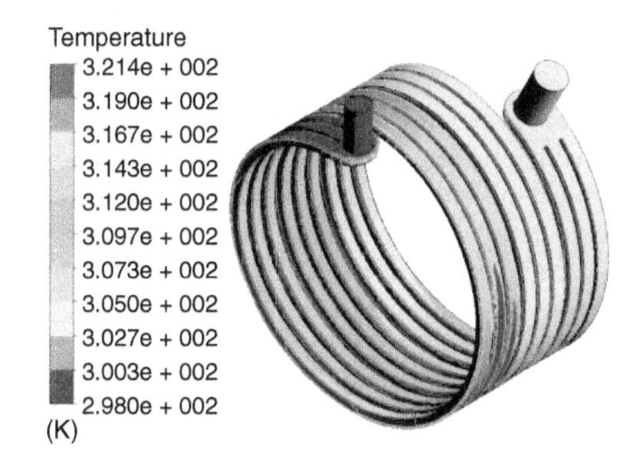

Figure 9.31 Temperature distribution contour plot of the fluid model.

9.3.1.3.3 The comparison between thermal current unidirectional and bidirectional coupling analysis

The analysis procedure for unidirectional coupling:

In this section, the traditional thermal unidirectional coupling analysis method is used as follows:

1. Assume the temperature of the fluid coupling (take the average value of the temperature on the motor coupling interface after the bidirectional coupling equilibrium is achieved); analyze the cooling liquid to acquire the distribution of heat transfer coefficients.
2. Output the distribution of the heat transfer coefficients on the fluid coupling interface and the temperature value.
3. Load the distribution of the heat transfer coefficients on the fluid coupling interface, and the temperature value on the structure heat analysis model; the other loads, boundary conditions,

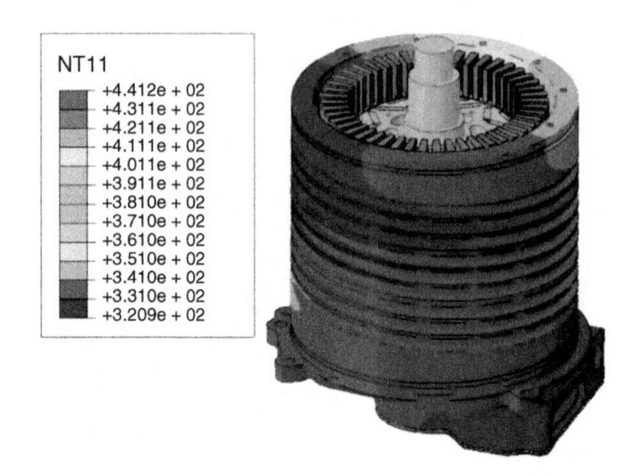

Figure 9.32 Temperature distribution contour plot of the motor in unidirectional coupling analysis.

are the same as the bidirectional coupling analysis; then, start the thermal steady-state analysis calculation.

The results of unidirectional coupling analysis:

After the unidirectional coupling analysis is done, the contour plot of temperature distribution in the motor model is presented in Figure 9.32. As shown in the motor model, the highest temperature (~441.2 K) occurs in the coils, and the lowest temperature (~320.9 K) occurs near the shell.

The comparison between unidirectional and bidirectional coupling:

The highest temperature T_{max} and the lowest temperature T_{min} obtained from unidirectional and bidirectional coupling method are shown in Table 9.10

Table 9.10 Comparison of temperature range of various components from unidirectional and bidirectional coupling method

Component	$T_{max, unidirectional}/K$, $T_{min, unidirectional}/K$	$T_{max, bidirectional}/K$, $T_{min, bidirectional}/K$	$\Delta T_{max}/\%$ $\Delta T_{min}/\%$
Stator	427.0 326.0	403.4 306.8	18.1 56.8
Rotor	401.1 377.3	382.1 362.8	17.4 16.1
Coil	441.2 409.7	417.2 388.0	16.6 18.9
Permanent magnet	398.7 386.7	380.2 370.6	17.3 16.5
Watercase	355.7 320.9	331.1 301.8	42.3 66.3
Shell 1	338.7 327.8	315.1 308.2	56.1 55.7
Shell 2	331.4 327.8	310.8 308.5	54.5 54.4
Shaft	380.6 371.2	365.4 357.8	16.5 15.8

Among them:

$$\Delta T_{\max} = \frac{(T_{\max,\text{unidirectional}} - 273) - (T_{\max,\text{bidirectional}} - 273)}{(T_{\max,\text{bidirectional}} - 273)}, \tag{9.15}$$

$$\Delta T_{min} = \frac{(T_{\min,\text{unidirectional}} - 273) - (T_{\min,\text{bidirectional}} - 273)}{(T_{\min,\text{bidirectional}} - 273)}. \tag{9.16}$$

According to Table 9.10, the temperature differences between the results from the unidirectional and bidirectional coupling methods is quite obvious. The maximum difference of the temperature is 66.3%. The reasons are as follows: (1) in unidirectional coupling analysis, the temperature distribution on the coupling interface is assumed to be uniform, but this is not conform to the actual situation; (2) in the unidirectional coupling, since the data transfer between the thermal solid and thermal fluid model is only carried out once, the temperature continuity condition and energy conservation across the coupling interface are not achieved. As a consequence, it is necessary to conduct bidirectional-coupling analysis for the design of electric motor.

9.3.1.4 Conclusions

1. The coupling among JMAG software, ABAQUS software, and FLUENT software is achieved for the first time by using the interface software INTESIM-ETFMotor which is developed by the INTESIM Company. Furthermore, the electromagnetic–thermal–fluid coupling analysis of the new PMSM is accomplished by automatic data transfer, and by process control in programs.
2. By the comparison of results from the unidirectional and bidirectional thermal coupling analysis, it is revealed that the traditional thermal unidirectional coupling analysis does not truly achieve the heat steady state equilibrium. The temperature continuity and energy conservation on the thermal coupling interface cannot be guaranteed. As a result, the thermal bidirectional coupling analysis is needed.
3. The proposed electromagnetic–thermal–fluid coupling analysis provides a convenient and efficient way to study and design PMSM, and other similar devices.
4. Experimental data is needed in future in order to confirm the accuracy of the bidirectional thermal-fluid coupling method; meanwhile, bidirectional electromagnetic-thermal coupling should also be implemented in the future.

9.3.2 A fully integrated solution for the coupling analysis of PMSM

In the previous section, an inter-solver coupling interface, INTESIM-ETFMotor, is developed and used for multiphysics simulation of temperature rise issue of a PMSM. In this approach, JMAG, ABAQUS, FLUENT are used for the electromagnetic, thermal-stress, and thermal-fluid analyses, respectively. This approach is complicated, and requires remarkable resources: simulation software, hardware, and professional engineers. This is impractical for many firms and research institutes.

INTESIM-Motor, as an integrated multiphysics software, is specially developed for the design and simulation of the electric motor. It provides a unified graphical user interface, and a unified solver system with the electromagnetic, thermal–stress, thermal–fluid solver included. The hybrid coupling methods that consist of the weak coupling method and strong coupling method are available in INTESIM-Motor which makes this software a convenient and powerful tool for PMSM design.

Compared with INTESIM-ETFMotor based multiple code coupling approach presented in Section 9.3.1, INTESIM-Motor uses strong coupling methods for thermal-stress and thermal-fluid coupling that guarantees a better convergence behavior. Besides, the bidirectional weak coupling method is used for thermal-stress and electromagnetic field coupling, that take into account the influences of the temperature to the material properties of the electromagnetic field. All these make INTESIM-Motor a convenient, robust, accurate, and low cost tool for electric motor design.

9.3.2.1 Coupling procedure

The weak coupling method is employed for the electromagnetic and the thermal field, and Gauss Seidel iteration is adopted. In this method, the average copper loss and iron loss are calculated from the electromagnetic field. The losses are then mapped to the corresponding part of the thermal field. The temperature acquired from the thermal field will affect the electromagnetic properties of materials. The flow chart of the bidirectional coupling analysis is demonstrated in Figure 9.33.

The MPC-based strong coupling method is used for the thermal-stress and thermal–fluid coupling analysis with unmatching interface meshes. As shown in Figure 9.33, the coupled thermal field, and the thermal-fluid field, are called the conjugate heat transfer problem.

9.3.2.2 Bidirectional coupling analysis of 3D simplified motor

1. Analysis model

   A simplified model of the PMSM is shown in Figure 9.34. This four-pole, twenty-four slots motor is composed of stator, rotor, coil, magnet, fluid, and air gap. It has a water-cooling system outside of the stator. There are also many vents on the rotor for maximizing the heat dissipation effects.

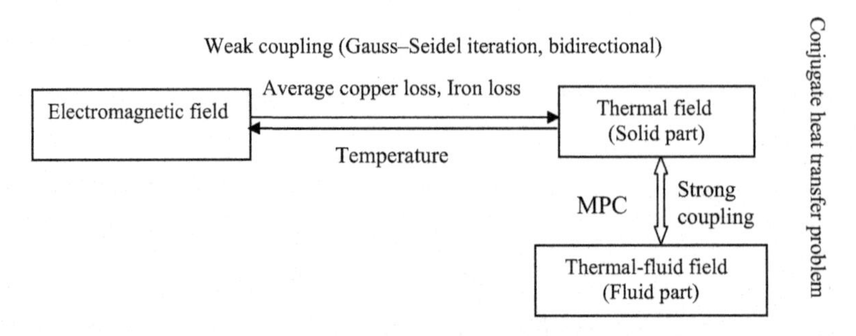

Figure 9.33 Flow chart of the bidirectional electromagnetic-thermal-fluid coupling algorithm.

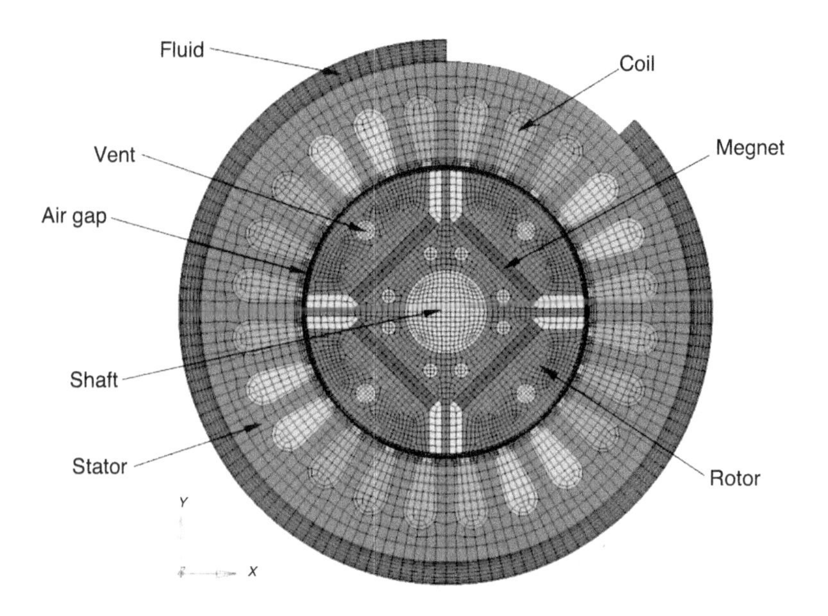

Figure 9.34 Analysis model of the PMSM.

The analysis model is set up based on the following hypotheses: the materials are considered as isotropic; there is no other heat source, except the copper loss and iron loss; all the parts of the motor are considered to be homogeneous heating elements, ignoring the axial (*Z*-axis, perpendicular to the paper) heat transfer, which means that the temperature distribution of each cross-section perpendicular to the axial direction is assumed to be identical; therefore, the model can be simplified as a short section along the axial direction. The incompressible fluid flow is assumed to be Newtonian laminar flow. Hexahedral elements are used for the mesh discretization of the model.

The material properties and the corresponding parts are listed in Table 9.11. The nonlinear B-H characteristic of material steel is given as Figure 9.35. The lamination factor of steel is 0.98, and the stacking direction is in *Z*-axis. In the bidirectional coupling analysis, the electrical resistivity of the coil depends on the temperature, which is expressed in the function $1.72\text{E}-7\,(1+0.00393\,(\text{Temp}-273))\,\Omega\cdot\text{m}$.

2. Analysis conditions

a. Electromagnetic field

The transient electromagnetic analysis model consists of stator, rotor, coil, magnet, and air gap. The excitation is achieved through a balanced three-phase current system. The phase sequence is U–W–V. The current distribution and magnetization orientation are shown in Figure 9.36.

The excitation given in terms of current density is listed as follows.

$$J_{\text{U}} = J_S \cdot \sin(2\pi f \cdot t + \theta \cdot \pi/180), \tag{9.17}$$

$$J_{\text{V}} = J_S \cdot \sin(2\pi f \cdot t + (\theta - 120) \cdot \pi/180), \tag{9.18}$$

Table 9.11 **Material properties of the simplified motor**

Materials	Steel	Copper	Magnet	Air	Water
Part	Stator, rotor	Coil	Magnet	Vent, shaft, air gap	Fluid
Relative permeability	BH curve (seen in Figure 9.35)	1	1.03	1	–
Coercive force (A/m)	–	–	920,000	–	–
Density (kg/m^3)	7850	8950	7500	1.293	998.2
Thermal conductivity (W/m · K)	23	380	20	0.02624	0.6
Specific heat (J/K · kg)	460	380	460	1012	4182
Dynamic viscosity (kg/m · s)	–	–	–	–	0.001003

$$J_W = J_S \cdot \sin(2\pi f \cdot t + (\theta + 120) \cdot \pi/180), \tag{9.19}$$

where $\boldsymbol{J}_U$, $\boldsymbol{J}_V$, $\boldsymbol{J}_W$ are the current density of the U, V, W phase; $\boldsymbol{J}_S$ is the amplitude of current density and is set to 2.0×10^6 A/m^2; $\boldsymbol{f}$ is the frequency, and is set to 50 Hz; $\boldsymbol{t}$ is the time; θ is the initial phase, and is set to 240°.

The rotational speed is set to 1500 rpm. The total analysis time of electromagnetic simulation is one time period of motor rotation, and the time step size is set to 0.001 s based on the rotational speed and the number of grids on the sliding face.

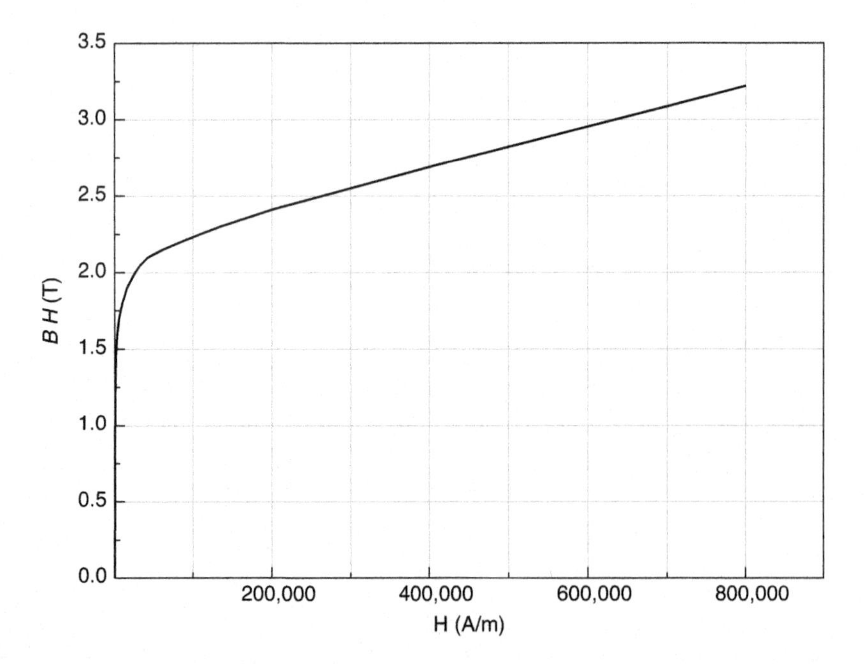

Figure 9.35 Nonlinear B-H characteristic of material steel.

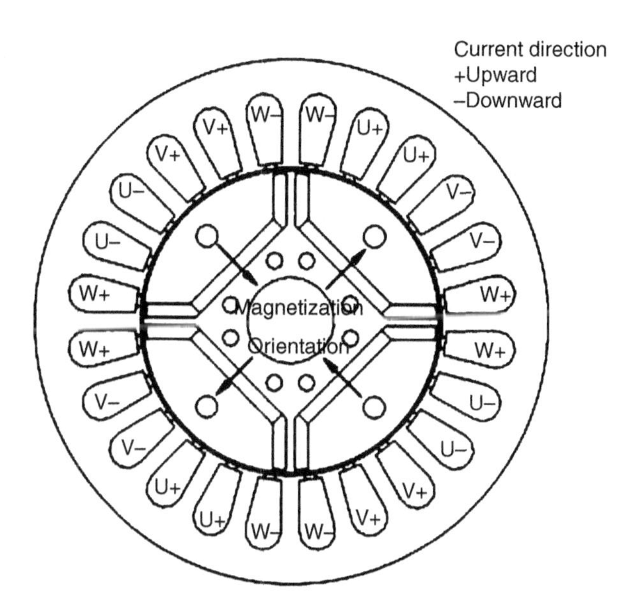

Figure 9.36 Diagram of the current distribution and magnetization orientation.

b. Thermal field
The thermal–field analysis model includes stator, rotor, coil, magnet, and air gap. There is a thermal resistance between permanent magnets and rotor, as well as the coils and stator, and the heat transfer coefficient is set for these thermal contacts by the penalty method. Although the heat transfer coefficient varies with ambient state, it is about 10 $W/m^2 \cdot K$ under nature convection. In the analysis, the heat transfer coefficient of nature convection is set to 10 $W/m^2 \cdot K$. The specific analysis conditions of the thermal–field analysis are listed in the Table 9.12.

c. Thermal–fluid field
The flow model for analysis is shown in Figure 9.37. The conditions of the thermal–fluid field are specified as follows. The inflow velocity at inlet is 0.01 m/s; the pressure at outlet is 0; the symmetry boundary is applied to the top and bottom of the fluid part (faces perpendicular to the *Z*-axis); no slip boundary is applied to the inside or outside of the fluid part (faces perpendicular to the *XY* plane).

Table 9.12 Analysis conditions of the thermal field

Object	Conditions	Value
Permanent magnets and rotor	Coupling setting, penalty method	Heat transfer coefficient: 500 W/($m^2 \cdot K$)
Coil and stator	Coupling setting, penalty method	Heat transfer coefficient: 270 W/($m^2 \cdot K$)
Air gap	Coupling setting, penalty method	Heat transfer coefficient: 142 W/($m^2 \cdot K$)
Surface in contact with air	Convection	Ambient temperature: 300 K; heat transfer coefficient: 10 W/($m^2 \cdot K$)

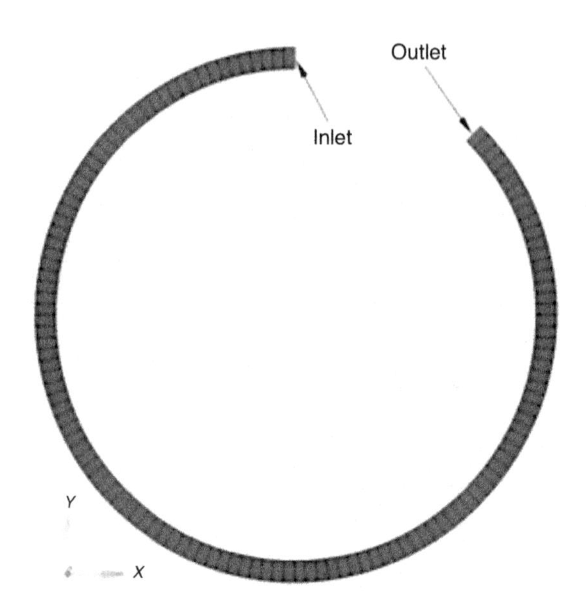

Figure 9.37 Grid model of the fluid part.

d. Coupling setting

The weak coupling condition is used for the data transfer between the electromagnetic field and the thermal field. The average copper loss and iron loss in one time period are transferred to the thermal field from the electromagnetic field; the temperature of the coil in the thermal field is transferred to the electromagnetic field simultaneously which affects the electrical resistivity of the coil. By setting the data transfer conditions through INTESIM, the bidirectional coupling analysis of the electromagnetic-thermal field can be achieved. The total analysis time is 0.02 s, and the coupling time is set to 0.02 s.

The thermal field and the thermal–fluid field mentioned above are set to strong coupling. The analysis type is set static.

3. Simulation Results

The results of the electromagnetic–thermal–fluid coupling analysis can be checked after iteration convergence. Figure 9.38 shows the magnet flux density distribution of the motor at $t = 0.02$ s. As shown in the figure, the area owning large magnet flux density is located on the magnetic bridge. The diagram of magnet flux density of the magnet can be seen in the Figure 9.39 at $t = 0.02$ s.

The time averaged copper loss and iron loss are transferred to the thermal field as the heat source. The contour plot of average copper loss in one time period is shown in Figure 9.40. The average copper loss in every coil is different, because the electrical resistivity depends on the temperature. The average copper loss of the coils near the outlet is larger than that of other coils, because of the relatively large resistivity caused by the relatively high temperature of coils.

The iron loss of the stator and rotor are calculated in the case. Figure 9.41 shows the iron loss distribution of the stator and rotor in one period. As shown in the figure, the area with high iron loss density is concentrated on the magnetic bridge of the rotor, and the rest areas of the rotor close to zero. Due to the rotation effect, the iron loss density of the stator is larger than that of the rotor. The iron loss distribution of the stator is uniform, and the iron loss density is approximately 1.5E4 W/m^3.

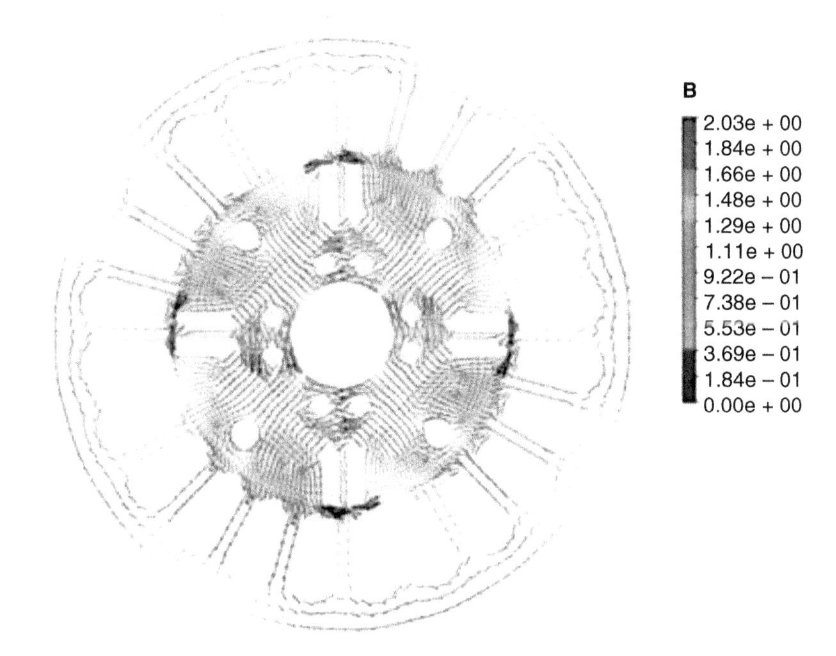

Figure 9.38 Vector plot of magnet flux density of the simplified motor.

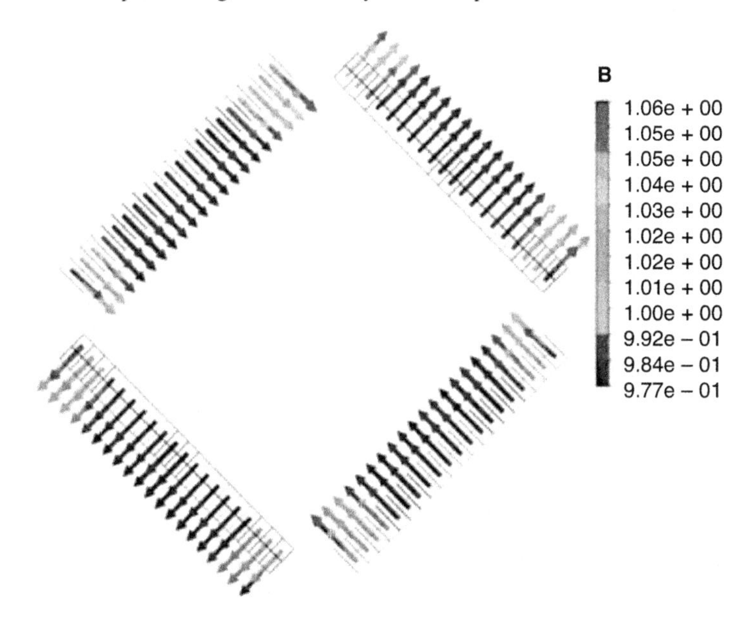

Figure 9.39 Vector plot of magnet flux density of the magnet part.

Figure 9.42 shows the contour plot of the temperature, after cooling. It is obvious that the high temperature occurs in the area between the inlet and outlet of the cooling water. The temperature distribution of the coils is approximately in the range of 313–317 K.

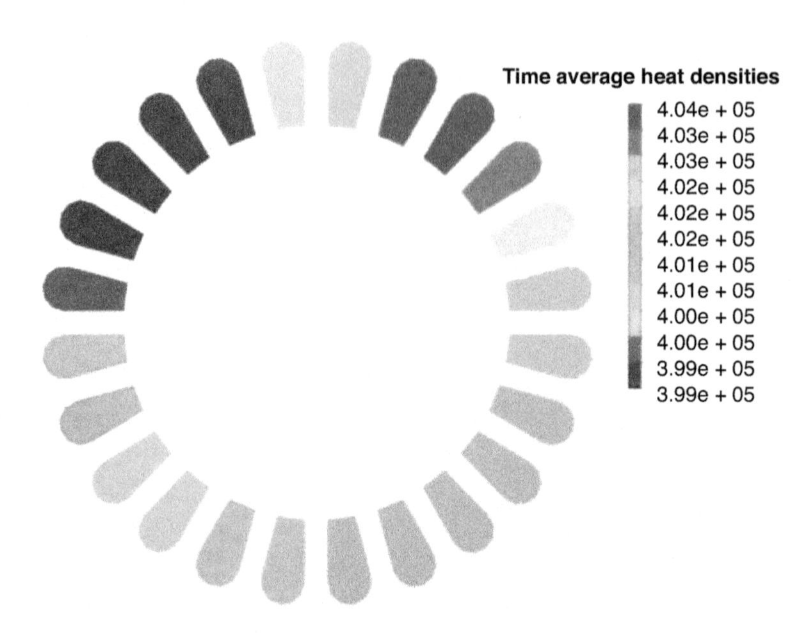

Figure 9.40 Contour plot of average copper loss of coils.

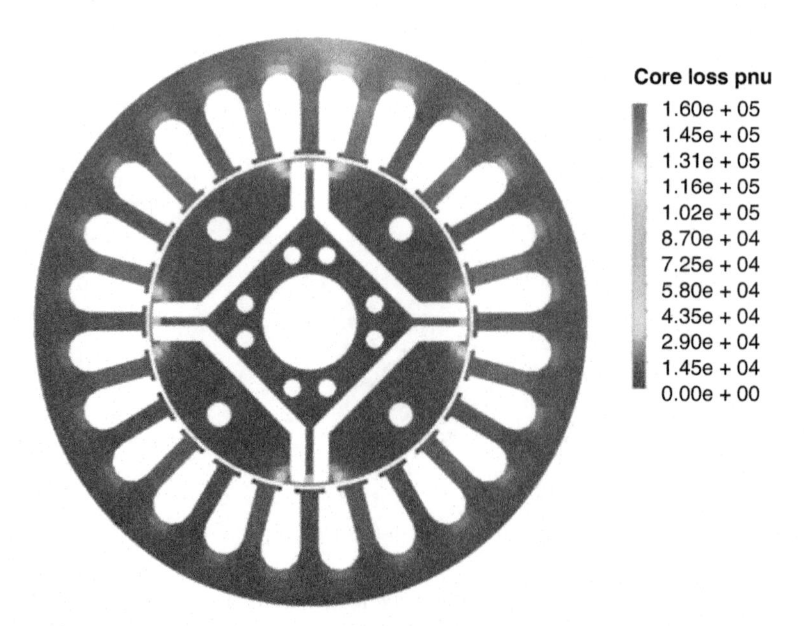

Figure 9.41 Contour plot of iron loss of the stator and rotor.

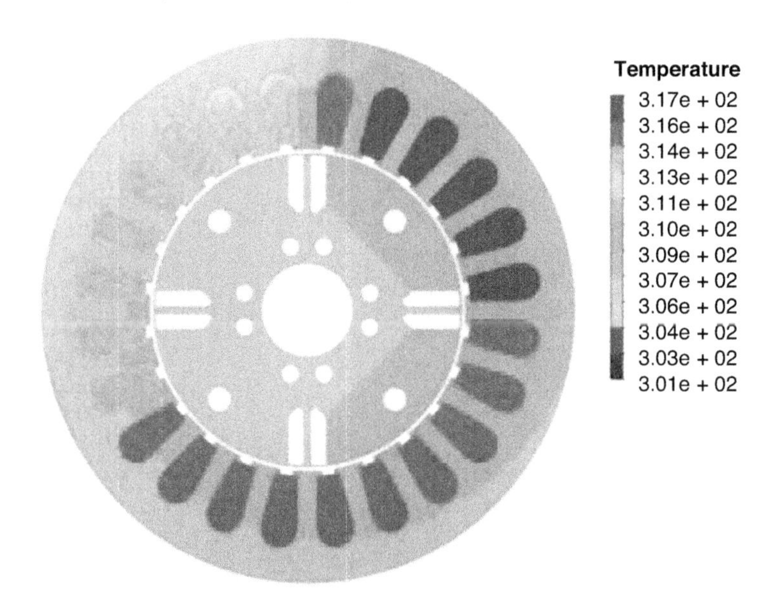

Figure 9.42 Contour plot of temperature after fluid cooling.

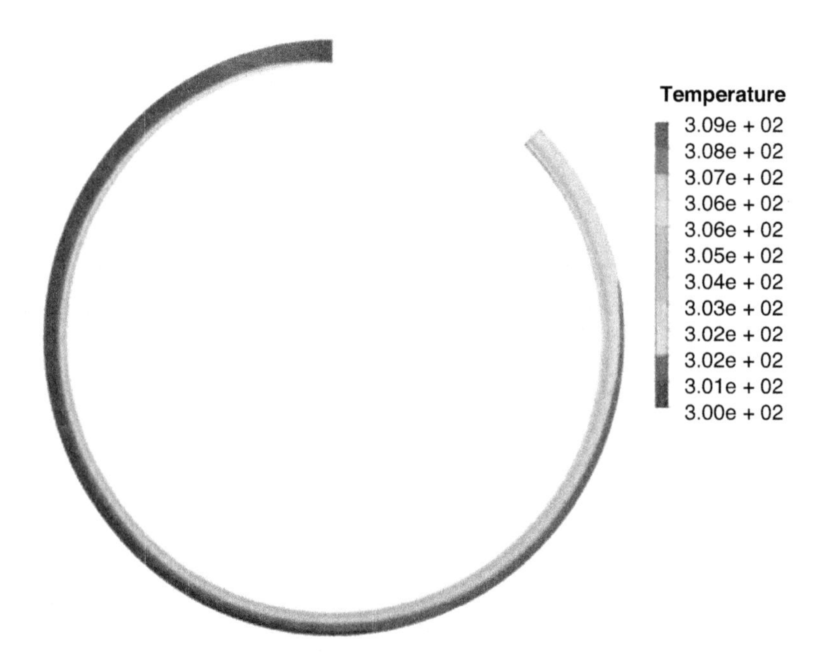

Figure 9.43 Contour plot of fluid temperature.

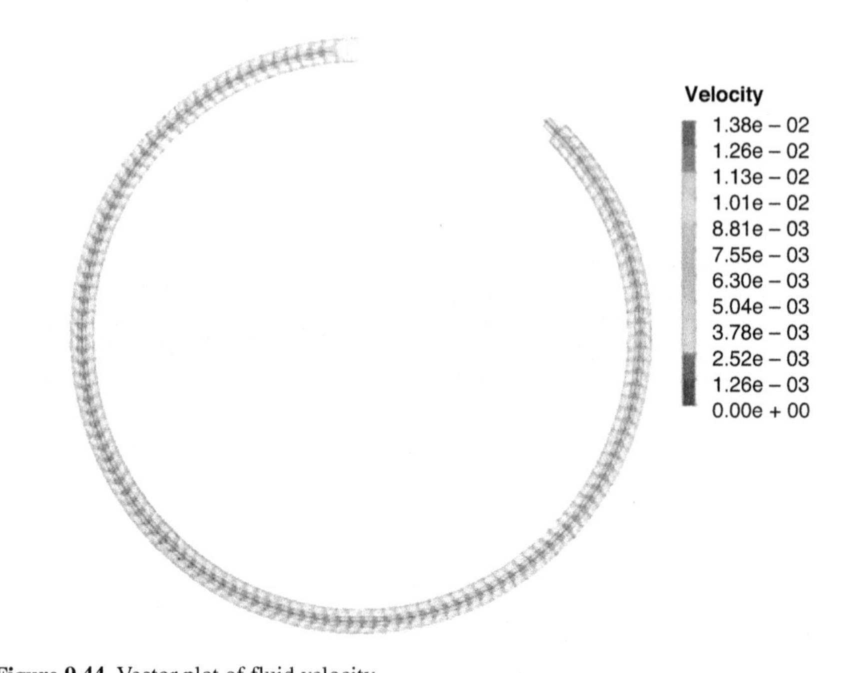

Figure 9.44 Vector plot of fluid velocity.

The temperature distribution of the cooling water is shown in Figure 9.43. From a common phenomenon, the closer the area is to the outlet, the higher the temperature becomes. Figure 9.44 shows the velocity of the fluid.

9.3.2.3 Conclusions

A customized software, INTESIM-Motor is developed for solving the temperature rise issue of PMSM. It consists of electromagnetic, thermal-stress, and thermal-fluid physical solvers, with both weak coupling and strong coupling methods available. With a unified graphical user interface, it is a convenient, robust, accurate, efficient tool for the design and multiphysics simulation of a motor. The procedure of electromagnetic-thermal–fluid coupling analysis is presented and demonstrated through the bidirectional coupling case of the simplified PMSM. Reasonable distributions of magnetic field strength, loss density, temperature, fluid velocity, and other physical quantities, can be obtained through analysis. The INTESIM-Motor software developed by the INTESIM company can help the design and optimization of motors.

Computational fluid dynamics in aerospace field and CFD-based multidisciplinary simulations

10

Chapter Outline

10.1 Application and development of computational fluid dynamics simulation in the aerospace field

Flight-testing is the most effective and direct way to verify aircraft performance. It takes thousands of hours to do flight testing for new types of aircrafts, and the designs need to be modified according to the test results. This design method of fly-and-fix is costly, which often leads to a prolonged developmental cycle of the aircraft, and may bring disastrous consequences as well as make the previous efforts meltdown. Literature (Salas, 2006) has mentioned that since the United States entered jet age, dozens of aircrafts needed frequent modifications during the flight-testing stage of initial development due to inaccurate prediction of the aerodynamic stability and control characteristics. Modification of design leads to increase in research funding, a prolonged research cycle, decrease in the flight performance index, and sacrifice of layout optimization. Several new concept aircrafts, such as the flexible wing aircraft Helios, crashed because of nonlinear coupling of flow, structure, and control system in flight-testing, which caused the project failure. Therefore, before flight-testing, a comprehensive study of aircraft's system characteristics under real flight conditions, is required to reduce cost and risk during the flight-testing phase.

Virtual flight-testing(VFT) is a research approach of aircraft aerodynamics stability and control law, initially developed from wind tunnel experiments. The US Air Force's Arnold Engineering Development Center (AEDC) has used tensile wire

Q. Zhang & S. Cen: Multiphysics Modeling. http://dx.doi.org/10.1016/B978-0-12-407709-6.00010-9

suspension technology and automatic control systems to carry out VFT research for missiles in the low-speed opening wind tunnel (Gebert et al., 2000; Lawrence and Mills, 2002), and coupling between 3-DOF (degrees of freedom) motion and a control system for missile has been achieved. Consequently, the stability and control law of missiles has been studied. China's Aerodynamics Research and Development Center has used 3-DOF motion combined with closed-loop control, to study wind tunnel VFT of one missile (Li et al., 2011). The experimental result shows that VFT can be used to study unsteady aerodynamics and control system coupling characteristics of aircrafts so as to reduce flight-testing cost. However, due to the limitations of wind tunnels, the experimental model can only achieve the motion of three rotational DOFs. The wind tunnel wall, support, and so on may also affect the wind tunnel model. Therefore, the measured response from the wind tunnel model is not equal to that of a real aircraft during flying.

With the development of computational fluid dynamics (CFD), the concept of numerical wind tunnel (NWT) has been paid more attention. Japan's National Aerospace Laboratory developed a numerical wind tunnel (Miyoshi et al., 1994; Iwamiya et al., 1996). A high-precision numerical wind tunnel (TH-Hi NWT) deployed on a Tianhe computer has been developed by Yan Zhenguo (2013) to solve complicated engineering problems using a high-precision differential method. Ren et al. (2000) have conducted research on numerical wind tunnel software systems that consist of structure dynamics, CFD solver, pre- and postprocessing, and an interactive interface module design. As a supplement of the physical wind tunnel, a numerical wind tunnel system takes advantage of CFD simulation's low cost, easy modeling, and rich information from calculation results, so that it is easy to generate a great number of aerodynamic simulation results with low cost. As an aerodynamic research tool, unsteady aerodynamic characteristics and control system effects have not been considered by numerical wind tunnels.

According to Salas (2006), digital flight has been taken as the last grand challenge in the development of CFD technology. In the literature, digital flying has been divided into flying through databases and flying through equations. Flying through equations, also called numerical flight simulation, can be considered numerical wind tunnel–based virtual flight-testing. Numerical flight simulation technology shows the development tendency of CFD simulation coupling with other disciplines. China's Aerodynamics Research and Development Center has adopted CFD to carry out virtual flight simulation (Tao et al., 2010; Da et al., 2012). The unsteady aerodynamic characteristics under free pitching, swinging, rolling, and maneuvering conditions for the missile were obtained. Simulating Aircraft Stability and Control (SimSAC) project was introduced in byRizzi (2011). This project carried out multidisciplinary modeling and simulation of stability and controlling problems during the aircraft conceptual design phase on the basis of CFD. The research on numerical virtual flight technology requires theories from multiple disciplines such as, CFD, structural dynamics, flight mechanics, and control theory. These theories could subsequently be used to develop computer software, which could carry out numerical simulation of complex multidisciplinary coupling problems with the help of supercomputers. Therefore, the research and application of CFD nowadays has confronted new challenges.

10.2 The research topic and its progress

10.2.1 Main problems overview

CFD has been developed from a single discipline simulation and gradually extended up to multidisciplinary coupling simulations, in which CFD and flight mechanics coupling becomes mature and widely applied. In addition to simulating 6-DOF motion of aircraft, rocket stage separation, aircraft's load delivery, the opening and closing of cabins, seat catapult and other single-/multibody separation dynamics processes, aircraft maneuvering under the action of control surface deflection can also be simulated. The aerodynamics simulation for problems listed earlier can be generalized as unsteady flow simulation with moving boundaries. Therefore, in addition to the studies on different scheme and turbulence models, additional specific problems about moving boundary simulation need to be resolved in CFD calculation.

Mesh morphing technology: in the numerical simulation of a flight, couplings among aerodynamics, flight mechanics, structural mechanics, and rudder control laws are included. This requires CFD models to be able to simulate 6-DOF motions of the aircraft, which is calculated by flight mechanics, the deformation from flexible structural dynamics, and multibody motion of the control surface after receiving the rotation and separation commands. Thus, an efficient and robust mesh-morphing algorithm needs to be developed to deal with complex moving boundary problems. For the unsteady flow caused by moving boundary simulations, because mesh shape varies with time, and mesh motion is coupled with fluid flow, the geometric conservation law needs to be satisfied during simulation.

Time discretization scheme and convective flux scheme: applying the *p*th-order time-accurate discrete scheme onto the dynamic grid cannot guarantee the same order of accuracy as on the Eulerian grid. The time discrete scheme on the deformed mesh needs to consider the geometric conservation law in the governing equation and construct the corresponding discrete scheme (Liu et al., 2009). Under the description of the Arbitrary Lagrangian Eulerian (ALE) formulation, this convective term is different from those on the Eulerian grid, thus the discretization scheme of convection term should also undertake certain corrections.

Mesh generation and remeshing: a mesh with good quality is crucial for improving the accuracy of calculation results. However, features of mesh generation and remeshing in the standard CFD software are not usually implemented. General practice is to generate the mesh data through a specialized mesh-generating software from a model entity, and then the CFD model is invocated by a flow field solver, and the mesh remains unchanged during the calculation process. Even for the CFD software with mesh generator (usually called preprocessing), the intercommunication between mesh generator and solver is very limited, and the mesh generation module is not called during flow field module processing. Interaction between the mesh generator and flow solver is needed while developing numerical simulation software because aircraft motion and deformation may cause mesh distortion and failure of the flow field. At present, communication of flow information between new and old mesh after remeshing, is processed by the interpolation schemes. Theory proves that this processing will bring calculation errors and decrease accuracy. Therefore, highly accurate information-passing algorithms are required.

Fluid–structure interaction (FSI) methods: usually two methods are adopted to solve the fluid–structure-coupling system namely, the strong coupling method and the weak coupling method. In the strong coupling method, the fluid control equations and structural equations are assembled as one systematic equation to be solved simultaneously. The formula derivation is complicated, and amount of calculation is relatively larger. It is difficult to be applied in actual engineering of FSI analysis of the multi-DOF structure field. The partitioned method is used to decompose coupling problems into two parts, flow fields and structures, which are solved separately. The coupling between flow field and structure is achieved by information exchanges across the fluid–structural interface, which is also called the conventional serial staggered (CSS) algorithm. Compared with strong coupling method, the weak coupling method can fully use the existing CFD and CSD calculation method and program, reducing the difficulty of program development and keeping program modularization. Therefore, the weak coupling method attracted great attention and has been extensively used since it was first proposed. However, the information exchanges on the coupling interface bring new problems. For example, a study found that after equation discretization, the fluid–structure interface cannot satisfy the condition of equal velocity and equal position simultaneously. If aerodynamics, flight mechanics, structural mechanics, rudder control law, and other multiple equations are solved simultaneously, the interface information exchange will be more abundant. Since interface-coupling algorithm is directly related to the accuracy of numerically simulated flight, detailed studies are required.

10.2.2 ALE formulation flow control equation

This section describes how to use ALE finite volume method to describe 3D NS equations of compressible flows in nondimensional form. For specific theoretical assumptions and the derivation process, refer the literature by Liu et al. (2009):

$$\frac{\partial}{\partial t}\int_{\Omega} \boldsymbol{Q}\,\mathrm{d}\Omega + \int_{\Gamma}\left[F_c(\boldsymbol{Q},\boldsymbol{x}_c) + F_v(\boldsymbol{Q})\right]\boldsymbol{n}\,\mathrm{d}\Gamma = 0 \tag{10.1}$$

where $\boldsymbol{Q}$ is the conservation variable:

$$\boldsymbol{Q} = \{\rho, \rho u, \rho v, \rho w, \rho e\}^{\mathrm{T}} \tag{10.2}$$

Here, u, v, and w are the velocities in x, y, and z directions, respectively. The total energy of per unit mass e of fluid consists of internal energy and kinetic energy. For ideal gas with a constant specific heat ratio $\gamma = 1.4$, the total energy per unit mass of fluid is:

$$e = \frac{p}{\rho(\gamma - 1)} + \frac{1}{2}\left(u^2 + v^2 + w^2\right). \tag{10.3}$$

In Equation (10.1), the viscous-irrelevant functions are written as follows:

$$F_c(\boldsymbol{Q}, \boldsymbol{x}_c) = F\boldsymbol{i} + G\boldsymbol{j} + H\boldsymbol{k}, \tag{10.4}$$

where $(\boldsymbol{i}, \boldsymbol{j}, \boldsymbol{k})$ are the unit vectors of three coordinate directions (x, y, z) in the Cartesian coordinate system. The flux of convective terms that contains relative motion can be written as:

$$\boldsymbol{F} = \begin{Bmatrix} \rho U \\ \rho U u + p \\ \rho U v \\ \rho U w \\ (\rho e + p)U + x_t p \end{Bmatrix}, \boldsymbol{G} = \begin{Bmatrix} \rho V \\ \rho V u \\ \rho V v + p \\ \rho V w \\ (\rho e + p)V + y_t p \end{Bmatrix}, \boldsymbol{H} = \begin{Bmatrix} \rho W \\ \rho W u \\ \rho W v \\ \rho W w + p \\ (\rho e + p)W + z_t p \end{Bmatrix}, \tag{10.5}$$

where the relative velocity of fluid flow under the ALE coordinate system is expressed as:

$$V_c = \{U, V, W\}^T = \{(u - x_t), (v - y_t), (w - z_t)\}^T. \tag{10.6}$$

In the literature about the finite volume method, sometimes $\boldsymbol{F}_c$ is called nonviscous flux, which is neither a vector nor a tensor. It is introduced as a symbol for brevity. Similarly, the "viscosity flux" caused by surface stress of the control body is expressed as:

$$\boldsymbol{F}_v(\boldsymbol{Q}) = F_\mu \boldsymbol{i} + G_\mu \boldsymbol{j} + H_\mu \boldsymbol{k}, \tag{10.7}$$

where the vectors related to viscous stress and heat conduction are:

$$\boldsymbol{F}_\mu = \frac{1}{Re}\begin{Bmatrix} 0 \\ \tau_{xx} \\ \tau_{xy} \\ \tau_{xz} \\ \varphi_x \end{Bmatrix}, \boldsymbol{G}_\mu = \frac{1}{Re}\begin{Bmatrix} 0 \\ \tau_{xy} \\ \tau_{yy} \\ \tau_{yz} \\ \varphi_y \end{Bmatrix}, \boldsymbol{H}_\mu = \frac{1}{Re}\begin{Bmatrix} 0 \\ \tau_{xz} \\ \tau_{yz} \\ \tau_{zz} \\ \varphi_z \end{Bmatrix}. \tag{10.8}$$

Herein,

$$Re = \rho_\infty^* u_\infty^* D / \mu_\infty^*, \tag{10.9}$$

where * denotes dimensional physical quantity and ∞ denotes free-stream reference physical quantity. For the laminar flow problem, the air viscosity coefficient is given by the Sutherland equation:

$$\mu_l = T^{3/2} \times \frac{1+(110.4/T_\infty^*)}{T+(110.4/T_\infty^*)}. \tag{10.10}$$

The energy equation includes the contribution of heat conduction and the contribution of work done by the stress:

$$\begin{aligned} \varphi_x &= u\tau_{xx} + v\tau_{xy} + w\tau_{xz} + q_x \\ \varphi_x &= u\tau_{yx} + v\tau_{yy} + w\tau_{yz} + q_y \\ \varphi_z &= u\tau_{xz} + v\tau_{yz} + w\tau_{zz} + q_z \end{aligned} \tag{10.11}$$

where the heat conduction components are expressed as:

$$q_x = k\frac{\partial T}{\partial x}, q_y = k\frac{\partial T}{\partial y}, q_z = k\frac{\partial T}{\partial z}. \tag{10.12}$$

The relationship between heat conduction coefficient k and viscosity coefficient μ_l is given by:

$$\begin{aligned} k &= \frac{\gamma}{\gamma-1}\frac{1}{Pr}\frac{\mu_l}{Re} \\ Pr &= u_\infty^* c_p^* \frac{D}{k_\infty} \end{aligned} \tag{10.13}$$

When setting up this equation, the definition of specific heat ratio $\gamma = c_p/c_v$ and the condition of universal gas constant are used implicitly. The use of specific heat ratio of diatomic molecule $\gamma = 7/5 = 1.4$ derived by the molecular kinetic theory can obtain relatively satisfying accuracy under most circumstances when it is used for air.

The equation above is not closed until the state equation and internal energy relational expression is specified:

$$p = T\rho \tag{10.14}$$

Note: different expressions of state equation are obtained if temperatures are getting from different dimensionless parameters; for example, expression of state equation $p = (T\rho)/\gamma$ will be obtained by using environment dimensionless temperature $T = T^*/T_\infty^*$, and the form of heat conduction coefficient will also change correspondingly.

If viscosity term flux is not taken into consideration, that is, $\boldsymbol{F}_v(\boldsymbol{Q}) = 0$, Equation (10.1) reduced to a 3D compressible unsteady dimensionless Euler equation by the description of the ALE finite volume method. When the numerical simulation technique in the framework of the finite volume method is adopted, there is no essential difference between solving the Euler and NS equations, using dynamic mesh. Therefore, the Euler equation is used in the following discussion for brevity. If there is no special explanation, all of the conclusions are also valid for the NS equation.

10.2.3 Discrete geometric conservation law

The concept of geometric conservation law was first put forward by Thomas and Lombard in 1978. In the literature(Thomas and Lombard, 1978), through model analysis and example demonstration, the following has been proved. Under the assumption of flow field parameter as constant, the geometric conservation law equation is the degeneration form of ALE formulation of fluid mechanics but is not an independent new constraint condition. Therefore, from the perspective of specification, adoption of the discrete geometric conservation law (DGCL) shall be more accurate. The essential mechanism of a nonphysical solution caused by unsatisfied geometric conservation is that, the volume increment is not equal to the volume change because of surface motion in the process of grid deformation

A triangular element is shown in Figure 10.1, with unit thickness in the direction that is perpendicular to the paper assumed, and infinitesimal volume and area for description used. From time step n to time step $n + 1$, infinitesimal volume has been changed from V^n to V^{n+1}, where the ith side S_i^n, moving at the speed of $\boldsymbol{x}_{ci}$, having been changed to S_i^{n+1}, at the same time, and the volume increment formed by the moving

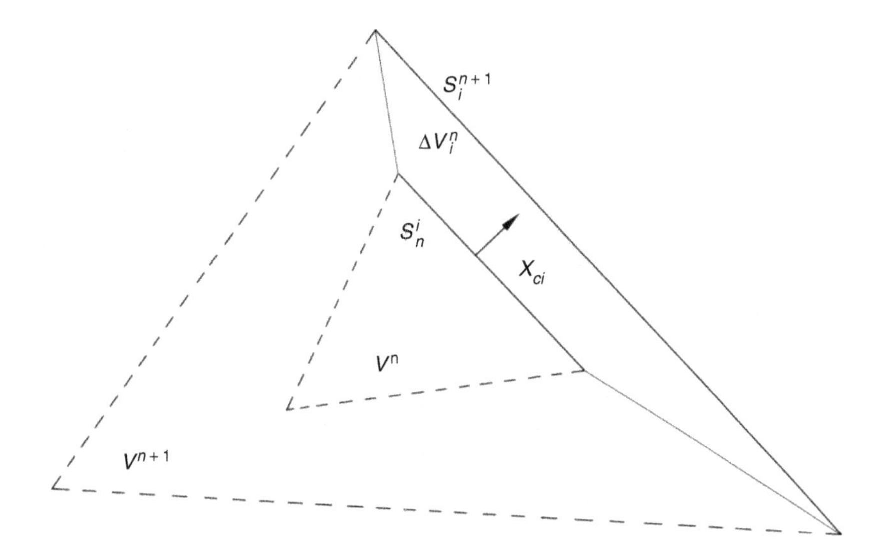

Figure 10.1 Grid motion diagram.

area, is taken as ΔV_i^n. Theoretically, volume change formed by the motion of control body surface is not equal to the actual volume increment, but things go differently after discretization. If the first-order explicit time integration scheme is used for the transient ALE formulation of fluid flow, parameter at time step n is used for calculating space flux integral. Approaching time step n to time step $n+1$, obviously there is:

$$\frac{V^{n+1}-V^n}{\Delta t} \neq \sum_{i=1}^{3} \boldsymbol{x}_{ci}\boldsymbol{n}_i S_i^n \text{ or } V_i^{n+1}-V_i^n \neq \boldsymbol{x}_{ci}\Delta t\, \boldsymbol{n}_i S_i^n \tag{10.15}$$

In the process of calculation, the geometric parameter is coupled together with the flow variable. This geometric inequality will cause the change of flow field. In some cases, this error would accumulate continuously along with time, which will lead to the failure of unsteady flow dynamic simulation or calculation process divergence. DGCL, which relates to the space–time discrete scheme of solving the flow field, is the method to eliminate this type of geometric error. Many kinds of algorithms have been constructed in the literature.

According to different correction parameters used, the geometric conservation law algorithms are divided into four types.

The first type is the modified area algorithm, which is mainly used for a two-step explicit time-stepping scheme. The flow field parameter at time step n is used for the calculation of flux integral (traditionally known as the right-hand side). However, the geometric parameter at time step $n+1$, is supplemented to calculate the average area:

$$S_i^{n+\frac{1}{2}} = \frac{1}{2}\left(S_i^n + S_i^{n+1}\right), \tag{10.16}$$

by replacing S_i^n at time step n by $S_i^{n+\frac{1}{2}}$ to carry out right-hand side flux integration. For the widely used Runge–Kutta explicit scheme, volume in the forecasting step is additionally required to use the same computational scheme as that in the flow field:

$$\begin{gathered} V^{(0)} = V^n \\ V'^{(j)} = V^{(0)} + \sigma_j \Delta t \left(\boldsymbol{n}_j S_j\right)^{n+\frac{1}{2}} \\ V^{n+1} = V'^{(m)} \\ \sigma_j = 1/(m-j+1) \quad j = 1,2,\ldots m \end{gathered} \tag{10.17}$$

This type of DGCL algorithm can realize the equal position condition of object surface. For calculation of flow equation, the boundary velocity of fluid model is calculated by the position of interface nodes of the structure. In the aforementioned figure, after the position of the vertex at time step $n+1$ is decided, the second-order accuracy velocity at the center of the edge is equal to the average velocity of vertexes, that is:

$$\boldsymbol{x}_{12} = \frac{1}{2}\left(\frac{\boldsymbol{x}_1^{n+1}-\boldsymbol{x}_1^n}{\Delta t} + \frac{\boldsymbol{x}_2^{n+1}-\boldsymbol{x}_2^n}{\Delta t}\right) \tag{10.18}$$

However, this type of DGCL algorithm has vague physical meaning. This is because the calculation of flow field flux at time step n involves the geometrical parameter at time step $n + 1$, and the new time step volume V^{n+1} will be obtained on the basis of flow field computational scheme. Theoretically, accumulative error could not be eliminated automatically, so it is difficult to be extended to a multi-time step scheme. For example, because the second-order implicit time scheme involves geometrical parameter at time step $n - 1$, the average area could not be defined. Therefore, it is inappropriate to use neither $S_i^{n+(1/2)}$ nor $S_i^{n-(1/2)}$.

The second type is the grid velocity modification algorithm. Flux integral of multitime step implicit scheme uses the geometrical parameter at time step $n + 1$. For example, when the fluid equation adopts second-order implicit time discretization scheme, the time derivative of the volume changes in discretization form as follows:

$$\frac{\partial}{\partial t}\int_{\Omega} \mathrm{d}\Omega = \frac{3V^{n+1} - 4V^{n} + V^{n-1}}{2\Delta t} \tag{10.19}$$

The integration area for the right-hand side flux is S_i^{n+1}; the grid velocity is $\boldsymbol{x}_{ci}$, which is calculated by geometrical relationship; and the geometrical nonconservation issue arises in this process. To satisfy the geometric conservation law, grid velocity $\boldsymbol{u}_g$ is introduced. The following grid velocity $\boldsymbol{u}_{gi}$ is used to replace $\boldsymbol{x}_{ci}$ in the process of flow field calculation, that is:

$$\boldsymbol{u}_{gi}\boldsymbol{n} = \frac{3\Delta V_i^{n} - \Delta V_i^{n-1}}{2\Delta t S_i^{n+1}}. \tag{10.20}$$

This algorithm cannot guarantee satisfaction of location conditions on the fluid–structure interface even though the velocity condition is not strictly satisfied. Farhat, who introduced the accurate CSS interface algorithm of energy analysis, also noticed this problem (Farhat et al., 2006). Therefore, he proposed a structure predictor to construct the new interface. The basic process is as follows:

1. Predict the displacement at time step $n + (1/2)$ on the basis of velocity $\boldsymbol{q}_t^n$.

$$\boldsymbol{q}^{(n+(1/2)),p} = \boldsymbol{q}^{n} + 0.5\boldsymbol{q}_t^{n}\Delta t \tag{10.21}$$

2. Advance Δt based on the flow field at time step $n - (1/2)$ to obtain the flow field at time step $n + (1/2)$. The second type of DGCL is used to calculate the fluid boundary grid velocity $\boldsymbol{u}_g^{n+(1/2)}$ according to $\boldsymbol{q}^{n-(1/2)}$ and $\boldsymbol{q}^{(n+1),p}$.
3. When the fluid load, $\boldsymbol{F}_{\mathrm{air}}^{n+(1/2)}$ is known, linear extrapolation is used to determine the aerodynamic force at time step, $n + 1$.

$$\boldsymbol{F}_{\mathrm{air}}^{n+1} = 2\boldsymbol{F}_{\mathrm{air}}^{n+(1/2)} - \boldsymbol{F}_{\mathrm{air}}^{n}. \tag{10.22}$$

On this basis, we can obtain $\boldsymbol{q}_t^{n+1}$ and $\boldsymbol{q}^{n+1}$ at time step $n + 1$ by calculating a structural equation with a second-order accuracy scheme.

According to the energy model analysis, this interface algorithm has second-order accuracy, which is called the high-resolution weak-coupling algorithm. However, after careful study of this algorithm, the following problems are found:

1. In the process of the coupling calculation, the fluid boundary location $\boldsymbol{x}^{n+1} = \boldsymbol{u}^{(n+1),p} \neq \boldsymbol{u}^{n+1}$ does not fully satisfy the equal position condition.
2. If first-order accuracy predicted value $\boldsymbol{u}^{(n+1),p}$ is used as the boundary condition of fluid flow, even though the inner iteration of the dual-time method has reached convergence, accuracy of the obtained fluid force $\boldsymbol{F}_{\text{air}}^{(n+(1/2)),p}$ is quite low. On this basis, the accuracy of aerodynamic force $\boldsymbol{F}_{\text{air}}^{n+1}$ and $\boldsymbol{u}^{n+1}$ obtained by extrapolation is difficult to verify. According to this interface algorithm, although the energy conservative transferred between fluid and structure reaches second-order accuracy, the precision of the transferred parameters themselves is not clear.

The third type is correcting velocity and area at the same time. When a semi-implicit scheme is used for the integration of flux, both S_i^{n+1} and S_i^n appear in the integrands. In order to solve this problem, the following geometric conservation equation

$$\frac{V^{n+1} - V^n}{\Delta t} = \sum_{i=1}^{nf} (\boldsymbol{u}_{gi}\boldsymbol{n}_i) S_i^{n+(1/2)}, \tag{10.23}$$

is used to obtained the grid velocity:

$$\boldsymbol{u}_{gi}\boldsymbol{n}_i = \frac{\Delta V_i}{\Delta t S_i^{n+(1/2)}} \tag{10.24}$$

In the literature(Liu et al., 2009), through the calculation examples and theory analysis, the "geometric conservation law is not the new constraint coming from the process of fluid–structure coupling, but an error produced by the inequality of volume increment and volume formed by the moving area at two different time steps," was demonstrated. Based on this mechanism, these three geometric conservation law algorithms are nothing more than using the modifying area or grid movement of the right-hand side (integrated flux) to satisfy geometrical equivalence. Now that we can modify the area or grid movement, why can't we modify volume? Based on such thinking, a new geometric conservation law algorithm is established in this chapter, which is called the volume correction algorithm.

The volume correction algorithm: in the flux calculation, practical physics value is used for all areas, that is, flux S_i^n at time step n and flux S_i^{n+1} at time step $n + 1$, and geometric conservation could be achieved through volume correction. There are many choices for specific treatment of this kind of algorithm.

For the Runge–Kutta explicit scheme given earlier, we can correct V^{n+1} or V^n.

1. In the process of time step n pushing forward to time step $n + 1$, replaces V^{n+1} with V':

$$V' = V^n + \Delta t \sum_{i=1}^{3} \boldsymbol{x}_{ci}\, \boldsymbol{n}_i S_i^n. \tag{10.25}$$

2. In the process of time step n pushing forward to time step $n + 1$, replaces V^n with V'':

$$V'' = V^{n+1} - \Delta t \sum_{i=1}^{3} \boldsymbol{x}_{ci}\, \boldsymbol{n}_i S_i^n. \tag{10.26}$$

No matter which kind of modification is chosen, when applying a coordinate at the corresponding time step to calculate V^n and V^{n+1}, there is no error accumulation.

For the second-order time implicit scheme earlier, there are three methods of correction:

1. Replaces V^{n+1} with V' for calculation:

$$V' = \frac{1}{3}\left(4V^{n+1} - V^{n-1} + 2\Delta t \sum_{i=1}^{3} \boldsymbol{x}_{ci} \boldsymbol{n}_i S_i^{n+1}\right). \tag{10.27}$$

2. Or replaces V^n with V''

$$V'' = \frac{1}{4}\left(3V^{n+1} + V^{n-1} - 2\Delta t \sum_{i=1}^{3} \boldsymbol{x}_{ci} \boldsymbol{n}_i S_i^{n+1}\right). \tag{10.28}$$

3. Or replaces V^{n-1} with V'''

$$V''' = 4V^n - 3V^{n+1} + 2\Delta t \sum_{i=1}^{3} \boldsymbol{x}_{ci} \boldsymbol{n}_i S_i^{n+1}. \tag{10.29}$$

For the volume correction algorithm, in addition to the advantages of clear physical meaning, convenient application, and small amount of calculation, it is more important that its mesh deformation velocity is calculated by the displacement according to the area central position of time step n and time step $n + 1$. A high-resolution interface algorithm can be built when the location condition is satisfied at the fluid–structure interface.

10.2.4 Mesh deformation algorithm

Fluid flow with moving interfaces such as FSI and free interface issues has widely existed in nature, human life, and production. At present, solving flow equations of the ALE form is one of the major methods for dealing with issues related to the moving interface. Using the ALE method with a mesh morphing scheme, to accommodate the movement and deformation of boundaries, mesh regeneration could be avoided, and the times of remeshing can be reduced as much as possible. Therefore, most of these flow issues with moving interfaces could be conducted conveniently. The dynamic mesh-morphing algorithm is a key factor for calculating flow issues related to moving interfaces by the ALE method. The effect of the dynamic mesh deformation algorithm on flow field calculation is mainly reflected in two aspects: the ability of mesh deformation and the computational efficiency of mesh morphing calculation. The remeshing in calculation will bring additional interpolation error, and precision of calculation will be reduced further. Times of local remeshing could be avoided or reduced when a good dynamic mesh-morphing algorithm is applied while the good quality of deformed mesh is maintained. This property plays an important role in ensuring precise calculations. On the other hand, to adapt to large-scale calculations in practical issues, mesh-morphing algorithms should have higher efficiency. But in fact, deformation ability and calculation efficiency could not be balanced in an easy way.

In the mesh-morphing method, under the condition of the known mesh displacement on the boundary, displacement of interior points by an interpolation technique could be obtained. Radial function is only the function of radial distance between any two points, which is not related to any data structure and fit for any type of mesh. In the computational domain, the displacement of one point can be interpolated by the radial basis function. And it can be expressed as follows:

$$s_i(\boldsymbol{X}) = \sum_{j=1}^{nb} \beta_{ji} \phi(\boldsymbol{X} - \boldsymbol{X}_{bj}) + p_i(\boldsymbol{X}) \tag{10.30}$$

where $s_i(\boldsymbol{X})$ is the displacement component in i direction at node x; $\boldsymbol{X}_{bj} = \{x_{bj}, y_{bj}, z_{bj}\}^T$ is the coordinate of boundary point with known displacement, and these mesh nodes are called centers, with nb indicating the number of boundary points; β_{ij} is the interpolation factor; ϕ represents the radial basis function; and $p_i(\boldsymbol{X})$ is the polynomial function. Usually first-order function is chosen, for example:

$$p_i(\boldsymbol{X}) = \gamma_{0i} + \gamma_{1i}x_1 + \gamma_{2i}x_2 + \gamma_{3i}x_3 \tag{10.31}$$

All displacements of the center points should be exactly described by Equation (10.30); therefore,

$$\boldsymbol{S}(\boldsymbol{X}_{jb}) = \boldsymbol{d}_{jb} \tag{10.32}$$

where, $\boldsymbol{d}_b$ is the known displacement of mesh nodes on the boundary. Equation (10.32) can be written in the matrix form:

$$d_{ib} = \begin{bmatrix} 1 & x_{i1} & x_{i2} & x_{i3} & \phi(\|x_i - x_1\|) & \phi(\|x_i - x_2\|) & \cdots & \phi(\|x_i - x_{nb}\|) \end{bmatrix} \cdot \begin{bmatrix} \gamma_{01} & \gamma_{02} & \gamma_{03} \\ \gamma_{11} & \gamma_{12} & \gamma_{13} \\ \gamma_{21} & \gamma_{22} & \gamma_{23} \\ \gamma_{31} & \gamma_{32} & \gamma_{33} \\ \beta_{11} & \beta_{12} & \beta_{13} \\ \vdots & \vdots & \vdots \\ \beta_{n1} & \beta_{n2} & \beta_{n3} \end{bmatrix} \quad (i = 1, 2, \cdots nb). \tag{10.33}$$

The above equation contains $3 \times (nb + 4)$ numbers unknowns, but only $nb \times 3$ number equations, so adding conditions to complete the equations is required. Additional equations are given below.

$$\sum_{i=1}^{n} \beta_{ij} = 0, \ \sum_{i=1}^{n} \beta_{ij} x_{ik} = 0 \quad (j, k = 1, 2, 3) \tag{10.34}$$

Simple translational and rotational deformation could be restored in view of the increased conditions. The equation set is completed after 12 equations are complemented, and the unique solution can be obtained. Then it can be written in full matrix form:

$$\begin{pmatrix} 0 \\ 0 \\ 0 \\ 0 \\ d_1 \\ \vdots \\ d_{nb} \end{pmatrix} = \begin{pmatrix} 0 & 0 & 0 & 0 & 1 & \cdots & 1 \\ 0 & 0 & 0 & 0 & x_{11} & \cdots & x_{n1} \\ 0 & 0 & 0 & 0 & x_{12} & \cdots & x_{n2} \\ 0 & 0 & 0 & 0 & x_{13} & \cdots & x_{n3} \\ 1 & x_{11} & x_{12} & x_{13} & \phi_{11} & \cdots & \phi_{1n} \\ \vdots & \vdots & \vdots & \vdots & \vdots & \ddots & \vdots \\ 1 & x_{n1} & x_{n2} & x_{n3} & \phi_{n1} & \cdots & \phi_{nn} \end{pmatrix} \begin{pmatrix} \gamma_{01} & \gamma_{02} & \gamma_{03} \\ \gamma_{11} & \gamma_{12} & \gamma_{13} \\ \gamma_{21} & \gamma_{22} & \gamma_{23} \\ \gamma_{31} & \gamma_{32} & \gamma_{33} \\ \beta_{11} & \beta_{12} & \beta_{13} \\ \vdots & \vdots & \vdots \\ \beta_{n1} & \beta_{n2} & \beta_{nn} \end{pmatrix} \tag{10.35}$$

The interpolation factor is obtained by solving above equations. LU decomposition is adopted to solve the equation. Then through the interpolation of Equation (10.30), the interior point displacement can be obtained.

De Boer et al. (2007) studied a variety of radial basis functions, and thought Wendland compactly supported functions and thin plate spline (TPS) functions have better precision and efficiency. Forms of these two functions are as follows:

Wendland function is:

$$\phi(\|r\|) = \begin{cases} (1-\alpha\|r\|)^4 (4\alpha\|r\|+1) & \|r\| \le 1/\alpha \\ 0 & \text{else} \end{cases} \tag{10.36}$$

TPS is:

$$\phi(\|r\|) = \|r\|^2 \, ln(\|r\|) \tag{10.37}$$

In Equations (10.36) and (10.37), $\|r\|$ is the Euclidian distance from the interpolation point to the center, and $1/\alpha$ is the support radius of compactly supported function. Generally, a larger support radius offers more precise interpolation. However, when the support radius gets larger, coefficient matrix also becomes denser. Conversely, if the support radius is smaller, the coefficient matrix will be relatively sparser, and equations get easily solved. The support radius should be chosen based on certain experiences. In the calculation of this chapter, TPS is used as the radial basis function.

Combining the radius basis function method(RBFs) with the motion subgrid algorithm(MSA), the RBFs–MSA mesh deformation method is proposed and described as follows:

1. After generation of the computational mesh, the motion subgrid also needs to be generated. In order to reduce the computational time, number of boundary points of the motion subgrid is smaller than the number of the entire computational mesh in general, and the number of interior nodes can be controlled depending on the specific problem. The number of motion subgrid nodes can be far less than that of computational mesh. The schematic of computational mesh and motion subgrid is shown in Figure 10.2.

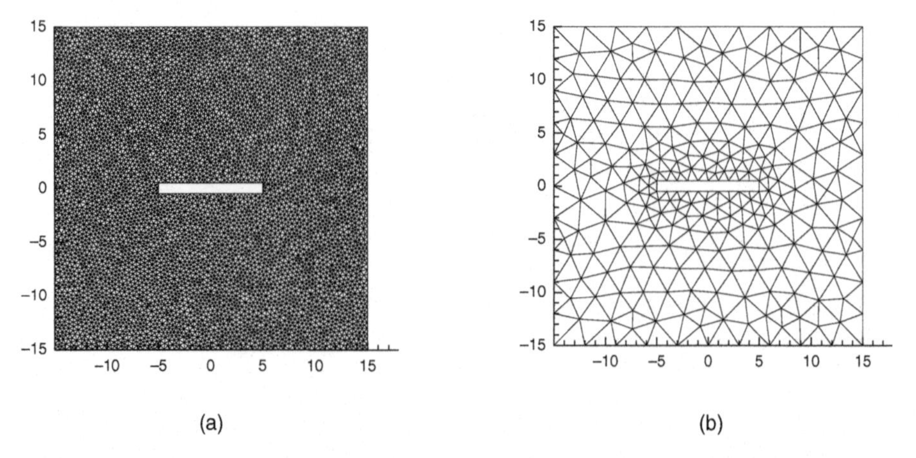

Figure 10.2 Initial computational mesh and motion subgrid. (a) Computational mesh and (b) motion subgrid.

2. In the element of motion subgrid, the interior points of the calculation mesh are located, the relative area or volume coordinates of the interior point in the element is determined, and location information and relative coordinates will also be stored.
3. Displacement of dynamic boundary is calculated, and the coordinates of the boundary nodes are also updated.
4. Based on the displacement on boundary nodes, using the interpolation factor, the coordinates of interior points of the motion subgrid will be obtained by radial basis function interpolation.
5. Using the area or volume coordinates of interior point in motion subgrid element and updated coordinate of the nodes, the interior point coordinate is obtained by interpolation.

Processes 1–5 shall be repeated until the end of the calculation.

The calculation method of relative area or volume coordinate of the interior points in a motion subgrid element is given below (Liu et al., 2006). The relative coordinate definition of one point in the triangle and tetrahedron element is described in Figure 10.3. For the 2D case, P is a point in $\triangle ABC$; the coordinate of P is $\boldsymbol{x}_p$; and the three vertex coordinates of $\triangle ABC$ are $\boldsymbol{x}_A$, $\boldsymbol{x}_B$, and $\boldsymbol{x}_C$, respectively. The area coordinate of P point relative to the area of $\triangle ABC$ can be obtained by the following equation:

$$e_i = \frac{S_i}{S}(i = 1,2,3) \tag{10.38}$$

where S is the area of $\triangle ABC$, and S_1, S_2 and S_3 represent the area of $\triangle PBC$, $\triangle$PCA, and $\triangle PAB$, respectively.

$$e_i = \frac{V_i}{V}(i = 1,2,3,4) \tag{10.39}$$

where, V is the volume of tetrahedron $ABCD$, and V_1, V_2, V_3, and V_4 are the volumes of $PBCD$, $PCDA$, $PDAB$, and $PABC$, respectively.

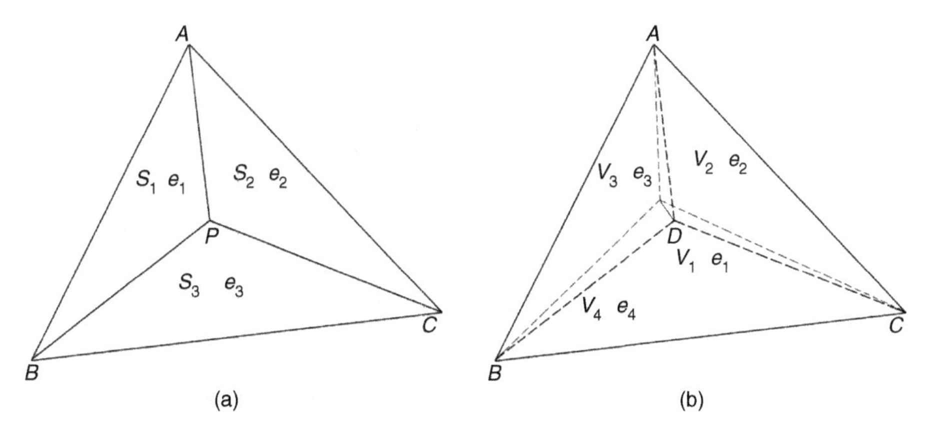

Figure 10.3 Area ratio coefficient and volume ratio coefficient. (a) Triangle element and (b) tetrahedron element.

The relative area or volume coordinate of point P in the motion subgrid elements could remain constant during grid deformation. Therefore, the coordinate P can be expressed as:

$$\begin{cases} x_p = \sum_i e_i x_i \\ y_p = \sum_i e_i y_i \\ z_p = \sum_i e_i z_i \end{cases}. \tag{10.40}$$

In these equations, (x_i, y_i, z_i) represent the vertex coordinate of the motion subgrid elements.

10.3 Example

10.3.1 Preposition and postposition double-action flapping wing

The example from the literature (Xiao et al., in press) is used in the dynamic mesh test. The front and back wings are a type of NACA0012 and the minimum horizontal distance between these wings is the length of chord C. Flapping laws of the front and back are given as follows:

Front wing:

$$h(t) = h_0 \sin(\omega t) \tag{10.41}$$

Back wing:

$$h(t) = h_0 \sin(\omega t + \varphi) \tag{10.42}$$

Flapping amplitude is $h(0) = 0.8C$. The phase between the flapping of front and the back wing is $\varphi = 180°$. Each flapping period is divided into 16 time steps. The computational grid and the motion subgrid are presented in Figure 10.4. The computational grid consists of 52,000 quadrilateral elements; 8,332 triangular elements; 740 boundary nodes; and 55,795 interior nodes. The motion subgrid consists of 1,142 triangular elements; 128 boundary nodes; and 506 interior nodes. The RBFs–MSA algorithm and the radial basis function method are used in this calculation. The definition of mesh quality indicator f_{size} refers to the literature (Knupp, 2003). The mesh at two positions of maximum relative motion in one period, is solved by the RBFs–MSA algorithm in Figure 10.5. In this figure, it is known that the mesh acquires good quality at the maximum relative displacement. To quantitatively analyze the quality

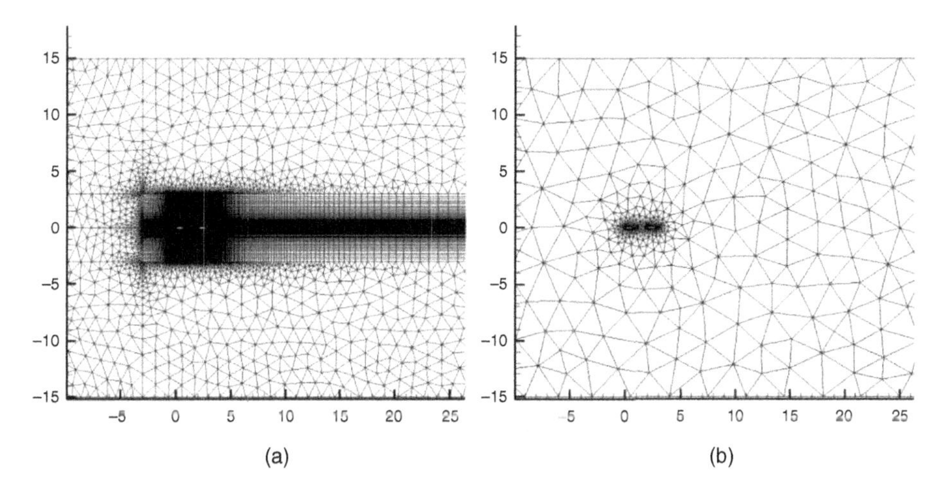

Figure 10.4 Computational grid and motion subgrid for preposition and postposition double-action flapping wing. (a) Computational grid and (b) motion subgrid.

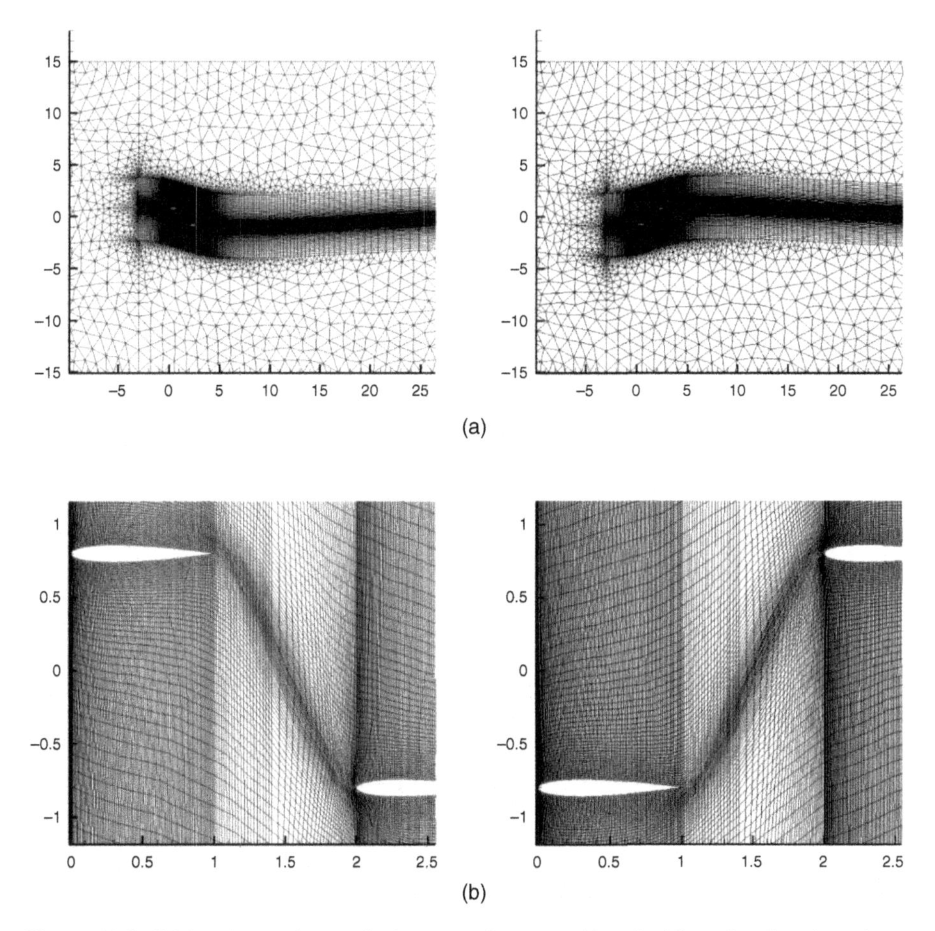

Figure 10.5 Grid at the maximum displacement for preposition double-action flapping wing. (a) Overall and (b) local.

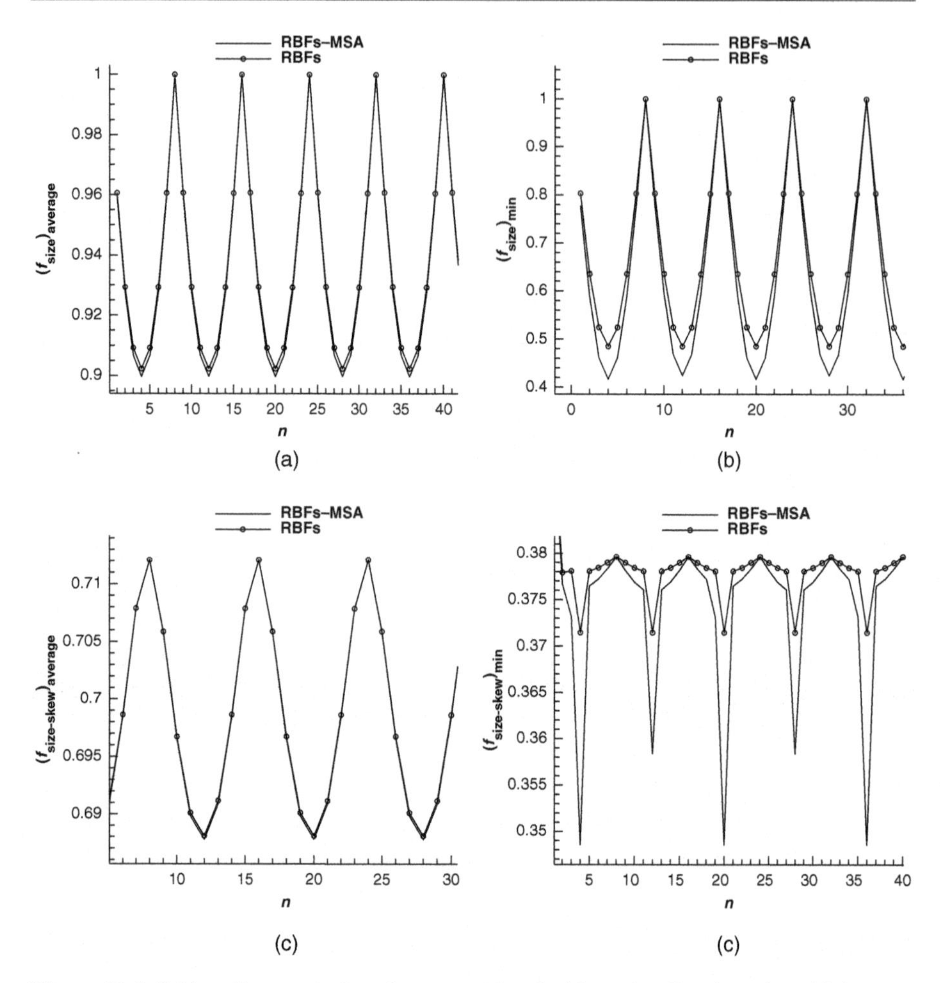

Figure 10.6 Grid quality comparison for preposition double-action flapping wing. (a) Average value of f_{size}; (b) minimum value of f_{size}; (c) average value of $f_{size-skew}$; and (d) minimum value of $f_{size-skew}$.

of mesh during deformation, the mesh quality from these two methods is compared in Figure 10.6. In Figure 10.6a, curve of average value of f_{size} during deformation is presented, and f_{size} reaches its minimum value when relative displacement is at the maximum point. But the minimum average values in both methods exceed 0.9, and f_{size} reaches the maximum value, that is close to 1.0, when the relative displacement is minimum. The minimum value curve of f_{size} is presented in Figure 10.6b. The value of f_{size} obtained by the RBFs–MSA algorithm is less than that obtained from the radial basis function method.

As the left wing reaches up to the maximum displacement, the right wing goes down the maximum displacement, and the difference of minimum value of f_{size} between these two methods becomes the largest. At this time, f_{size} is around 0.41, as predicted by

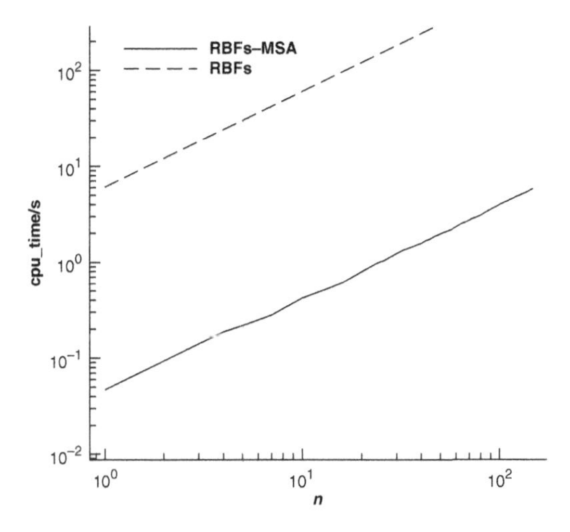

Figure 10.7 CPU time comparison during grid deformation for preposition double-action flapping wing.

the RBFs–MSA algorithm and around 0.47, as predicted by the radial basis function method. The average value of $f_{size-skew}$ is shown in Figure 10.6c, which represents the comprehensive mesh quality index of the quadrilateral element. It is observed that the difference between these two methods is quite small, and the variation of average $f_{size-skew}$ during the whole deformation, is less than 5%. The minimum value curve of $f_{size-skew}$ is shown in Figure 10.6d; the difference between these two methods is below 8%. These results show that grid orthogonality remains well during mesh deformation, which is important for the calculation of flow in the boundary layer.

For comparison of the computational efficiency between the RBFs–MSA algorithm and the radial basis function method, curve of accumulation of the calculation time in CPU varying with time step n, is shown in Figure 10.7. The CPU time and speed-up ratio in each average step of these two methods are shown in Table 10.1. In this example, the computational time of the RBFs–MSA algorithm is two magnitudes quicker than that of the radial basis function method. The reason is that boundary nodes and interior nodes of the motion subgrid are far more less than that in calculation grid. The number of boundary nodes in motion subgrid, is less than 1/5th of that in calculation grid, and number of interior nodes in motion subgrid is less than 1/100th of that in calculation grid. Therefore, the amount of computational time on the solving coefficient equations in the radial basis function method, and the interpolation of the interior nodes displacement needed by the motion subgrid is far less than that of the entire calculation grid. Meanwhile, storage of the motion subgrid is decreased dramatically. After the position of the motion, the subgrid is updated, the interpolation calculation of the interior nodes according to the relative area coordinate holds a very small portion of the whole amount of the calculation. Therefore, as compared to the radial basis function method, the amount of computational time could be reduced dramatically by the RBFs–MSA algorithm.

Table 10.1 Comparison of CPU time and speed-up ratio between RBFs–MSA and RBFs

Method	CPU time (s)	Speed-up ratio
RBFs	6.076	1.0
RBFs–MSA	0.039	155

It is shown that, in this example, the mesh quality by the RBFs–MSA algorithm is close to that of the deformed grid in radial basis function method. Meanwhile, the computational time is greatly reduced.

10.3.2 External store separation

To test and verify the capability and accuracy of developed software platforms in solving flow around the moving boundary, external store separation below the 3D wing is simulated and compared with the results in experiments and the literature. This example is a standard model in testing and verifying the flow around the moving boundary with public experimental designate parameters and measurement data in the wind tunnel, which is used in verification of the calculation accuracy in multibody separation in many literatures.

The simulation model in this experimental example is shown in Figures 10.8–10.12, including wing, pylon, and store. The axis direction of the inertial coordinate system is demonstrated in Figure 10.9. It is shown that, in Figure 10.11, the wall grid of the wing and pylon includes 37,216 triangular elements. A local remeshing strategy is adopted in this calculation. The red frame in Figure 10.10 is the boundary of local remeshing region, including the whole wing. While generating the mesh, the mesh inside the remeshing region is finer than that the outer part; when the quality of mesh

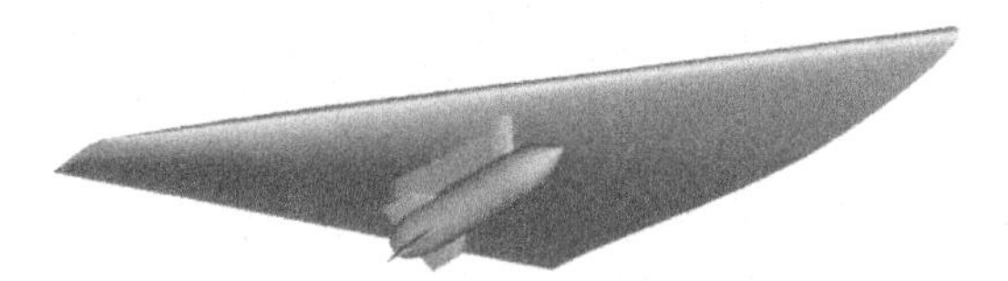

Figure 10.8 Store-dropped model under the wing.

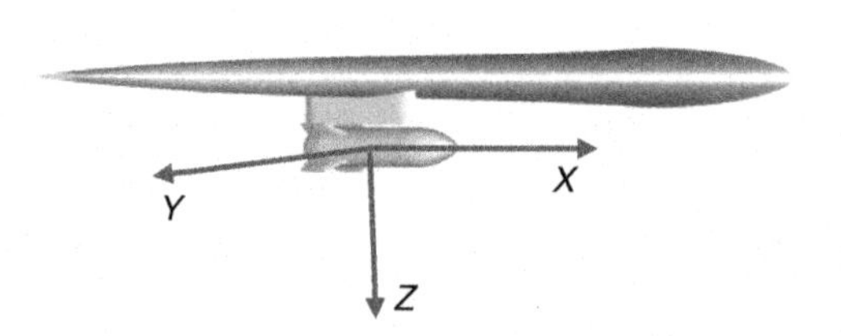

Figure 10.9 Coordinate definition.

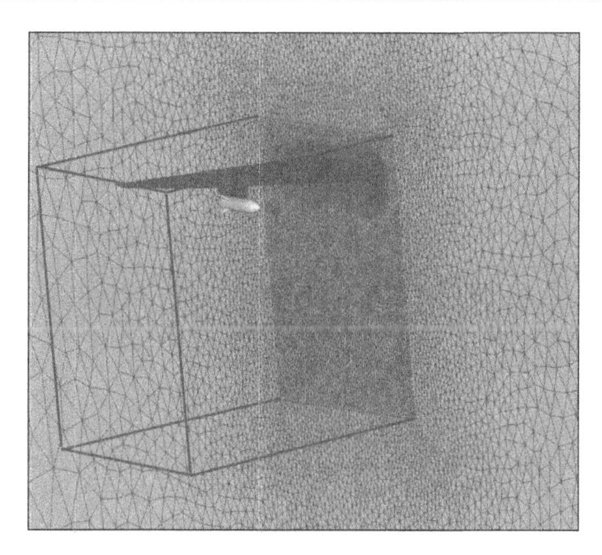

Figure 10.10 Surface grid of the wing and the pylon.

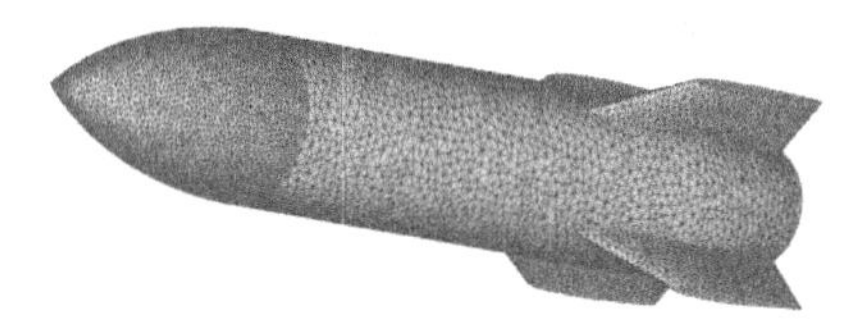

Figure 10.11 Store surface grid.

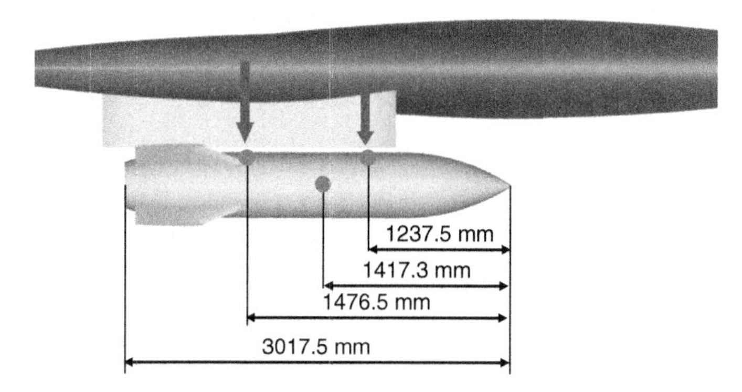

Figure 10.12 Position of the ejection force.

drops below the critical level due to the motion of the store, the meshes in this remeshing region need to be regenerated. Wall grid of the store is displayed in Figure 10.11, including 20,250 triangular elements. The grid of the whole solving area contains about 1,800,000 tetrahedron elements and about 1,200,000 tetrahedron elements in the remeshing region.

Calculation parameters in this example: Ma = 1.2; angle of attack is 0; separation height is 11,600 m; incoming flow density is 0.33217 kg/m^3; and incoming flow pressure is 20,657 Pa. Before separation, the width between external store and pylons is 3.66 cm. Because the initial velocity of the store is 0, the ejection force should be applied during separation. Position and value of the ejection force on the store is presented in Figure 10.12 and Table 10.2. The ejection force remains constant during calculation and disappears after 0.054 s. Parameters of the store are shown in Table 10.2.

A four-CPU workstation is adopted in the parallel calculation. Each CPU is an eight-core Intel Xeon X7550 with a basic frequency of 2.0 GHz and a memory of 64 GB. Grid partition software package METIS is used to generate grid partitioning automatically. Two layers of the grid are overlapped between the partitions for data transmission.

Computational efficiency under the condition of different numbers of processors is shown in Table 10.3. Computational efficiency is effectively improved by parallel computing, and parallel efficiency is 82.5% in 16 processors and 74.4% in 32 processors, which indicates that the parallel algorithm employed here has good parallel efficiency. However, with the number of processors rising, the number of meshes in one partition got reduced accordingly, and the ratio of overlapped meshes needed recalculation is

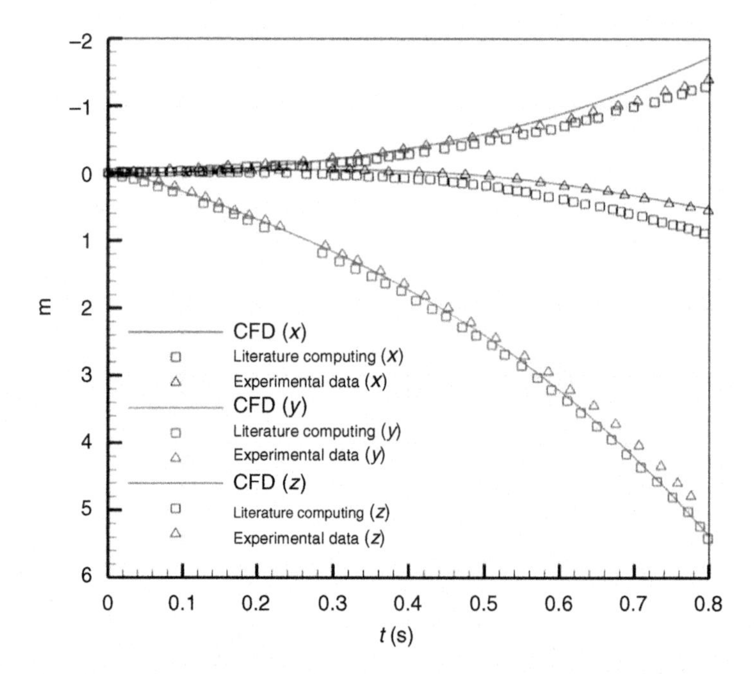

Figure 10.13 Centroid position curve of the store.

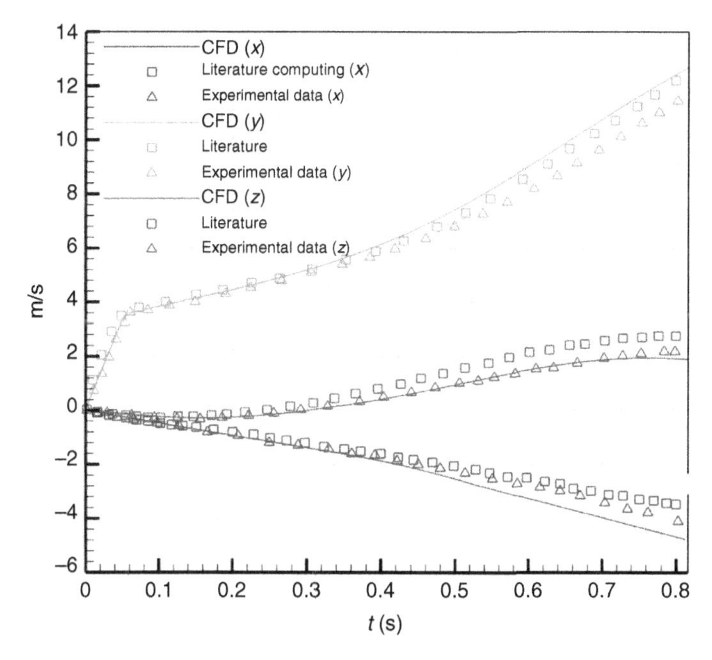

Figure 10.14 Velocity curve of the store centroid position.

Table 10.2 Parameter of store and ejection force

Parameter	Unit	Value
Mass	Mg	907
Moment of inertia I_{xx}	kg · m^2	27
Moment of inertia I_{yy}, I_{zz}	kg · m^2	488
Centroid position in X direction	mm	1417.3 (to the front apex of store)
Forward ejection position in tX direction	mm	1237.5 (to the front apex of store)
Value of the forward ejection force	kN	10.7
Backward ejection position in X direction	mm	1746.5 (to the front apex of store)
Value of the backward ejection force	kN	42.7
Action time	s	0.054

Table 10.3 Comparison of the parallel calculation efficiency

Number of processors	Average grid elements in each course (×1000)	Time of 100 steps in explicit calculation (s)	Speed-up ratio	Parallel efficiency
1	1800	1190	1	–
2	910	600	1.98	99%
4	460	310	3.8	95%
8	230	167	7.1	88.8%
16	120	90	13.2	82.5%
32	~60	50	23.8	74.4%

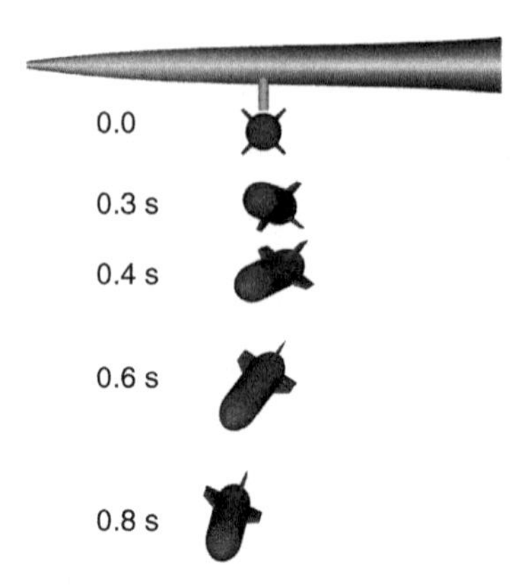

Figure 10.15 Store separation position change.

increased. Meanwhile, the communication traffic across processors rises, so that parallel scalability goes down.

The end time of the calculation is 0.8 s for this example. Time history of the position of centroid of the store is presented in Figure 10.13, and it is shown that the result in Z direction conforms well to that of the experiment and literature. This is because the effect of ejection force and gravity in this direction, is larger than that of aerodynamic forces, while the separation parameter is little affected by the fluid field. In X and Y directions, calculation results conform well to the results of the experiment and the literature, so precision of the related algorithm and program is verified.

Velocity curve of the store centroid position is presented in Figure 10.14, which shows that the results conform well to those of the experiment and the literature, and high precision of algorithm and program is further confirmed. In addition, the effect of ejection forces in Z direction is shown clearly in the figure. After separation, the velocity of store is swiftly increased under the effect of the ejection force. The ejection force disappears after about 0.05 s. Store velocity rising in Z direction is mainly attributable to the effects of gravity. The separation position of the store is indicated in Figure 10.15. The store position and morphological change after separation is observed more visually. These results shows that while dealing with the simulation of multibody separation, the developed dynamic mesh technique has high precision and reliability, which ensures its application in practical engineering.

10.3.3 AGARD445.6 flutter of aerofoil

Wind tunnel testing on the flutter property of AGARD445.6 aerofoil has been completed in the transonic dynamic tunnel by the NASA Langley Research Center (Yates, 1988). Serial model from this experiment is the standard type of AGARD445.6, which is usually used for the property inspection of transonic aerofoil flutter. The AGARD445.6 standard model consists of five groups in all, including four groups of strong models and one group of weak model. At present, the AGARD445.6 weak aerofoil model is usually taken as a standard example for computational program testing of

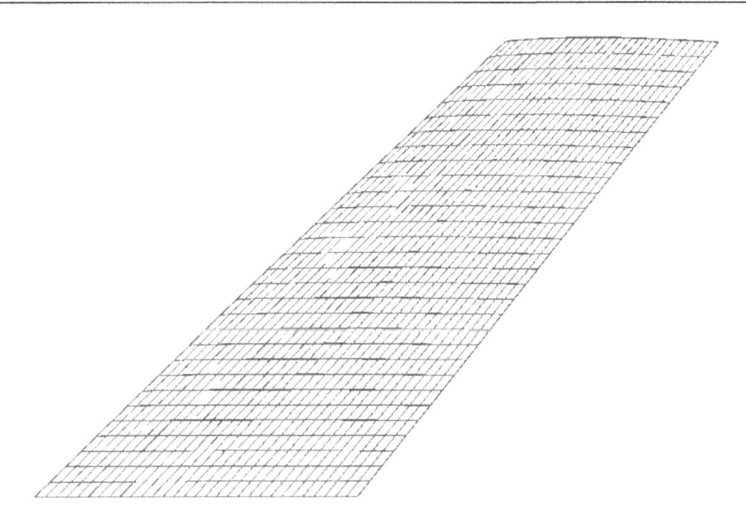

Figure 10.16 Wing surface mesh.

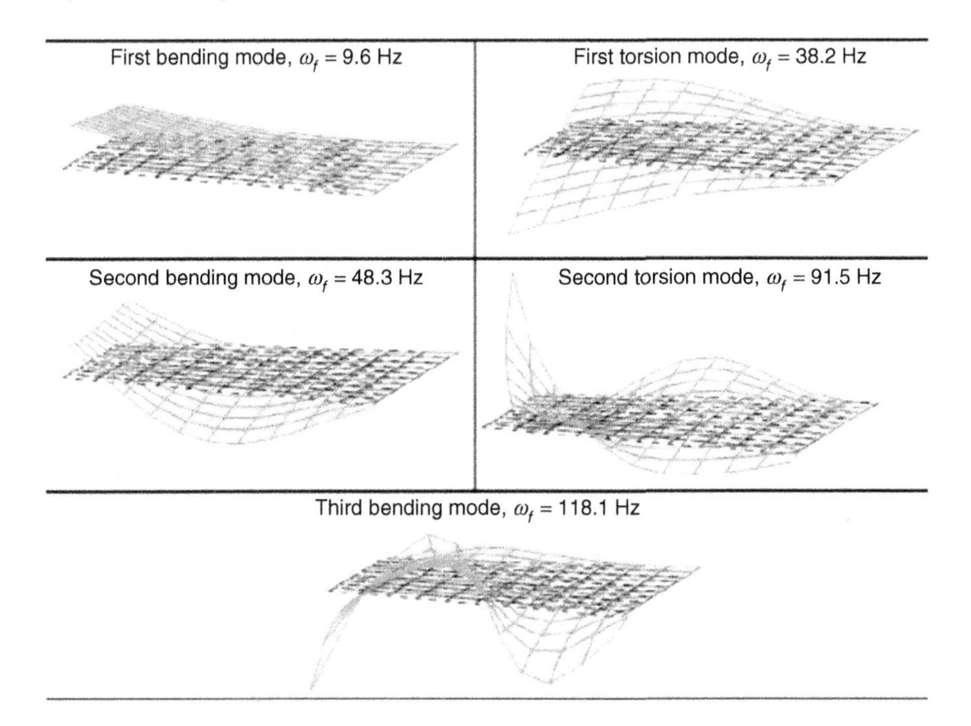

Figure 10.17 AGARD445.6 wing vibration mode.

the transonic flutter property. The wing type of the AGARD445.6 weak wing model is NACA64A004. The geometry configuration and dimension of the wing are shown in Figure 10.16, the chord length of the wing's end is 21.96 in., wingspan is 30 in., and the sweep angle of the leading edge is 45°. Vibration mode of each stage is shown in Figure 10.17.

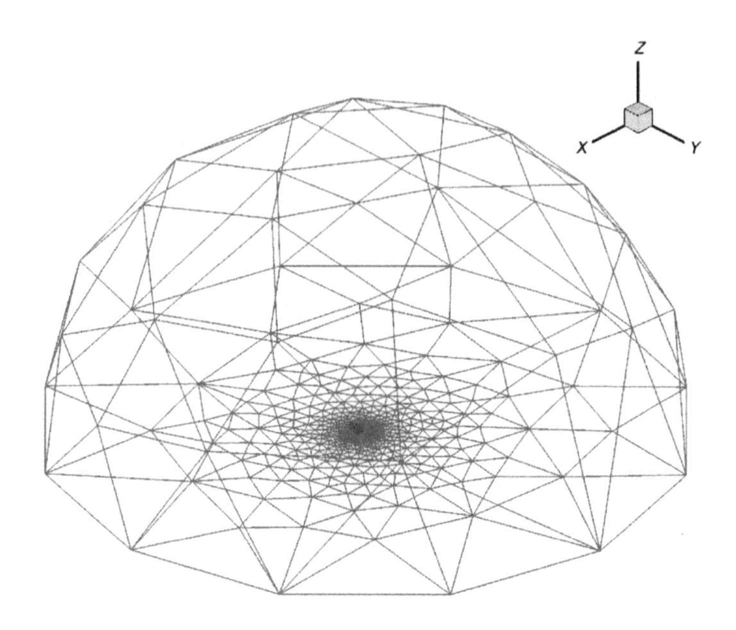

Figure 10.18 Calculative region.

For the case of transonic velocity calculation, to reduce the boundary effect, the outer boundary of a calculative region is usually specified farther. Computational region of this example is shown in Figures 10.18 and 10.19, the radius of the calculative region is 600 in.. Quadrilateral meshes are used in the wing surface, and the number of grids is $31 \times 31 \times 2$. Triangle meshes are adopted in the symmetry plane and outer boundary. Pyramid-type volume mesh is first grown on the surface of the wing, and then tetrahedral meshes are formed in the whole calculative region. Calculative mesh has 36,651 solid elements, 7,900 mesh nodes, and 3,818 boundary surfaces.

A total of seven different Mach numbers are calculated namely, 0.678, 0.900, 0.950, 0.954, 0.960, 0.980, and 1.072, respectively. First, steady fluid field under each Mach number is calculated, as shown in Figure 10.20. Because NACA64A004 is one type of symmetry wing and the angle of attack is zero, the flow situations adjacent to the upper and lower wing surfaces are identical. Steady fluid field calculation is first conducted, and then, on this basis, the unsteady flutter calculation starts. The initial conditions are as: generalized acceleration and displacement of all modes, generalized velocity of the first and second mode, and generalized velocity of other three modes at the initial stage are all zero.

The simulation procedure is as follows: first, the flutter dynamic pressure obtained from the test is taken as a reference value; then the calculation of the rough range of flutter dynamic pressure is carried out for each Mach number; then the

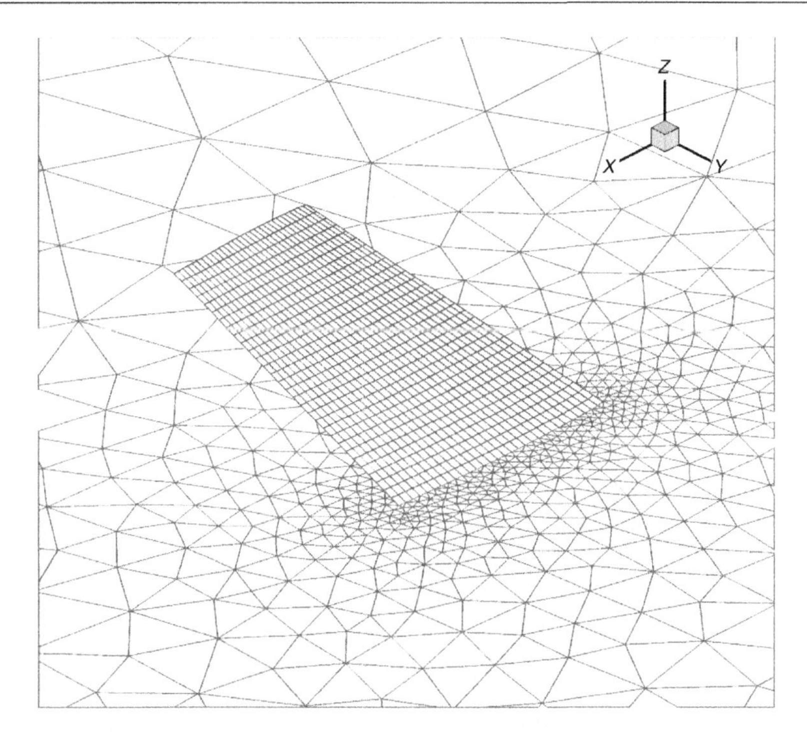

Figure 10.19 AGARD445.6 calculative mesh.

range is reduced gradually, and the flutter dynamic pressure is finally determined. The numbers of calculation samples for different Mach numbers are not equal, but the total is 40. The partial computational results under each Mach number are presented in Figures 10.21–10.27. Horizontal coordinate indicates time, and longitudinal coordinate indicates the generalized displacement in each mode. When Ma = 0.96

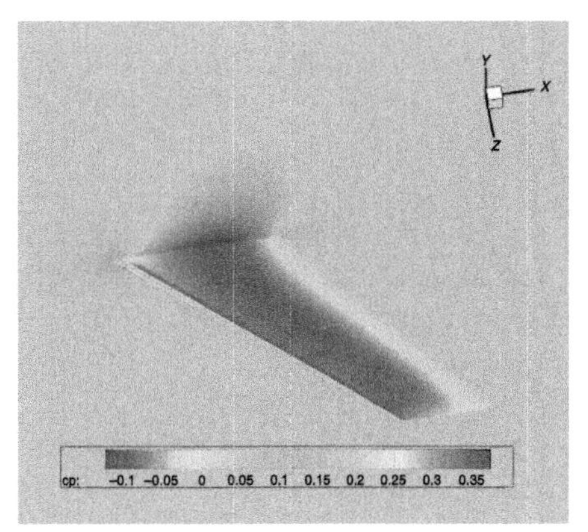

Ma = 0.678

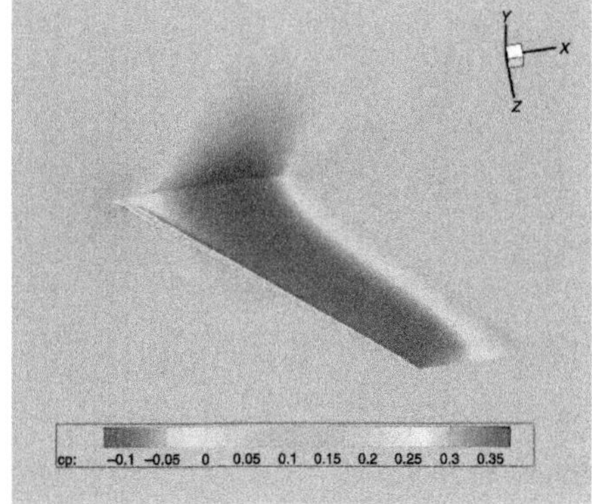

Ma = 0.900

Figure 10.20 Steady flow.

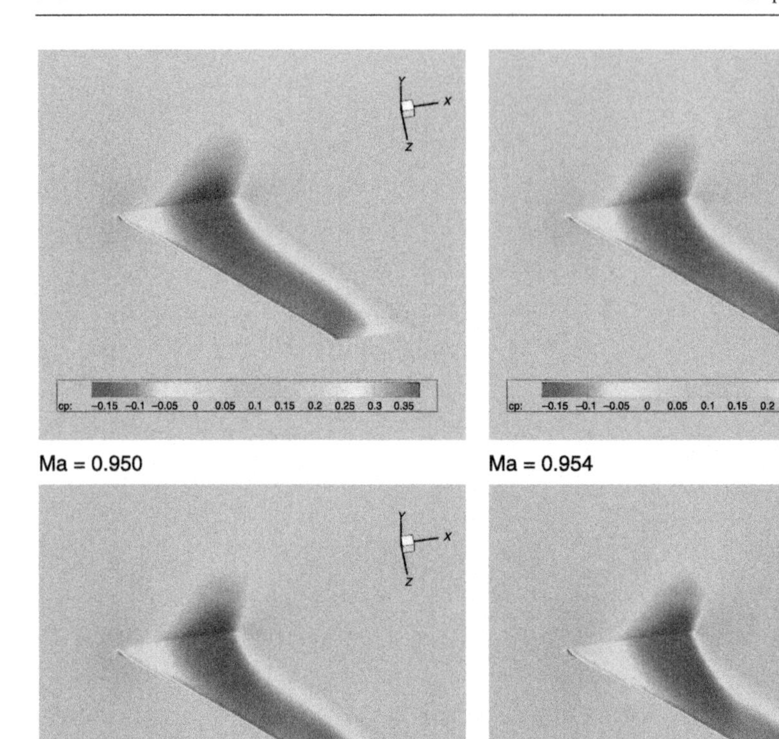

Ma = 0.950

Ma = 0.954

Ma = 0.960

Ma = 0.980

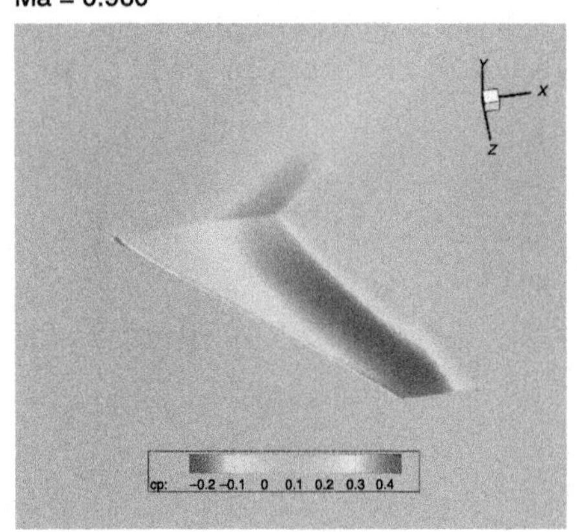

Ma = 1.072

Figure 10.20 *(cont.)*

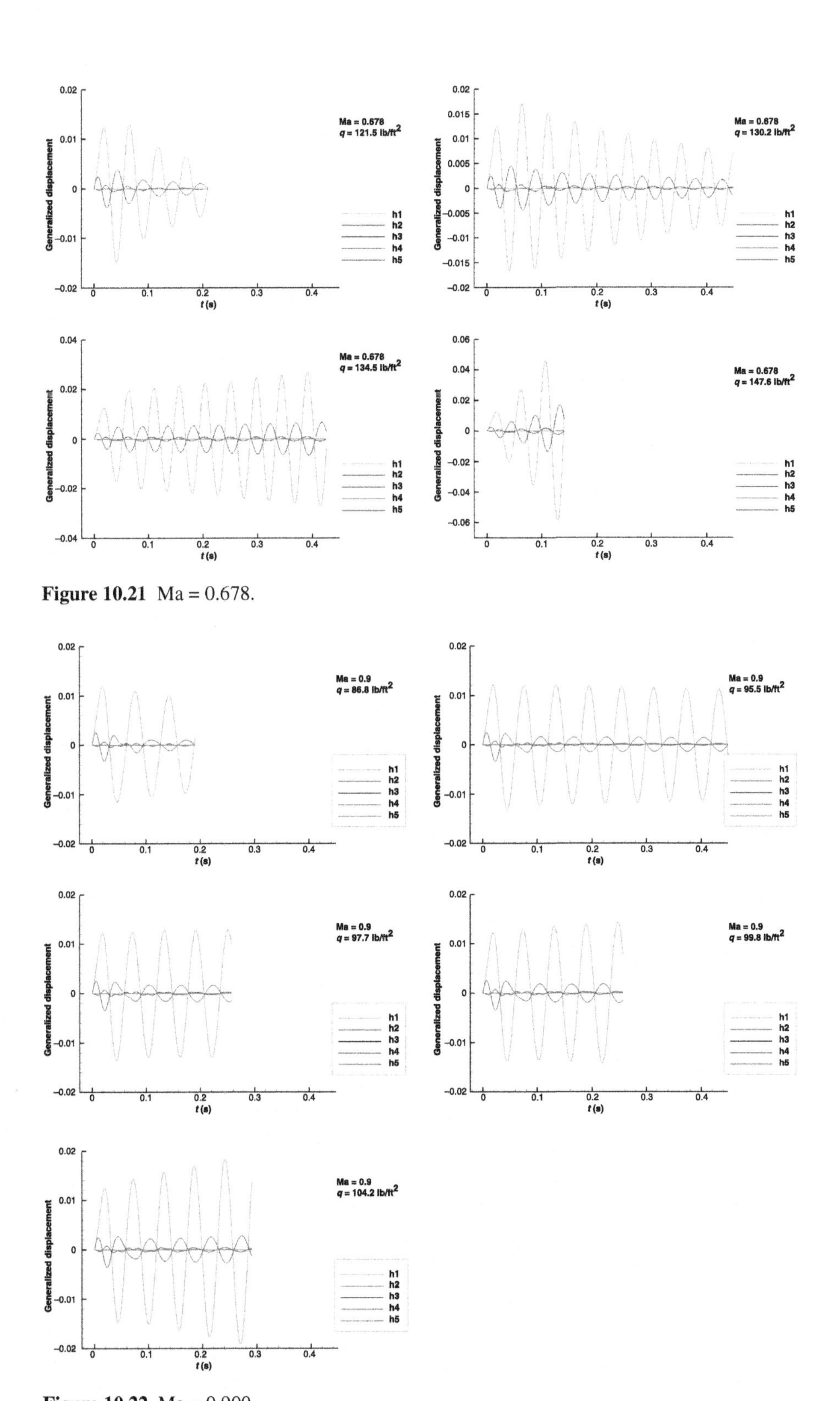

Figure 10.21 Ma = 0.678.

Figure 10.22 Ma = 0.900.

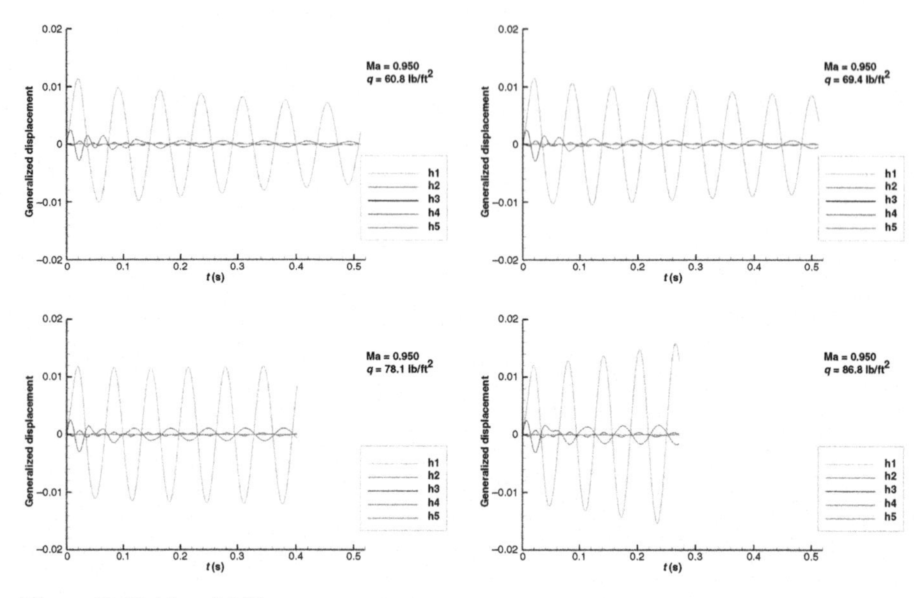

Figure 10.23 Ma = 0.950.

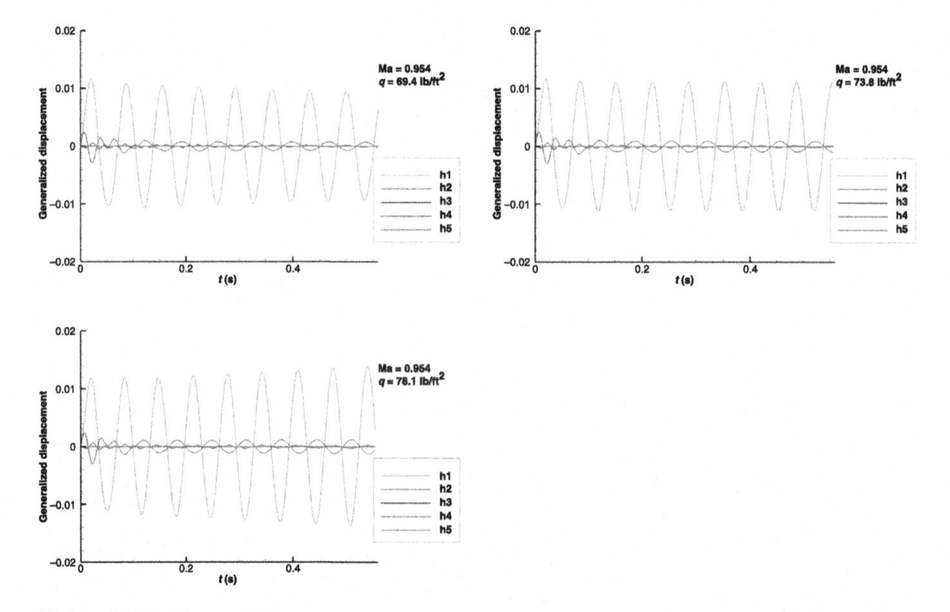

Figure 10.24 Ma = 0.954.

and q = 78.1 lb/ft.^{2}, the wing deformation diagram at different times, is shown in Figure 10.28. Because actual deformation in the wing is small, the wing deformation gets amplified 20 times over.

After calculating flutter pressure under each Mach number, the flutter pressure is compared with the test value. It is observed from Figure 10.29 that when the Mach

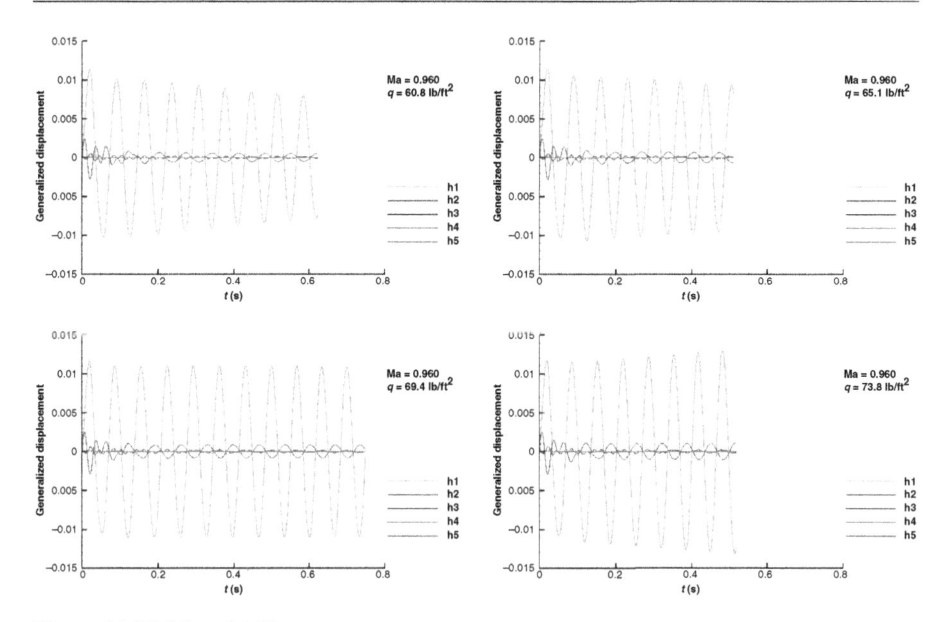

Figure 10.25 Ma = 0.960.

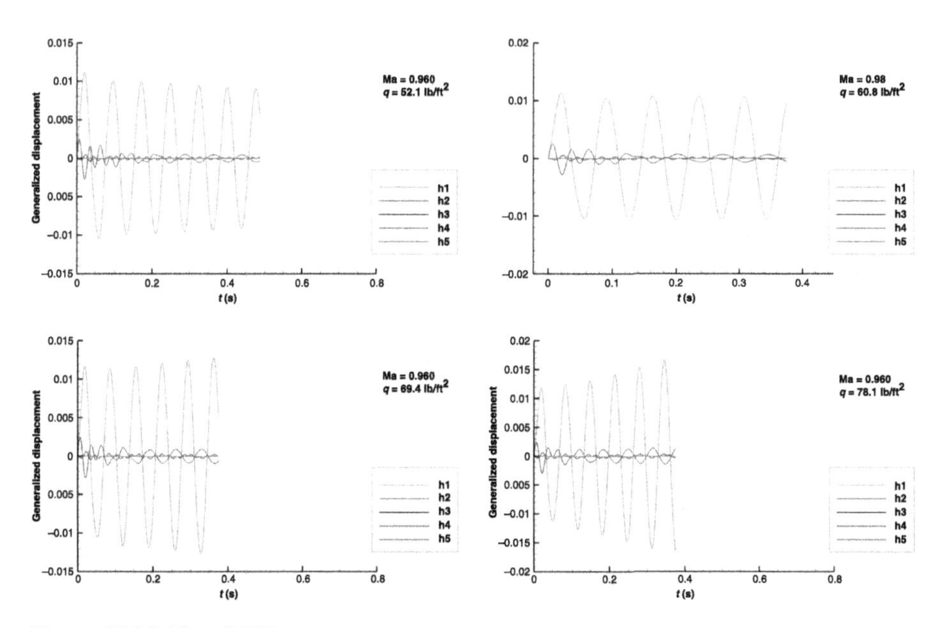

Figure 10.26 Ma = 0.980.

number is less than 0.98, the flutter pressure is in inverse proportion with the Mach number. And when the Mach number exceeds 0.98, the flutter pressure becomes larger as the Mach number rises. Calculative results show that as Ma = 0.98, flutter pressure becomes minimum. But as Ma = 0.954, flutter pressure becomes minimum in the experiment, and results of the calculation and experiment have a slight difference.

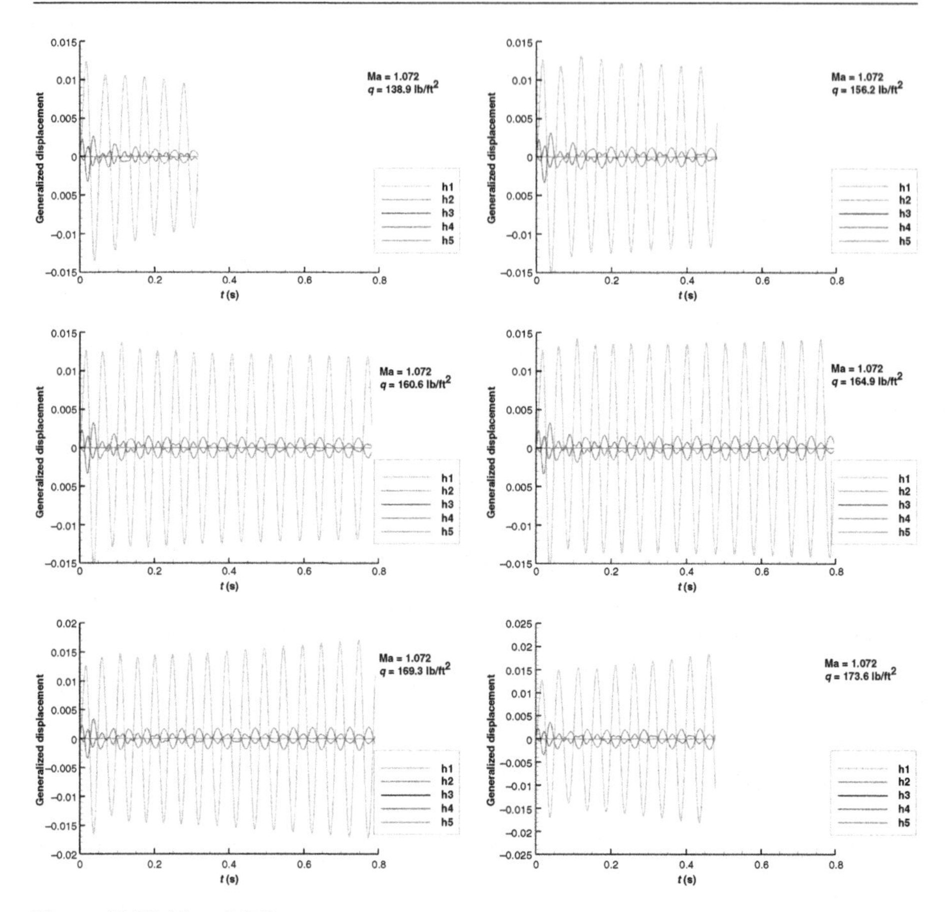

Figure 10.27 Ma = 1.072.

However, the value of the minimum flutter pressure 60.76 lb/ft.^{2} is close to experimental value 60.6 lb/ft.^{2}

The mesh discretization in this example is relatively coarse, with the viscous effect not taken into consideration. When the viscosity effect exists, the position of the shock wave is changed by the boundary layer, which leads to changes in the aerodynamic force and torque of the wing compared with the nonviscous condition. From Figure 10.29, the flutter dynamic pressure error is larger under supersonic conditions. The results show that although viscosity is not taken into consideration, the calculation result of the minimum flutter dynamic pressure is consistent with that of experimental result, and the "transonic dimple phenomenon" is reproduced (Chen et al., 2010). Therefore, the effectiveness of coupling calculation method adopted in this case is verified.

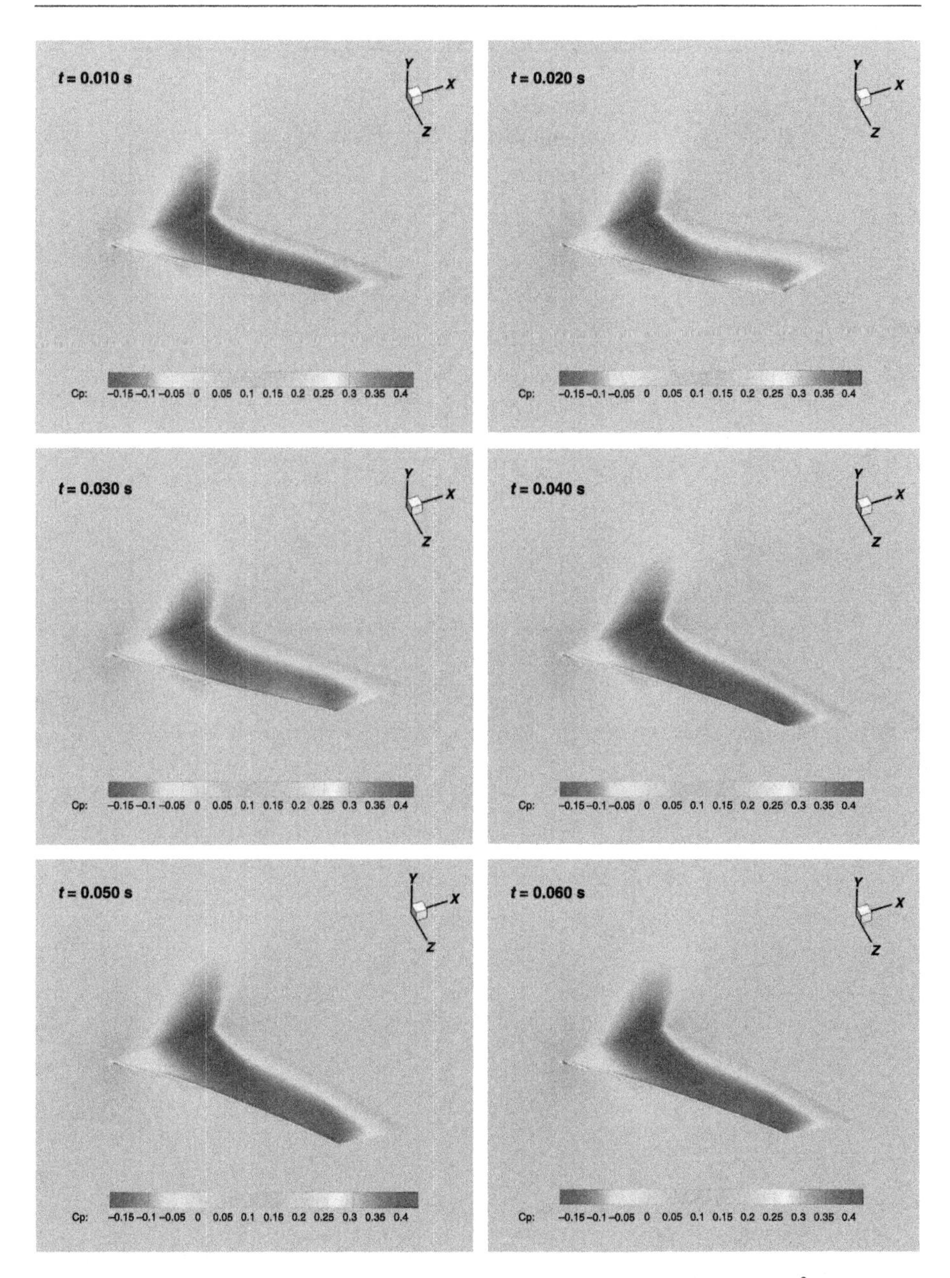

Figure 10.28 Wing deformation at different time steps. Ma = 0.96; q = 78.1 lb/ft.^{2}. The displacement gets amplified 20 times over.

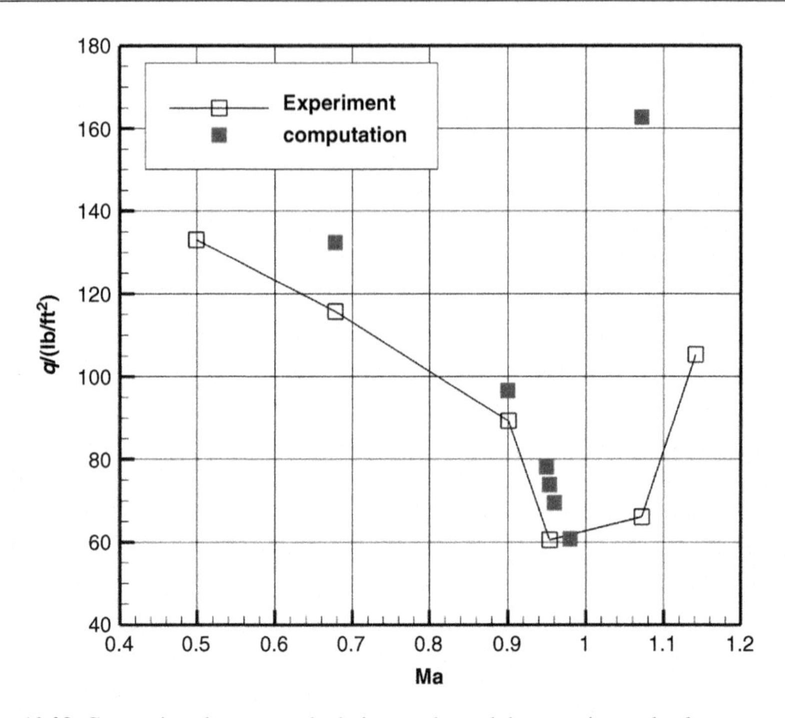

Figure 10.29 Comparison between calculative results and the experimental value.

Multiphysics simulation of microelectro-mechanical systems devices

Chapter Outline

11.1 Introduction to MEMS

A microelectro-mechanical system (MEMS) is a system in which microcircuits and micromachinery are integrated on a chip by function, and its scale is usually within millimeters or even micrometers. Due to the interdisciplinary features, MEMS devices are typical systems, in which multiphysics simulations are used. A related study may cover many branches of physics, including mechanics, electrics, optics, magnetism, and so on. The coupling between several scientific fields that takes place in MEMS is unavoidable and causes a main challenge in numerical simulation.

11.2 Micropump

Microfluidics has emerged from MEMS technology as an important research field and a promising market. The micropump is one of the most important microfluidic component. This case presents one kind of mechanical micropump, which utilizes a piezoelectric patch bonded to a silicon membrane for driving the fluid in the fluid chamber.

11.2.1 Physical model

This device consists of a chamber filled with fluid and a flexible membrane on the top, which could deform when stimulated by piezoelectric layer. The dimension of the

Q. Zhang & S. Cen: Multiphysics Modeling. http://dx.doi.org/10.1016/B978-0-12-407709-6.00011-0

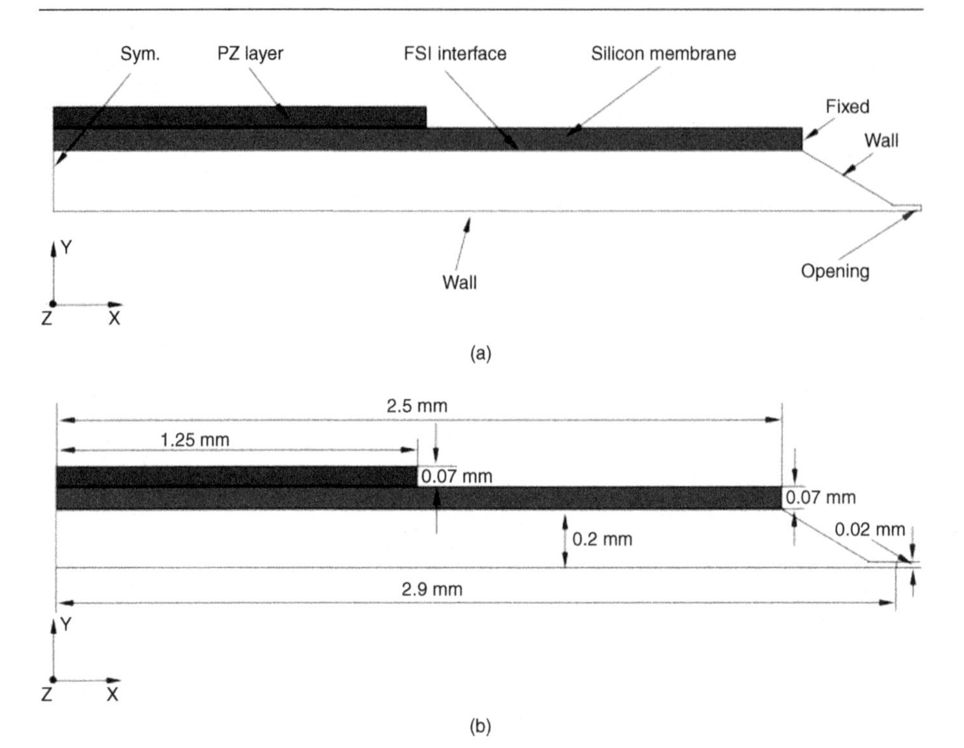

Figure 11.1 Model of the micropump. (a) Structure and (b) dimensions.

device is given in Figure 11.1, and the thickness in *Z* direction (perpendicular to the paper) is 0.2 mm. For estimating the damping and inertia force, a simplified loading history is considered. Obviously, this is a typical multiphysics strong coupling problem including piezoelectric field, solid field, and fluid field, which need to be solved at the same time.

The flexible membrane is made of silicon, and the piezoelectric patch is PZT4. Their basic material properties used in simulation, are listed in Table 11.1. The polar axis of the piezoelectric patch is along the thickness direction (*Y* axis), and the corresponding dielectric permittivity is 659.7, which in other directions are both 804.6. The piezoelectric stress constant matrix is as follows:

$$[e]=\begin{bmatrix} 0 & -4.1 & 0 \\ 0 & 14.1 & 0 \\ 0 & -4.1 & 0 \\ 10.5 & 0 & 0 \\ 0 & 0 & 10.5 \\ 0 & 0 & 0 \end{bmatrix}\left(\mathrm{C/m^2}\right) \tag{11.1}$$

Table 11.1 **Material properties used in simulation**

Material properties	Silicon	PZT4
Mass density, ρ (kg/m^3)	2329	7500
Young's modulus, E (Pa)	1.689e11	1.0e11
Poisson's ratio, v	0.3	0.3

The mesh discretization is demonstrated in Figure 11.2, and the total number of elements is 640.

11.2.2 Analysis conditions

INTESIM supplies a specialized element type, accounting for the piezoelectric field analysis, which works as a combination of structure element and an electrostatic element and further expansion. The degrees of freedom (DOFs) between piezoelectric pad and silicon membrane are coupled directly because they share the common nodes. And the coupled DOFs are displacements. However, the nodes of silicon membrane and fluid chamber do not match with each other on the fluid–solid coupling interface. Therefore, a coupling method based on constraint equations is used. The coupling type is mechanical coupling. So the piezoelectricity–solid–fluid strong coupling field analysis can be achieved. And this approach is the most efficient method for this problem. A constant voltage is applied to the piezoelectric patch during the whole analysis process, with the topside −0.175 V and the bottom side 0 V. The fluid domain adopts a laminar model for calculation. Analysis type is set to transient analysis, and the time integration method is the PMA scheme. The total analysis time is set to 5e-4s, and the time step size is 1e-5s.

11.2.3 Results and discussions

The results of the piezoelectricity–solid–fluid fields can be checked separately when the solution is done. The results are presented at $t = 3.97\text{e-}4$s. Figure 11.3 gives the distribution of voltage showing a result as expected, the voltage increasing from −0.175 V at one side to 0 V at the other side, gradually. Figure 11.4 presents the deformation of the silicon membrane and the piezoelectric patch and the mesh morphing of the fluid field. It is noted that the color transition from the solid zone to the fluid zone is smooth, indicating that the movement of the fluid mesh follows the solid deformation closely on the fluid–solid coupling interface, which is the very reflection of the strong coupling analysis that has been carried out. The maximum deformation occurs

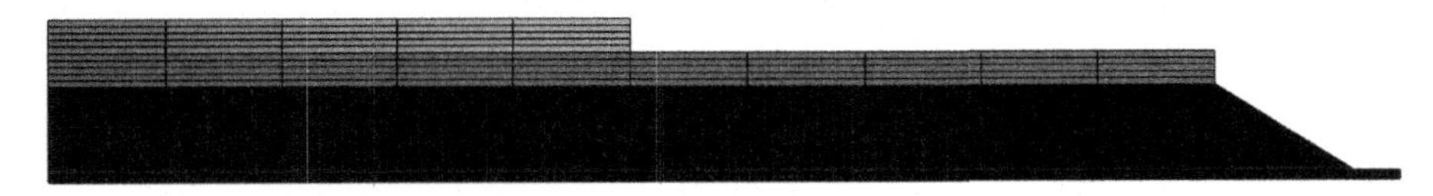

Figure 11.2 Mesh discretization of 3D model.

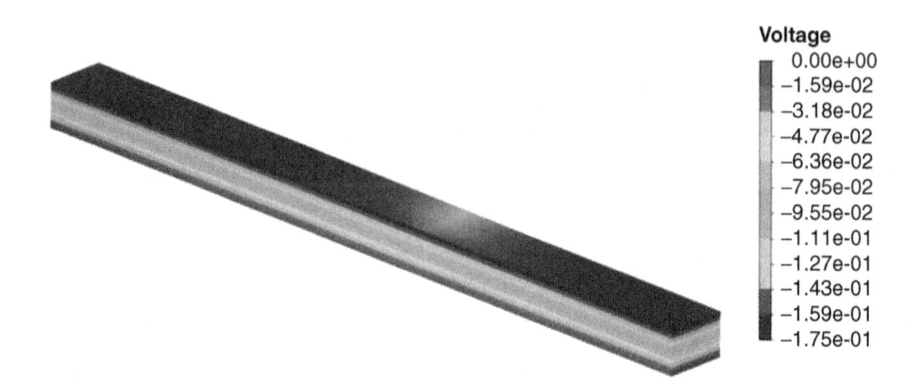

Figure 11.3 Distribution of voltage (unit: V).

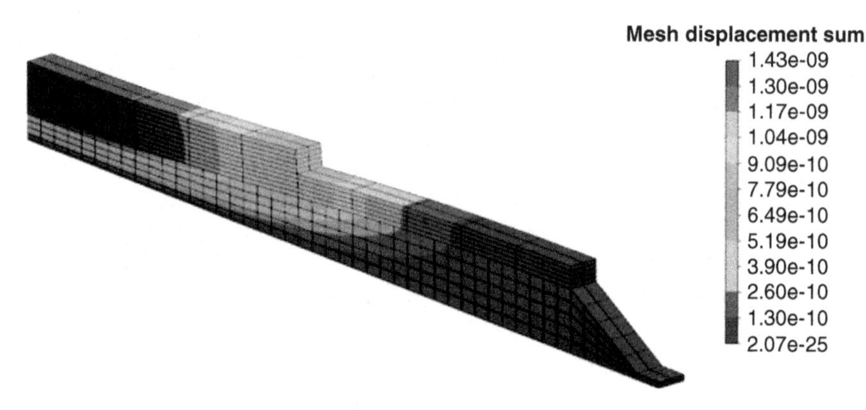

Figure 11.4 Contour plot of displacement of solid field and mesh morphing of fluid field (unit: m).

at the silicon membrane tip with a value of 1.43e-9m. The piezoelectric layer bends toward the silicon membrane due to the applied voltage load. Meanwhile, the silicon membrane is forced to deliver a pressure to the fluid chamber, driving the fluid to flow out the outlet.

Figure 11.5 is the time history of displacement of one node at the silicon tip during the whole analysis. It is noted that there is a big fluctuation at the beginning, which is caused by the stimulation of the piezoelectric patch. Then with the effect of fluid damping, displacement gradually levels off.

As presented in Figure 11.6, maximum pressure appears at the center part of the chamber, and the maximum displacement of the silicon membrane also takes place. The vector diagram of the fluid velocity can be seen in Figure 11.7, which shows that the velocity in the chamber is small and uniform, but it increases sharply when coming to the outlet. The results mentioned earlier basically agree with the actual physical phenomena.

So we can see that INTESIM software is applicable to the studies of numerical simulation of micropumps, in a fully coupled way.

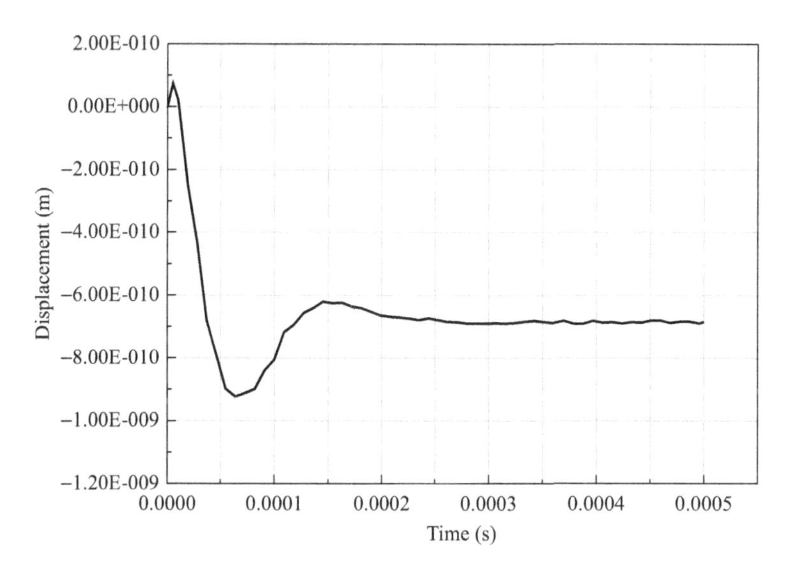

Figure 11.5 Diagram of displacement history (unit: m).

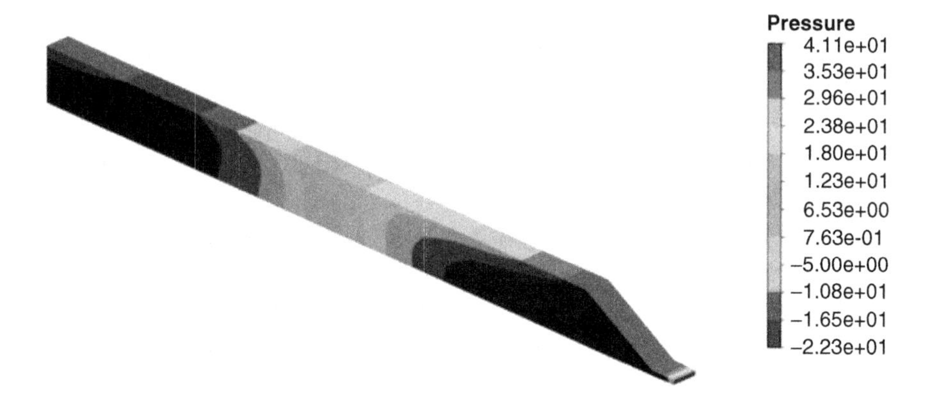

Figure 11.6 Contour plot of fluid pressure (unit: Pa).

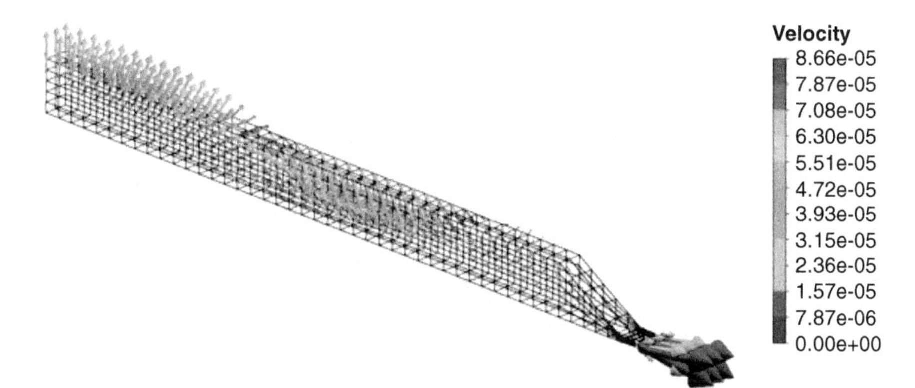

Figure 11.7 Vector diagram of fluid velocity (unit: m/s).

11.3 Natural convection cooling of a microelectronic chip

The simulation of natural convection that happens in a microelectronic chip, is featured by its microscale and multiphysics problem. The capacity of INTESIM for solving this problem is demonstrated in this case.

11.3.1 Physical model

This example presents a device composed by a coil, a metal plate, and the air around them, as shown in Figure 11.8. The coil is heated by voltage loads applied on the ends, the surfaces of which act as hot walls in the air box during convective heat transfer to the surroundings. Meanwhile, the initial thermal field of the plate will also be changed under the influence of natural convection. All the walls of the air box are made of a heat-insulated material. The detailed dimensions are shown in Figure 11.9. Gravity acts downward and the objective is to compute the flow and temperature distribution in the box, as well as the temperature distribution of the plate.

The coil and plate are all made of steel. The air is assumed to be an ideal gas. Basic material properties used in the simulation are listed in Table 11.2.

Mesh discretization with an all-hexahedral element is demonstrated in Figure 11.10, and the total number of elements is 17,279.

11.3.2 Analysis conditions

The problem itself is multiphysics, involving three different types of physical fields namely: electric field, thermal field, and fluid field. Moreover, interface coupling methods are used differently for solving the domain of each physical field either overlapped in space or only surface interacted. The scheme that combines weak coupling

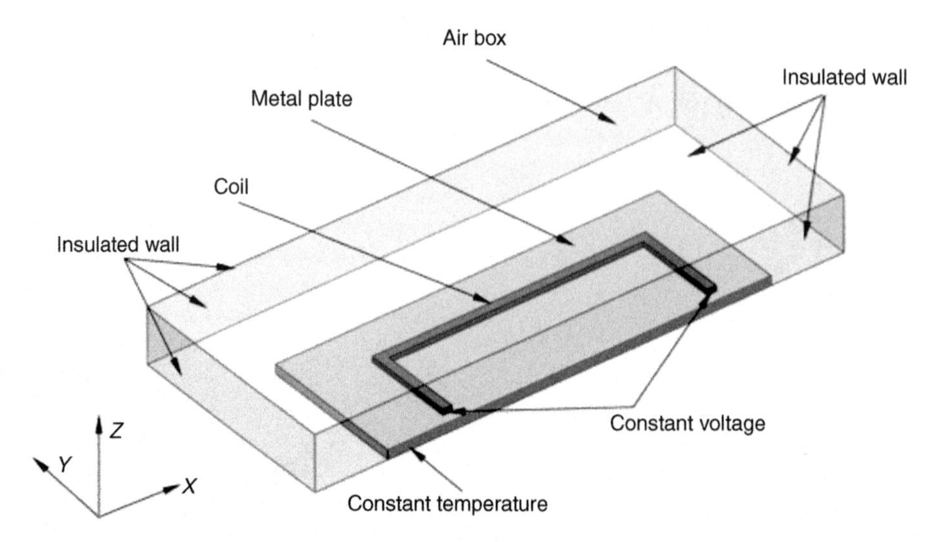

Figure 11.8 Internal constitution and boundary conditions of the device.

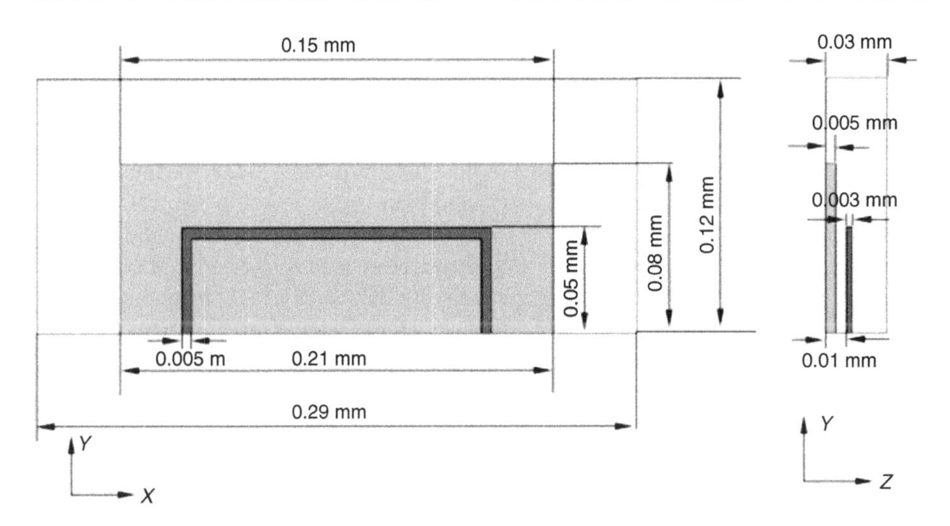

Figure 11.9 Dimensions of the device.

Table 11.2 **Material properties used in simulation**

Material properties	Steel	Air
Mass density, ρ (kg/m^3)	7800	1.21
Specific heat, C (J/kg·K)	400	1005
Thermal conductivity, k (W/(m K))	50.2	0.0257
Resistivity, r (Ω m)	1.7e-7	–
Dynamic viscosity, μ (Pa s)	–	1.511e-5

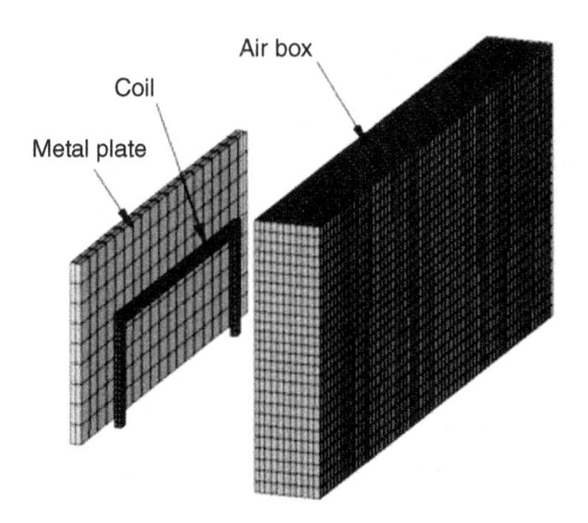

Figure 11.10 Mesh discretization of 3D model.

and strong coupling methods could be more efficient. Three solving domains of the physical field, D1, D2, and D3, are created. D1 comprises of a coil grid and is used to solve the heat generation problem of electric fields. D2 consists of the coil, air, and metal plate mesh, and it is used to solve the problem of pure thermal field. D3 comprises of air grid, and it is used to solve the flow field problems. In D2, the temperature DOFs on the interfaces of coil, air, and metal plate are coupled by the MPC-based strong coupled method. D2 accepts heat generation from D1, to complete the calculation of temperature field, and the resulting temperature load is passed into D3 for flow field calculation. D3 sends the calculated velocity data back to D2, repeating it until an iteration convergence (Figure 11.11). Stable voltages are applied on both the ends of the coils, 0.03 and 0 V, respectively. At the bottom surface of the metal plate, a constant temperature is maintained at 22°C. The initial temperature of the entire computational domain is set to be 22°C. Laminar flow model is employed in flow field analysis, and the analysis type is the steady state.

11.3.3 Results and discussions

The simulation result of each physics model can be obtained after the iteration convergence. Figure 11.12 shows the heat generation rate in the coils at a given voltage in the electric field analysis. Figure 11.13 shows the temperature distribution in the medium of air. The temperature near the coil is high and forms a disperse-state distribution to the surrounding, which is under the interaction of heat conduction and convection effect.

Velocity distribution of the airfield is shown in Figure 11.14. It is clear that in the flow field region, there are two large vortices, which are formed on the left and right

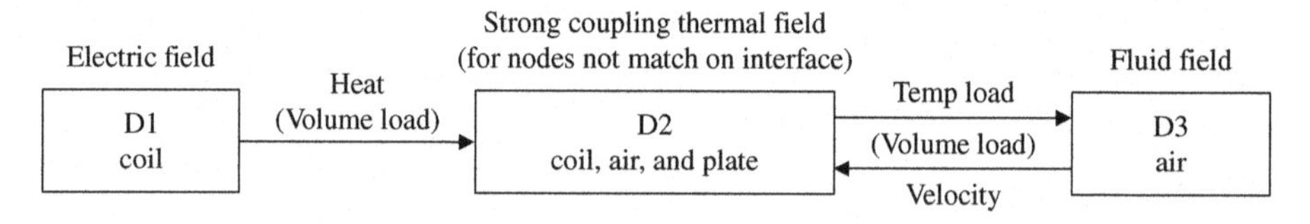

Figure 11.11 The data flow between multisolvers.

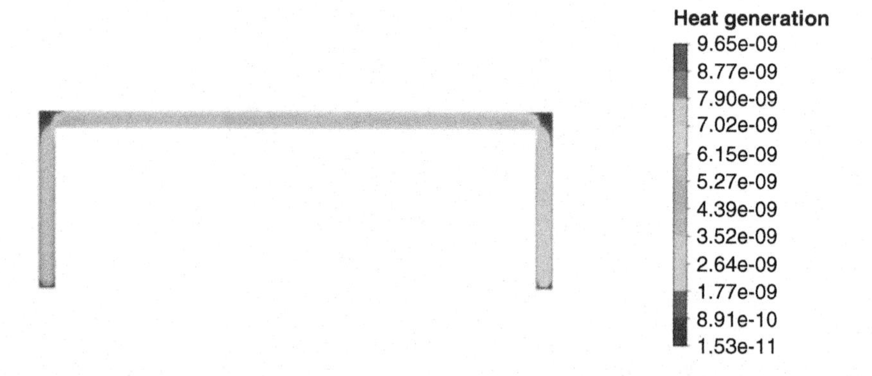

Figure 11.12 Contour plot of heat generation of the coil (unit: W).

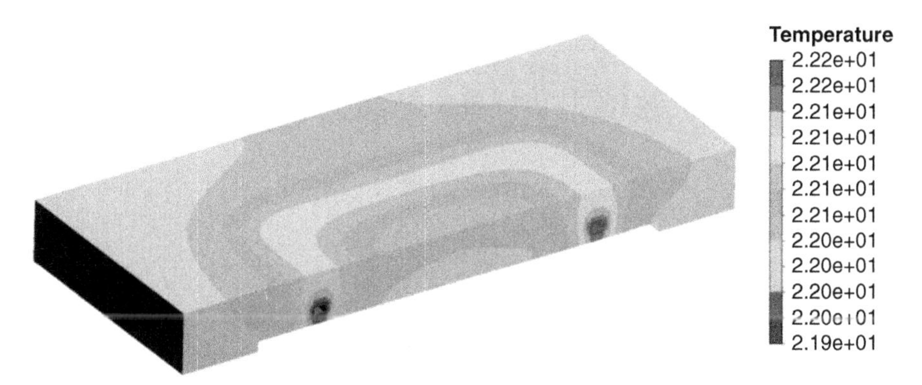

Figure 11.13 Contour plot of the air temperature (unit: °C).

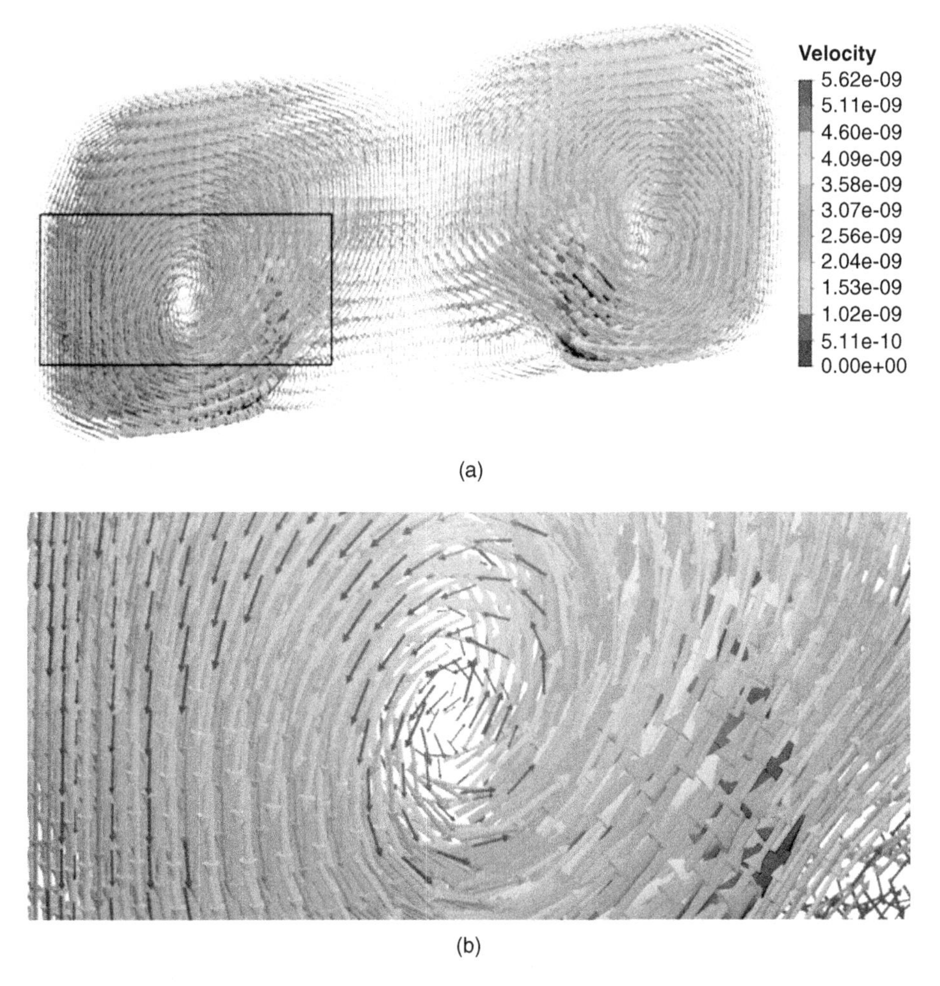

Figure 11.14 Vector diagram of the fluid velocity (unit: m/s). (a) Global view and (b) partial enlarged view of the box zone marked in view (a).

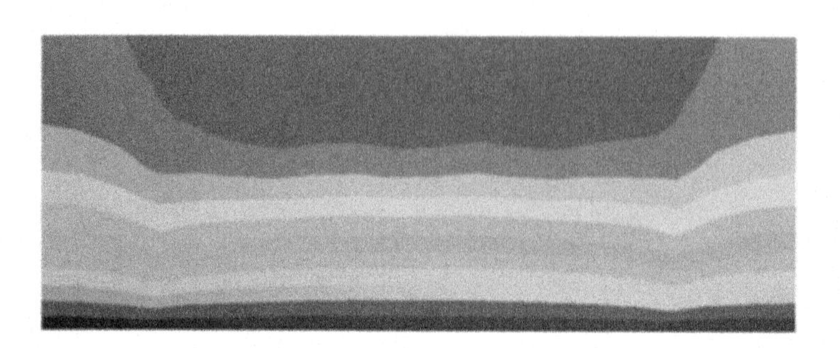

Figure 11.15 Distribution of the plate temperature (unit: °C).

sides. The reason is that, increasing temperature causes a reduction of gas density near the coil, and under the action of buoyancy, gas in this part rises upward. An airflow is formed and ultimately, a large vortex is formed in each half of the entire flow field. Figure 11.15 shows the temperature distribution of the metal plate because the bottom is applied with temperature constraint, which maintains at the initial temperature, and the temperature of the upper portion increases slightly by the fluid-heating effect.

This analysis shows that the simulation of natural convection results in reasonable calculation of the microelectronic chips. The key technologies such as the strong coupling between interfaces of adjacent physical fields are employed, so as the body load bidirectional transfer between different physical fields, which shows INTESIM's efficient calculation and analysis capabilities in electrical, thermal, and flow multiphysics coupling simulation under microscale.

Bidirectional multiphysics simulation of turbine machinery

12

Chapter Outline

12.1 The fluid–structure–thermal bidirectional coupling analysis on the rotor system of turbo expander

The turbo expander is the key device for air separation and liquefied natural gas separation, which are frequently encountered in metallurgical, petrochemical, thermal power, and other important industrial fields. With a huge demand and significant development in these fields, the development of turbo expanders gets confronted with the challenge of ultra-large-scale and low energy consumption. And higher technical requirements are put forward on the aspects of expander design and manufacture, reliability analysis, failure diagnosis, and nonlinear dynamics analysis. Among these aspects, the fluid–structure–thermal coupling analysis for rotor systems of turbo expanders is a key comprehensive project for a technical breakthrough. Under complicated working conditions, rotor system of the turbo expander endures aerodynamic force of flow field, centrifugal force, and thermal loads. In addition, factors such as oil film force, sealing force, and impact force of droplets are also need to be considered in certain circumstances. To design the turbo expander of large and ultra scale, it is essential to analyze the working condition and stress level of a rotor system. Multiphysics simulation for the impeller of turbo expander would be needed to promote the development and design optimization of a large-scale expander. The statistical data

Q. Zhang & S. Cen: Multiphysics Modeling. http://dx.doi.org/10.1016/B978-0-12-407709-6.00012-2

also shows that more than 65% of the blade failure in turbo machinery is attributable to the mechanical failure, which indicates that coupling stress at the impeller may have a direct effect on failure of the blade. Therefore, research on fluid–solid–thermal coupling could provide some solutions for the failure analysis of turbo expander. In this section, the rotor system of turbo expander is presented as an example to provide some analysis solutions for engineers on the design of turbo expanders.

12.1.1 The physics model

Structure of the rotor system of a typical turbo expander is shown in Figure 12.1. The turbo expander is mainly composed of flow passage, turbocharger, and body. The flow passage is mainly used to get low-temperature gas. The gas is uniformly distributed to the spray, after it enters the volute through the pipeline of the expander. In the spray nozzle, the gas undergoes expansion and transforms its enthalpy to the kinetic energy, part of which is specially used to drive the expansion impeller to output external work. When finishing expansion, gas is discharged through a diffuser into the low-temperature pipeline. The turbocharger is the power-consuming component of the turbo expander, and mechanical work is consumed by compressing the gas by the turbo impeller through spindle six. The brake air is inhaled through the inlet pipe of the end cover and is then compressed by the turbo impeller. After that, it is diffused by the vaneless diffuser and turbocharger volute nine and finally discharged into the

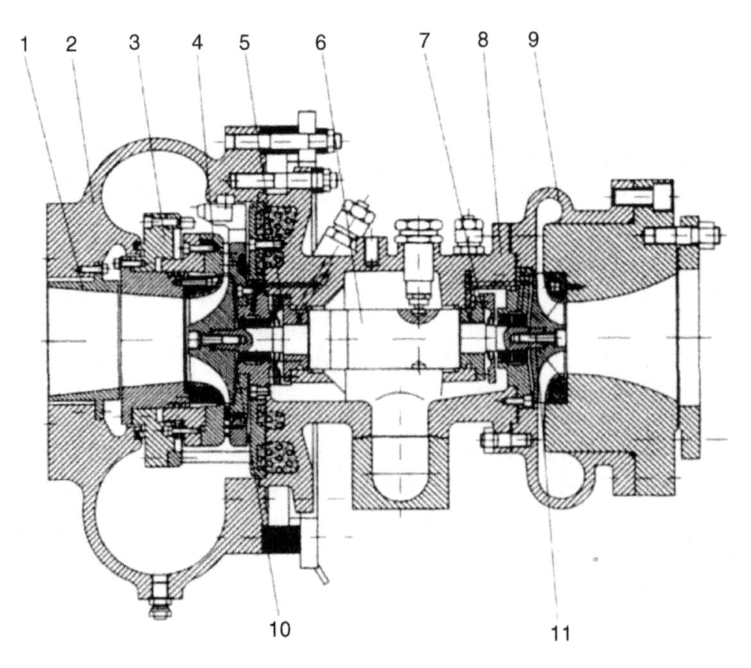

Figure 12.1 Structure of the turbo expander. (1) Diffuser, (2) volute, (3) nozzle-clamping mechanism, (4) spray nozzle, (5) bearing, (6) spindle, (7) bearing, (8) body, (9) turbocharger volute, (10) expansion impeller, and (11) turbo impeller.

outlet pipeline. The rotor system is composed of expansion impeller, turbo impeller, spindle, and other rotating parts.

This section focuses on fluid–structure–thermal coupling analysis for the rotor system of the turbo expander. It is assumed that the influences of oil film force, sealing force, droplet impact force, and other forces on the rotor system are negligible. The structure model, including expansion impeller, shaft, and turbo impeller, of the rotor system, is shown in Figure 12.2. The expansion impeller consists of 13 blades, and the turbo impeller consists of 17 blades. The expansion impeller and the turbo impeller are made of aluminum alloy, and the spindle is made of steel. The material properties are shown in Table 12.1. The fluid analysis model of the rotor system is presented in Figure 12.3. To ensure the convergence of fluid calculation and the stability of the

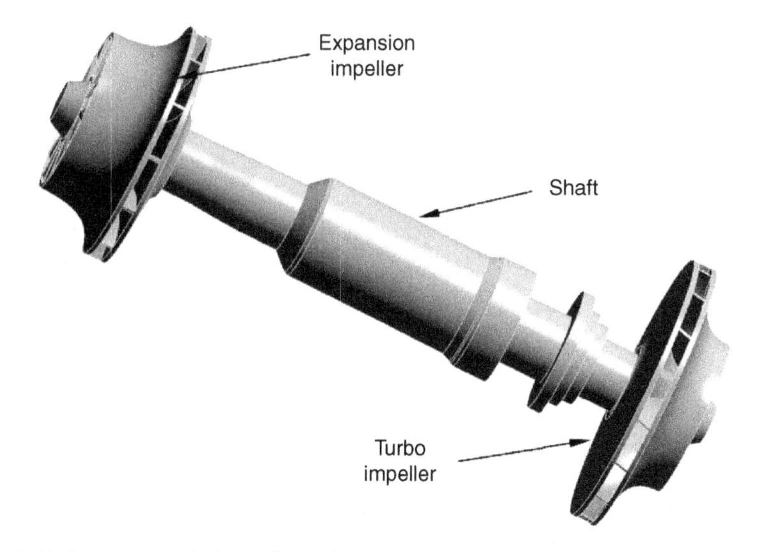

Figure 12.2 Structure analysis model of the rotor system.

Table 12.1 **Material properties**

Material	Density (kg/m^3)	Young's modulus (Pa)	Poisson's rate	Specific heat (J/kg°C)	Heat conductivity (W/m°C)	
					Temperature (°C)	Heat conductivity (W/m°C)
Aluminum alloy	2770	7.1e + 10	0.33	875	−100	114
					0	144
					100	165
					200	175
Steel	7850	2.0e + 11	0.3	434	60.5	

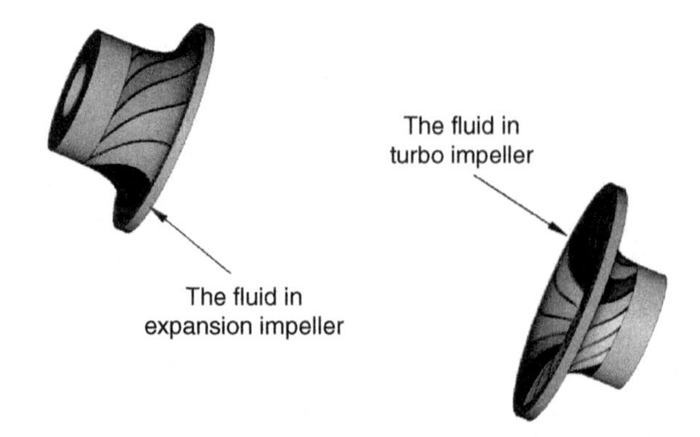

Figure 12.3 Fluid analysis model of the rotor system.

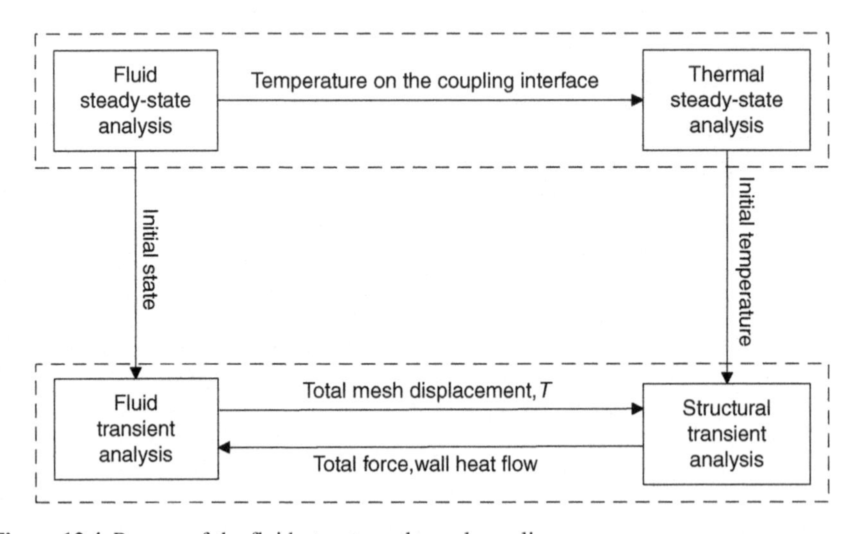

Figure 12.4 Process of the fluid–structure–thermal coupling.

internal channel, the length of the inlet and outlet in the impeller is extended appropriately. Ideal gas is assumed for fluid analysis.

The analysis procedure of the fluid–structure–thermal coupling of the rotor system is as follows: The steady-state analysis of the thermal–fluid model is first performed. The resulting temperature on the coupling interface of the thermal–fluid domain is then transferred to the solid domain to carry out the steady-state structure thermal analysis. When the unidirectional steady coupling analysis is finished, the bidirectional transient fluid–structure–thermal coupling analysis that includes the mechanical and thermal coupling will be started with the initial conditions from the previous steady-state analysis for both the fluid and structure domain. The outline of the coupling simulation process is shown in Figure 12.4. The ANSYS-CFX weak coupling method is adopted for this case.

12.1.2 Steady-state analysis

12.1.2.1 The conditions of steady-state analysis

The system is analyzed in steady-state condition, and the body-attached rotating frame is used. For the fluid region of the expansion impeller, the rotating speed of the whole fluid region is specified as 22,700 rpm, and the reference pressure is 0. As shown in Figure 12.5, the mass flow rate of the inlet is 14.36 kg/s. The relative angle of incidence is 40.675, that is, the radial component is cos θ = 0.7584, the circumferential component is sin θ = 0.6518, and the axial component is 0. The total temperature at inlet is 170 K, and the static pressure at outlet is 1 MPa. Other boundaries are specified as adiabatic wall.

At the turbo impeller side, the rotating speed of the whole fluid region is the same as the speed of the expansion impeller, and the reference pressure is also 0. As shown in Figure 12.6, the mass flow rate at inlet is 14.36 kg/s with vertical incidence. The specified total temperature at the inlet is 314.5 K, and the static pressure at the outlet is 3.95 MPa. Other boundaries are specified as adiabatic wall boundary. In addition, when conducting the steady-state analysis, the fluid regions belonging to the expansion impeller and the turbo impeller are built in the same fluid analysis model to perform fluid–structure–thermal coupling simulation.

When the steady-state analysis of the fluid is done, the resulting temperature distribution on the coupling interface is used as the boundary condition to carry out the subsequent steady-state thermal analysis for the structure model of the rotor system. Finally, the resulting temperature distribution is used as the initial temperature loads for the structural transient stress analysis.

12.1.2.2 The result of the steady-state analysis

The results of steady-state analysis of fluid flow are presented. The contour plots of temperature distribution are shown in Figures 12.7 and 12.8. The streamline

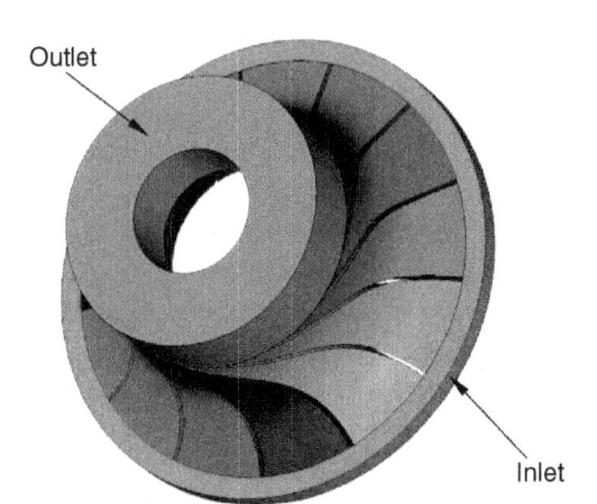

Figure 12.5 The inlet and outlet of the expansion impeller.

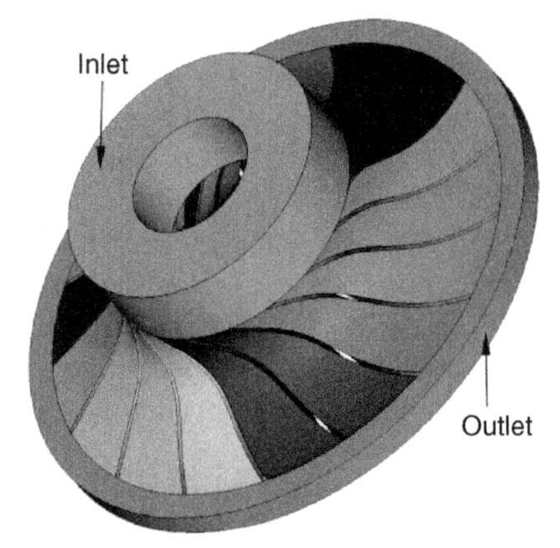

Figure 12.6 The inlet and outlet of the turbo impeller.

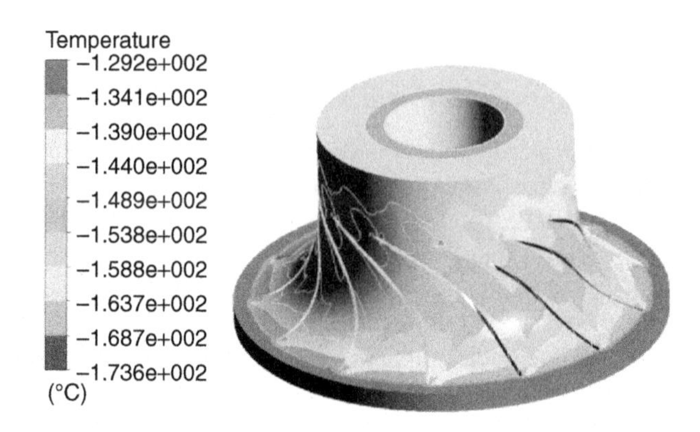

Figure 12.7 Contour plot of temperature distribution of the fluid inside expansion impeller.

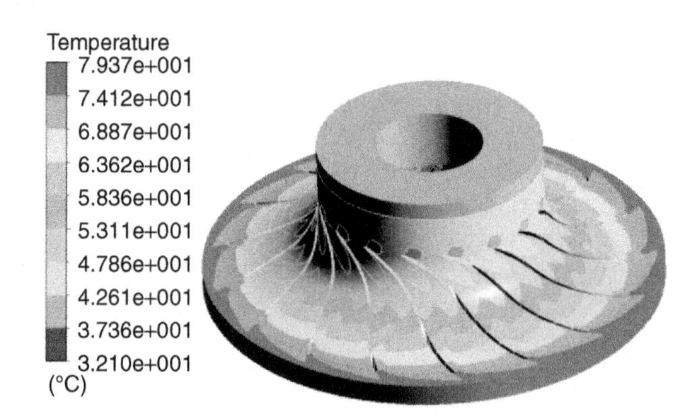

Figure 12.8 Contour plot of the temperature distribution of the fluid inside the turbo impeller.

diagrams of the fluid flow insides the expansion impeller and turbo impeller are shown in Figures 12.9 and 12.10, respectively. The contour plots of pressure distribution are shown in Figures 12.11 and 12.12. From these pictures, it is known that the expansion impeller is working in the condition of low temperature, which would affect the mechanical performance of the impeller structure. As the expanding gas flows in the pipeline, the pressure in the expansion impeller decreases gradually, and the loads on the impeller structure reduce accordingly. Due to the compression effect from the turbo impeller, temperature and pressure gradually increase along the flow direction.

The temperature distribution obtained from structural thermal steady-state analysis is shown in Figure 12.13. The unit is Celsius in Figure 12.13. It can be seen that the expansion impeller is in a low-temperature state, and the turbo impeller is in a high-temperature state.

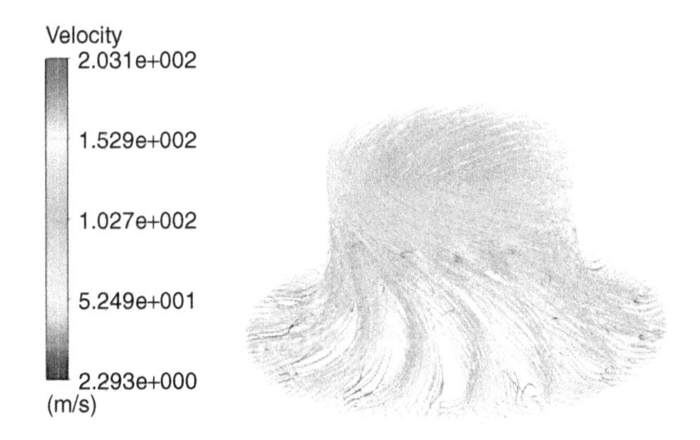

Figure 12.9 Streamline diagram of the fluid velocity inside expansion impeller.

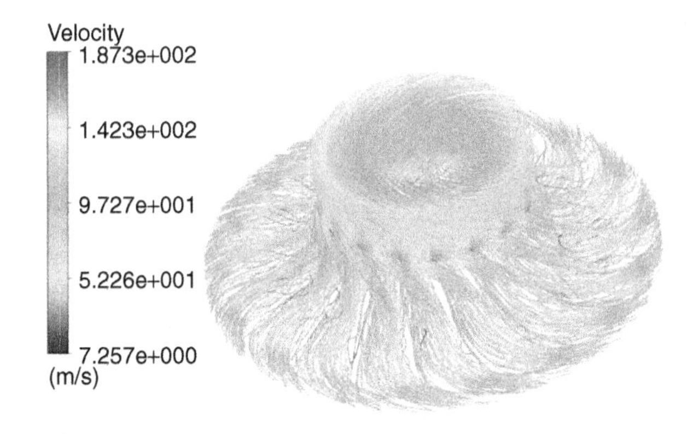

Figure 12.10 Streamline diagram of the fluid velocity inside the turbo impeller.

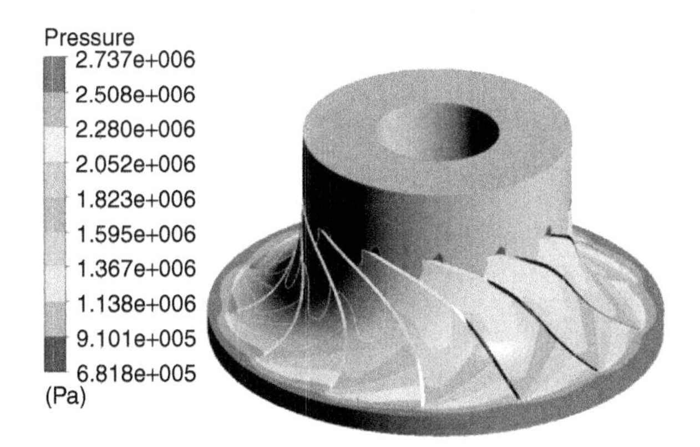

Figure 12.11 Pressure distribution contour of the fluid part inside the expansion impeller.

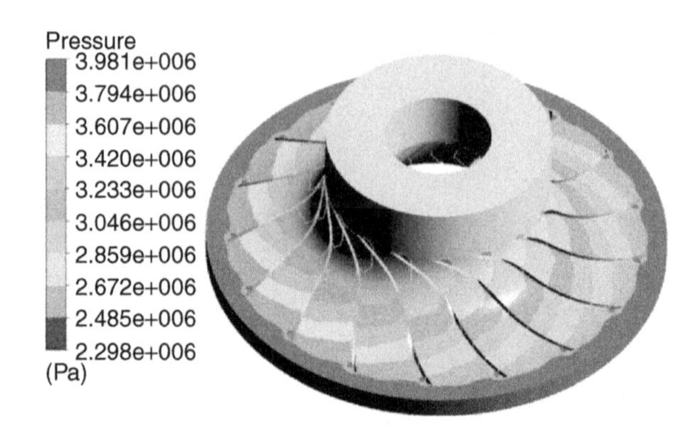

Figure 12.12 Pressure distribution contour of the fluid part inside the turbo impeller.

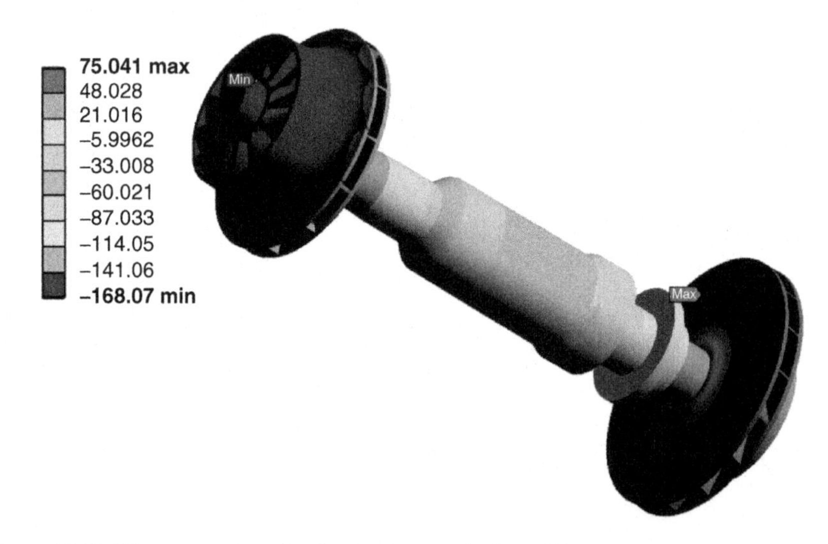

Figure 12.13 The temperature distribution contour in the solid rotor system.

12.1.3 Transient coupling analysis

12.1.3.1 Conditions of transient coupling analysis

The time step size for coupling analysis is specified as 7.77e-6 s, which is the time for the turbo impeller to rotate 1/20 pitch. The total analysis time is 4.66e-4 s, which is the time for the turbo impeller to rotate three pitches. The results from this fluid steady-state analysis are considered as the initial condition of the expansion impeller and the turbo impeller. Except for the coupling interface, other boundary conditions are consistent with those from the steady-state analysis. In addition, the torque is monitored at the coupling interface between the turbo impeller and the expansion impeller to understand the steady state of the whole rotor system.

As for the structure model, the temperature obtained previously, by structure thermal steady-state analysis is considered as the initial condition in the transient state of structure analysis. The rotational speed in the whole system is specified as 22,700 rpm, which is consistent with that in the fluid region. To prevent the rotor system from axial drifting by the fluid force, the middle position A is fixed in the shaft, and the z-direction displacement is fixed at the end interface of the expansion impeller in the shaft (position B). Then the x- and y-directions displacement is fixed in the shaft at position C, as shown in Figure 12.14.

12.1.4 The results of the transient coupling analysis

Since the transient analysis is based on the steady-state analysis, the whole variation range of the temperature and pressure is not very large. So the temperature and pressure distribution at $t = 4.66\text{e-}4$ s are presented in Figures 12.15 and 12.16. It is expected that the expansion impeller remains low temperature all the time, which is similar to the results from the steady-state analysis. Since the gas is expanded along the flow pipeline, pressure reduces gradually, and the force acting on the impeller reduces accordingly. On the contrary, since the gas in the turbo impeller is compressed by it, the temperature and pressure gradually increase along the flow direction.

Figure 12.17 shows the contour plot of temperature distribution in the solid part of the rotor system at $t = 4.66\text{e-}4$ s. Because of the low temperature, the structure temperature in the expansion impeller is close to the gas temperature at the end of the expansion impeller. The temperature of the shaft varies from the low-temperature zone in the expansion impeller to the high-temperature zone in the turbo impeller zone. The

Figure 12.14 The position diagram of the boundary condition in the structural transient analysis.

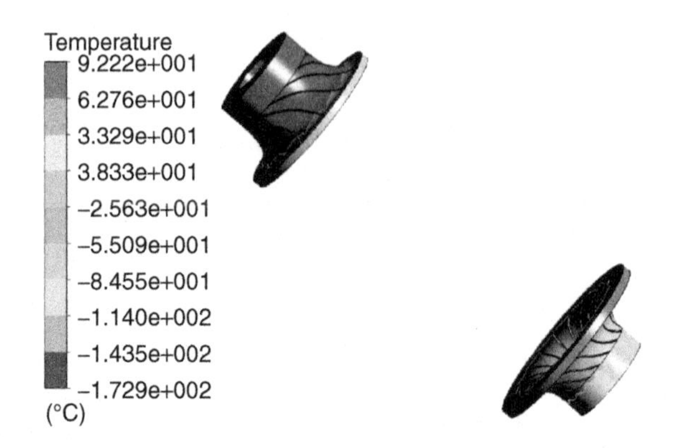

Figure 12.15 The temperature distribution contour of fluid domain at t = 4.66e-4 s.

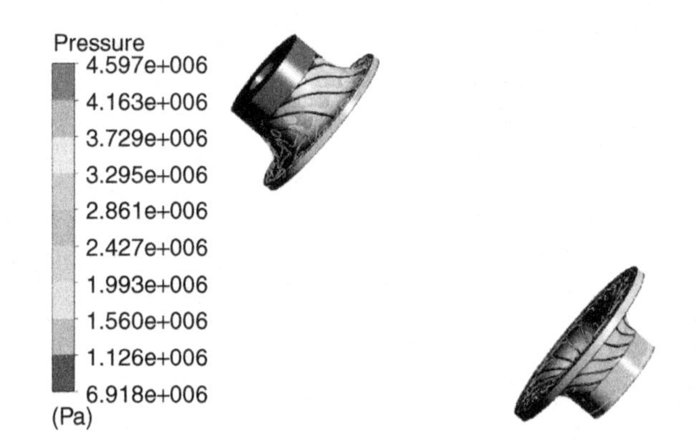

Figure 12.16 The pressure distribution contour of fluid domain at t = 4.66e-4 s.

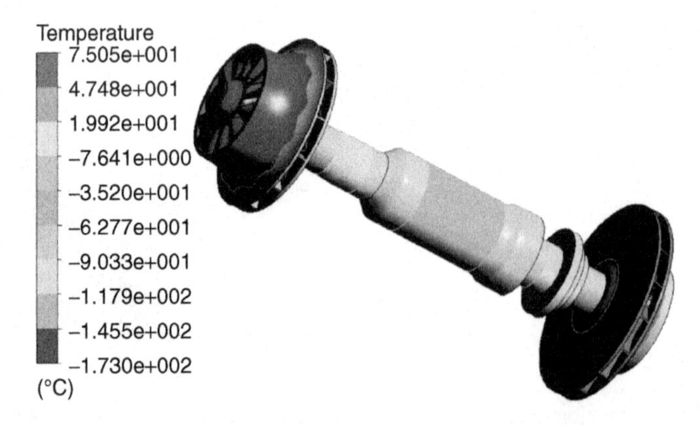

Figure 12.17 The temperature distribution contour of the structural domain at t = 4.66e-4 s.

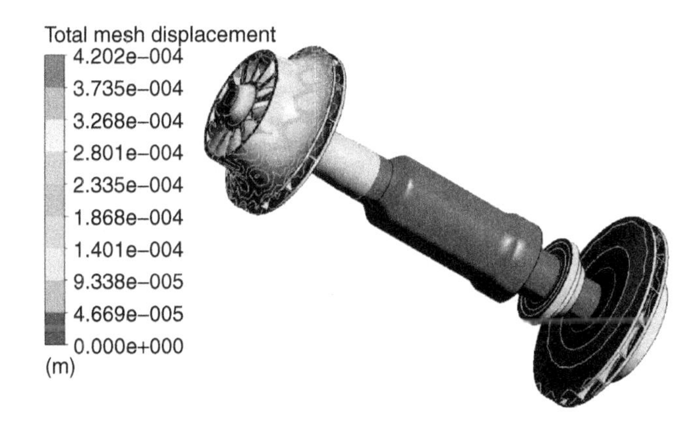

Figure 12.18 The deformation contour of the structural domain at t = 4.66e-4 s.

contour plot of displacement of the structure at t = 4.66e-4 s is shown in Figure 12.18. Due to the fluid force, the displacement at the expansion impeller and the turbo impeller is relatively large, while the displacement at the fixed part in the shaft is 0.

12.1.5 Conclusions

In this case, the steady fluid–structure–thermal coupling analysis is applied to the rotor system of the turbo expander. The oil film force, sealing force, and impact force of droplets are assumed to be negligible. The temperature distribution of the structural rotor system and the temperature and pressure distribution of the fluid regions that belong to the expansion impeller as well as the turbo impeller are obtained. Then a transient fluid–structure–thermal coupling analysis is performed based on the results from the steady-state analysis. The method presented here provides an alternative way for engineers to analyze devices such as turbo expanders. The next step is to improve the physical models by including the effects of deposition patterns of droplets and gas–liquid phase change.

12.2 The fluid–structure coupling analysis of the turbine blade

12.2.1 The physics model

The steam turbine that transforms steam power into mechanical work is a kind of rotary power machine. Steam turbines can be used to drive electrical machine as well as various kinds of pumps, fans, compressors, and ship propellers. The blade structure of the turbine is the most commonly used component in the turbine. The reliability of steam turbine directly relates to the working performance of the turbine. Therefore, the analysis and research of the blade vibration of the turbine is the key to the design and study of the turbine.

The mesh used to discretize the turbine blade is shown in Figure 12.19. Here, the commercial software ANSYS and CFX are used in the bidirectional fluid–structure coupling analysis of the blade to acquire the vibration property of the blade. The results are compared with the natural vibration frequency of the blade to verify whether the blade structure satisfy the requirement of the vibration characteristics.

Since the middle of the blade structure shown in Figure 12.19 has lashing wire structure, the fluid modeling will be more complicated. Thus, appropriate simplifications are made in the fluid model. As shown in Figure 12.20, the corresponding

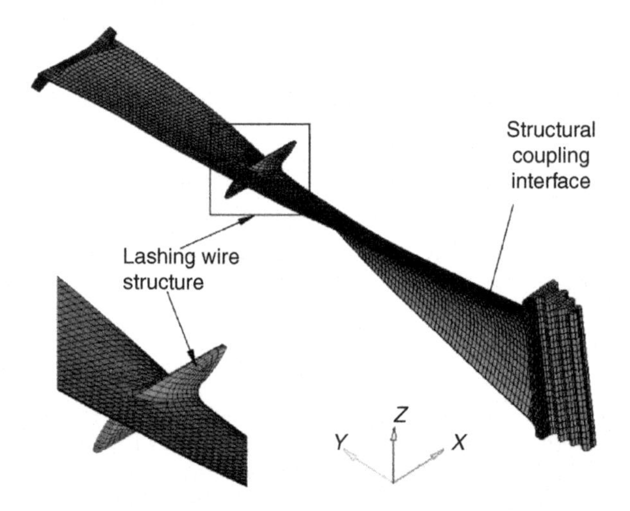

Figure 12.19 The mesh model of the blade.

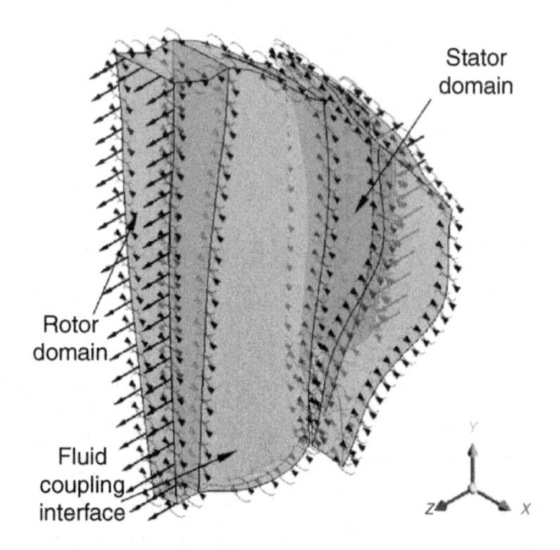

Figure 12.20 Model of the fluid region.

positions of lashing wire structure (highlights further) are removed from the coupling interface in the fluid model for a smooth transition, provided that this simplification will not affect the overall solution of the fluid flow.

As shown in Figure 12.19, the material of the turbine blade is steel, with density 7771.4 kg/m^3, Young's modulus 1.93e+11 Pa, and Poisson's ratio 0.27. In the fluid model of Figure 12.20, the pointed stator domain is the region of inlet flow, and the pointed rotor domain is the region of outlet flow. Also, the stator and rotor domains are all one pitch. Considering the impact of gas–liquid phase changes on the blades, the components of the fluid are defined as the uniform mixture of the liquid water and wet steam, the properties of which are specified by the IAPWS IF97. The lowest temperature is 273.15 K, and the highest temperature is 450 K. The lowest pressure is 620 Pa, and the highest pressure is 1.0e+5 Pa.

12.2.2 Analysis procedure

When analyzing the fluid–structure coupling in the blade model shown in Figures 12.19 and 12.20, the analysis process consists of three steps: the steady-state analysis (including structure and fluid steady-state analysis), the transient analysis of the single field, and the fluid–structure coupling analysis. As shown in Figure 12.21, first the steady analysis is conducted, which provides initial condition for the single-field transient analysis and the fluid–structure-coupling analysis. Then the single-field transient analysis is conducted. The force acting on the fluid-coupling interface is monitored. After the profile of the monitored force reaches a periodic and stable state, the fluid field results at an arbitrary time are taken as the initial condition for the coupling analysis. Finally, the transient bidirectional fluid–structure-coupling analysis is

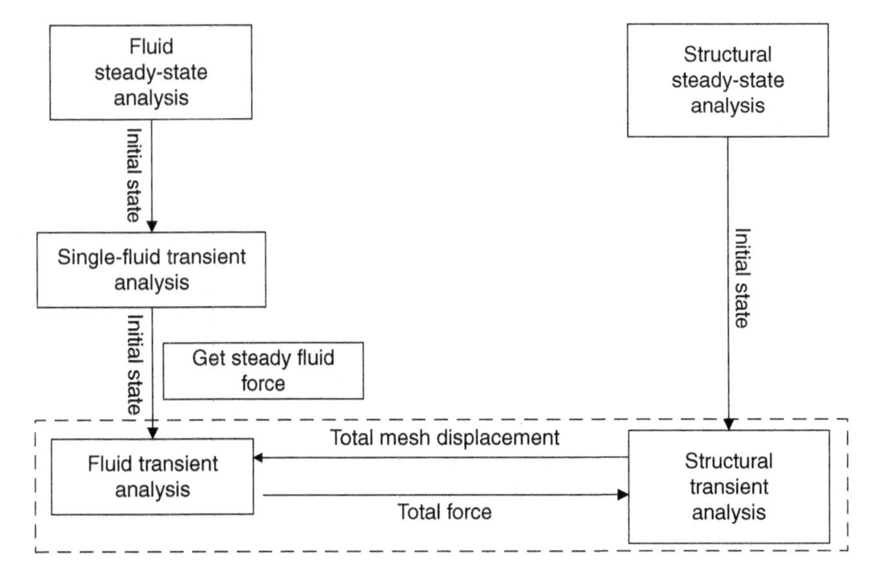

Figure 12.21 The procedure of coupling analysis.

conducted; the data are transferred between the structure and fluid interface. When the data transfer at the current time step is converged and each physical field meets convergence criteria, then the next time step calculation is started. After the total time for the coupling analysis is reached, the calculation is finished.

12.2.3 Steady-state analysis

12.2.3.1 Conditions for the steady-state analysis

As for the structure model, the analysis and load are conducted in the rotating reference frame. The rotational speed is –6000 rpm, the rotary shaft is z-axis, and roots of the blades are fixed as shown in Figure 12.22. Also, the rotational periodic boundary conditions are specified on the top, root, and positions of the blade lashing wire as shown in Figure 12.23.

For the steady-state analysis of the fluid model, the shear stress transport turbulence model is applied. It is assumed that the phase change of gas–liquid takes place quickly at the interaction between the fluid field and the structure blade where both the gas and liquid have an equal temperature.

As shown in Figure 12.20, the fluid domain consists of rotor and stator regions. As for the rotor region, the designed rotational speed is –6000 rpm, and the rotary shaft is z-axis. The fluid velocity, the pressure, and the enthalpy loading at the inlet are specified according to experimental data. The periodic rotating boundary condition about the z-axis is applied at the two boundaries in the rotor region pitch, and other boundaries are defined as no-slip and adiabatic wall boundary conditions.

The stator region is indicated as a nonrotational region, the outlet of which is indicated as pressure boundary. At the pressure outlet, the profile from experimental data is specified. The periodic rotating boundary condition about the z-axis is also applied

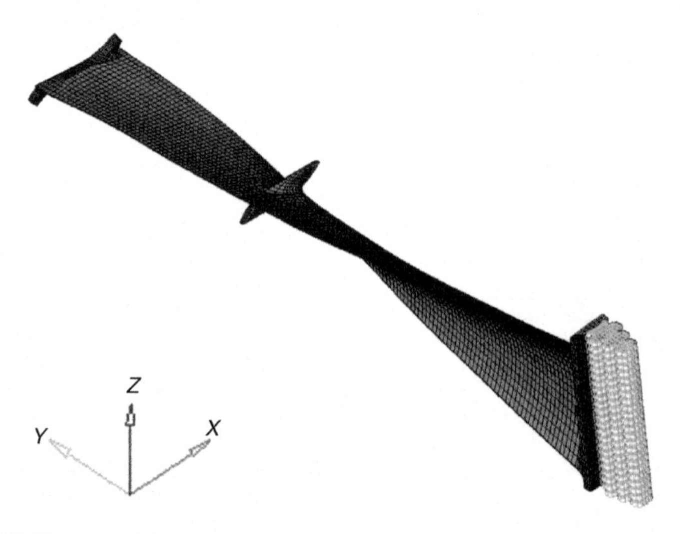

Figure 12.22 The root of the blade is fixed.

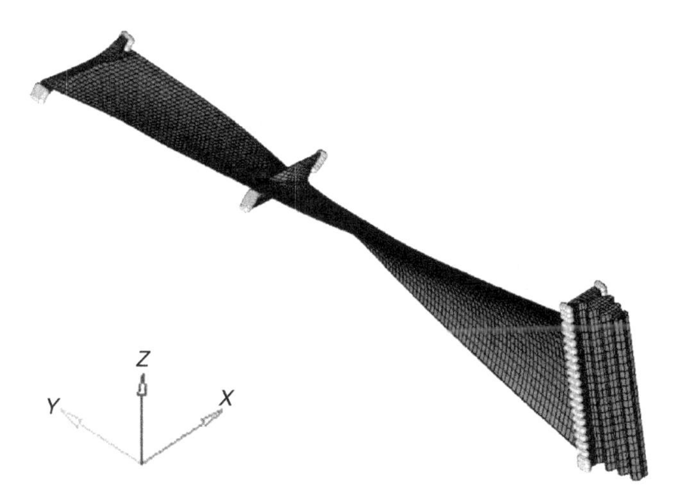

Figure 12.23 The load position of periodic boundary condition.

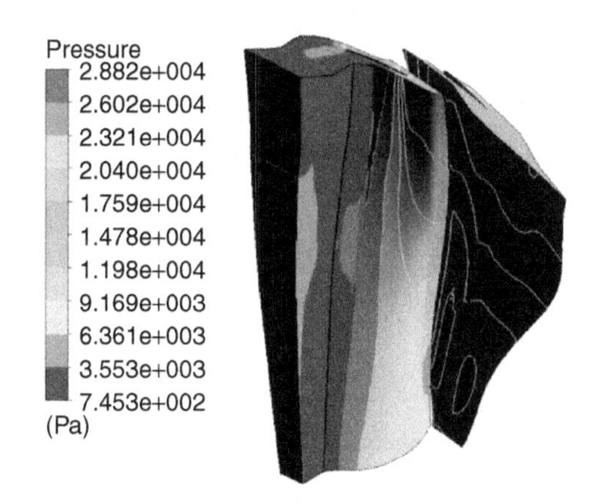

Figure 12.24 The pressure distribution contour of the fluid.

at the two boundaries in the stator region pitch, and other boundaries are defined as nonslip and adiabatic wall boundary conditions.

The sliding interface between the rotor and stator is defined as a frozen rotor stator, and the pitch change is defined by pitch angles.

12.2.3.2 Steady-state analysis results

After the steady-state analysis, the results are presented as follows. In Figure 12.24, the pressure distribution of the fluid steady-state analysis shows that the pressure reduces gradually from the stator region to the rotor region. The distribution of contour plot of the flow velocity is shown in Figure 12.25 in the stator coordinate.

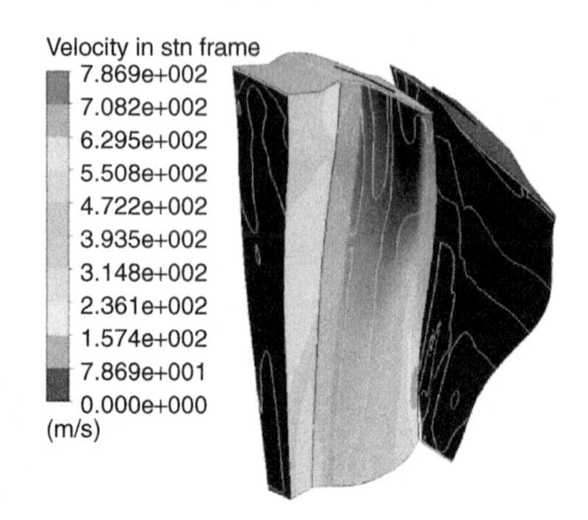

Figure 12.25 The velocity distribution contour of the fluid.

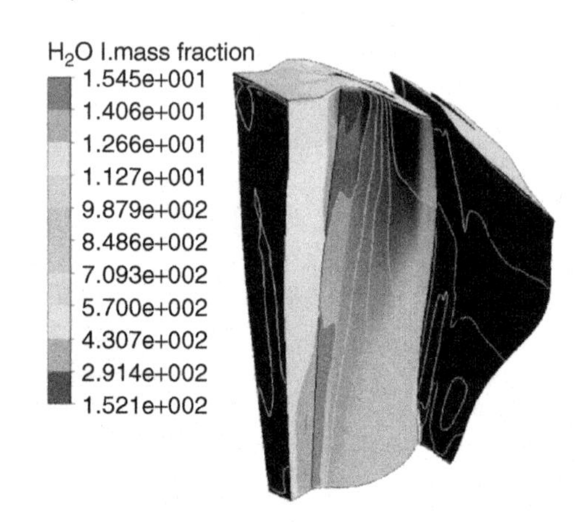

Figure 12.26 The mass fraction of the liquid water in the fluid.

The mass fraction of the liquid water and steam in the fluid is shown in Figures 12.26 and 12.27, respectively. From the distribution, it can be seen that the fluid is mostly in the state of gas phase.

The structural results of the steady-state analysis are presented as follows. In Figure 12.28, the contour plot of the structural displacement shows that the position of the largest displacement is near the tip of leaf blade. In Figure 12.29, it is known that the position of maximum local stress is at the lashing wire structure. Therefore, appropriate mesh refinement at this place should be used to acquire more accurate results.

12.2.4 The transient analysis of the single-flow field

In the transient analysis of the single fluid field, the steady-state analysis result is taken as the initial condition, and the analysis time step is 1.5625e-5 s, which is 0.1 of one pitch time in the fluid rotor region. The total analysis time is 0.005 s, which is the

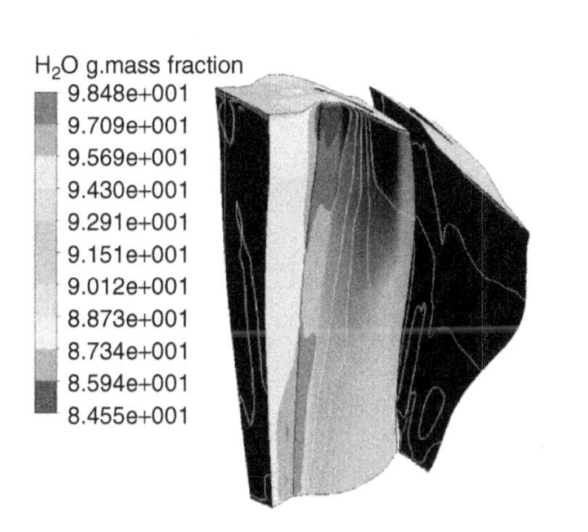

Figure 12.27 The mass fraction of the steam in the fluid.

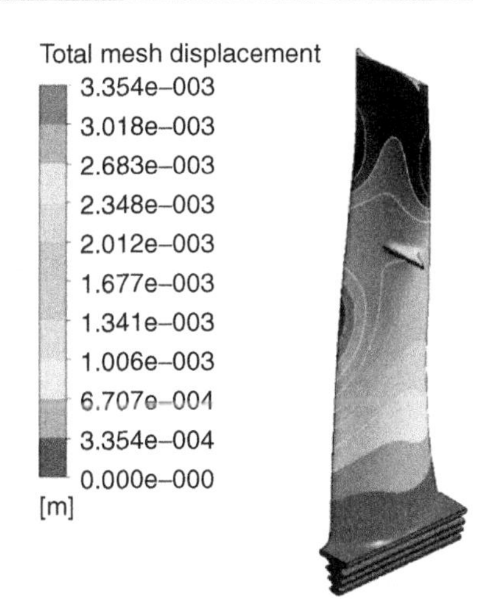

Figure 12.28 The contour plot of the structural displacement.

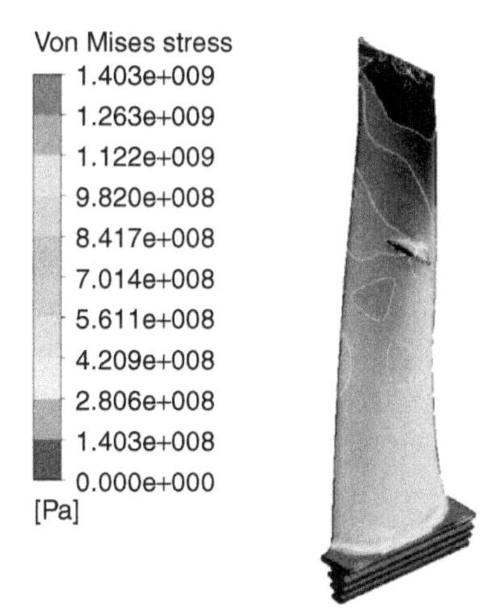

Figure 12.29 The contour plot of the structural von Mises stress.

half period time in calculation. The sliding interface is applied in the way of transient rotor stator, and the other conditions remain consistent with those in the steady-state analysis.

The time history of the fluid force acting on the coupling interface is shown in Figure 12.30. From this figure, it can be seen that the fluid force acting on the coupling interface has steady periodicity after the 155th step. Thus, the analysis results at any time after this time step could be taken as the initial state of the fluid in the coupling analysis. In this case, the transient analysis results (the fluid force is minimum) in the 210th step is taken as the initial condition in the coupling analysis.

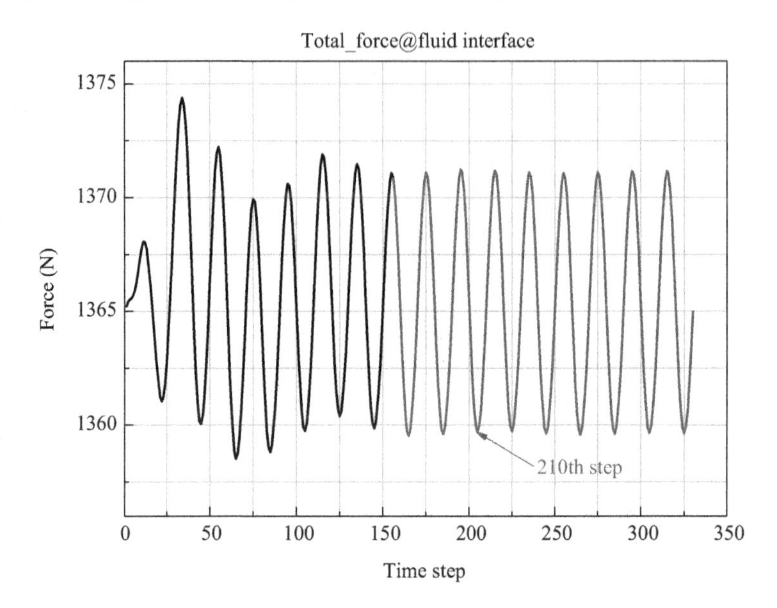

Figure 12.30 The time history curve of the fluid force on coupling interface.

The contour plots of fluid pressure and the flow velocity in the static coordinate are shown in Figures 12.31 and 12.32, respectively.

12.2.5 Fluid–structure bidirectional coupling analysis

For the structure model, the total analysis time is 0.015 s, which is 1.5 periods of calculation, and the time step size is 1.5625e-5 s, which is consistent with that in the fluid transient analysis. Then the results of the structure steady-state analysis are taken as the initial condition, and the other conditions remain as same as those in the

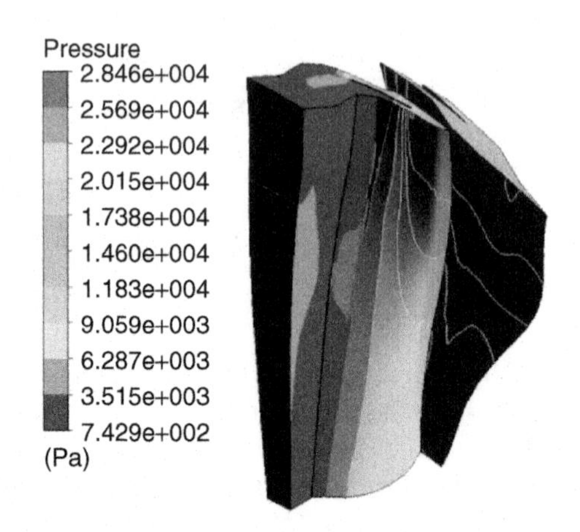

Figure 12.31 The contour plot of fluid pressure.

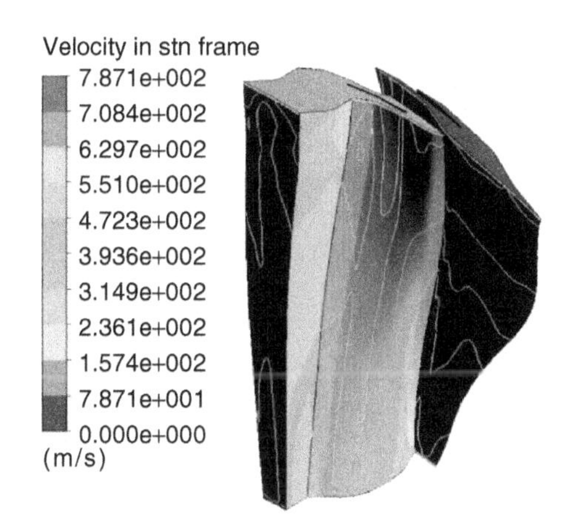

Figure 12.32 The contour plot of fluid velocity.

structure steady-state analysis. In addition, the outer interface of the blade is defined as the structure-coupling interface to conduct the data transfer with the fluid coupling interface.

As for the fluid model, the total analysis time is 0.015 s, and the time step size is 1.5625e-5 s. Then the 210th results in the transient analysis of the single fluid are taken as the initial condition. As for data transfer on the coupling interface, the fluid coupling interface is defined to transfer the total force to the structure coupling interface, and the total mesh displacement of the structural coupling interface is transferred to the fluid coupling interface. Other boundaries are defined as the same as those in the single transient analysis.

After the analysis, the results are presented as follows. At $t = 0.015$ s, the contour plots of the fluid pressure and the flow velocity in the stator coordinate are shown in Figures 12.33 and 12.34, respectively.

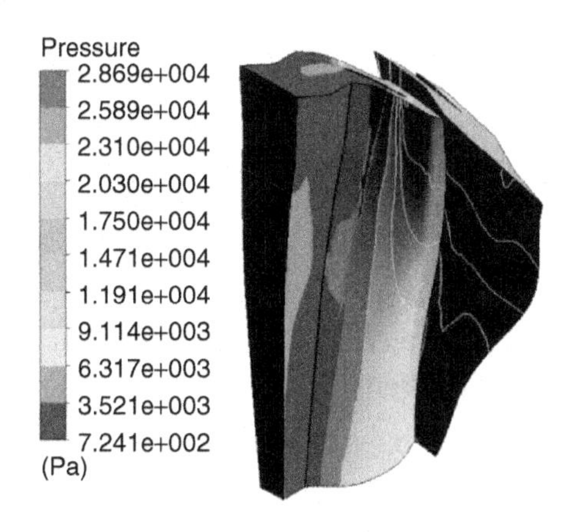

Figure 12.33 The contour plot of the fluid pressure at $t = 0.015$ s.

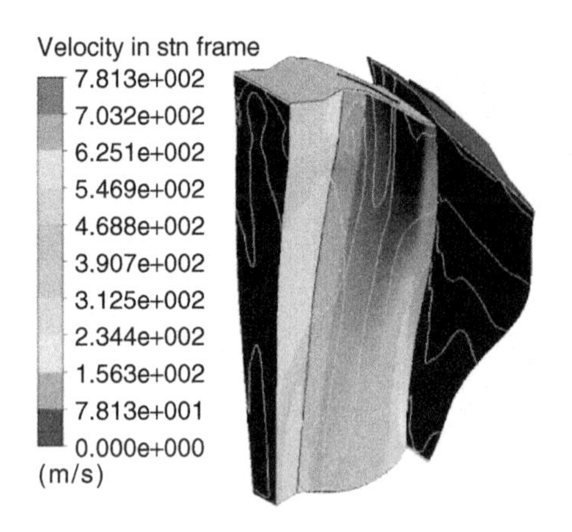

Figure 12.34 The contour plot of the flow velocity in the stator coordinate at $t = 0.015$ s.

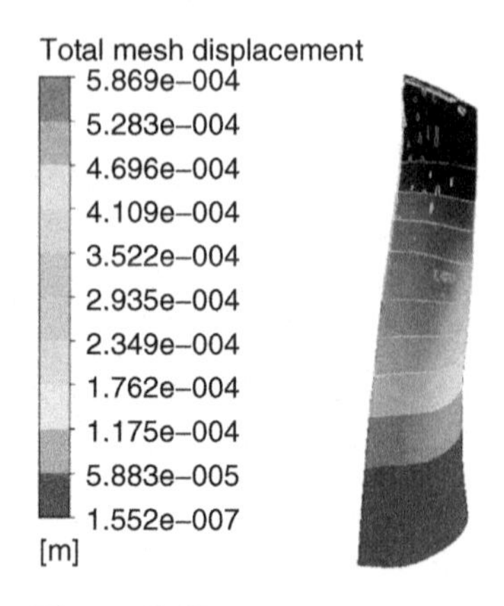

Figure 12.35 The contour plot of mesh displacement on the fluid coupling interface at $t = 0.015$ s.

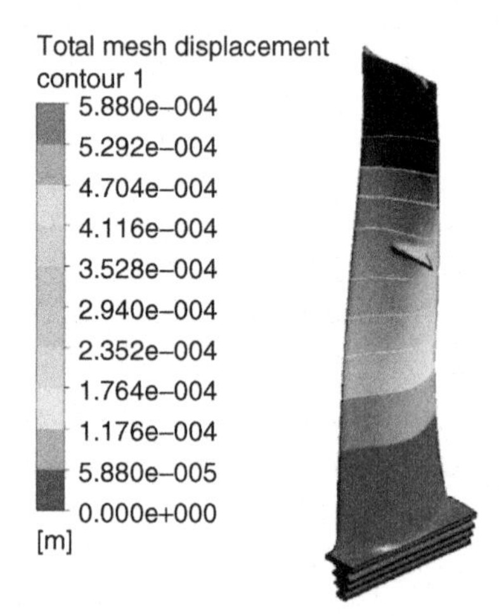

Figure 12.36 The contour plot of mesh displacement in the structure at $t = 0.015$ s.

The displacement on the fluid-coupling interface is presented in Figure 12.35. From this figure, it can be seen that the deformation is larger near the tip of leaf blade, which is consistent with the mesh displacement result in the structure shown in Figure 12.36.

12.2.6 Comparison of the fluid–structure unidirectional and bidirectional coupling analysis

12.2.6.1 The unidirectional coupling analysis

To demonstrate the necessity of the bidirectional coupling analysis, the unidirectional transient coupling analysis is also conducted for this model. The unidirectional transient coupling analysis only considers the effect of the fluid forces on the blade structure, while the effect of the structural deformation on the fluid is excluded. In the setup of analysis, data transfer is only demanded from the fluid force to the structure on the coupling interface, and other conditions remain the same as in the bidirectional coupling method. After the analysis, the fluid force on the fluid-coupling interface is compared with that in the bidirectional coupling analysis shown in Figure 12.37.

As shown in Figure 12.37, when comparing the unidirectional and bidirectional fluid-structure coupling, fluid forces differ comparatively largely on fluid coupling interface. In the unidirectional coupling analysis, the profile of the fluid force has a single frequency $f = 48/(0.15\ \text{s}) = 3200$ Hz; while using the bidirectional fluid-structure coupling method, the profile of fluid force on the fluid coupling interface shows another new frequency $f = 2/(0.15\ \text{s}) = 133.33$ Hz. This new frequency may be caused by the structural deformation. Therefore, the unidirectional coupling analysis could not yield the reasonable stress in the blade structure, and the bidirectional coupling analysis should be adopted in this case.

12.2.7 Modal analysis for blade structure

To identify the reason for the large circle of fluid force on the coupling interface and to confirm the flutter in the blade structure, the modal analysis is carried out for the blade model in Figure 12.19. The results of the steady-state analysis in the structure are taken as the prestressed force condition in the modal analysis, and the first 10 order modes are calculated in this case.

After the analysis, frequency distribution of the first 10 order modes is shown in Table 12.2.

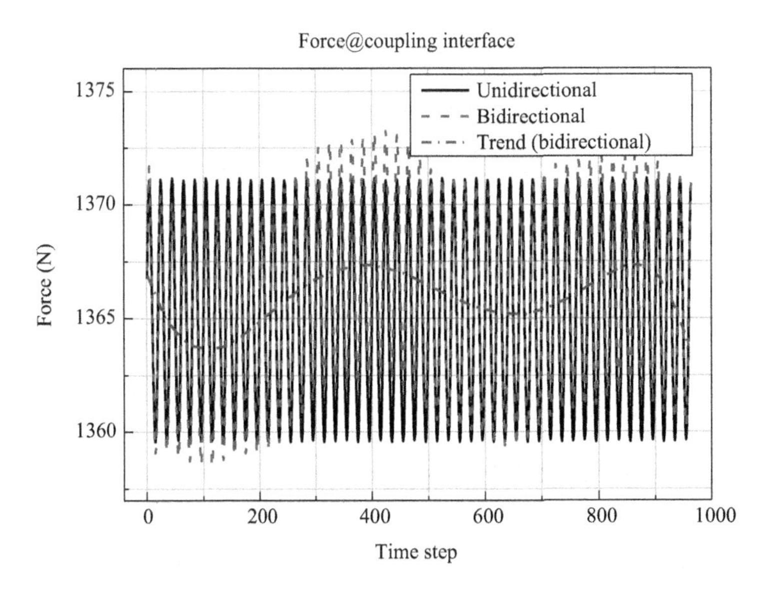

Figure 12.37 The comparison of the fluid forces on the fluid coupling interface.

Table 12.2 **The frequency distribution of the first 10 order modes**

Order	Frequency (Hz)
1	91.14
2	136.90
3	139.14
4	145.88
5	151.88
6	153.44
7	156.68
8	160.52
9	163.85
10	167.02

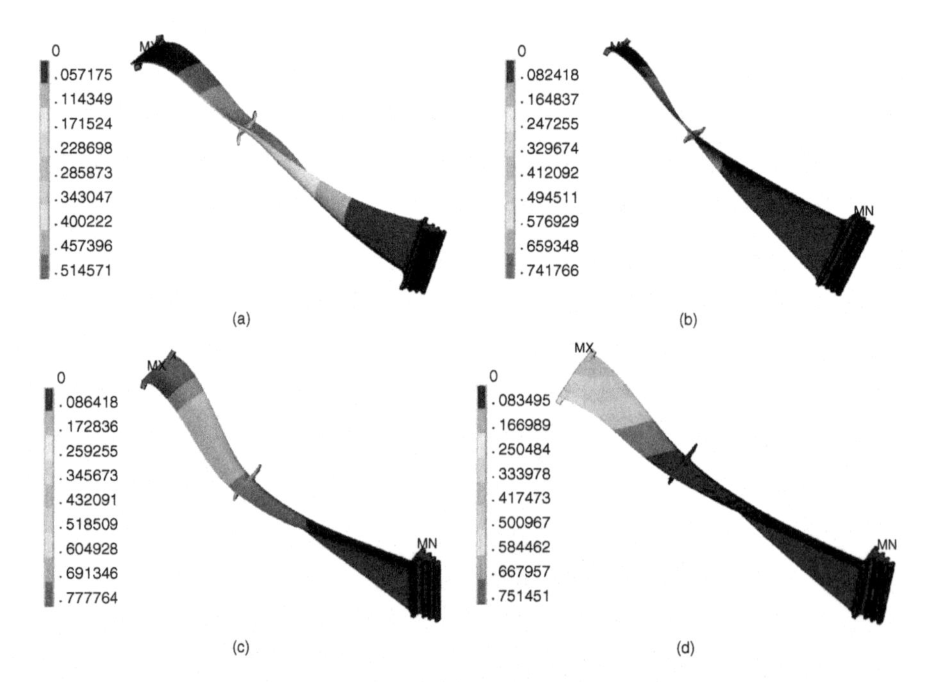

Figure 12.38 First four order modes of blade structure. (a) First-order frequency, 91.14 Hz; (b) second-order frequency 136.90 Hz; (c) third-order frequency, 139.14 Hz; and (d) fourth-order frequency, 145.88 Hz.

The first four order modes are shown in Figure 12.38.

From the results of modal analysis, frequency of the second-order mode is close to the newly appeared frequency $f = 133.33$ Hz of the fluid force in the fluid–structure coupling analysis. Thus, the periodic varying of the fluid forces on the coupling interface is influenced by the structural deformation, and it is the effect of the superimposition of the structural natural vibration and the periodic effects of fluid force. Therefore, the flutter happens in the blade structure and might cause damage to the blade structure. So the structure should be carefully designed to avoid the flutter.

12.2.8 Conclusions

In this case, bidirectional fluid–structure-coupling analysis is used to acquire the vibration characteristics of the steam blade. The results are compared with those from the unidirectional coupling analysis to verify the necessity of usages of the bidirectional fluid–structure-coupling analysis. At the same time, the comparison is made between the vibration frequency of the fluid force on the coupling interface and the mode frequency of the blade structure. The results demonstrate capability of predicting resonance of the blade by bidirectional fluid–structure-coupling analysis. It provides a useful coupling simulation procedure and engineering insights for the designer to improve the reliability of turbine blades.

Multiphysics modeling for biomechanical problems

13

Chapter Outline

13.1 Numerical analysis of a 3D simplified artificial heart

For fluid–structure interaction (FSI) problems involving structural buckling and extremely large domain changes, advanced theory and numerical methods are required (Zhang and Hisada, 2001). A 3D finite element analysis program can be used to solve the proposed problem. Simulation of the pulsation of a 3D simplified artificial heart with structural buckling and large domain changes is demonstrated to show the capability of a strong coupling method method with stabilization schemes in space and time. In the present analysis, Arbitrary Lagrangian–Eulerian (ALE) finite element formulation is used for the fluid, and total Lagrangian formulation is used for the structure.

13.1.1 Numerical model

The blood pump of a totally implantable pulsatile artificial heart has the same efficacy as the ventricle of a human heart. Under the action of oil pressure, a round diaphragm is pushed to discharge blood from the artificial heart and then pulled back to draw blood into the heart. The artificial heart is hence filled with blood again. This is the working principle of the artificial heart.

In this analysis, the material properties of blood are as follows: mass density, $\rho = 1.06 \times 10^3$ kg/m^3; dynamic viscosity, $\mu = 4.71 \times 10^{-3}$ Pa·s; and bulk modulus, $B = 2.06 \times 10^9$ Pa. The material properties of the diaphragm are Young's modulus, $E = 2.0 \times 10^7$ Pa; Poisson's ratio, $\nu = 0.45$; mass density, $\rho = 1.13 \times 10^3$ kg/m^3; and thickness, $t = 0.2$ mm. The initial configuration and geometrical dimensions of a 3D simplified artificial heart are shown in Figure 13.1.

Q. Zhang & S. Cen: Multiphysics Modeling. http://dx.doi.org/10.1016/B978-0-12-407709-6.00013-4

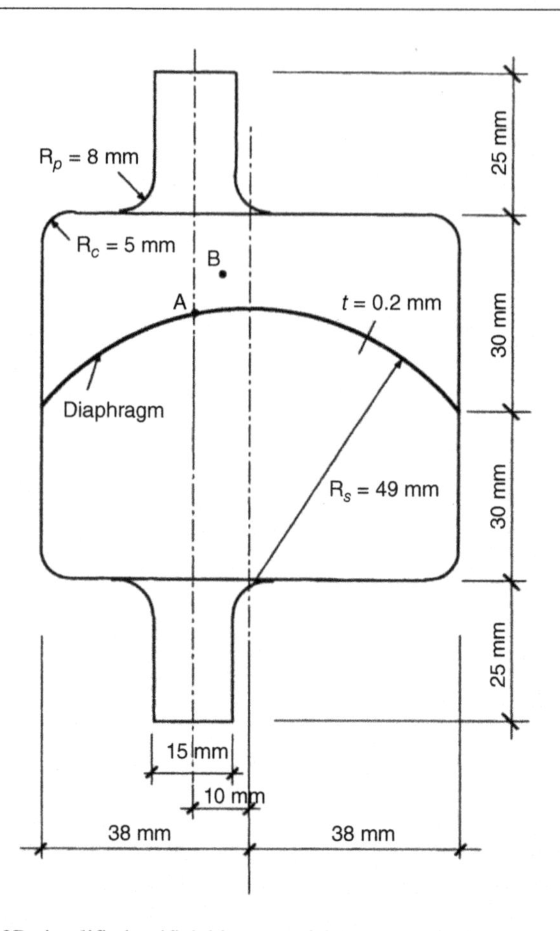

Figure 13.1 The 3D simplified artificial heart model (cross-section).

The mesh of the 3D model is shown in Figure 13.2a. A total of 10,624 fluid elements and 640 shell elements are used for the mesh discretization. The total number of nodes is 22,245, which include 657 interface nodes, and total number of degrees of freedom is 41,296. In the case, the Q1/P0 (eight nodes for the bilinear velocity interpolation/the constant pressure field) mixed hexahedral elements are used for the fluid, and structure is discretized by the four-node MITC shell elements.

13.1.2 Analysis conditions

Initial conditions are velocity, $\mathbf{v} = \mathbf{0}$ and pressure, p = hydrostatic pressure. The initial stress of the membrane is assumed to be zero. The pulsation procedure of the artificial heart is divided into two phases as follows:

1. Contraction phase ($t = 0.0$ s$\rightarrow t = 0.5$ s): Downward velocity is prescribed on the inlet boundary (the upper tube end), and its history is assumed to follow the fourth-order polynomial

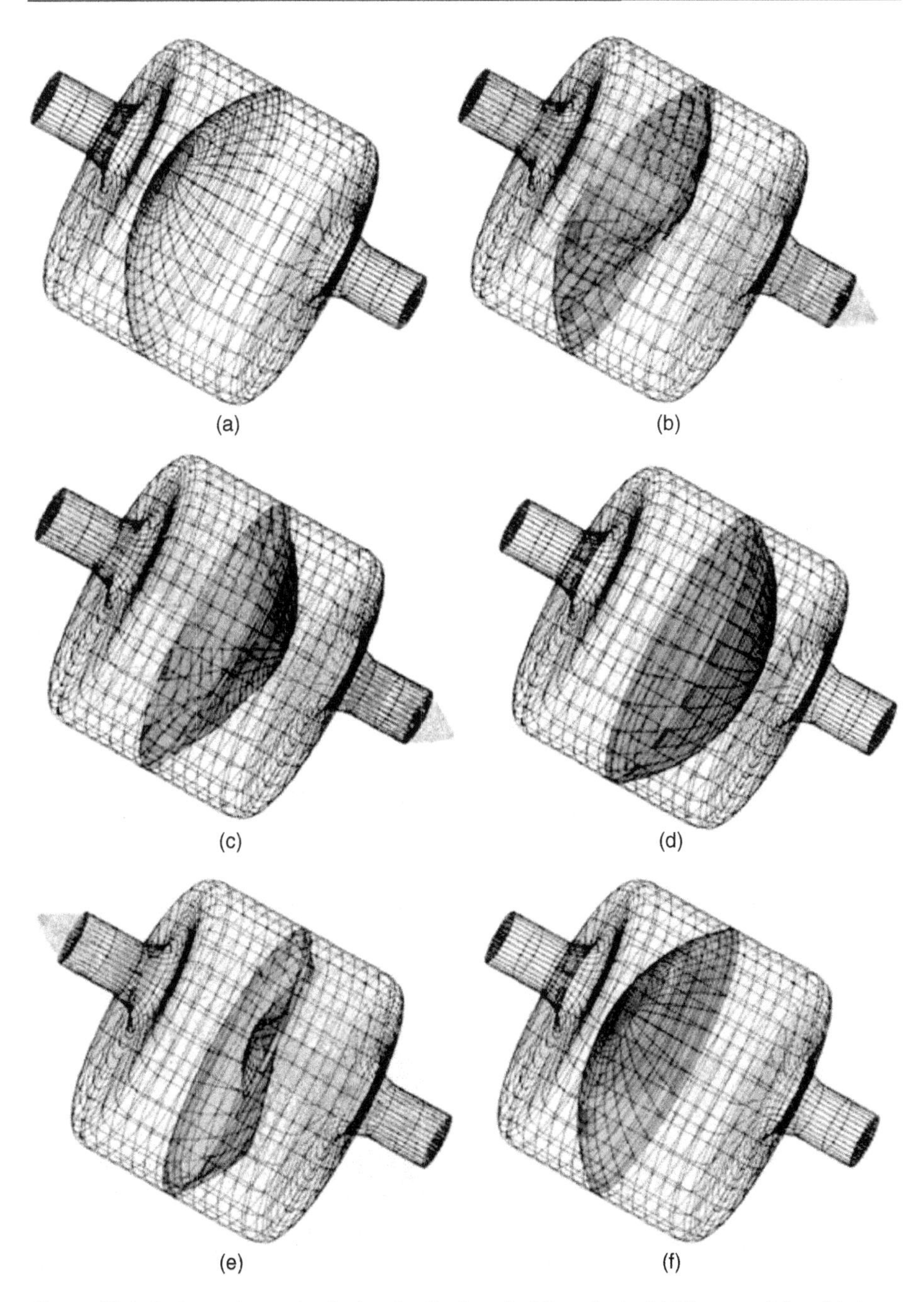

Figure 13.2 Deformation and velocity distributions in 3D analysis. (a) Time: $t = 0.0$ s; (b) time: $t = 0.2$ s; (c) time: $t = 0.3$ s; (d) time: $t = 0.5$ s; (e) time: $t = 0.7$ s; and (f) time: $t = 1$ s.

function of time ($V_n = At^4 + Bt^3 + Ct^2 + Dt + E$). The distribution of velocity is given by assuming laminar flow in a cylinder, and the coefficients A, B, C, and D are decided by setting the flow to increase up to 0.1×10^{-3} m^3 smoothly. The initial hydrostatic pressure is applied to the outlet open boundary constantly over time. In the horizontal direction, the stress-free conditions are given for both inlet and outlet.

2. Relaxation phase ($t = 0.50$ s$\rightarrow t = 1.0$ s): Upward velocity is prescribed on the outlet boundary (the upper tube end). The same velocity distribution and history are given as those for the contraction phase. The initial hydrostatic pressure is applied to inlet open boundary constantly over time. To stabilize the spurious oscillations of velocity at the inlet boundary, a parabolic velocity distribution is assumed. Stress-free conditions are applied in the horizontal direction for both inlet and outlet.

The proportional mesh control method is used according to the change of fluid domain and geometrical configuration. To cope with the instability in time integration, numerical damping is introduced by setting, $\gamma = 0.6$ and $\beta = 0.3025$. Strong coupling method is adopted in this analysis.

13.1.3 Simulation results

As mentioned earlier, one of the significant features of this analysis is that the problem includes structural buckling. In this section, mechanical and numerical characteristics of the FSI problem with structural buckling and large domain changes are clarified. Figure 13.2b–f illustrates deformation of the shell and velocity distributions of the blood flow. Histories of acceleration, velocity, and displacement are given in Figure 13.3 for the selected node on the interface (point A in Figure 13.1). It is observed in Figure 13.3a,b that the oscillations of acceleration and velocity of the node on the interface are induced during some periods. The oscillations of acceleration and velocity are also observed in most of the fluid domain during the same periods. On the other hand, at the beginning of the contraction phase ($t = 0.0$–0.05 s) and in the fully turning over period ($t = 0.45$–0.62 s), the oscillations of responses of the coupled system are small. The determinant of the tangent stiffness matrix of the structure is kept positive for the first 0.05 s; however, it changes the sign frequently after this. This means that the matrix becomes singular, that is, the structural buckling occurs after $t = 0.05$ s. It is thus understood that the oscillation is caused by the buckling. During the period from 0.45 to 0.62 s, the diaphragm fully turns over (Figure 13.3c), and the membrane is mostly in the tensile stress state. The oscillation is therefore stabilized during this period.

Through the successful calculation for a 3D model, it is demonstrated that the strong coupling method and numerical stabilization methods are effective for FSI problems with structural buckling large domain changes.

13.2 FSI simulation of a vascular tumor

13.2.1 Physical model of a vascular tumor

Vascular tumor is a prevalent disease that is of significant concern because of the morbidity associated with the continuing expansion of abdominal aorta and its ultimate

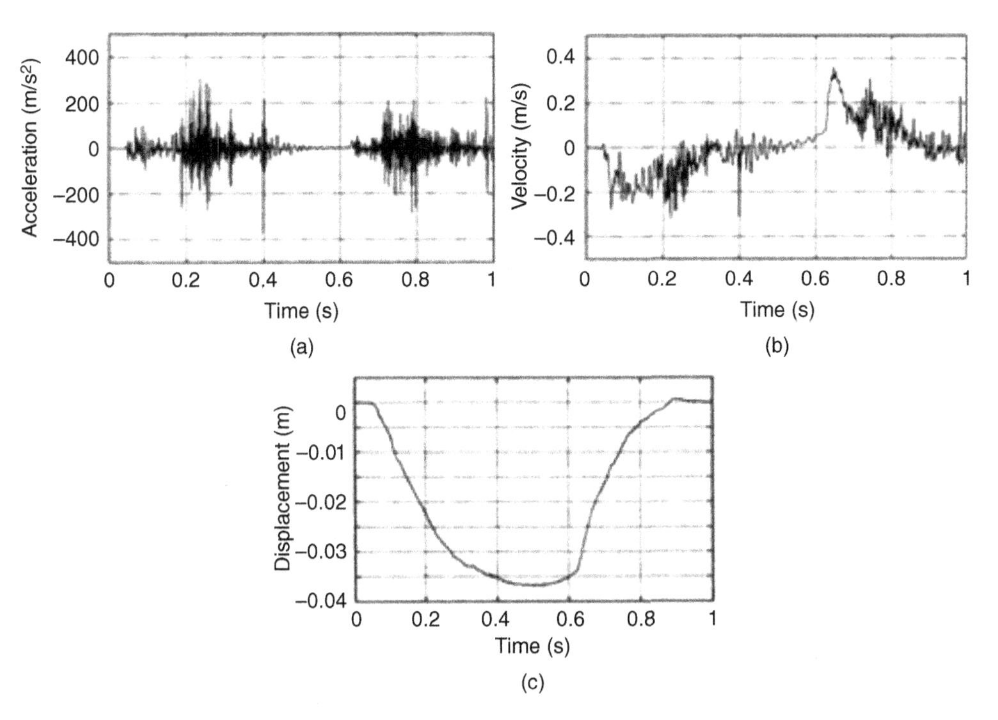

Figure 13.3 Responses of point A. (a) Acceleration; (b) velocity; and (c) displacement.

rupture. The transient interaction between blood flow and the vessel wall contributes to wall stress, which when exceeds the failure strength of the dilated arterial wall, leads to rupture of the aneurysm.

In this example, we describe the complex interaction of blood flow and a compliant vascular tumor wall by using a time-dependent, fully coupled FSI methodology to determine effects on mechanical stresses and vortex dynamics in INTESIM software.

The model consists of a fluid domain, Ω^F, representing the aorticlumen and a structural domain, Ω^S, representing the vascular tumor wall. The fluid domain is characterized by circular cross-sections parallel to the x–y plane, with a nonuniform diameter, $d = 20$ mm, at the inlet and outlet sections and a maximum diameter, $D = 3\ d$, at the midsection of the vascular tumor. The geometry of the fluid domain is given by Equation (13.1), which defines the diameter of each cross-section, (z),

$$\phi(z) = \begin{cases} d\left(\cos\left(\dfrac{\pi}{3d}(z-6d)\right)+2\right), 3d \le z \le 9d \\ d, 0 \le x \le 3d \text{ and } 9d \le z \le 12d \end{cases} \tag{13.1}$$

The geometry of structural domain is given by a vascular tumor wall with a uniform thickness. The wall models thickness of the vascular tumor as material extruding normally from the surface enclosing the lumen. The structural model has a thickness that is given by $t = 1.5$ mm. The fluid domain and structural domain are shown in Figure 13.4a,b separately. Scotti et al. (2005) simulated the same problem with ADINA-FSI.

The material properties of blood are: density, ρ_f = 1050 kg/m^3; dynamic viscosity, μ = 0.00385 Pa·s; and a bulk modulus, $B = 2.15 \times 10^9$ Pa. The vascular tumor wall is assumed to be isotropic, linear, and an elastic solid with density, ρ_s = 2.000 kg/m^3; Young's modulus, $E = 2.7 \times 10^6$ Pa; and Poisson's ratio, $\upsilon = 0.45$.

In this example, we make use of linear hexahedral, eight-node elements to discretize the fluid and structural domains. The FSI models are composed of 17,600 hexahedral elements (19,521 nodes) for the fluid and 5,760 hexahedral elements (8,784 nodes) for the structural domains. Figure 13.5 shows the fluid and structural meshes.

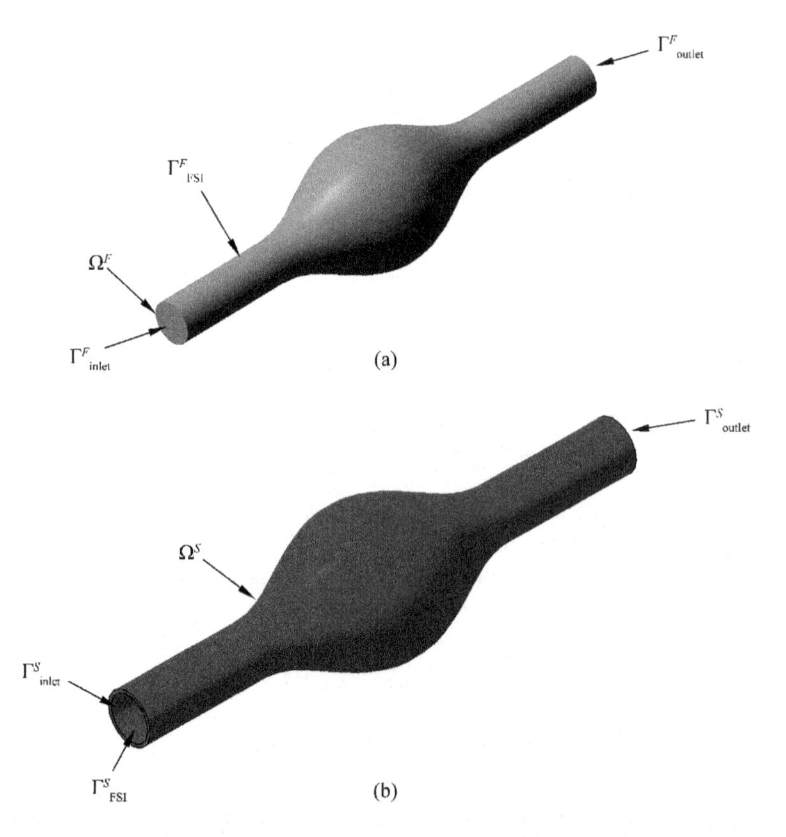

Figure 13.4 CAD geometry of a vascular tumor. (a) Fluid domain and (b) structural domain.

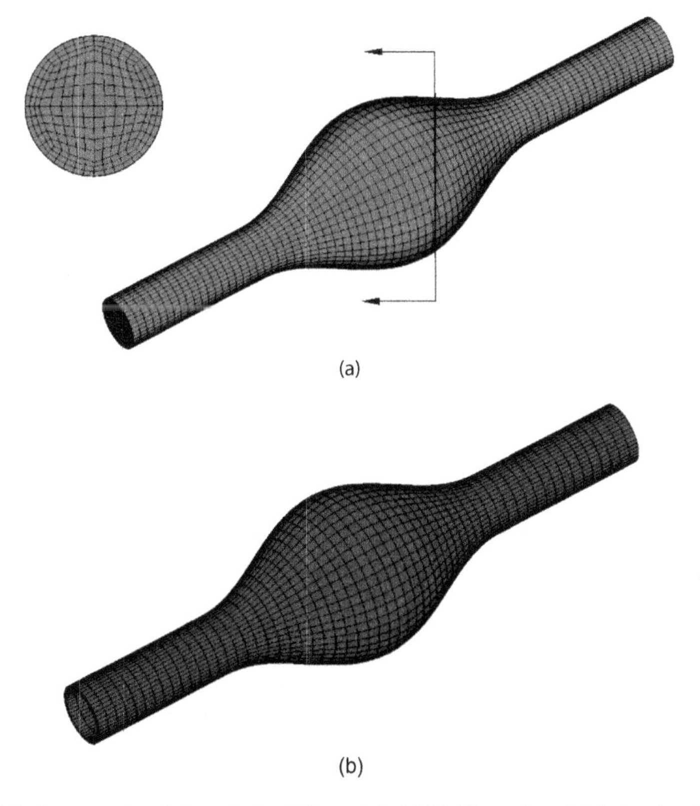

(a)

(b)

Figure 13.5 Computational domain for FSI model. (a) Fluid mesh and (b) structural mesh.

13.2.2 Analysis conditions

The boundary of the fluid domain is divided into the following regions for the assignment of boundary conditions: inlet (Γ^F_{inlet}), outlet (Γ^F_{outlet}), and the FSI interface (Γ^F_{FSI}), as shown in Figure 13.4a. The applied boundary conditions on the non-FSI regions are: (1) time-dependent fully developed velocity profile on Γ^F_{inlet} which is presented by Equation (13.2), and (2) time-dependent luminal pressure on Γ^F_{outlet}.

$$v(t,x,y) = -2u(t)\left\{1-\left(x^2+y^2\right)\right\}, \quad (x,y)\in\Gamma^F_{\text{inlet}} \tag{13.2}$$

where $u(t)$ is the time-dependent velocity waveform shown in Figure 13.6a, and x, y are the cross-section coordinates on Γ^F_{inlet}. The time-dependent pressure wave form on Γ^F_{outlet} is shown in Figure 13.6b.

Figure 13.4b shows the boundary of structural domain divided into inlet (Γ^S_{inlet}), outlet (Γ^S_{outlet}), and fluid–structure interface (Γ^S_{FSI}) regions. The FSI interfaces Γ^F_{FSI} and Γ^S_{FSI} are identical, coupling the fluid and structural domains. The non-FSI regions of the structural domain are imposed zero transformation on the ends.

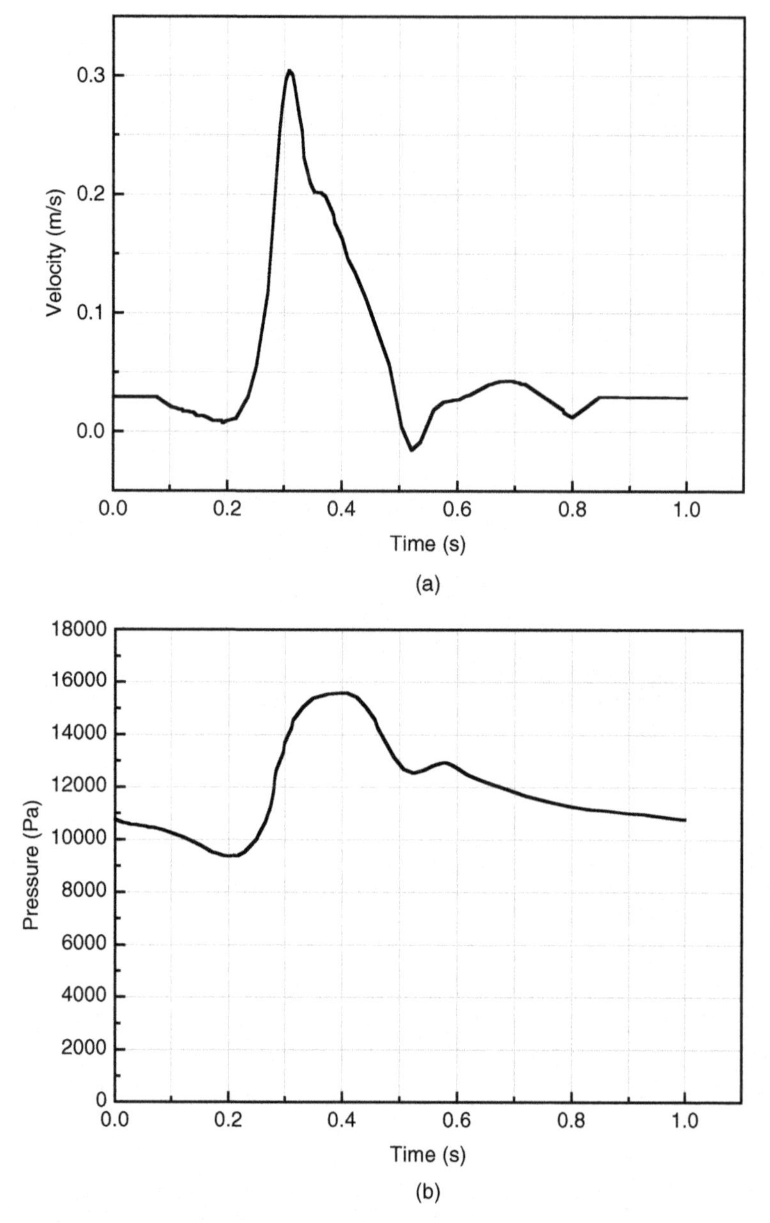

Figure 13.6 The boundary conditions applied up on the fluid non-FSI regions. (a) Velocity waveform on Γ^F_{inlet} and (b) pressure waveform on Γ^F_{outlet}.

13.2.3 Results and discussion

In this study, INTESIM is used for numerical simulation of FSI between the vessel wall and blood flow.

The blood flow pattern at $t = 0.4$ s (peak systolic pressure in reference to the outlet pressure boundary condition) is shown in Figure 13.7a. The velocity vectors illustrate a streamlined profile absent of vortices, a flow path customarily associated with a condition of systolic acceleration. Also, the wall pressure distribution in Figure 13.7b is nearly identical at the longitudinal cross-section, which is given by the *YZ* plane.

Figure 13.8 shows the displacement magnitude and stress distributions for the structural wall model. Significant gradients occur at inflection points of the aneurysm curvature. The changes in this curvature result in higher displacements and increased stress, suggestive of the gradual expansion and contraction effects of the flow through the wall.

Besides, we can have the time history curves of displacement and von Mises stress on the wall where the maximum value occurs during the whole period of time. The curves are shown in Figure 13.9.

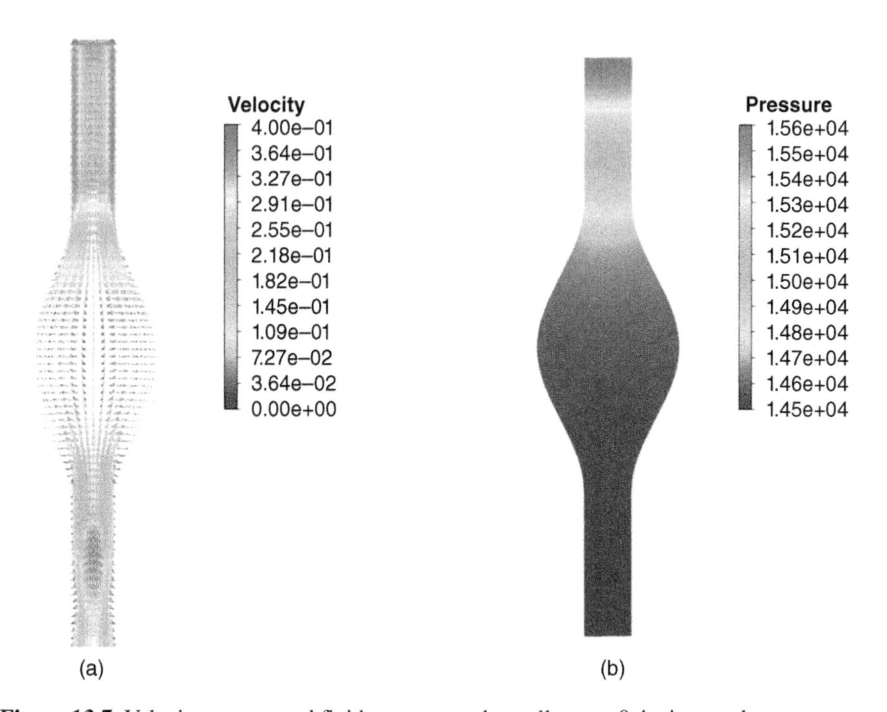

Figure 13.7 Velocity vectors and fluid pressure at the wall at: $t = 0.4$ s in y–z plane. (a) Velocity vector and (b) fluid pressure.

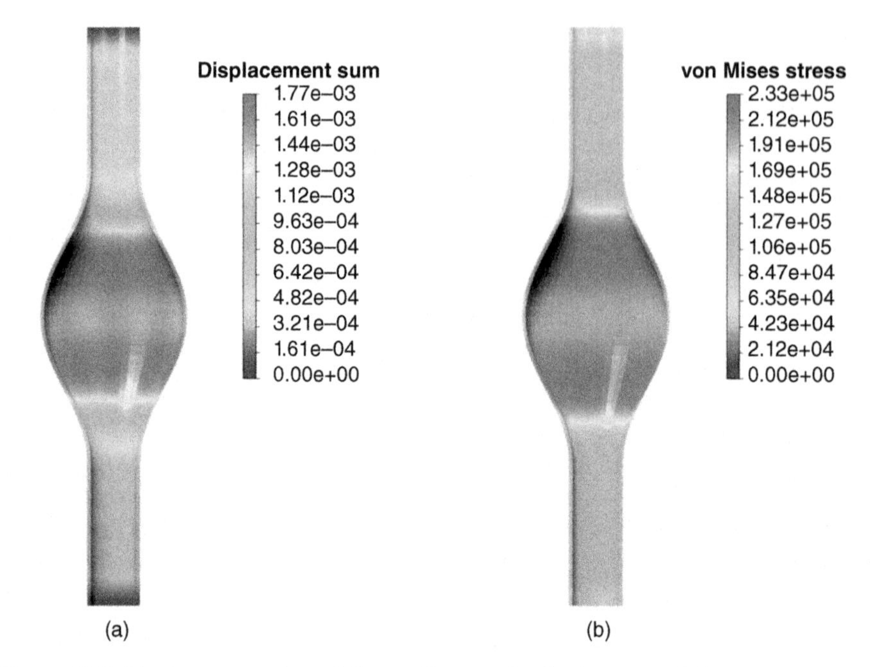

Figure 13.8 Displacement and von Mises wall stress distributions at $t = 0.4$ s. (a) Displacement and (b) von Mises wall stress.

Table 13.1 Maximum values, which occurred in the transient time domain

	Max von Mises stress (N/cm^2)	Max displacement (cm)
INTESIM	23.1	0.17
ADINA	23.8	0.18

Finally, the analysis results obtained by INTESIM are compared with those by ADINA-FSI to verify the accuracy of FSI simulation. The maximum values shown in Table 13.1 that occurred in the transient time domain, are observed for comparison. It can be seen from this table that INTESIM and ADINA-FSI produced a very close solution of the maximum stress.

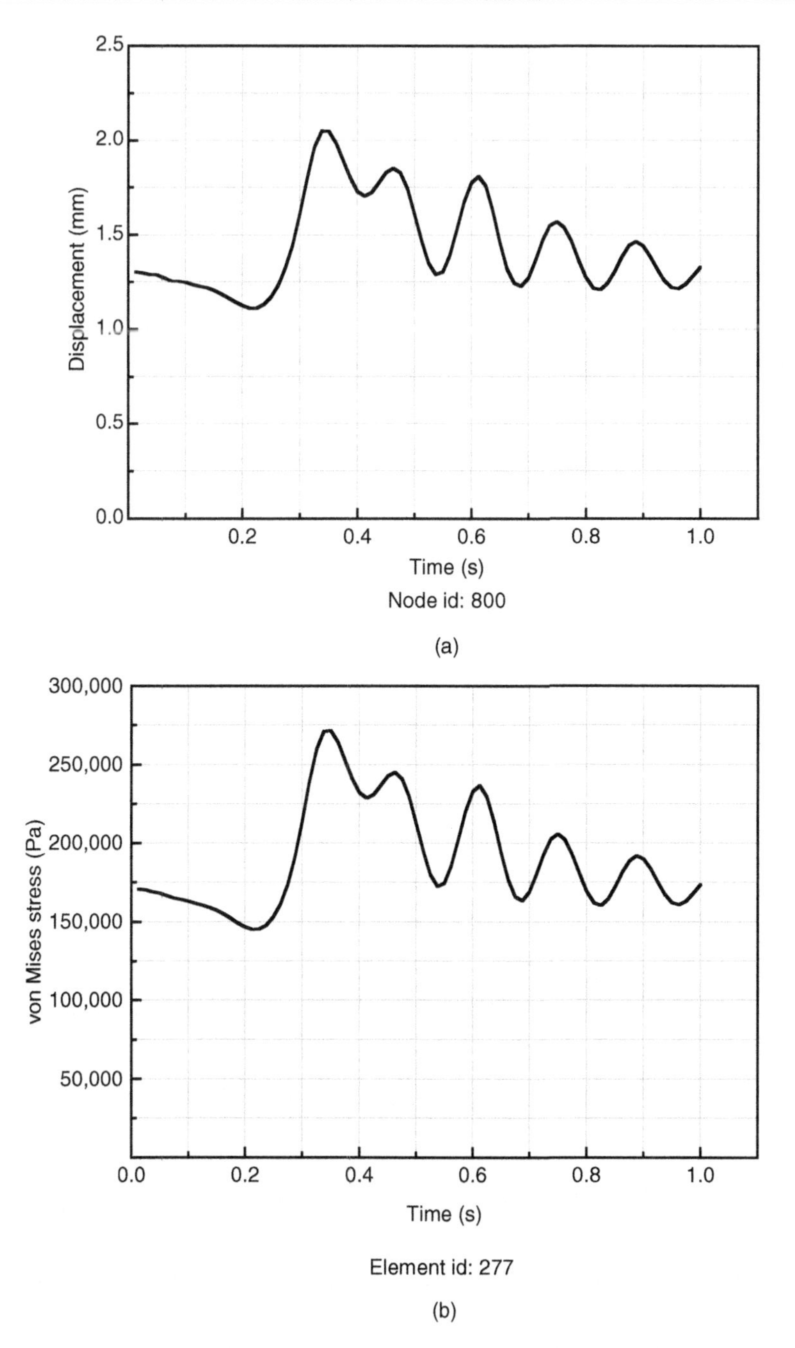

Figure 13.9 Displacement and von Mises time history curves where maximum value occurs. (a) Time history of displacement where the max value occurs. (b) Time history of von Mises stress where the maximum value occurs.

Other multiphysics applications

14

Chapter Outline

14.1 FSI simulation of a sensor device in civil engineering

Sensor is a kind of a measuring device that transforms the measurement quantities into electrical signals or other forms of information using certain correlations. In civil engineering, the sensor is widely used to measure stress, strain, displacement, and other physical quantities. In this section, the miniature sensor model originating from civil engineering is chosen as an example. Through fluid–structural coupling analysis of this model, dynamic characteristics of the sensor are clarified.

14.1.1 The physics model

As shown in Figure 14.1, the sensor model for civil engineering consists of four structural components labeled as solid 1, solid 2, solid 3, and solid 4 and one fluid domain labeled as fluid. The fluid part is enclosed by the structural components. By exerting sinusoidal displacement on the top of solid 1, the structural components of solid 1, solid 2, and solid 3 vibrate and squeeze the fluid domain. The successive fluid forces around the structure of solid 4 deform it. Then, the strain value is obtained through the embedded sensor element in solid 4. In this way, we can get the investigating physical quantities from this sensor.

From Figure 14.1, it is known that the global size of a model is as small as the micrometer level, and thickness in the direction that is perpendicular to the paper is 0.5 μm. In addition to the tiny order of magnitude of the model size, material of the structure in this model is comparatively extreme. As shown in Figure 14.1, relatively hard materials are used in solid 1, while soft materials are used in solid 2 and solid 3 with a small Young's modulus, but the Poisson's ratio is, $\nu = 0.49$,

Q. Zhang & S. Cen: Multiphysics Modeling. http://dx.doi.org/10.1016/B978-0-12-407709-6.00014-6

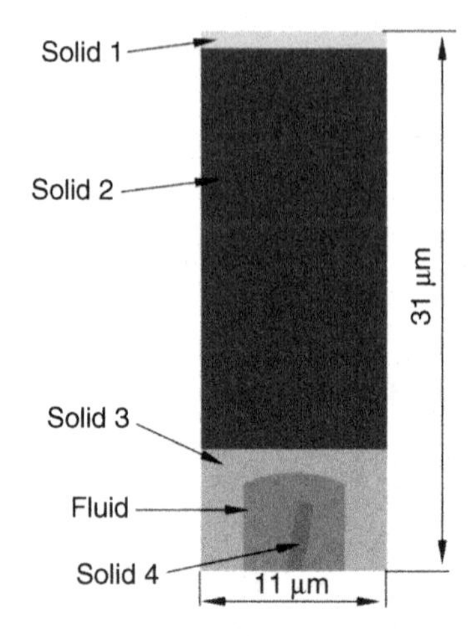

Figure 14.1 Dimensions of the whole model.

which is close to the incompressible material. The material in solid 4 also has a small Young's modulus, which is used with density, $\rho = 1.0 \times 10^{-15}$ kg/μm^3; viscosity, $\mu = 1.0 \times 10^{-9}$ kg/μm·s; and bulk modulus, B = 2.0×10^3 MPa. Due to the coupling between incompressible fluid in a closed space and structure, solution cannot converge using the weak coupling method. Therefore, the fluid–structural strong coupling analysis built into the INTESIM software is adopted in this example. The description on the applicable range of strong and weak coupling method is given in Chapter 3 (Table 14.1).

14.1.2 Analysis conditions

The total coupling analysis time is 0.1 s, and the time step size is 0.0001 s. The analysis condition of the structural model is shown in Figures 14.2 and 14.3. Time-varying displacement condition, $u_x = 1.0 \sin(2\pi ft)$μm, is specified in *X* direction on the top line boundary of solid 1, as shown in Figure 14.2. Symmetry conditions are applied on the top, front, and back surfaces of the model. No-slip wall condition is applied at the bottom surface. The position contacting the fluid is defined as the structural coupling interface. Furthermore, a monitoring point, point 1, located at (0.2085, 3.996, 0.5) is built on solid 4, to monitor the deformation in the *X* direction in order to provide needed observation data for civil engineering.

Table 14.1 Material properties of structural components

Components	Densities (kg/μm^3)	Young's modulus (MPa)	Poisson's ratio
Solid 1	2e-12	2e4	0.45
Solid 2	1e-15	1e-6	0.49
Solid 3	1e-15	1e-6	0.49
Solid 4	1.35e-15	1e-2	0.35

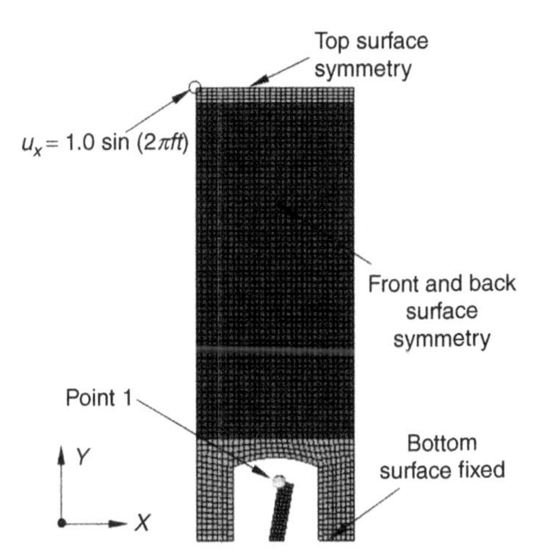

Figure 14.2 Boundary condition of the structural model.

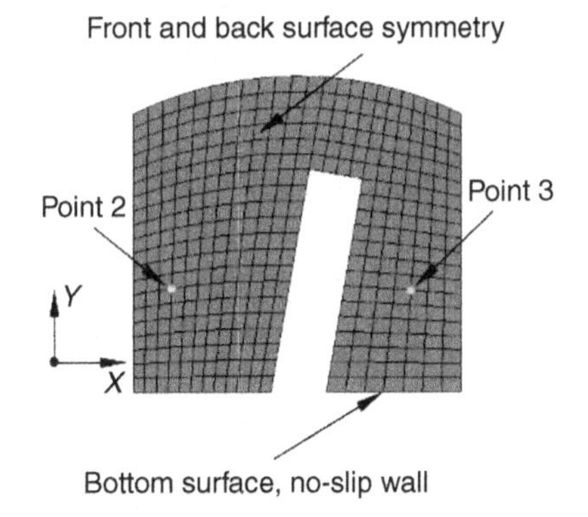

Figure 14.3 Boundary condition of the fluid model.

The analysis condition of the fluid model is shown in Figure 14.3, where symmetry boundary is applied on the front and back surfaces and no-slip wall boundary condition is applied on the bottom surface, while the other boundaries are considered as a fluid–solid coupling interface. The monitoring point 2 and point 3 are located at (2.297, 1.871, 0.5) and (2.084, 1.832, 0.5) at initial state, respectively, as shown in Figure 14.3 to monitor the time history of fluid pressure. As for the coupling conditions, strong coupling method is used for the structural and fluid coupling interface by the multipoint constraint (MPC)-based method, which can be referred to in Section 3.2.3.

14.1.3 Simulation results

We check the simulation results at frequency point $f = 15$ Hz. The time histories of fluid pressure at the monitoring points are presented in Figure 14.4. The black line

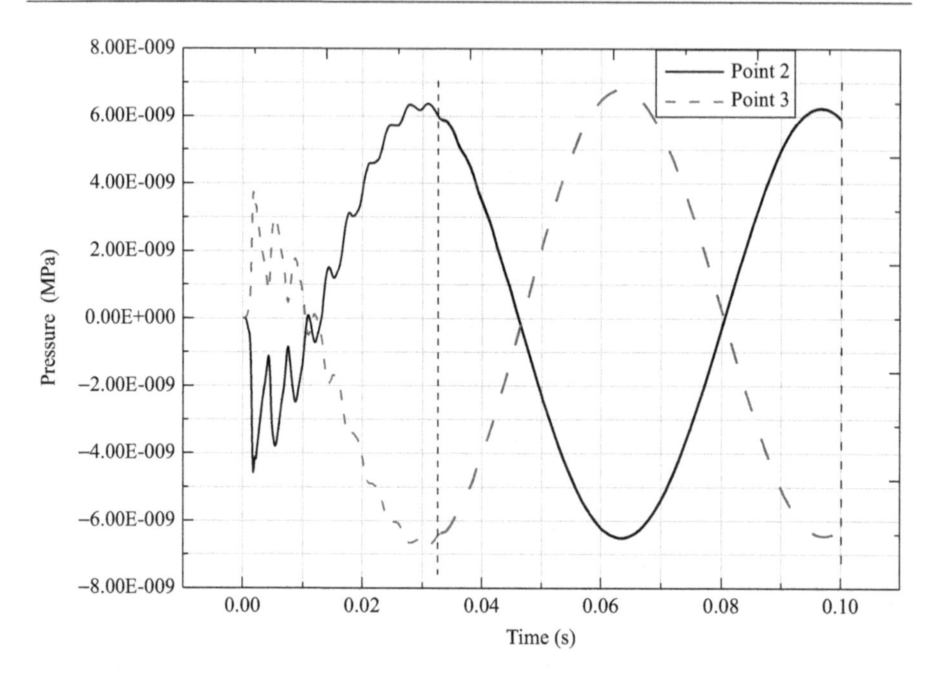

Figure 14.4 Time histories of the fluid pressure on the fluid monitor point.

represents the time history of pressure at point 2, and the red line represents the time histories of pressure at point 3. From this diagram, it is found that during the initial period, resulting pressure of the fluid starts with oscillation, and it soon turns to stable. Then a periodic stable solution is formed, and the vibration frequency is close to frequency of imposed displacement on top of the structure.

The structural component, solid 4, deforms under the effect of fluid force. The required monitoring data is provided for the application in civil engineering. Due to the effect of fluid–solid bidirectional coupling, the time history of displacement in the X direction at monitoring point 1, is presented in Figure 14.5. From this diagram, because of the unstable fluid force in the beginning of the analysis, the structural deformation also starts with oscillation and soon turns to periodic stable state. The vibration frequency of the structure is almost the same with the frequency of oscillating fluid pressure, as well as the frequency of imposed displacement on top of solid 1.

Take stable period $t = 0.033 - 0.1$ s in Figure 14.6 as an example. The contour plot of structural displacement is fully examined. As shown in Figure 14.6, the contour plot of structural displacement is presented at $t = \frac{T}{4}, \frac{T}{2}, \frac{3}{4}T$, and T with unit μm. In this period, imposed displacement in the X direction appears as the cosine function of time and as shown in the figure, the structure deforms in the negative direction of X at, $t = \frac{T}{4}$ and gradually transforms to the positive direction. So the structure deforms

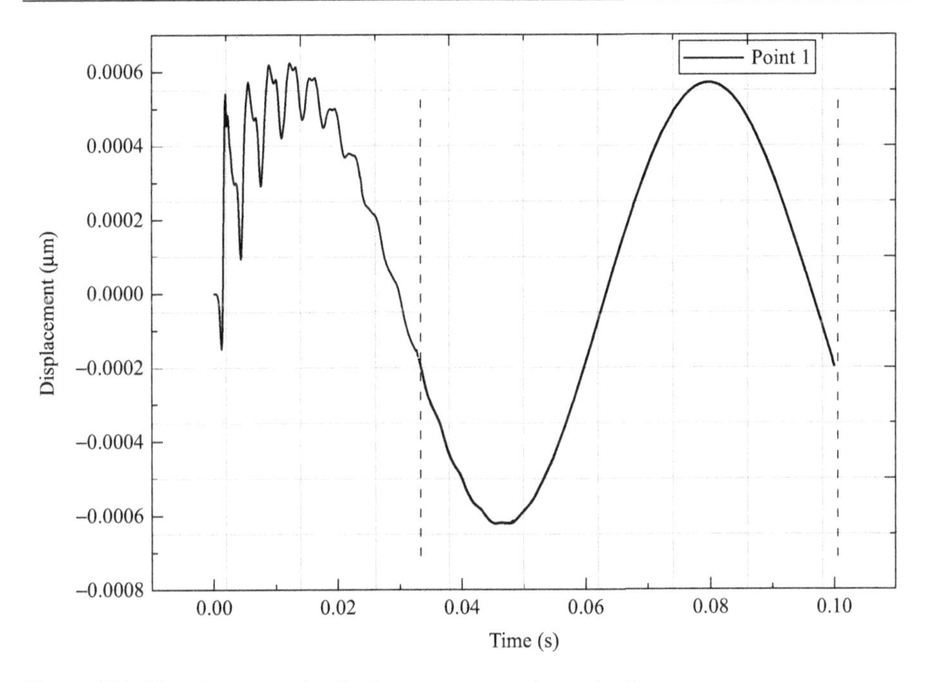

Figure 14.5 Time history of the displacement on monitor point 1.

with the direction of the imposed displacement, there is no obvious phase change between them.

The vector diagrams of flow velocity distribution at $t = \frac{T}{4}, \frac{T}{2}, \frac{3}{4}T$, and T in one time period, $t = 0.033 - 0.1$ s, are presented in Figure 14.7. It is found that the interface coupling in the flow direction is driven by, and follows with deformation of the structure.

14.1.4 Conclusions

The MPC-based strong coupling method implemented in the INTESIM software is applied to solve the sensor model that involves interaction between enclosed incompressible fluid flow and flexible structure. Time histories of the structure displacement at the monitoring points under certain load frequency are obtained. From the fluid pressure-monitoring curve, it is found that initial oscillation of the fluid pressure becomes stable shortly. Consequently, it is advised that at least two analysis time periods should be carried out to get a stable periodic response.

In addition, because of the nature of the interaction between enclosed incompressible fluid flow with a flexible structure, the weak coupling method is unable to get a converged solution. Therefore, only the strong coupling method can be used to analyze this kind of a problem.

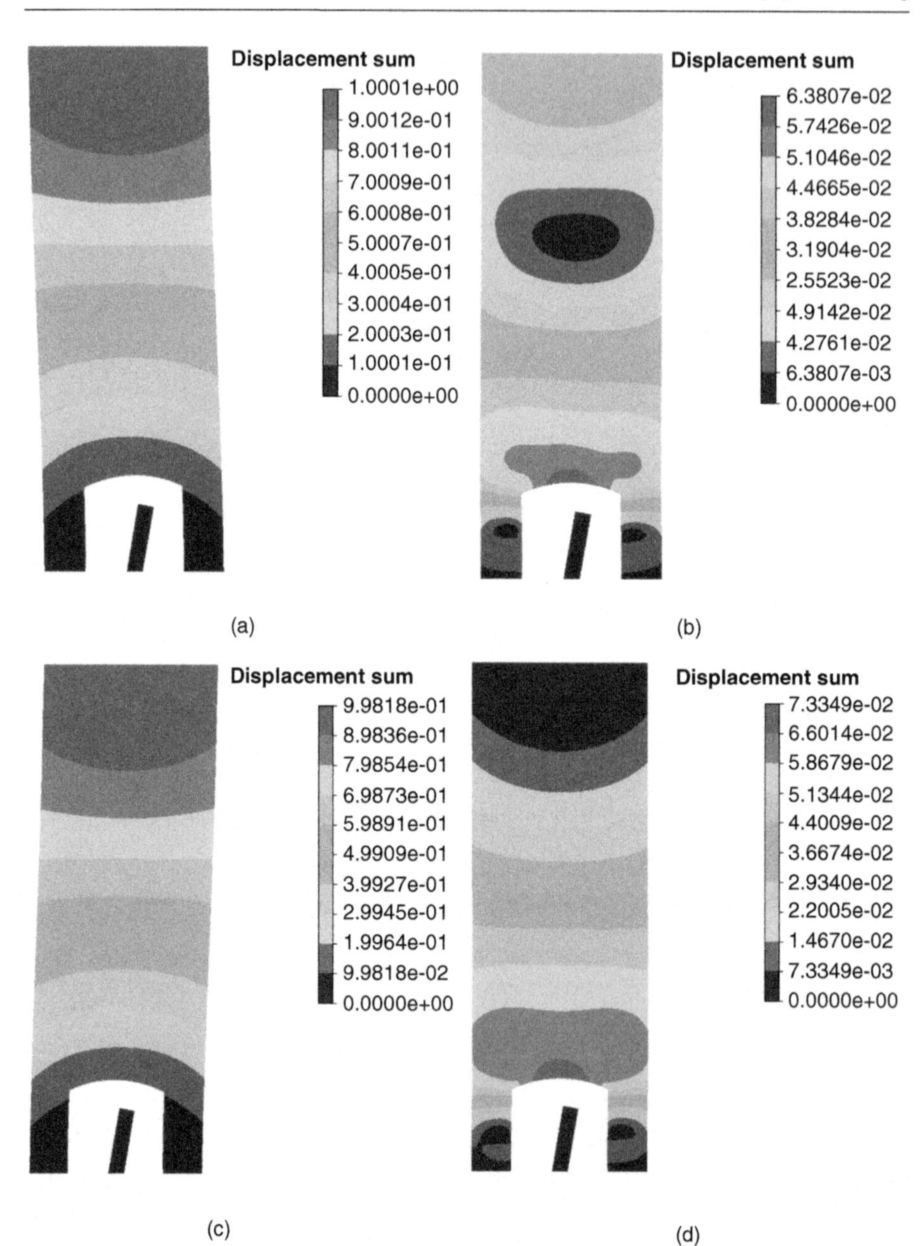

Figure 14.6 Contour plot of the structural displacement. (a) $t = \frac{T}{4}$; (b) $t = \frac{T}{2}$; (c) $t = \frac{3}{4}T$; and (d) $t = T$.

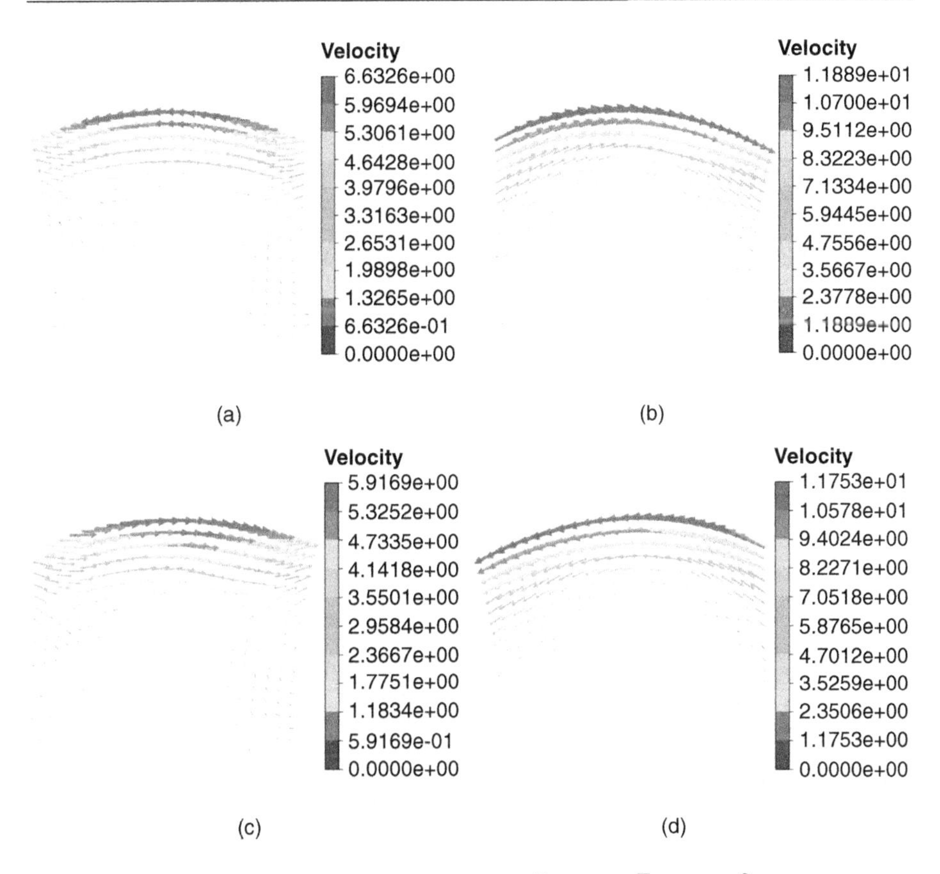

Figure 14.7 Velocity vector of the fluid domain. (a) $t = \frac{T}{4}$; (b) $t = \frac{T}{2}$; (c) $t = \frac{3}{4}T$; and (d) $t = T$.

14.2 Acoustic structural coupling case

14.2.1 Physical model of the acoustic structural coupling model

In most cases, the coupling relation between the acoustic field and structure could be ignored, and the separate calculation of the acoustic field will not affect the computational accuracy. However, under some special circumstances, the coupling relation between the acoustic field and the structure must be considered. Otherwise, the simulation result will be inaccurate. In the case of fluid medium with a large density or the size of the structure being large and the material being soft, the sound counter reflects on the structure could be strong, and the coupling relation between fluid and structure needs to be considered.

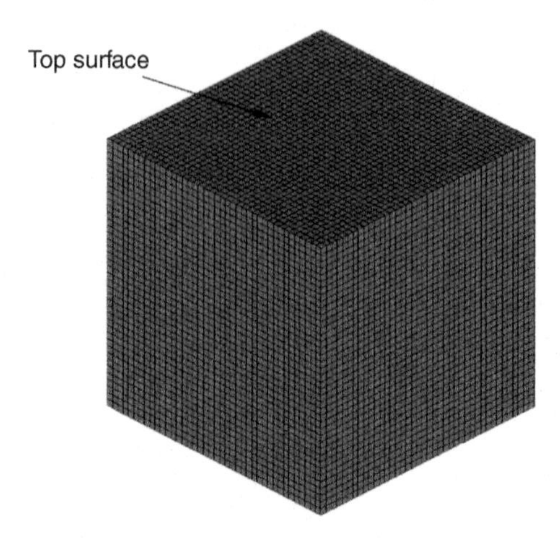

Figure 14.8 The grid model of the closed metal box.

Consider the problem of a hollow cube box enclosed by thin aluminum plates whose side length and thickness are 1 and 0.002 m, respectively. The box is filled with water. Inside the box, a structural excitation of simple harmonic motion is exerted at the center of the bottom surface where the four baselines are fixed. The structural and acoustic field responses are analyzed under the coupling effects of structural and acoustic fields driven by the structural excitation.

The hexahedral solid elements are adopted in meshing, and the mesh size is chosen as 0.025 mm, as shown in Figure 14.8. The structure's material is linear elastic, density is 2710 kg/m^3, Young's modulus is 7.0E + 10 Pa, and Poisson's ratio is 0.346.

Hexahedral elements are also adopted in the acoustic mesh inside the closed box, and the element size is the same as the structural mesh. The sound propagation velocity of the acoustic material of water is set to be 1500 m/s, and the density of the water is 1000 kg/m^3.

14.2.2 Analysis conditions

First of all, modal analysis of the acoustic-structural coupling system composed of the box and the inside fluid is carried to find out the resonance frequency and modal shape of the system. The simple harmonic motion with amplitude of 0.01 m in the *Z* direction and the first resonance frequency in the coupling system are placed in a square range of 0.1 m $\times$ 0.1 m at the center of the box's bottom surface. Then the structural and acoustic responses are analyzed under the coupling effects of structural and acoustic field driven by the structural excitation. In this analysis, displacement of the four base lines is fixed, and the nodes on the inner surface of the structural mesh are coupled with the outside surface of the acoustic field. The loads and constraint conditions on the acoustic-structural coupling in the metal box are presented in Figure 14.9.

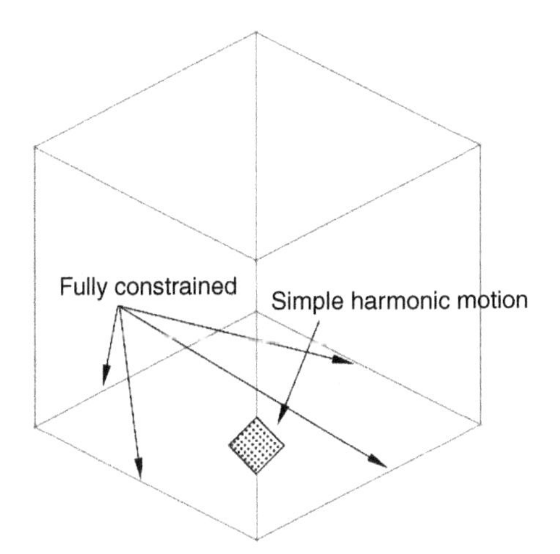

Figure 14.9 The loads and constraint conditions on the acoustic-structural coupling in the box.

14.2.3 Results and discussion

The first 10 order resonance frequencies of the acoustic-structural coupling systems of the box are shown as Table 14.2.

The structure deformation contour plot of the first-order resonance modal in the acoustic-structural coupling system is shown in Figure 14.10.

As shown in Figure 14.10, larger deformation is produced on the top surface of the box than on the four sidewalls, which is the result of the fully fixed constraints of the four baselines of the box.

Distribution contour plot of the acoustic pressure in the first-order resonance modal of the acoustic-structural coupling system is shown in Figure 14.11:

Table 14.2 **Resonance frequencies of the acoustic-structural coupling system of the box**

Modals	Frequencies (Hz)
1	5.1207
2	12.2475
3	12.8362
4	12.8381
5	13.8266
6	17.8138
7	32.2373
8	32.2397
9	33.8133
10	35.7645

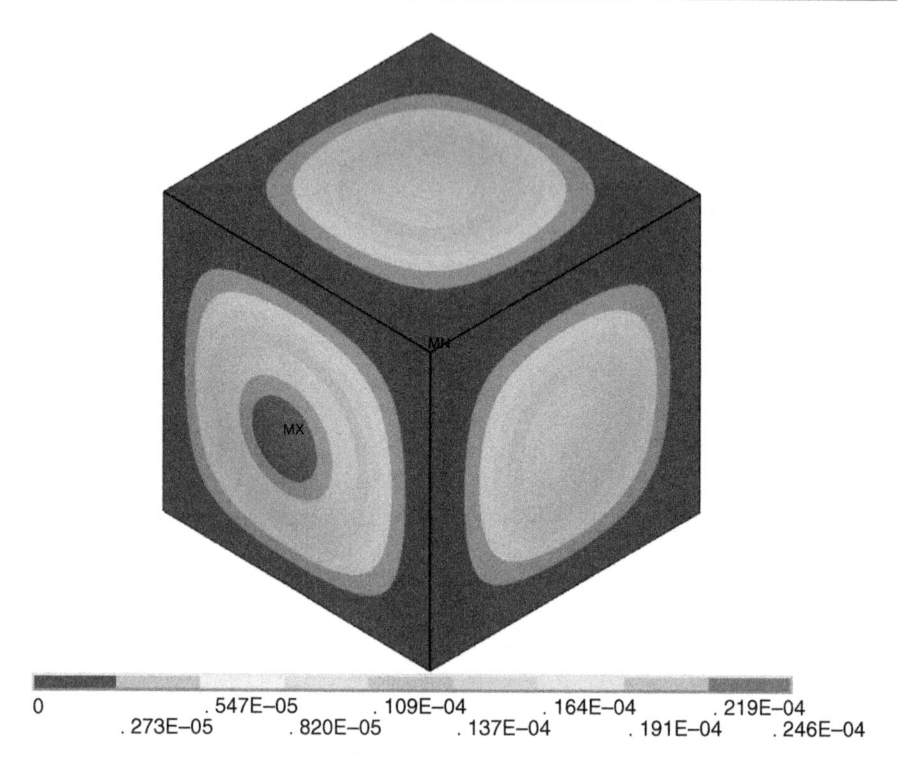

Figure 14.10 The structural deformation contour plot of the first-order resonance modal in the acoustic-structural coupling system.

According to Figure 14.11, a larger acoustic pressure distribution is generated near the roof of the box in the internal acoustic field, and it becomes smaller at the internal acoustic field close to the bottom.

The contour plots of the structural displacement of the coupling system are demonstrated in Figure 14.12. A first-order harmonic excitation, with resonance frequency 5.1207 Hz and amplitude of the displacement value 0.01 m, is exerted at the bottom wall. Compared to the contour plot of structural deformation from modal analysis in Figure 14.10, both results show the same pattern, but the excitation phases have a difference of 180°. The distributions of the acoustic pressure of the coupling system are shown Figure 14.13. Compared with the results in Figure 14.11, the same pattern can be found, but the excitation phases also have a difference of 180°. These results prove that after applying a harmonic excitation with 5.102 Hz frequencies on the bottom of the box, the structural field resonates with the acoustic field through coupling. The resonance mode has the same shape as that of the modal analysis at a frequency of 5.102 Hz. However, it is in the opposite direction because of the 180° difference of the excitation phase.

By coupling the structural and acoustic fields, energy produced by structural excitation could spread from the bottom surface to the top surface through sound fields.

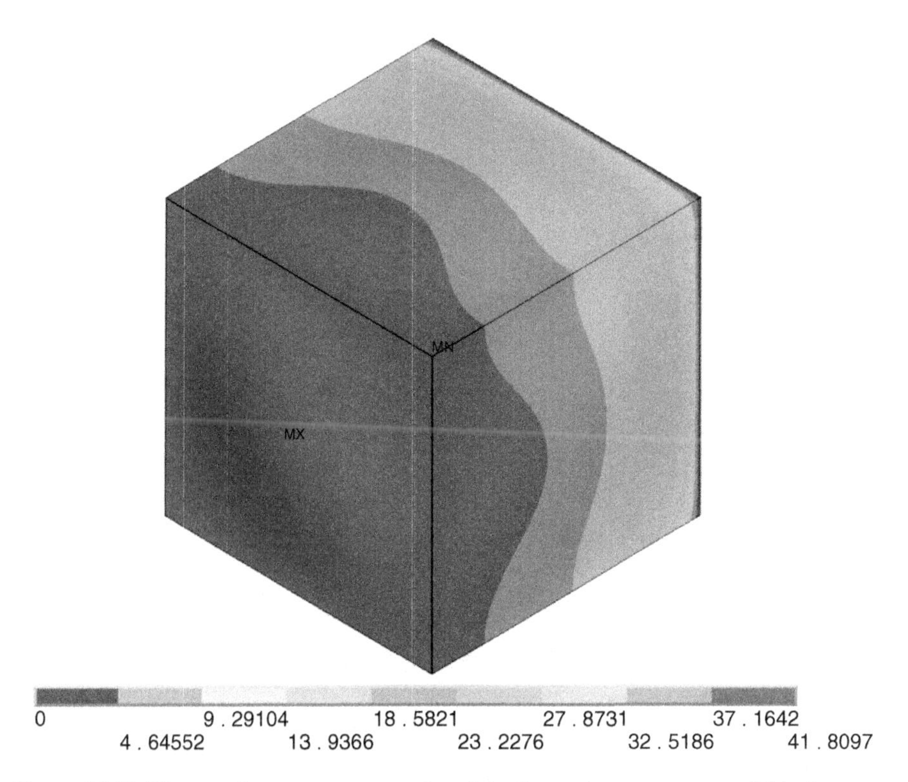

Figure 14.11 The sound pressure contour plot of the first-order resonance modal in the structural–sound coupling system.

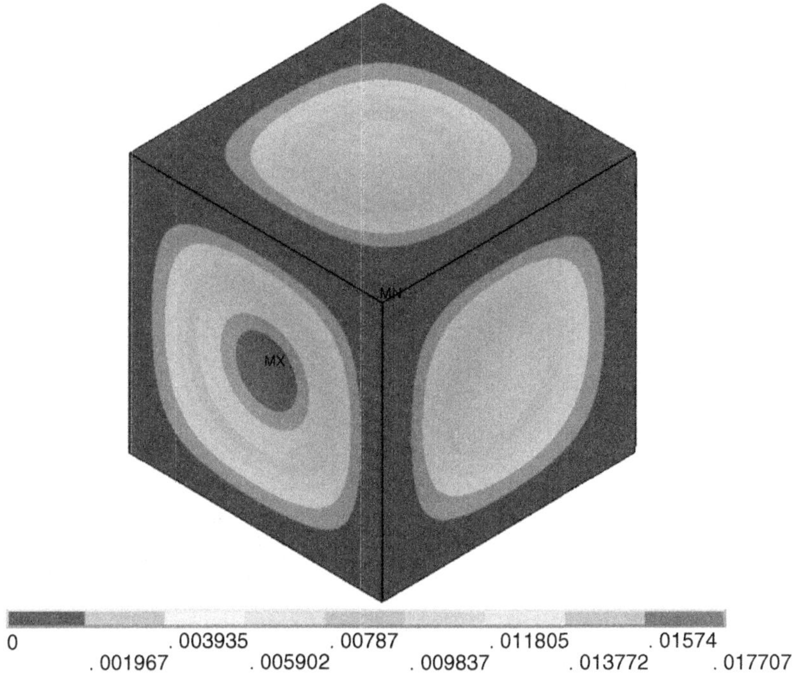

Figure 14.12 The structural deformation contour plot under the simple harmonic motion of the structural–sound coupling system.

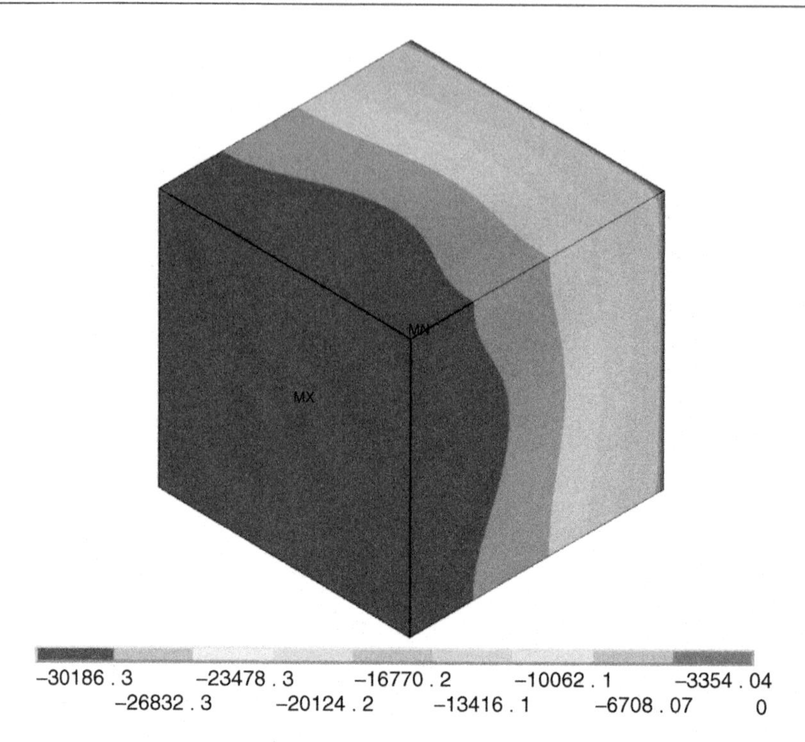

Figure 14.13 The acoustic pressure contour plot under the simple harmonic motion of the structural–sound coupling system.

Consequently, larger deformation and acoustic pressure occur on the top surface of the structure.

This also proves the necessity in the analysis of this kind of acoustic-structural coupling problem. If the acoustic field or the structural field is calculated alone and the coupling effect is neglected, the result would no longer be accurate.

Code implementation of multiphysics modeling

15

Chapter Outline

15.1 Overview of commercial CAE software for multiphysics

15.1.1 Major trends in commercial coupling physics simulations

- Intersolver coupling through a third-party coupling interface.
- Single set of codes to do multiphysics simulation (e.g., ADINA, INTESIM): strong coupling but less flexibilities and efficiency.
- Unified GUI for all physics models.
- Integrated CAD/CAE/PDM and design optimization platform with multiphysics simulation capability.
- Multiphysics simulation for large-scale real engineering problems.

15.1.2 ANSYS

1. Largest CAE software and technology focused company in the world.
2. Currently state of the art solver technology for each individual physics: structure/solid mechanics (ANSYS mechanics), CFD (Fluent, CFX), and electromagnetic (ANSYS-Emag, ANSOFT), etc.
3. Unified CAD, meshing, and design tools in the workbench environment.
   - **a.** Not unified GUI in CAE pre- and postprocessing, user needs to learn several software programs to carry out multiphysics simulation for different physics types.
   - **b.** No strong coupling method to deal with strongly coupled fluid-structure problems.
   - **c.** Expensive and complicated.

Q. Zhang & S. Cen: Multiphysics Modeling. http://dx.doi.org/10.1016/B978-0-12-407709-6.00015-8

15.1.3 COMSOL (Comsol, 2013)

1. Developed by a multiphysics simulation software company in Sweden, partial differential equations-based multiphysics solver.
2. Considerable marketing efforts make the company and their products one of the most influential multiphysics simulation products.
3. Unified user interface for all physics modules.
4. A wide range of physics modules.
   a. Based on partial differential equations directly, certain mathematical and physics background is needed for advanced users.
   b. No promising capability for fluid–structure interaction problems.
   c. Less efficient for large-scale problems.
   d. Most of the users are from academic fields, fewer users are from the industrial fields.

15.1.4 ADINA

1. Research-oriented multiphysics simulator.
2. Both strong coupling and weak coupling methods (single set of codes) are available for fluid–structure problems.
   a. Not a very friendly user interface.
   b. Structure, CFD, and heat transfer modules and recently piezoelectric modules have been developed.

15.1.5 INTESIM

1. Advanced physics modules that include control surface mapping, CFD, heat flow, electrostatic, electric, piezoelectric, electromagnetic, and acoustics modules
2. A variety of coupling technologies: coupled elements at element matrices and degrees of freedom (DOFs) level; direct interface DOF coupling; constraint equation–based interface strong coupling; Lagrange multiplier–based strong coupling; general weak coupling interface
3. Unified graphical user interface for all physics problems and physics couplings
4. Arbitrary Lagrangian Eulerian–based morphing and automatic remeshing
5. Object-oriented C++ code
   a. Missing some important features in each individual physics models compared with most popular general CAE solvers
   b. Importing major CAD files; newly released CAD modeling tools
   c. Relatively new products

15.1.6 Others

ABAQUS has started to add simple CFD features into structure analysis code. The CD-Adapco Group is adding mechanics features into CFD code.

MpCCI still has a stand-alone code-coupling interface; no physics solver is provided.

Algor has not had many new development activities since being purchased by Autodesk.

15.2 Code implementation for multiphysics modeling

Most of the numerical methods and capabilities listed in Chapters 1–14 have been implemented in INTESIM software. This section provides simple ideas about how to implement a multiphysics simulation code.

15.2.1 Infrastructure of strong coupling

```
Initialize:
        1.  Setup default value
        2.  Read input file into Solver database
        3.  Model preparation: Generate interior nodes; Create External node id to
            internal node id map array; Reset node id to internal id for element
            connectivity and BCs
        4.  Initialization: physics models,  mesh  initialize,  and coupling conditions
        5.  Node reordering (if not sparse solver)
        6.  Solver setup

Restart setup: Initial Mesh quality measurement;
               If <Remesh Key is on>
                      Solution interpolation:
                           •  Read restart file, get the solutions on old mesh
                           •  Do Mapping between Old Mesh and New Mesh
                           •  Interpolation: Use interpolation routine for New Mesh
                              region, directly (node to node map) get value  for un-
                              remeshed region
                           •  Move New Mesh to Old Mesh, mesh quality initialization
                           •  Set Remesh Key to 0
               <Else> Read previous result file as a normal restart

Time step loop
* Setup Timestep, Time
* Solve coupling
       Backup solutions and variables at previous time step.
       1. Set Global Dirichlet BCs
       1a. Set Dirichlet BCs for Mesh motion
       2. PMA predictor (in case of PMA time integration) and updates FSI related
       variables
       2a. Updates FSI variable: (Updates stress, grid coordinates, Solve mesh
morphing etc)
       3. Nonlinear iteration loop
              a.  Begin Nonlinear stagger loop
              b.  Calculate physics models
              c.  Setup Neumann BCs for all physics models
              d.  Calculate residual vector; convergence check
              e.  Solver coupled equation
              f.  PMA corrector
              g.   Update FSI variable: (Updates stress, grid coordinates, Solve mesh
                  morphing etc)
              h.   End of stagger loop
```

```
      4. Update dt back to previous step or go to next step
* Output results:
If remesh feature is on:
        Mesh quality measurement (Check element quality for Old Mesh)
        •  Update Coordinates for Old Mesh
        •  Mesh quality check for Old Mesh, Setup Remesh Key
        If Remesh Key is needed
        •  Prepare remesh data (e.g., setup boundary meshes and write necessary
           files for mesh generator);
        •  Do Remesh
        •  Read new mesh file, reset exterior node ids and element ids for new mesh
           group, make it consistent with the boundary node ids of the old mesh
        •  Create a new input file: Delete and add elements, interior nodes,
           components (nodal and element type for current release); and BCs if
           necessary. Modify Coordinates of un-remeshed region.
        •  Cleanup most Solver memories;
        •  Go to Initialize 2
 Else go to next time step
 End of time

 Finalize
```

15.2.2 The definition of the coupled element in INTESIM

15.2.2.1 Object oriented data structure for coupled elements

It is convenient to use object oriented concept (C++) to define coupled physics elements. The base class for element data is BaseElementData, which includes all common element information for all physics types (e.g., element properties, element type, element node connectivity).

Following is one example of the class definition of the thermal fluid element that is inherited from the fluid element and thermal element. Both the fluid and thermal element classes are inherited from the base element class.

BaseElementData: with properties, type, node connectivity
StructureElementData: + BaseElementData
ThermalElementData: + BaseElementData
ThermalStructure: FluidElementData + ThermalElementData (Figure 15.1)

```
class CElementData
{
protected:
        int * _nodes;               // node index
        int _properties[ELEM_END];// element properties
        double *_grid;              // element nodal coordinates
        int _ngauss;                // number of integration points
};
CElementData_Therm::CElementData_Therm () : CElementData()
{
}
```

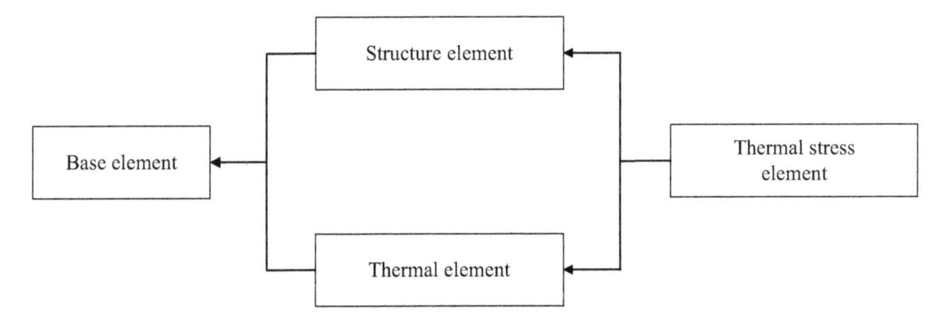

Figure 15.1 Element inheritance structure for thermal–stress element (C++ style).

```
CElementData_Fluid::CElementData_Solid ():CElementData()
{
}

CElementData_ThermalSolid::CElementData_ThermalSolid ():CElementData_Solid(),
CElementData_Therm()
{
}
```

15.2.2.2 Calculation procedure for the thermal–stress element

Loop over all thermal–stress element

1. Calculate the pure structure related matrices and vectors.
2. Calculate the pure thermal related matrices and vectors.
3. Calculate the coupling matrices and vectors.
4. Assemble the matrices and vectors from each physics and coupling terms into the coupled matrices and vectors

End of element calculation

15.2.2.3 Management of nodal DOFs for multiphysics element

One of the features of the multiphysics code is allowed to freely define node types and add DOFs into existing nodes. This makes it possible to combine any two, three, four base element types or base node types into a coupled element or coupled node.

Example eight-node hexahedral thermal stress coupled element:

Structural node: $\{U_x, U_y, U_z\}$

Temperature node: $\{T\}$

For thermal structure node $\{U_x, U_y, U_z\} + \{T\} = \{U_x, U_y, U_z, T\}$

The nodal DOFs list for eight-node hexahedral thermal element are given:

```
/* Thermal: 8-node hex*/
{ {TEMP}, {TEMP}, {TEMP}, {TEMP},
{TEMP}, {TEMP}, {TEMP}, {TEMP} }
```

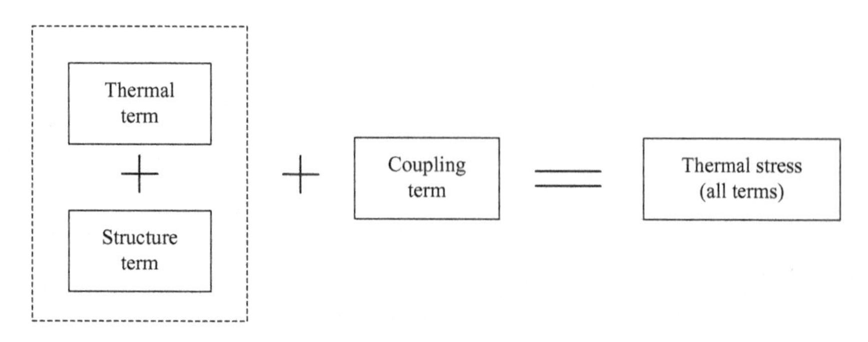

Figure 15.2 Element assembly for the thermal stress.

The eight-node hexahedral structure element:

```
/* CFD: 8-node  hex*/
{ {UX, UY, UZ}, {UX, UY, UZ}, {UX, UY, UZ}, {UX, UY, UZ},
{UX, UY, UZ}, {UX, UY, UZ}, {UX, UY, UZ}, {UX, UY, UZ} }
```

The eight-node thermal–stress element:

```
/* Thermal stress: 8-node  hex*/
{ {UX, UY, UZ, TEMP}, {UX, UY, UZ, TEMP}, {UX, UY, UZ, TEMP}, {UX, UY, UZ, TEMP},
{UX, UY, UZ, TEMP}, {UX, UY, UZ, TEMP}, {UX, UY, UZ, TEMP}, {UX, UY, UZ, TEMP} }
```

where TEMP represents the nodal temperature and {UX, UY, UZ} is the displacement vector (Figure 15.2).

15.2.2.4 The matrix and vector assembly for interface strong coupling method

The implementation of the matrix assembly for the interface strong coupling method.

1. Eliminate the slave DOFs on the coupling interface.
   The DOFs of the slave node on the coupling interface need to be constrained on the coupling interface.
2. Assemble the matrix and vector from the slave DOFs into the master DOFs based on the mapping result.
   The residual vector of the slave DOFs need to be distributed into the corresponding master DOFs. Map the slave nodes into the master mesh and distribute the nodal value on the slave into the related nodes on the master.

$$R_i^S = \sum_{k=1}^{n^{S(i)}} w_{k,i} R_k^M \tag{15.1}$$

where, $\boldsymbol{w_{k,i}}$, is the weight factor of slave node $\boldsymbol{i}$ corresponding master node $\boldsymbol{k}$ and $\sum_{k=1}^{n^{S(i)}} w_{k,i} = 1$. $\boldsymbol{n^{S(i)}}$, is the number of corresponding master nodes for slave $\boldsymbol{i}$.

On the contrary, the residual for mater node k due to the contribution from corresponding slave nodes will be:

$$R_k^M = R_k^{M(M)} + \sum_{i=1}^{n^{s(k)}} w_{k,i} R_i^S \tag{15.2}$$

where, $\boldsymbol{n^{S(k)}}$, is the number of corresponding slave nodes of master node $\boldsymbol{k}$.

$R_k^{M(M)}$, is the residual vector of the master DOFs contributed by the elements on the master side. R_i^S, can be expressed by matrix and vector production form:

$$R_i^S = \sum_{j=1}^{ne} k_{ij}^S \varphi_j^S \rightarrow \tag{15.3}$$

where ne, is the number of nodes of slave element; φ_j^S, is the solution of node j slave mesh; and k_{ij}^S, is the element matrix.

Substituting Equation (15.3) in Equation (15.2), one has:

$$R_k^M = R_k^{M(M)} + \sum_{i=1}^{n^{S(k)}} w_{k,i} \sum_{j=1}^{ne} k_{ij}^S \varphi_j^S = R_k^M + \sum_{j=1}^{ne} \left(\sum_{i=1}^{n^{s(k)}} w_{k,i} k_{ij}^S \right) \varphi_j^S \tag{15.4}$$

If we define the matrix contribution of slave to master:

$$k_{kj}^{M(S)} = \sum_{i=1}^{n^{s(k)}} w_{k,i} k_{ij}^S \tag{15.5}$$

Then the residual contribution from slave node to master node can be expressed as:

$$R_k^M = \sum_{j=1}^{ne} k_{kj}^{M(S)} \varphi_j^S. \tag{15.6}$$

3. Solve the coupled equations for master DOFs only.
4. Update the slave DOFs by the constraint conditions on the coupling interface.

$$\varphi_i^S = \sum_{k=1}^{n^{M(i)}} w_{k,i} \varphi_k^M$$

where $n^{s(k)}$ and $n^{M(i)}$, are the number of slave nodes for master node k and the number of master nodes for slave node i, respectively.

15.2.3 Code implementation for intersolver coupling

15.2.3.1 Set up the synchronization points

While using the third-party code coupling interface such as MpCCI, each and every individual code, needs to be linked with the coupling library and inserted into the appropriate function calls at the synchronization points where data transfer need to be done (Figure 15.3).

15.2.3.2 Set up the get and put functions

Several get and put functions that include (but are not limited to) get and put the mesh data and get and put the solution need to be implemented on each physics solver.

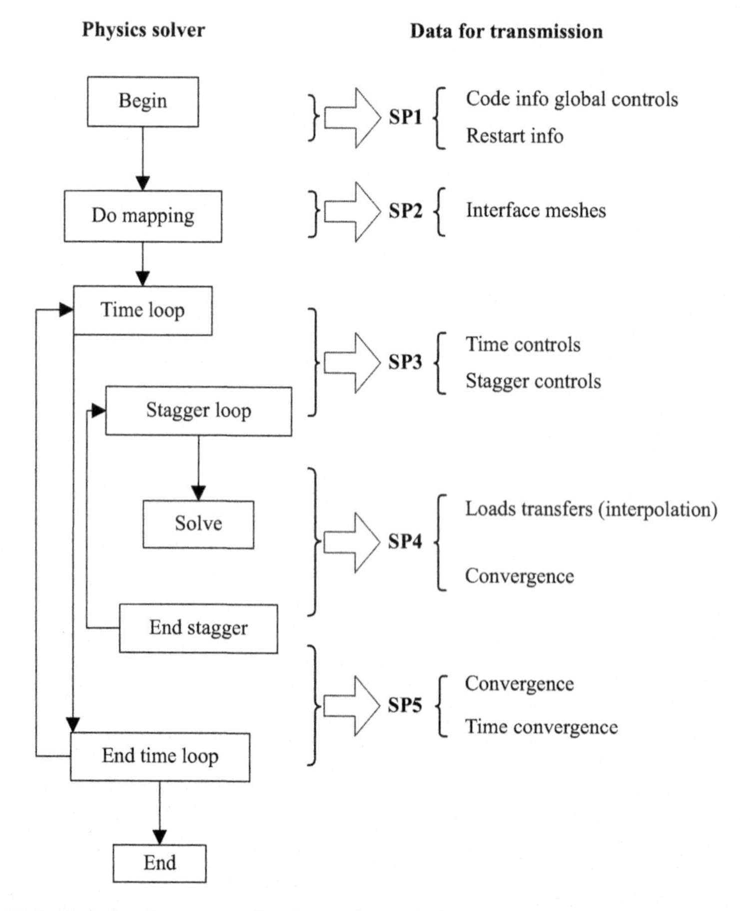

Figure 15.3 Coupling loop control and data transmission in MpCCI.

Table 15.1 Description of the mesh data and arguments in getMesh function

Names	Type	Descriptions
where	string	The name of the location where mesh is located
RegionType	string	The region type: surface, volumetric
iDim	int	Dimension of the mesh
nElem	int	Number of elements
NNMAX	int	Maximum number of nodes per element
NSUMAX	int	Maximum number of nodes on element faces
elemIds	int*	Element connectivities
ElemType	int*	Element type
NodePerElem	int*	Number of nodes for each element
ElemNodes	int*	Node connectivities for each elements
nNode	int	Number of nodes
nodeIds	Int*	Node ID list
grid	Double*	Coordinates of nodes

getMesh function is used to get the mesh data from the other physics for data transfer.

```
int getMesh (char *where, char * RegionType, MeshData * mesh ) { }
```

The data structure that holds the mesh data MeshData (Table 15.1):

```
class MeshData {
protected:
        int iDim, int nElem, int NNMAX,  int NSUMAX,
        int *elemIds, int *ElemType,  int * NodePerElem, int *ElemNodes,
        int * nNode,  int ** nodeIds, double ** grid
}
```

getSolution is the function to get the solution array from the current physics solver.

```
int getSolution (char * What, char *where, char *  when, char * csysidchar, char * FrameType,
                 int *varType, int *num, int *varDim, double ** value )
```

putSolution is the function to put the received data array as the solution of the current physics solver (Table 15.2).

```
int putSolution (char * What, char *where, char * when, char *csysidchar, char * FrameType,
double * value )
```

Table 15.2 The description of arguments in a function of getSolution and putSolution

Names	Type	Descriptions
What	string	The name of the value
where	string	The name of the location where the value is located
csysidchar	string	Coordinate system ID
FrameType	string	The type of reference frame: body-attached rotating; global stationary
varType	int	Type of value
num	int	Number of value sets
varDim	int	Dimension of one value set
elemIds	int*	Element connectivities
value	double*	Value of the solution array

15.2.3.3 Make the code support repeatable iteration

If you want to do bidirectional coupling and need a converged solution during coupling iteration, then your code needs to support repeatable time loops and iterations. This means, during the coupling iteration, each physics solver should be able to repeat the time increment calculation Δt_{cp} inside the coupling stager loop, as shown in Figure 15.4.

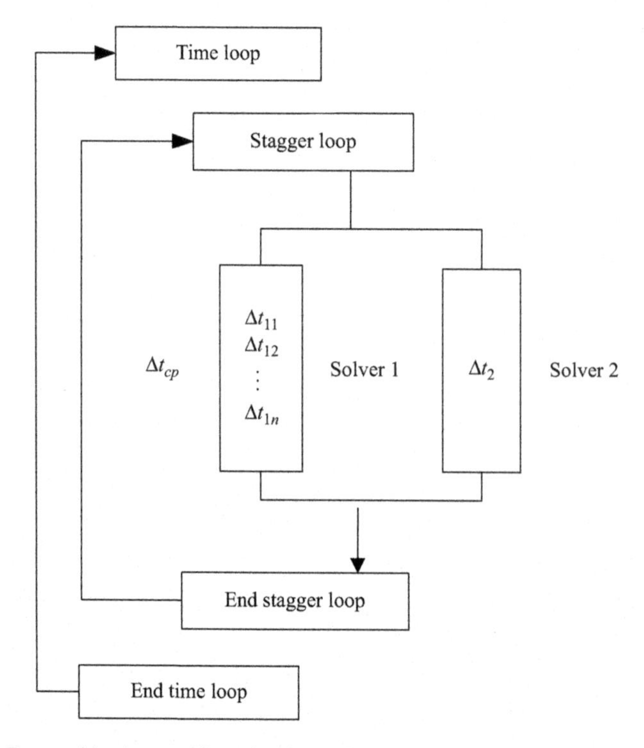

Figure 15.4 Repeatable time and iteration loops for physics solver.

References

Abdoulaev, G., Kuznetsov, Y., Pironneau, O., 1996. The numerical implementation of the domain decomposition method with mortar finite elements for a 3D problem. Preprint, Laboratoire d'Analyse Numerique, Univ. Pierre et Marie Curie, Paris.

Achdou, Y., Kuznetsov, Y.A., 1995. Substructuring preconditioners for finite element methods on nonmatching grids. East West J. Num. Math. 3 (1), 1–28.

Achdou, Y., Maday, Y., Widlund, O., 1996. Méthode itérative de sous-structuration pour les éléments avec joints. C. R. Acad. Sci. 1 Math. 322 (2), 185–190.

Albanese, R., Rubinacci, G., 1990. Magnetostatic field computations in terms of two-component vector potentials. Int. J. Num. Method. Eng. 29 (3), 515–532.

Allen, M., Maute, K., 2005. Reliability-based shape optimization of structures undergoing fluid–structure interaction phenomena. Comput. Method. Appl. Mech. Eng. 194 (30), 3472–3495.

Allik, H., Hughes, T.J., 1970. Finite element method for piezoelectric vibration. Int. J. Num. Method. Eng. 2 (2), 151–157.

Alonso, A., Valli, A., 1996. Some remarks on the characterization of the space of tangential traces of H (rot; Ω) and the construction of an extension operator. Manuscripta Mathematica 89 (1), 159–178.

Alonso, A., Valli, A., 1999. An optimal domain decomposition preconditioner for low-frequency time-harmonic Maxwell equations. Math. Comput. Am. Math. Soc. 68 (226), 607–631.

Amestoy, P.R., Duff, I.S., L'Excellent, J.Y., 2000. Multifrontal parallel distributed symmetric and unsymmetric solvers. Comp. Method. Appl. Mech. Eng. 184 (2), 501–520.

Aminpour, M.A., 1992. An assumed-stress hybrid 4-node shell element with drilling degrees of freedom. Int. J. Num. Method. Eng. 33 (1), 19–38.

ANSYS Mechanical APDL Theory Reference, 2013. Release 15.0.

ANSYS CFX-Solver Theory Guide, 2013. Release 15.0.

ANSYS Fluent Theory Guide, 2013. Release 15.0.

Antunes, O.J., Bastos, J.P., Sadowski, N., Razek, A., Santandrea, L., Bouillault, F., Rapetti, F., 2005. Using hierarchic interpolation with mortar element method for electrical machines analysis. IEEE Trans. Magnet. 41 (5), 1472–1475.

Antunes, O.J., Bastos, J.P.A., Sadowski, N., Razek, A., Santandrea, L., Bouillault, F., Rapetti, F., 2006. Torque calculation with conforming and nonconforming movement interface. IEEE Trans. Magnet. 42 (4), 983–986.

Atluri, S.N., 1992. Computational nonlinear mechanics in aerospace engineering, vol. 146, Reston: Aiaa.

Badiansky, B., 1974. Theory of buckling and post-buckling behavior of elastic structure. Chia-Shun, Yih (Ed.), Advances in Applied Mechanics, vol. 14, Academic Press, New York, pp. 2–63.

Baek, H., Karniadakis, G.E., 2012. A convergence study of a new partitioned fluid–structure interaction algorithm based on fictitious mass and damping. J. Comput. Phy. 231 (2), 629–652.

Barth, T.J., Jespersen, D.C., 1989. The design and application of upwind schemes on unstructured meshes.

Bastos, J.P.A., Sadowski, N., 2003. Electromagnetic Modeling by Finite Element Methods. Boca Raton: CRC press.

Q. Zhang & S. Cen: Multiphysics Modeling. http://dx.doi.org/10.1016/B978-0-12-407709-6.00021-3

Bathe, K.J., 1982. Finite element procedures. Upper Saddle River: Prentice Hall.

Bathe K.J., 1996. Finite Element Procedures, USA Prentice-Hall. Flexible Rotors. Engineering Computations (S0264-4401). 1 (1), 52–64.

Bathe, K.J., 2006. Finite element procedures. Klaus–Jurgen Bathe.

Bathe, K.J., Baig, M.M.I., 2005. On a composite implicit time integration procedure for nonlinear dynamics. Comp. Struct. 83 (31), 2513–2524.

Bathe, K.J., Chaudhary, A., 1985. A solution method for planar and axisymmetric contact problems. Int. J. Num. Method. Eng. 21 (1), 65–88.

Bathe, K.J., Dvorkin, E.N., 1985. A four – node plate bending element based on Mindlin/Reissner plate theory and a mixed interpolation. Int. J. Num. Method. Eng. 21 (2), 367–383.

Bathe, K.J., Wilson, E.L., 1973. Stability and accuracy analysis of direct integration methods. Earthquake Eng. Struct. Dyn. 1 (3), 283–291.

Bathe, K.J., Zhang, H., Ji, S., 1999. Finite element analysis of fluid flows fully coupled with structural interactions. Comp. Struct. 72 (1), 1–16.

Batoz, J.L., Tahar, M.B., 1982. Evaluation of a new quadrilateral thin plate bending element. Int. J. Num. Method. Eng. 18 (11), 1655–1677.

Bazilevs, Y., Calo, V.M., Tezduyar, T.E., Hughes, T.J., 2007. YZβ discontinuity capturing for advection-dominated processes with application to arterial drug delivery. Int. J. Num. Method. Fluids 54, 593–608.

Bazilevs, Y., Takizawa, K., Tezduyar, T.E., 2013. Computational fluid-structure interaction. Wiley, United Kingdom.

Beak, H., Karniadakis, G.E., 2012. A convergence study of a new partitioned fluid-structure interaction algorithm based on fictitious mass and damping. J. Comput. Phy. 231, 629–652.

Belgacem, B.F., Maday, Y., 1997. The mortar and the primal hybrid mortar finite element method of the class H. France.

Belytschko, T., Liu, W.K., Moran, B., Elkhodary, K., 2013. Nonlinear finite elements for continua and structures. New York: John Wiley & Sons.

Ben Belgacem, F., Maday, Y., 1997. The mortar element method for three dimensional finite elements. RAIRO-Modélisation mathématique et analyse numérique 31 (2), 289–302.

Bíró, O., 1999. Edge element formulations of eddy current problems. Comp. Method. Appl. Mech. Eng. 169 (3), 391–405.

Bossavit, A., 1998. Computational electromagnetism: variational formulations, complementarity, edge elements. New York: Academic Press.

Bossavit, A., 2004. Electromagnetisme, en vue de la modelisation, vol. 14, Springer Science & Business Media.

Braess, D., Dahmen, W., 1998. Stability estimates of the mortar finite element method for 3-dimensional problems. East West J. Numer. Math. 6, 249–264.

Brooks, A.N., Hughes, T.J.R., 1982. Stream upwind/Petrov-Galerkin formulation for convection dominated flows with particular emphasis on the incompressible Navier-Stokes equation. Comput. Method. Appl. Mech. Eng. 32, 199–259.

Budiansky, B., 1973. Theory of buckling and post-buckling behavior of elastic structures (No. DEAP-S-5). Cambridge Mass: Harvard University Division of Engineering and Applied Physics.

Buffa, A., Ciarlet, P., 2001. On traces for functional spaces related to Maxwell's equations Part I: an integration by parts formula in Lipschitz polyhedra. Math. Method. Appl. Sci. 24 (1), 9–30.

Buffa, A., Maday, Y., Rapetti, F., 2000. Calculation of eddy currents in moving structures by a sliding mesh-finite element method. IEEE Trans. Magnetics 36 (4), 1356–1359.

Buffa, A., Maday, Y., Rapetti, F., 2001. A sliding mesh-mortar method for a two dimensional eddy currents model of electric engines. ESAIM Math. Model. Num. Anal. 35 (2), 191–228.

Buffa, A., Costabel, M., Sheen, D., 2002. On traces for H (curl, Ω) in Lipschitz domains. J. Math. Anal. Appl. 276 (2), 845–867.

Cao, Z.L., Li, J., Guo, K.H, Zhang, Q., 2012. Simulations of dynamic characteristics of hydraulic engine mounts by FEM with strong fluid-structure coupling. J. Vibration Shock 31 (10), 4–8.

Carpenter, N., Stolarski, H., Belyschko, T., 1986. Improvements in 3-node triangular shell elements. Int. J. Num. Method. Eng. 23, 1643–1667.

Cen, S., Soh, A.K., Long, Y.Q., Yao, Z.H., 2002a. A new 4-node quadrilateral FE model with variable electrical degrees of freedom for the analysis of piezoelectric laminated composite plates. Compos. Struct. 58 (4), 583–599.

Cen, S., Long, Y., Yao, Z., 2002b. A new hybrid-enhanced displacement-based element for the analysis of laminated composite plates. Comp. Struct. 80 (9), 819–833.

Cen, S., Soh, A.K., Long, Y.Q., Yao, Z.H., 2002c. A new 4-node quadrilateral FE model with variable electrical degrees of freedom for the analysis of piezoelectric laminated composite plates. Compos. Struct. 58 (4), 583–599.

Cen, S., Long, Y.Q., Yao, Z.H., Chiew, S.P., 2006. Application of the quadrilateral area co-ordinate method: a new element for Mindlin–Reissner plate. Int. J. Num. Method. Eng. 66 (1), 1–45.

Cen, S., Fu, X., Long, Y., Li, H., Yao, Z., 2007. Application of the quadrilateral area coordinate method: a new element for laminated composite plate bending problems. Acta Mechanica Sinica 23 (5), 561–575.

Chan, T.F., Smith, B.F., Zou, J., 1996. Overlapping Schwarz methods on unstructured meshes using non-matching coarse grids. Numerische Mathematik 73 (2), 149–167.

Chen, W.J., Cheung, Y.K., 1999. Refined non-conforming triangular elements for analysis of shell structures. Int. J. Num. Method. Eng. 46, 433–455.

Chen, Z., Du, Q., Zou, J., 2000. Finite element methods with matching and nonmatching meshes for Maxwell equations with discontinuous coefficients. SIAM J. Num. Anal. 37 (5), 1542–1570.

Chen, Y.L., Cen, S., Yao, Z.H., Long, Y.Q., Long, Z.F., 2003. Development of triangular flat-shell element using a new thin-thick plate bending element based on SemiLoof constrains. Struct. Eng. Mech. 15 (1), 83–114.

Chen, G.B., Yang, C., Zou, C.Q., 2010. Basics of aero-elasticity design, second ed. Beijing: Beihang University Press. 2nd edition.

Cheng, S.W., Dey, T.K., Shewchuk, J., 2012. Delaunay mesh generation. Boca Raton: CRC Press.

Chessa, J., Belytschko, T., 2004. Arbitrary discontinuities in space-time finite elements by level sets and X-FEM. Int. J. Num. Method. Eng. 61 (15), 2595–2614.

Chung, J., Hulbert, G.M., 1993. A time integration algorithm for structural dynamics with improved numerical dissipation: the generalized-α method. J. Appl. Mech. 60 (2), 371–375.

Ciarlet, P.G., 2002. The Finite Element Method for Elliptic Problems, vol. 40, Philadelphia: Siam.

Ciarlet, P. Jr., Zou, J., 1999. Fully discrete finite element approaches for time-dependent Maxwell's equations. Numerische Mathematik 82 (2), 193–219.

Clough, R.W., Johnson, C.P., 1968. A finite element approximation for the analysis of thin shells. Int. J. Solid. Struct. 4 (1), 43–60.

Cnem, W.J., Advances in finite element with high performances. In: Yuan, M.W., Sun, S.L. (Eds.), 2001. Computational Mechanics in Engineering and Science. Proceedings of the Chinese Conference on Computational Mechanics'2001, Guangzhou, China, Peking University Press, 136–151 (in Chinese).

COMSOL Multiphysics User's Guide, Version 4.4, 2013.

Corsini, A., Rispoli, F., Santoriello, A., Tezduyar, T.E., 2006. Improved discontinuity-capturing finite element techniques for reaction effects in turbulence computation. Comput. Mech. 38 (4–5), 356–364.

Da, X.Y., Tao, Y., Zhao, Z.L., 2012. Virtual flight simulation based on predictor corrector and nested mesh. J. Aviation 33 (6), 977–983, (in Chinese).

Davat, B., Ren, Z., Lajoie-Mazenc, M., 1985. The movement in field modeling. Magn. IEEE Trans. 21 (6), 2296–2298.

De Boer, A., Van Zuijlen, A.H., Bijl, H., 2007. Review of coupling methods for non-matching meshes. Comp. Method. Appl. Mech. Eng. 196, 1515–1525.

Farhat, C., Van Der Zee, K.G., Geuzaine, P., 2006. Provably second-order time-accurate loosely-coupled solution algorithms for transient nonlinear computational aeroelasticity. Comp. Method. Appl. Mech. Eng. 195, 1973–2001.

Feng, M.L., Li, J., Han, P.B., 2007. Fluid-solid coupling thermal analysis of flameproof-diesel exhaust manifold cooling system. Meikuang Jixie (Coal Mine Machinery) 28 (2), 65–67.

Fish, J., Belytschko, T., 1992. Stabilized rapidly convergent 18 - degrees - of - freedom flat shell triangular element. Int. J. Num. Method. Eng. 33 (1), 149–162.

Flemisch, B., Wohlmuth, B.I., 2006. Nonconforming discretization techniques for coupled problems. Multifield Problems in Solid and Fluid Mechanics. Springer, Berlin, Heidelberg, pp. 531–560.

Fraunhofer, S.C.A.I., 2007. MpCCI Documentation, Part V, User Manual. Fraunhofer Institute for Algorithms and Scientific Computing SCAI, Germany.

Geisberger, A., Khajepour, A., Golnaraghi, F., 2002. Non-linear modelling of hydraulic mounts: theory and experiment. J. Sound Vibration 249 (2), 371–397.

Gebert, G., Kelly, J., Lopez, J., et al., 2000. Wind tunnel based virtual flight testing. AIAA, 2000-0829.

Genta, G., 2007. Dynamics of Rotating Systems. Berlin: Springer Science & Business Media.

Ghosal, S., 1996. An analysis of numerical errors in large-eddy simulations of turbulence. J. Comput. Phy. 125 (1), 187–206.

Giannakopoulos, A.E., 1989. The return mapping method for the integration of friction constitutive relations. Comput. Struct. 32 (1), 157–167.

Glenn, G., Joy, K., Juan, L., 2000. Wind tunnel based virtual flight testing.

Glowinski, R., Pan, T.W., Periaux, J., 1994. A fictitious domain method for external incompressible viscous flow modeled by Navier-Stokes equations. Comp. Method. Appl. Mech. Eng. 112 (1), 133–148.

Glowinski, R., Pan, T.W., Hesla, T.I., Joseph, D.D., 1999. A distributed Lagrange multiplier/fictitious domain method for particulate flows. Int. J. Multiphase Flow 25 (5), 755–794.

Grisvard, P., 2011. Elliptic problems in nonsmooth domains, vol. 69, Philadelphia: SIAM.

Guan, Y., Tang, L., 1992. A quasi-conforming nine-node degenerated shell finite element. Finite elements in analysis and design 11 (2), 165–176.

Guo, J.Z., Wang, M.Z., 2006. Analysis method based on Petri net. Network Inf. Technol. 25 (7), 53–55, in Chinese.

Guo, D., Chu, F.L., Zheng, Z.C., 2001. The influence of rotation on vibration of a thick cylindrical shell. J. Sound Vibration 242 (3), 487–505.

Harold, D.N., 1998. Rotordynamic modeling and analysis procedures: A review. Machine Elements and Manufacturing. JSME Int. J. Series C Mech. Sys. 41 (1), 1–12.

Hibbitt, Karlsson, Sorensen., Inc., 1998. ABAQUS/Standard User's Manual, Version 5.8, Rawtucket, Rhode Island, USA.

Hoppe, R.H., 1999. Mortar edge element methods in R^ 3. East West J. Num. Math. 7, 159–174.

Hu, Q., Zou, J., 2004. Substructuring preconditioners for saddle-point problems arising from Maxwell's equations in three dimensions. Math. Comput. 73 (245), 35–61.

Huerta, A., Liu, W.K., 1988. Viscous flow with large free surface motion. Comp. Method. Appl. Mech. Eng. 69 (3), 277–324.

Hughes, T.J.R., Taylor, R.L., Kanokunkulchai, W., 1977. A simple and efficient finite element for plate bending. Int. J. Numer. Method Eng. 11, 1529.

Hughes, T.J.R., Cohen, M., Haroun, M., 1978. Reduced and selective integration techniques in the finite element analysis of plates. Nucl. Eng. Design 46, 203–222.

INTESIM2010 Theory Manual, Dalian, China, www.intesim.cn.

Iwamiya, T., Nakamura, T., Yoshida, M., 1996. Numerical wind tunnel (NWT) project. Progress of Theoretical Physics Supplement 122, 57–66.

Jansen, K., Shakib, F., Hughes, T.J., 1992. Fast projection algorithm for unstructured meshes. Computational Nonlinear Mechanics in Aerospace Engineering. American Institute of Aeronautics and Astronautics, Inc., Washington, United States, p. 175.

Jansen, K.E., Whiting, C.H., Hulbert, G.M., 2000. A generalized-α method for integrating the filtered Navier–Stokes equations with a stabilized finite element method. Comp. Method. Appl. Mech. Eng. 190 (3), 305–319.

Jeffcott, H., 1919. The lateral vibration of loaded shafts in the neighborhood of a wiring speed-the effect of want of balance. Philos. Mag. J. 37 (6), 304–314.

Jin, J.M., 2014. The finite element method in electromagnetics. New York: John Wiley & Sons, New York.

John, V., Knobloch, P., 2006. On Discontinuity—Capturing Methods for Convection—Diffusion Equations. Numerical mathematics and advanced applications. Springer, Berlin Heidelberg, pp. 336–344.

Johnson, A.A., Tezduyar, T.E., 1994. Mesh update strategies in parallel finite element computations of flow problems with moving boundaries and interfaces. Comp. Method. Appl. Mech. Eng. 119 (1), 73–94.

Johnson, A.A., Tezduyar, T.E., 1996. Simulation of multiple spheres falling in a liquid-filled tube. Comp. Method. Appl. Mech. Eng. 134 (3), 351–373.

Kahn, H., 1955. Use of different Monte Carlo sampling techniques. Santa Monica: Rand Corporation.

Khadra, K., Angot, P., Parneix, S., Caltagirone, J.P., 2000. Fictitious domain approach for numerical modelling of Navier–Stokes equations. Int. J. Num. Method. Fluids 34 (8), 651–684.

Kim, C., Lazarov, R.D., Pasciak, J.E., Vassilevski, P.S., 2001. Multiplier spaces for the mortar finite element method in three dimensions. SIAM J. Num. Anal. 39 (2), 519–538.

Kim, J., Yoon, J.C., Kang, B.S., 2007. Finite element analysis and modeling of structure with bolted joints. Appl. Math. Model. 31 (5), 895–911.

Knobloch, P., 2006. Improvements of the Mizukami–Hughes method for convection–diffusion equations. Comp. Method. Appl. Mech. Eng. 196 (1), 579–594.

Knupp, P.M., 2003. Algebraic mesh quality metrics for unstructured initial meshes, finite elements in analysis and design. SIAM J. Sci. Comp. 39 (3), 217–241.

Kong, Q.L., Liu, Q.B., 1999. Calculation and Strengthening of Railway Transport Capacity. Beijing: China Railway Publishing House, Beijing, China.

Kuczmann, M., 2009. Potential formulations in magnetics applying the finite element method. Lecture notes, laboratory of electromagnetic fields, Széchenyi István University, Gyor, Hungary.

Kuhl, D., Crisfield, M.A., 1999. Energy-conserving and decaying algorithms in non-linear structural dynamics. Int. J. Num. Method. Eng. 45 (5), 569–599.

Küttler, U., Wall, W.A., 2008. Fixed-point fluid–structure interaction solvers with dynamic relaxation. Comput. Mech. 43 (1), 61–72.

Kuzmin, D., Turek, S., 2004. High-resolution FEM-TVD schemes based on a fully multidimensional flux limiter. J. Comput. Phy. 198 (1), 131–158.

Lai, M.C., Peskin, C.S., 2000. An immersed boundary method with formal second-order accuracy and reduced numerical viscosity. J. Comput. Phy. 160 (2), 705–719.

Lalanne, M., Ferraris, G., 1998. Rotordynamics Prediction in Engineering. Hoboken: Wiley.

Landau, L.D., Bell, J.S., Kearsley, M.J., Pitaevskii, L.P., Lifshitz, E.M., Sykes, J.B., 1984. Electrodynamics of Continuous Media, vol. 8, Amsterdam: elsevier.

Lange, E., Henrotte, F., Hameyer, K., 2010. A variational formulation for nonconforming sliding interfaces in finite element analysis of electric machines. Magnetics, IEEE Trans. 46 (8), 2755–2758.

Laursen, T.A., Simo, J.C., 1993. A continuum-based finite element formulation for the implicit solution of multibody, large deformation-frictional contact problems. Int. J. Num. Method. Eng. 36 (20), 3451–3485.

Lawrence, F.C., Mills, B.H., 2002. Status update of the AEDC wind tunnel virtual flight testing development program. In: The Fortieth AIAA Aerospace Sciences Meeting & Exhibit, pp. 168.

Lee, S.H., 2009. Efficient finite element electromagnetic analysis for high-frequency/high-speed circuits and multiconductor transmission lines. Doctoral dissertation, University of Illinois at Urbana-Champaign.

LeVeque, R.J., 2002. Finite Volume Methods for Hyperbolic Problems, vol. 31, London: Cambridge University Press.

Li, J.T., Wang, Y., 2010. Modeling and analysis of microscopic simulation of railway marshalling yard operation. J. Transport. Sys. Eng. and Inf. 10 (6), 150–155, (in Chinese).

Li, H., Zhao, Z.L., Fan, Z.L., 2011. Study on the simulation method of virtual flight experiment in wind tunnel. Experimental Fluid Mech. 25 (6), 72–76, (in Chinese).

Lin, C., 1997. Model and approximate performance analysis of a resource sharing system. J. Comp. Sci. 20 (10), 865–871.

Lin, C., 2005. Stochastic Petri Net and System Performance Evaluation. Publishing House of Tingshua University, Beijing.

Liu, G.R., Liu, M.B., 2003. Smoothed Particle Hydrodynamics: A Meshfree Particle Method. Singapore: World Scientific.

Liu, X., Han, J.H., Li, J., 2006. Analysis of resource competition based on stochastic Petri nets. J. HeFei University Technol. 29 (4), 399–402, (in Chinese).

Liu, X., Qin, N., Xia, H., 2006. Fast dynamic grid deformation based on Delaunay graph mapping. J. Comput. Phy. 211, 405–423.

Liu, J., Bai, X. Z., Guo, Z., 2009. Computational method of unstructural dynamic mesh and applications in flow simulations containing moving boundaries.

Liu, Y., Lo, S.H., Guan, Z.Q., Zhang, H.W., 2014. Boundary recovery for 3D delaunay triangulation. Finite Elements in Analysis and Design 84, 32–43.

Liu, J, Bai, X.Z., Zhang, H.X., et al., Discussion about GCL for deforming grids. Aeronautical Computer Technique, 39 (4), 1–5.

Loeven, G.J.A., 2005. Efficient uncertainty quantification in computational fluid-structure interaction, Master of Science Thesis, Delft University of Technology.

Long, Z.F., 1993. Two generalized conforming plate elements based on SemiLoof constraints. Comput. Struct. 47 (2), 299–304.

Long, Z.F., 1993. Generalized conforming quadrilateral plate element by using semi – loof constraints. Comm. Num. Method. Eng. 9 (5), 417–426.

Long, Z.F., Cen, S., 2001. New monograph of finite element method: principle, programming, developments. China Hydraulic and Water-power Press, Beijing, (in Chinese).

Long, Y.Q., Huang, M.F., 1988. A generalized conforming isoparametric element. Appl. Math. Mech. 9 (10), 929–936.

Long, Y.Q., Xin, K.G., 1989. Generalized conforming element for bending and buckling analysis of plates. Finite Elem. Anal. Design 5 (1), 15–30.

Long, Y.Q., Xu, Y., 1993. Generalized conforming triangular flat shell element. Eng. Mech. 10 (4), 1–7, (in Chinese).

Long, Y.Q., Xu, Y., 1994a. Generalized conforming quadrilateral membrane element with vertex rigid rotational freedom. Finite elements in analysis and design. Comput. Struct. 52 (4), 749–755.

Long, Y.Q., Yin, X., 1994. Generalized conforming flat rectangular thin shell element. Chinese J. Comput. Mech. 2.

Long, Y.Q., Bu, X.M., Long, Z.F., Xu, Y., 1995. Generalized conforming plate bending elements using point and line compatibility conditions. Comput. Struct. 54 (4), 717–723.

Long, Y.Q., Cen, S., Long, Z.F., 2009. Advanced finite element method in structural engineering. Beijing: Tsinghua University Press.

Lu, Y., Chang, J.H., Guan, Z.Q., 2012. Enhanced 3D optimal Delaunay triangulation optimization method for tetrahedral mesh quality. J. Comput.-Aided Design Comput. Graphics 7, 014.

Lwamiya, T., Nakamura, T., Yoshida, M., 1996. Numerical wind tunnel (NWT) project. PThPS 122, 57–66.

Macneal, R.H., Harder, R.L., 1985. A proposed standard set of problems to test finite element accuracy. Finite Elements Anal. Design 1 (1), 3–20.

Manges, J.B., Cendes, Z.J., 1995. A generalized tree-cotree gauge for magnetic field computation. IEEE Trans. Magnetics 31 (3), 1342–1347.

Marechal, Y., Meunier, G., Coulomb, J.L., Magnin, H., 1992. A general purpose tool for restoring inter-element continuity. IEEE Trans. Magnetics 28 (2), 1728–1731.

Masud, A., 1993. A Space–Time Finite Element Method for Fluid–Structure Interaction, PhD Thesis, Palo Alto: Stanford University.

Meunier, G. (Ed.), 2010. The Finite Element Method for Electromagnetic Modeling, vol. 33, New York: John Wiley & Sons.

Mittal, R., Iaccarino, G., 2005. Immersed boundary methods. Annual Rev. Fluid Mech. 37, 239–261.

Miyoshi, H., Fukuda, M., Iwamiya, T., Nakamura, T., Tuchiya, M., Yoshida, M., Kishimoto, M. 1994. Development and achievement of NAL numerical wind tunnel (NWT) for CFD computations. In: Proceedings of the 1994 ACM/IEEE conference on Supercomputing. IEEE Computer Society Press. pp. 685–692.

Monaghan, J.J., 2005. Smoothed particle hydrodynamics. Reports on Progress in Physics 68 (8), 1703.

Monk, P., 1992. A finite element method for approximating the time-harmonic Maxwell equations. Numerische Mathematik 63 (1), 243–261.

Monk, P., 2003. Finite element methods for Maxwell's equations. London: Oxford University Press.

Montgomery, D.C., 1984. Design and analysis of experiments, vol. 7, Wiley, New York.

Nédélec, J.C., 1980. Mixed finite elements in R3. Numerische Mathematik 35 (3), 315–341.

Nédélec, J.C., 1986. A new family of mixed finite elements in R3. Numerische Mathematik 50 (1), 57–81.

Nedelec, 1980. A new family of mixed finite elements in R3. Numerische Mathematik 35, 315–341.

Nelson, H.D., 1980. A finite rotating shaft element using Timoshenko beam theory. J. Mech. Design 102 (4), 793–803.

Nelson, H.D., Mcvaugh, J.M., 1976. The dynamics of rotor-bearing systems with internal damping. ASME J. Eng. Power 98 (2), 593–600.

Newmark, N.M., 1959. A method of computation for structural dynamics. J. Eng. Mech. Division 85 (3), 67–94.

Ney, J.F., 1957. Physical properties of crystals: Their Representation by Tensors and Matrices. OUP Oxford University Press, Oxford.

Nye, J.F., 1985. Physical properties of crystals: their representation by tensors and matrices. Oxford University Press, Oxford.

Olson, M.D., Bearden, T.W., 1979. A simple flat triangular shell element revisited. Int. J. Num. Method. Eng. 14 (1), 51–68.

Parisch, H., 1979. A critical survey of the 9-node degenerated shell element with special emphasis on thin shell application and reduced integration. Comput. Method. Appl. Mech. Eng. 20 (3), 323–350.

Peskin, C.S., 2002. The immersed boundary method. Acta Numerica 11, 479–517.

Preis, K., Bardi, I., Biro, O., Magele, C., Renhart, W., Richter, K.R., Vrisk, G., 1991. Numerical analysis of 3D magnetostatic fields. IEEE Trans. Magnetics 27 (5), 3798–3803.

Providas, E., Kattis, M.A., 2000. An assessment of two fundamental flat triangular shell elements with drilling rotations. Comput. Struct. 77 (2), 129–139.

Qiu-yang, F.U., 2009. CFD & FEA coupling analysis application in exhaust manifold analysis. J. Hefei Univ. Technol. (Nat. Sci.), S1.

Rankine, W.M., 1869. On the centrifugal force of rotating shafts. Engineer 27, 249.

Rao, J.S., 2011. History of rotating machinery dynamics, vol. 20, Berlin: Springer Science & Business Media.

Rao, J.S., Sreenivas, R., 2003. Dynamics of asymmetric rotors using solid models. In: Proceedings of the International Gas Turbine Congress, pp. 2–7.

Reh, S., Khor, E.H., 2005. Behind the Scenes of Response Surface Methods, The Twenty-third CADFEM Users' Meeting Germany.

Ren, Z., 1996. Influence of the RHS on the convergence behaviour of the curl-curl equation. IEEE Trans. Magnetics 32 (3), 655–658.

Ren, B.N., Zhu, P.H., Wei, S.H., 2000. Study on numerical wind tunnel software system. J. Northwestern Polytech. Univ.

Riesen, D.V., Monzel, C., Kachler, C., Schlensok, C., Henneberger, G., 2004. iMOOSE: an open-source environment for finite-element calculations. IEEE Trans. Magnetics 38 (2), 613–616.

Riesen, D., Monzel, C., Kaehler, C., Schlensok, C., Henneberger, G., 2004. iMOOSE: an open-source environment for finite-element calculations. IEEE Trans. Magnetics 40 (2), 1390–1393.

Rispoli, F., Corsini, A., Tezduyar, T.E., 2005. Finite Element Computation of M Turbulent Flows with the Discontinuity-Capturing Directional Dissipation. Italy Mechanical Engineering, Rice University, Houston.

Rispoli, F., Corsini, A., Tezduyar, T.E., 2007. Finite element computation of turbulent flows with the discontinuity-capturing directional dissipation (DCDD). Comput. Fluids 36 (1), 121–126.

Rizzi, A., 2011. Modelling and simulating aircraft stability and control – the SimSAC project. Progr. Aerospace Sci. 47, 573–588.

Rodger, D., Lai, H.C., Leonard, P.J., 1990. Coupled elements for problems involving movement, switched reluctance motor. IEEE Trans. Magnetics 26 (2), 548–550.

Roger, W.P., 2011. Multiphysics Modeling Using Comsol. Infinity Science Press, Sudbury, Mass.

Rong, H.H., 2005. The history Haoshan. A based on stochastic Petri net resource sharing system performance analysis method. Appl. Comput. 25 (4), 881–882.

Rugonyi, S., Bathe, K.J., 2001. On finite element analysis of fluid flows fully coupled with structural interactions. CMES 2 (2), 195–212.

Ruhl, R.L., Booker, J.F., 1972. A finite element model for distributed parameter turborotor systems. J. Manufacturing Sci. Eng. 94 (1), 126–132.

Salas, M.D., 2006. Digital flight: the last CFD aeronautical grand challenge. J. Sci. Comput. 28 (2–3), 479–505.

Scotti, C.M., Shkolnik, A.D., Muluk, S.C., Finol, E.A., 2005. Fluid-structure interaction in abdominal aortic aneurysms: effects of asymmetry and wall thickness. BioMed. Eng. Online 4 (1), 64.

Shangguan, W.B., Lu, Z.H., 2004. Experimental study and simulation of a hydraulic engine mount with fully coupled fluid–structure interaction finite element analysis model. Comput. Struct. 82 (22), 1751–1771.

Shewchuk, J.R., 1998. Tetrahedral mesh generation by Delaunay refinement. In: Proceedings of the Fourteenth Annual Symposium on Computational Geometry, ACM pp. 86–95.

Shewchuk, J.R., 2002. Delaunay refinement algorithms for triangular mesh generation. Comput. Geometry 22 (1), 21–74.

Shinozuka, M., 1983. Basic analysis of structural safety. J. Struct. Eng.

Simo, J.C., Hughes, T.J., 2006. Computational Inelasticity, vol. 7, Springer Science & Business Media.

Simo, J.C., Laursen, T.A., 1992. An augmented Lagrangian treatment of contact problems involving friction. Comput. Struct. 42 (1), 97–116.

Soh, A.K., Long, Z.F., Cen, S., 1999. A new nine DOF triangular element for analysis of thick and thin plates. Comput. Mech. 24 (5), 408–417.

Soh, A.K., Long, Y.Q., Song, C., 2000. Development of eight-node quadrilateral membrane elements using the area coordinates method. Comput. Mech. 25 (4), 376–384.

Soh, A.K., Cen, S., Long, Y.Q., Long, Z.F., 2001. A new twelve DOF quadrilateral element for analysis of thick and thin plates. European J. Mech.-A/Solids 20 (2), 299–326.

Stein, K., Benney, R., Kalro, V., Tezduyar, T.E., Leonard, J., Accorsi, M., 2000. Parachute fluid–structure interactions: 3-D computation. Comput. Methods Appl. Mech. Eng. 190 (3), 373–386.

Stodola, A., 1927. Steam and gas turbines. McGraw-Hill, New York.

Sukumar, N., Chopp, D.L., Moës, N., Belytschko, T., 2001. Modeling holes and inclusions by level sets in the extended finite-element method. Comput. Methods Appl. Mech. Eng. 190 (46), 6183–6200.

Sun, J., Long, Z., Long, Y., Zhang, C., 2001. Geometrically nonlinear stability analysis of shells using generalized conforming shallow shell element. Int. J. Struct. Stability Dynamics 1 (03), 313–332.

Sundararajan, C.R., 2012. Probabilistic Structural Mechanics Handbook: Theory and Industrial Applications. Springer Science & Business Media, Berlin, Heidelberg.

Tao, Y., Fan, Z.L., Wu, J.F., 2010. Longitudinal virtual flight simulation of a square missile based on CFD. J. Mech. 42 (2), 169–176.

Taylor, C.A., Hughes, T.J., Zarins, C.K., 1998. Finite element modeling of blood flow in arteries. Comput. Methods Appl. Mech. Eng. 158 (1), 155–196.

Tezduyar, T., Sathe, S., 2003a. Stabilization parameters in SUPG and PSPG formulations. J. Comput. Appl. Mech. 4 (1), 71–88.
Tezduyar, T., Sathe, S., 2003b. Stabilization parameters in SUPG and PSPG formulations. J. Comput. Appl. Mech. 4 (1), 71–88.
Tezduyar, T.E., Senga, M., 2007. SUPG finite element computation of inviscid supersonic flows with YZβ shock-capturing. Comput. Fluids 36 (1), 147–159.
Thomas, P.D., Lombard, C.K., 1978. Geometric Conservation law and application to flow computations on moving grids. AIAA J. 17 (10), 1030–1037.
Verardi, S.L.L., Cardoso, J.R., Motta, C.C., 1998. A solution of two-dimensional magnetohydrodynamic flow using the finite element method. IEEE Trans. Magnetics 34 (5), 3134–3137.
Versteeg, H.K., Malalasekera, W., 1995. An Introduction to Computational Fluid Dynamics. Longman Scientific & Technical, England.
Wang, X.C., 2002. Finite Element Method. Tsing Hua University Press, Beijing, (in Chinese).
Wang, H.L., 2010. Modeling and Analysis of the Railway Marshalling Station Technical Operation Simulation System. Dalian Jiaotong University, Dalian, (in Chinese).
Wanji, C., Cheung, Y.K., 1999. Refined non-conforming triangular elements for analysis of shell structures. Int. J. Num. Methods Eng. 46 (3), 433–455.
Watanabe, H., 1995. Study of Mixed Finite Element Analysis for Incompressible Hyper-Elastic Materials, Doctoral dissertation, PhD. Thesis, University of Tokyo.
Wohlmuth, B.I., 1999. A residual based error estimator for mortar finite element discretizations. Numerische Mathematik 84 (1), 143–171.
Wohlmuth, B.I., 1999. Hierarchical a posteriori error estimators for mortar finite element methods with Lagrange multipliers. SIAM J. Num. Anal. 36 (5), 1636–1658.
Wohlmuth, B.I., 2000. A mortar finite element method using dual spaces for the Lagrange multiplier. SIAM J. Num. Anal. 38 (3), 989–1012.
Wohlmuth, B.I., 2002. A comparison of dual Lagrange multiplier spaces for mortar finite element discretizations. ESAIM: Math. Model. Num. Anal. 36 (6), 995–1012.
Wriggers, P., Van, T.V., Stein, E., 1990. Finite element formulation of large deformation impact-contact problems with friction. Comput. Struct. 37 (3), 319–331.
Wu, J.H., 1994. Design and optimization of railway marshalling station system. China Railway Publishing House, Beijing.
Xiao, L.Z., Line, P., 2006. Speed of marshalling station subsystem internal matching and coordinating relationship between China Railway Science 27 (5), 118–121, (in Chinese).
Xiao, T.H., Ang, H.S., Tong, C., 2008. A new dynamic mesh generation method for large movements of flapping-wings with complex geometries. Acta Aeronautica et Astronautica Sinica 29 (1), 41–48.
Xu, Y., Long, Z.F., 1996. Advances of membrane and thin shell elements with the generalized conforming approach. In: Si, Y., Zhiliang, M. (Eds.), New Developments in Structural Engineering, 218–223.
Xu, Y., Long, Y.Q., Long, Z.F., 1994. A triangular shell element with drilling freedoms based on generalized compatibility conditions. In: Proceedings of WCCM III. Japan, 1234–1235.
Xu, Y., Long, Y.Q., Long, Z.F., 1999. A generalized conforming triangular flat shell element with high accuracy. In: Long, Y.Q., (Ed.), The Proceedings of the First International Conference on Structural Engineering. KunMing, China, 700–706. (in Chinese)
Yagawa, G., Yoshimura, S., Nakao, K., 1995. Automatic mesh generation of complex geometries based on fuzzy knowledge processing and computational geometry. Integr. Comput.-Aided Eng. 2 (4), 265–280.

Yan, Z.G., 2013. The role of high precision numerical wind tunnel in aircraft design. Aviation Sci. Technol.

Yao, W., 2010. Finite element analysis of 3D electric machine problems. Doctoral dissertation, University of Illinois at Urbana-Champaign.

Yates, E.C.J., 1988. AGARD standard aeroelastic configurations for dynamic response candidate configuration I. – wing 445.6. NASA-TM-100492.

Yuan, C.Y., 2005. Principle and application of Petri nets. Beijing: Publishing House of Electronics Industry.

Zeng, J.L., Xu, Y., Han, Y.P., Li, G.H., Zhang, Q., 2014. Cooling simulation analysis based on multi-field coupling technology of permanent magnet synchronous motor. J. Shanghai Jiaotong Univ. 48 (9), 1246–1251, (in Chinese).

Zhang, Q., 1999. Analysis of structure-fluid interaction problems with structural buckling and large domain changes by ALE finite element method. Doctoral dissertation, PhD Thesis, Department of Mechano-Informatics, University of Tokyo.

Zhang, Q., Hisada, T., 1998. Theoretical and applied mechanics. Proceedings of the Forty-seventh Japan National Congress for Applied Mechanics 47, 167–174.

Zhang, Q., Hisada, T., 2001. Analysis of fluid–structure interaction problems with structural buckling and large domain changes by ALE finite element method. Comput. Method. Appl. Mech. Eng. 190 (48), 6341–6357.

Zhang, Q., Toshiaki, H., 2004. Studies of the strong coupling and weak coupling methods in FSI analysis. Int. J. Num. Method. Eng. 60 (12), 2013–2029.

Zhang, X., Wang, T.S., 2007. Computational dynamics. Tsinghua University Press, Beijing, (in Chinese).

Zhang, N., Zhang, Q.S., 2001. Research on the modeling and analysis of marshalling station operation system using stochastic Petri nets. J. China Railway 23 (1), 13–18, (in Chinese).

Zhang, Q., Zhu, B.S., 2012. An integrated coupling framework for highly nonlinear fluid-structure problems. Comput. Fluids 60, 36–48.

Zhang, Q., Lu, M., Kuang, W.Q., 1998. Geometric non-linear analysis of space shell structures using generalized conforming flat shell elements—for space shell structures. Comm. Num. Method. Eng. 14 (10), 941–957.

Zhang, Q., Lu, M., Kuang, W.Q., 1999. Application of the generalized conforming flat shell element to geometrical non-linear analysis for composite stiffened shell structures. Comm. Num. Method. Eng. 15 (6), 399–412.

Zhang, Y.X., Cao, S.L., Zhu, B.S., 2005. An implicit SMAC method for incompressible 3D turbulence. J. Tsing Hua Univ. 45 (11), 1561–1564.

Zhang, Z.T., Cui, B.M., Qu, C.J., 2010. Study on the efficiency of the disintegration of the marshalling station. Traffic Standard. 8 (1), 130–133.

Zhang, L., Liu, T., Gao, W.F., Lin, W.X., 2012. Performance comparison of four kinds of turbulence models for two-dimensional backward flow to a step flow simulation. J. Yunnan Normal Univ. 32 (4), 1007-9793, (in Chinese).

Zhang, Q., Zhu, R.B., Han, Y.P., 2013. Strong coupled fluid-thermal analysis of engine exhaust pipe. Comput. Aided Eng. 22 (3), 34–38.

Zienkiewicz, O.C., Taylor, R.L., 1991. The Finite Element Method. Fourth edition. Volume 2 – Solid and fluid mechanics & dynamics and non-linearity. McGraw-Hill- Book Company, London.

Zorzi, E.S., Nelson, H.D., 1977. Finite element simulation of rotor-bearing systems with internal damping. J. Eng. Gas Turbines Power 99 (1), 71–76.

Zorzi, E.S., Nelson, H.D., 1980. The dynamics of rotor-bearing systems with axial torque – a finite element approach. J. Mech. Design 102 (1), 158–161.

Index

F

N

O

P

Y

Z

Printed in the United States
By Bookmasters